Lecture Notes in Computer Science   **9712**

Commenced Publication in 1973
Founding and Former Series Editors:
Gerhard Goos, Juris Hartmanis, and Jan van Leeuwen

More information about this series at http://www.springer.com/series/7407

Ying Tan · Yuhui Shi
Ben Niu (Eds.)

Advances in Swarm Intelligence

7th International Conference, ICSI 2016
Bali, Indonesia, June 25–30, 2016
Proceedings, Part I

 Springer

Editors
Ying Tan
Peking University
Beijing
China

Ben Niu
Shenzhen University
Shenzhen
China

Yuhui Shi
Xi'an Jiaotong-Liverpool University
Suzhou
China

ISSN 0302-9743   ISSN 1611-3349 (electronic)
Lecture Notes in Computer Science
ISBN 978-3-319-40999-3   ISBN 978-3-319-41000-5 (eBook)
DOI 10.1007/978-3-319-41000-5

Library of Congress Control Number: 2016942017

LNCS Sublibrary: SL1 – Theoretical Computer Science and General Issues

Printed on acid-free paper

This Springer imprint is published by Springer Nature
The registered company is Springer International Publishing AG Switzerland

Preface

This book and its companion volumes, LNCS vols. 9712 and 9713, constitute the proceedings of the 7th International Conference on Swarm Intelligence (ICSI 2016) held during June 25–30, 2016, in Bali, Indonesia.

The theme of ICSI 2016 was "Serving Life with Intelligence and Data Science." ICSI 2016 provided an excellent opportunity and/or an academic forum for academics and practitioners to present and discuss the latest scientific results and methods, innovative ideas, and advantages in theories, technologies, and applications in swarm intelligence. The technical program covered all aspects of swarm intelligence and related areas.

ICSI 2016 was the seventh international gathering in the world for researchers working on all aspects of swarm intelligence, following successful events in Beijing (ICSI-CCI 2015), Hefei (ICSI 2014), Harbin (ICSI 2013), Shenzhen (ICSI 2012), Chongqing (ICSI 2011), and Beijing (ICSI 2010), which provided a high-level academic forum for participants to disseminate their new research findings and discuss emerging areas of research. It also created a stimulating environment for participants to interact and exchange information on future challenges and opportunities in the field of swarm intelligence research. ICSI 2016 was held in conjunction with the International Conference on Data Mining and Big Data (DMBD 2016) at Bali, Indonesia, for sharing common mutual ideas, promoting transverse fusion, and stimulating innovation.

Bali is a famous Indonesian island with the provincial capital at Denpasar. Lying between Java to the west and Lombok to the east, this island is renowned for its volcanic lakes, spectacular rice terraces, stunning tropical beaches, ancient temples and palaces, as well as dance and elaborate religious festivals. Bali is also the largest tourist destination in the country and is renowned for its highly developed arts, including traditional and modern dance, sculpture, painting, leather, metalworking, and music. Since the late 20th century, the province has had a big rise in tourism. Bali received the Best Island Award from *Travel and Leisure* in 2010. The island of Bali won because of its attractive surroundings (both mountain and coastal areas), diverse tourist attractions, excellent international and local restaurants, and the friendliness of the local people. According to BBC Travel released in 2011, Bali is one of the world's best islands, rank second after Greece!

ICSI 2016 received 231 submissions from about 693 authors in 42 countries and regions (Algeria, Australia, Austria, Bangladesh, Bolivia, Brazil, Brunei Darussalam, Canada, Chile, China, Colombia, Fiji, France, Germany, Hong Kong, India, Indonesia, Italy, Japan, Kazakhstan, Republic of Korea, Macao, Malaysia, Mexico, The Netherlands, New Zealand, Nigeria, Oman, Pakistan, Peru, Portugal, Russian Federation, Serbia, Singapore, South Africa, Spain, Chinese Taiwan, Thailand, Tunisia, Turkey, United Arab Emirates, USA) across six continents (Asia, Europe, North America, South America, Africa, and Oceania). Each submission was reviewed by at least two reviewers, and on average 2.9 reviewers. Based on rigorous reviews by the Program Committee members and reviewers, 130 high-quality papers were selected for publication in this proceedings volume, with an acceptance rate of 56.28 %. The papers are organized in 22 cohesive sections covering major topics of swarm intelligence and related areas.

In addition to the contributed papers, the technical program of ICSI 2016 included three plenary talks by Prof. Gary G. Yen (Oklahoma State University, USA), Prof. Kay Chen Tan (National University of Singapore, Singapore), and Prof. Tshilidzi Marwala (The University of Johannesburg, South Africa). Besides the regular parallel sessions, the conference also had four special sessions on hot-spot topics, four tutorials, as well as poster sessions.

On behalf of the Organizing Committee of ICSI 2016, we would like to express sincere thanks to Peking University and Xian Jiaotong-Liverpool University for their sponsorship, and to Shenzhen University and Beijing Xinghui Hi-Tech Co. for its co-sponsorship, as well as to the IEEE Computational Intelligence Society, World Federation on Soft Computing, and International Neural Network Society, and IEEE Beijing Section for their technical co-sponsorship. We would also like to thank the members of the Advisory Committee for their guidance, the members of the international Program Committee and additional reviewers for reviewing the papers, and the members of the Publications Committee for checking the accepted papers in a short period of time. We are particularly grateful to the proceedings publisher Springer for publishing the proceedings in the prestigious series of *Lecture Notes in Computer Science*. Moreover, we wish to express our heartfelt appreciation to the plenary speakers, session chairs, and student helpers. In addition, there are still many more colleagues, associates, friends, and supporters who helped us in immeasurable ways; we express our sincere gratitude to them all. Last but not the least, we would like to thank all the speakers, authors, and participants for their great contributions that made ICSI 2016 successful and all the hard work worthwhile.

May 2016

Ying Tan
Yuhui Shi
Ben Niu

Organization

General Chairs

Ying Tan   Peking University, China
Russell C. Eberhart   Indiana University Purdue University Indianapolis, IUPUI, USA

Advisory Committee Chairs

Gary G. Yen   Oklahoma University, USA
Jun Wang   City University of Hong Kong, SAR China

Program Committee Chair

Yuhui Shi   Xi'an Jiaotong-Liverpool University, China

Technical Committee Co-chairs

Haibo He   University of Rhode Island Kingston, USA
Martin Middendorf   University of Leipzig, Germany
Xiaodong Li   RMIT University, Australia
Hideyuki Takagi   Kyushu University, Japan
Ponnuthurai Nagaratnam Suganthan   Nanyang Technological University, Singapore
Kay Chen Tan   National University of Singapore, Singapore

Special Sessions Co-chairs

Shi Cheng   Nottinggham University Ningbo, China
Ben Niu   Shenzhen University, China
Yuan Yuan   Chinese Academy of Sciences, China

Invited Speakers Session Co-chairs

Liangjun Ke   Xi'an Jiao Tong University, China
Komla Folly   University of Cape Town, South Africa

Publications Co-chairs

Radu-Emil Precup   Politehnica University of Timisoara, Romania
Swagatham Das   Indian Statistical Institute, India

Plenary Session Co-chairs

Nikola Kasabov   Auckland University of Technology, New Zealand
Rachid Chelouah   EISTI, France

Tutorial Co-chairs

Milan Tuba   University of Belgrade, Serbia
Dunwei Gong   China University of Mining and Technology, China
Li Li   Shenzhen University, China

Symposia Co-chairs

Maoguo Gong   Northwest Polytechnical University, China
Yan Pei   University of Aziz, Japan

Publicity Co-chairs

Yew-Soon Ong   Nanyang Technological University, Singapore
Carlos A. Coello Coello CINVESTAV-IPN, Mexico
Pramod Kumar Singh   Indian Institute of Information Technology
                          and Management, India
Yaochu Jin   University of Surrey, UK
Fernando Buarque   Universidade of Pernambuco, Brazil
Eugene Semenkin   Siberian Aerospace University, Russia
Somnuk Phon-Amnuaisuk   Institut Teknologi Brunei, Brunei

Finance and Registration Co-chairs

Andreas Janecek   University of Vienna, Austria
Chao Deng   Peking University, China
Suicheng Gu   Google Corporation, USA

Program Committee

Mohd Helmy Abd Wahab   Universiti Tun Hussein Onn, Malaysia
Lounis Adouane   LASMEA, France
Ramakrishna Akella   University of California, USA
Miltiadis Alamaniotis   Purdue University, USA
Rafael Alcala   University of Granada, Spain
Peter Andras   Keele University, UK
Esther Andrés   INTA, Spain
Helio Barbosa   Laboratório Nacional de Computação Científica, Brazil
Anasse Bari   New York University, USA
Carmelo J.A. Bastos Filho University of Pernambuco, Brazil
Christian Blum   IKERBASQUE, Basque Foundation for Science, Spain

Vladimir Bukhtoyarov	Siberian State Aerospace University, Russia
David Camacho	Universidad Autonoma de Madrid, Spain
Bin Cao	Tsinghua University, China
Jinde Cao	Southeast University
Kit Yan Chan	Curtin University, Australia
Chien-Hsing Chen	Ling Tung University, Taiwan
Liang Chen	University of Northern British Columbia, Canada
Walter Chen	National Taipei University of Technology, Taiwan
Shi Cheng	The University of Nottingham Ningbo, China
Manuel Chica	European Centre for Soft Computing, Spain
Carlos Coello Coello	CINVESTAV-IPN, Mexico
Jose Alfredo Ferreira Costa	UFRN – Universidade Federal do Rio Grande do Norte, Brazil
Micael S. Couceiro	Polytechnic Institute of Coimbra, Portugal
Prithviraj Dasgupta	University of Nebraska, Omaha, USA
Kusum Deep	Indian Institute of Technology Roorkee, India
Mingcong Deng	Tokyo University of Agriculture and Technology, Japan
Ke Ding	Baidu Corporation, China
Yongsheng Dong	Henan University of Science and Technology, China
Haibin Duan	Beijing University of Aeronautics and Astronautics, China
Mark Embrechts	Rensselaer Polytechnic Institute, USA
Andries Engelbrecht	University of Pretoria, South Africa
Jianwu Fang	Xi'an Institute of Optics and Precision Mechanics of CAS, China
Shangce Gao	University of Toyama, Japan
Ying Gao	Guangzhou University, China
Beatriz Aurora Garro Licon	IIMAS-UNAM, Mexico
Maoguo Gong	Xidian University, China
Amel Grissa	Ecole Nationale d'Ingénieurs de Tunis, Tunisia
Shenshen Gu	Shanghai University, China
Yinan Guo	Chinese University of Mining and Technology, China
Fei Han	Jiangsu University, China
Haibo He	University of Rhode Island, USA
Shan He	University of Birmingham, UK
Lu Hongtao	Shanghai Jiao Tong University, China
Mo Hongwei	Harbin Engineering University, China
Reza Hoseinnezhad	RMIT University, Australia
Jun Hu	Chinese Academy of Sciences, China
Teturo Itami	Hirokoku University, Japan
Andreas Janecek	University of Vienna, Austria
Yunyi Jia	Clemson University, USA
Changan Jiang	Ritsumeikan University, Japan
Mingyan Jiang	Shandong University, China
Licheng Jiao	Xidian University, China

Colin Johnson	University of Kent, UK
Matthew Joordens	Deakin University, Australia
Ahmed Kattan	Umm Al-qura University, Saudi Arabia
Liangjun Ke	Xi'an Jiaotong University, China
Arun Khosla	National Institute of Technology, Jalandhar, Punjab, India
Slawomir Koziel	Reykjavik University, Iceland
Thanatchai Kulworawanichpong	Suranaree University of Technology, Thailand
Rajesh Kumar	MNIT, India
Hung La	University of Nevada, USA
Germano Lambert-Torres	PS Solutions, Brazil
Xiujuan Lei	Shaanxi Normal University, China
Bin Li	University of Science and Technology of China, China
Xiaodong Li	RMIT University, Australia
Xuelong Li	Chinese Academy of Sciences, China
Andrei Lihu	Politehnica University of Timisoara, Romania
Fernando B. De Lima Neto	University of Pernambuco, Brazil
Bin Liu	Nanjing University of Post and Telecommunications, China
Ju Liu	Shandong University, China
Qun Liu	Chongqing University of Posts and Communications, China
Wenlian Lu	Fudan University, China
Wenjian Luo	University of Science and Technology of China, China
Chengying Mao	Jiangxi University of Finance and Economics, China
Michalis Mavrovouniotis	De Montfort University, UK
Mohamed Arezki Mellal	M'Hamed Bougara University, Algeria
Bernd Meyer	Monash University, Australia
Martin Middendorf	University of Leipzig, Germany
Sanaz Mostaghim	Institute IWS, Germany
Krishnendu Mukhopadhyaya	Indian Statistical Institute, India
Ben Niu	Shenzhen University, China
Yew-Soon Ong	Nanyang Technological University, Singapore
Feng Pan	Beijing Institute of Technology, China
Jeng-Shyang Pan Pan	National Kaohsiung University of Applied Sciences, Taiwan
Quan-Ke Pan	Nanyang Technological University, Singapore
Shahram Payandeh	Simon Fraser University, Canada
Yan Pei	The University of Aizu, Japan
Somnuk Phon-Amnuaisuk	Institut Teknologi, Brunei
Ghazaleh Pour Sadrollah	Monash University, Australia
Radu-Emil Precup	Politehnica University of Timisoara, Romania
Kai Qin	RMIT University, Australia
Quande Qin	Shenzhen University, China

Jun Zhang	Waseda University, Japan
Junqi Zhang	Tongji University, China
Lifeng Zhang	Renmin University of China, China
Mengjie Zhang	Victoria University of Wellington, New Zealand
Qieshi Zhang	Waseda University, Japan
Qiangfu Zhao	The University of Aizu, Japan
Shaoqiu Zheng	Peking University, China
Yujun Zheng	Zhejiang University of Technology, China
Zhongyang Zheng	Peking University, China
Cui Zhihua	Complex System and Computational Intelligence Laboratory, China
Guokang Zhu	Shanghai University of Electric Power, China
Xingquan Zuo	Beijing University of Posts and Telecommunications, China

Additional Reviewers

Bari, Anasse	Jia, Guanbo	Portugal, David
Chen, Zonggan	Jing, Sun	Shan, Qihe
Cheng, Shi	Lee, Jie	Shang, Ke
Dai, Hongwei	Li, Junzhi	Tao, Fazhan
Dehzangi, Abdollah	Li, Mengyang	Wan, Ying
Ding, Sanbo	Li, Yaoyi	Wang, Junyi
Feng, Jinwang	Lian, Cheng	Weibo, Yang
Ghafari, Seyed Mohssen	Lin, Jianzhe	Yan, Shankai
Guo, Xing	Liu, Xiaofang	Yao, Wei
Guyue, Mi	Liu, Zhenbao	Yu, Chao
Hu, Jianqiang	Lv, Gang	Zhang, Yong
Hu, Weiwei	Lyu, Yueming	Zi-Jia, Wang
Hu, Zihao	Paiva, Fábio	

Contents – Part I

Hybrid Search Optimization

Particle Swarm Optimization

PSO Applications

Ant Colony Optimization

Brain Storm Optimization

Fireworks Algorithms

Multi-Objective Optimization

Large-Scale Global Optimization

Biometrics

Contents – Part II

Clustering Algorithm

Classification

Image Classification and Encryption

Data Mining

Sensor Networks and Social Networks

Neural Networks

Swarm intelligence in Management Decision Making and Operations Research

Robot Control

Swarm Robotics

Intelligent Energy and Communications Systems

Intelligent Interactive and Tutoring Systems

Trend and Models of Swarm Intelligence Research

Swarm Intelligence in Architectural Design

Sebastian Wiesenhuetter$^{(\boxtimes)}$, Andreas Wilde,
and Joerg Rainer Noennig

Laboratory of Knowledge Architecture, Faculty of Architecture,
TU Dresden, Dresden, Germany
{sebastian.wiesenhuetter,andreas.wilde}@tu-dresden.de,
joerg.noennig@mailbox.tu-dresden.de

Abstract. This paper investigates the application of swarm intelligence in the field of architecture. We seek to distinguish different fields of application by regarding swarm intelligence as a potential tool to support the design process, to improve architectural use and further create novel building systems, based on self-organization principles. In architectural applications, swarm intelligence offers a high potential of resilience, and solutions that are fit to the task. We analyze two case studies, one concerning adaptive buildings with intelligent behavior, and one in the field of algorithmic design which makes use of agents during the planning process. Regarding their potentials and deficits, we propose a broader perspective on agent based architectural design. By integrating self-organized construction processes that are related both to the design process and to the usage, we propose combining the different tendencies to a more resilient system that covers a buildings ontogeny from beginning to end.

Keywords: Multi-agent design · Adaptive building · Evolutionary design · Distributed construction · Programmable matter · Architectural ontogeny

1 Introduction

This paper investigates the application of swarm intelligence in the field of architectural design. The scope of investigation includes both the design processes, as well as the resulting behavior of buildings in their physical environment. We seek to distinguish different fields of application by regarding swarm intelligence as a potential tool to support the design process and to improve architectural use and further create novel building systems. The benefits of this approach rely mainly on the principle of Self Organization. Self-Organization is given in cases of local interactions that lead to emergent order in an overall form or resulting structure. This is not only found in Swarm Intelligence but also in Evolutionary Algorithms [1]. Architectural design processes are enormously complex as it is a constant challenge to unite different simultaneously demanding tasks in one building without intriguing another. It has been found that these approaches can lead to emergent qualities in design. Thus both methods have been used in the field of architecture before [2, 3].

Whereas swarm intelligence achieved a remarkable relevance in mathematical and technological fields, it is still widely untapped in architectural contexts especially in professional practice. In architectural applications, swarm intelligence offers a high

© Springer International Publishing Switzerland 2016
Y. Tan et al. (Eds.): ICSI 2016, Part I, LNCS 9712, pp. 3–13, 2016.
DOI: 10.1007/978-3-319-41000-5_1

potential of resilience, time- and material- efficiency and solutions that are fit to the task. The problem of lacking "task fitness" and the subsequent deficit in architectural methodology has been addressed by Alexander [4] and others [5].

In recent years, mainly since 1970s two main strategies have been established in order to address such problems by help of distributed intelligence. On the one hand a new approach can be seen in terms of adaptive buildings with intelligent behavior, that are often directly inspired by natural processes [6]. They seek to provide better fit to changing tasks or environmental conditions through a range of possible solutions. On the other hand the field of algorithmic design has been established [7], which makes use of Swarm Intelligence strategies during the design and planning process.

The natural self-assembly processes which can be seen in bees and ants results in extreme adaptability, complexity and regularity within their construction [8].

2 Scientific Background

Architectural thinking is a highly creative process that goes beyond finding solutions to specific problems. The deficit of current design approaches is their limitation on mental capabilities and personal convictions of the individual designer who bases his decisions on tacit knowledge and personal experiences. Thus, the traditional approach, usually understood as creative process cannot compete with the possibilities of parametric design, especially if it comes to reaching a larger variety of options [6]. The author of a design program has the task to conceptualize rules, and in case, behavior of agents. Criteria of fitness, set by the author himself, can replace or support the designer's decision for a certain solution. The common attribute of any computer aided design is that both human designer and computer need to interact and communicate in order to reach a design solution. Therefore, the externalization of design methodology is crucial. Verbalization allows translation, automatization and a further leading progression of solutions. Today within the professional practice of architecture, these parametric design techniques are only implemented to a small extent. Those applications, however, rarely exceed the boundaries of parametric modeling, and even then, often stay object centered [5] and static design approaches. The design process of architecture usually does not make use of codified methodology, as it can be found in most fields of science.

Parametric design can be understood as a general verbalization of design processes, which enhances progress in complex design [10]. Parametric modeling itself does not use the full potential of parametric design. Additionally verbalizing requirements or methods instead, is necessary to gain new and fit-to-task solutions. Still, architecture is a field that can, until today not be fully abstracted into parametric values, and even simple abstractions already lead to an enormous complexity that is not yet in full control.

Historically, relations between society and architecture were often stable and only evolving simultaneously over many generations. Single buildings were usually not endangered to lose functional relevance over the years. Sustainable architecture was possible and ordinary within those boundaries, as buildings adapted to their function. This also led to relatively long life cycles or efficient renewal features that are often missing in contemporary architecture. Our environment today is more dynamic than it used to be. Innovations, social changes happen within shorter time spans, and buildings are usually planned for one certain recent use, no matter what following generations might make of it. The two potential reactions on a more dynamic environment could either be to plan with shorter buildings ontogenies, which is a strategy that is currently in use. Buildings today are usually planned for only 30 years, and indeed, buildings often change their function after only a few decades. The other option would be to follow strategies of adaptive architecture that is not static but dynamic itself, changing with the demands of society, climate, political system or simply a new user. This strategy is technically demanding, but allows a better sustainability concerning energy-resource- and time efficiency [4]. Christopher Alexander proved that architectural design can be seen as multi objective optimization, but it must be noted that the requirements that a building is supposed to acquire are constantly changing [4]. Therefore we have to see the context of architecture as a dynamic one. In its consequence, we assume the problems of architectural design as dynamic multi-objective problems (DMOP) [9].

We assume that there are comprehensive principles of Swarm Intelligence that can support the whole process of architecture. Recent progresses in computational design support research on adaptive features of architecture. The aim of this adaptive architecture is to precisely match to buildings requirements. The key in this new approach is parametric design. Parametric design enables application of iterative processes like Evolutionary Algorithms [1] or Swarm Intelligence algorithms that are often the key way of comparing and finding solutions. The use of Swarm intelligence - enabled by parametric design - extends the boundaries of human cognition, which then allows the finding of design solutions that are close to the designer's original intentions but are not directly accessible.

Today, we can see Swarm Intelligent approaches in two main tendencies in architectural design:

(A) Adaptive structures are real-time responsive and dynamic by reacting to environmental influences in a physical environment. This usually takes place after a construction that defines the boundaries of parameters or states that the building can adopt.

(B) In the design process, Swarm Intelligent applications can be used to optimize a defined solution to a design problem. A building can for example be optimized concerning light, structural design, layout and so forth, or according to multiple criteria. One application that could lead to emergent qualities could be Evolutionary Algorithms.

3 Approach

It is possible to design a building that makes use of Swarm Intelligence both for the design process itself and in ways of adaptive architecture during use. Those two phases of a building are usually separated by a materialization process, namely the construction (Fig. 1): The building is initiated as a space for specific purposes. After the design process the plan is converted into an embodiment. The building is then used until function and building do not correlate to a reasonable extent anymore. Restoration/renovation can prolong this time span by multiple decades. Adapting an expression from biology referring to the development of single organisms, we define a building's process from initial design to its use, adaption (for example renovations), and final demolition as ontogeny.

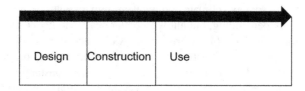

Fig. 1. 3 phases of a buildings ontogeny in their known order

3.1   Case Studies

We would like to discuss two case studies of Swarm Intelligence in architectural design as they represent the two main approaches. The first is an application during the use of buildings and is represented by a responsive folding structure. The second is the application during design phase and is exemplified by an emergent design approach.

3.1.1   Responsive Building Structure

Case Study I: Modular Multi-agent Folding Robot
    This study combines two biologically inspired principles:

(a) The principle of folding which enables animals and plants to change surface properties and allows flexible movement in general. Such deployability often results in higher energy-efficiency e.g. the blooming of buds and leafs allows the process of photosynthesis from the very beginning of unfolding.
(b) The second principle is the way of controlling complex movements through swarm behavior based on local interactions.

    The intention is to obtain a responsive building structure e.g. a roof which can adapt to the environmental conditions. Hence it can achieve better performance e.g. in terms of building climatic properties such as maximizing solar gains in cold surrounding while providing optimum shading without blinding. A possible scenario could include optimizing the airflow around and within a building in accordance with the current locally measured wind circulation in order to provide cooling or fresh air in hot surroundings. This example poses a complex problem since turbulences in airflow respond dramatically to even slight changes in the conditions and hence are mostly hard to predict and cope with through predefined or non-responsive configurations. The proposed strategy can adaptively "learn" and respond to unforeseen constellations. The setup uses irregular folding structures in order to achieve a higher range of possible shapes. Thus instead of precalculated configurations new solutions can emerge from the behavior. The potential of deployable folding structures for engineering and architectural purposes has been investigated and highlighted by Tachi [13]. A crucial point in rigid-foldable structures is to ensure finite motion. In our case the transition between the folding states of the global structure is constrained by the foldability at each node of the mesh structure and the current fold angles around the neighboring nodes at the same time. Hence the motion planning is rather complex and the search space for behavioral strategies is huge. Since the actual target states of movement are also subject to optimization and not predetermined it seems natural to include the task of motion planning along with this process. Thus it becomes rather a process of motion search and finding. This case study intends to deal with the complexity of these tasks by regarding each fold angle as an agent in a multi-agent-system which is physically embodied within a swarm robotic building [as seen in Fig. 2], [14]. The experimental setup consists of 22 agents, equipped with sensors and actuators respectively. The setup was developed as a testbed for different PSO algorithms.

However this approach is limited by the design of the folding pattern which has been embodied into the structure beforehand. Hence this robotic building can only behave within the range of possible states. To overcome this limitation it would require to redesign and possibly extend its folding pattern and to perform physical construction processes.

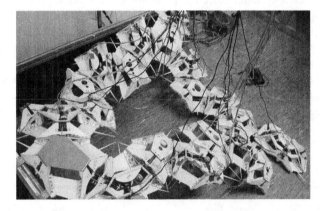

Fig. 2. Shape robot: multi-agent folding for adaptive behavior, B. Felbrich (2014)

3.1.2 Emergent Design Approach

Case Study II: Evolutionary Cellular Design Engine

The study investigates the potential application of agent based modules to Architectural Design. The objective is to utilize the emergent qualities of self-organized systems for the search of design solutions. Therefore, we formulate an experimental setup for design approaches in architecture that are generated in an evolutionary process [Fig. 3]. The setup is a virtual simulation only that is set up in Grasshopper, which is a Plugin for the CAD Software Rhino.

Research on the analogies between natural principles has led to the introduction of principles of self-organization in the architectural field. Especially swarm building behavior of termites, just as informal settlements offer a conversion from biological processes to design approaches in architecture. As a study, a generator is set up that is based on a cellular automaton. Common or uncommon rules of design are hereby integrated into the agent simulation that finally defines the configuration of modules. The specific application works with the task of layout design in case of apartment housing. This task still represents a major part of classic architectural design work. Systems Theory just like architectural theories like structuralism [15] support the understanding of buildings as systems of interrelated elements and their organization. Those modules are defined as simple room configurations like 'Garden', '2 rooms', 'Living Room' and so on. Rules describe pleasant or unpleasant configurations of rooms, open spaces, directions, and the number of appearances in one house that then affect the overall fitness of the design solution. Especially interesting is the appearance of the Von-Neumann-neighborhood, both in modular space design and three-dimensional cellular automata. This similarity is used in the program. Different rooms with different states are set in relation to their neighboring modules. Those relations are finally evaluated. The applied evolutionary solver then either selects the most potent option or proposes different solutions that can be found on the pareto front.

The generator finds very performant solutions to single problems, however it's performance concerning multi-objective problems often lacks quality. To select possible design options from a pareto front with up to 20 dimensions, the different requirements often are rebalanced in their resulting effect on an overall fitness. However, the generator does find designs that are not intrinsically found, and asymmetric, very unlike from classic housing designs.

As far as the simulation goes, we can find very suitable solutions for a virtual setup which is defined by the author. However, this simulation is configured based again on tacit knowledge and beliefs of the author that might not work accordingly in reality.

Fig. 3. Simulation engine for evolutionary optimized cellular housing design

Therefore, a Swarm Intelligent process, including the physical setup and real application could gain further knowledge and therefore update the virtual model and respond to newly emerging needs during application. Different visions of metabolist architecture sketched such settings of potentially adaptive modular housing. It is therefore an ambitious approach, but not unrealistic in the future of architecture.

4   Synthesis

As we can see Swarm Intelligence can influence architecture heavily, but those examples are very specific and do not change the principle structure of architectural ontogeny.

Swarm Intelligent applications however could change the way that buildings evolve in the construction process itself. Those processes could be self-responsive, taking place in a physical environment, being optimized in ways of efficiency in work distribution and a very short term reaction process between possible changes in the requirements of the building and the physical structure. One potential field is distributed construction by robots, like distributed 3D printing [11].

The limitation of optimizing in different phases of the ontogeny could be overcome by Swarm Intelligent processes going beyond the three different phases of a building. They could interconnect those phases and change the static order of Design, Construction and Use. Here we propose a comprehensive setup for future building systems, that interlinks strongly throughout the three phases of development [Fig. 4]. Natural processes as growth, adaptation and evolution can serve as useful analogies.

In nature, the design of an individual is firstly shaped by evolution and secondly by morphologic adaptations, namely growth processes that are based on the individual's situation and thirdly on movements, such as tropisms and nastic or rapid movements in plants. Growth and function are linked in homogenous processes. Obviously a tree is growing with nutrition it gets from its daily working functions (use), adapting to its

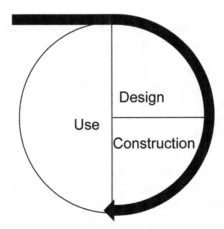

Fig. 4. 3 phases of a buildings ontogeny linked in an iterative system

very situation and place (design), growing taller and fit to its tasks (materialization). Biological design processes hence demonstrate the overlap of those 3 fields as the most potent option.

We propose an iterative time scale: The intentional design is not a final design but rather a proposition that is further evolved during construction and use. Therefore the process needs constant cycles through the states of the buildings ontogeny.

5 Conclusion

In theory, an approach of combining swarm intelligence in design, construction and use in one process could even lead to a paradigm shift in the relation of a building and time. Looking at this relation in different scales we can see 3 periods in which adaptation can take place, and in some cases Swarm Intelligence is already a part or architectural reality:

(a) Within a period of use, or situation, usually minutes to hours. One example of this would be the change of weather or the absence or presence of people in specific places.
(b) Within a buildings ontogeny, usually between one and hundreds of years. An example would be the changed requirements towards a building because of new clients or a different number of client members.
(c) Within ages of change, meaning changes that can take place in hundreds of years, like the evolution from an agricultural to the industrial age.

A major difference between (a) and (b) towards (c) is that the adaptation over hundreds of years does usually not take place within one building but can be understood in generations of buildings. Churches, for example, are in many ways still very alike to the first churches of roman times. However it is fairly easy to see significant differences between different centuries. Historically, case (c) is a case in which we can find swarm behavior shaping design, without the conscious intention to do so. Every human being can be seen as a potential agent that observes, processes and releases design.

Another scale on which different types of swarm adaptation can be found is the scale of architecture. Swarm intelligence can or could affect:

(a) the chemical configuration of a buildings material
(b) the material properties including smart properties, texture, e.g.
(c) technical components
(d) rooms or buildings
(e) networks of buildings like cities
(f) global or universal scale

Swarm Intelligent structures that already exist are especially found in the big scale of villages, cities and so on, and eventually in a very small scale, concerning chemical configurations that are rarely studied. This is due to the size and complexity of a single agent. Human agents can be found in the cases (e) and (f) but not in smaller structures.

What we can see in cities as a case model of agent based design is that the stages of the cities ontogeny are not happening after another but simultaneously, very alike to the morphology of organisms. This feature, the ontogeny without separate phases, could possibly be adapted to an architectural ideal model. This results in structures that evolve in an integral process including design, construction and use.

Hence, we propose an extension of this current state of Swarm Intelligence in Architectural Design. By integrating self-organized construction processes that are closely related to the design process and to the usage, we combine the different tendencies to a more resilient system that covers a buildings ontogeny from beginning to end [Fig. 5].

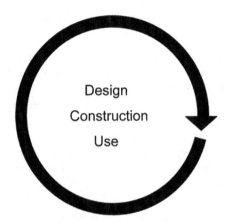

Fig. 5. 3 phases of a buildings ontogeny taking place simultaneously

However we have to note considerable differences between biological organisms and a house in terms of adaptation that lead to the conclusion that natural design is not an essentially suitable model for physical architectural adaptation. It is a very own feature of evolution to fail in single instances while leading to emergent solutions in the scale of species and ecosystems. Architecture needs to avoid failure at any cost. Failing could either mean a lack of fitness of a given solution, or fitness criteria that do not correlate to actual needs or interests. It has to be noted that those interests or needs, are, even in a fully agent-based emergent design theory, still defined by human beings. The proposed model cannot exclude the verbalization of a design target, and doesn't answer the question of how to verbalize architectural needs.

If we refer to a virtual simulation of physical spaces, individual solutions that lack quality are welcome, and even necessary for optimization. We have to note that virtual simulation is inevitable in architectural adaptation if we want to make use of evolutionary optimization. In those cases, optimization must take place prior to the process of materialization. In an overall swarm intelligent system however, this optimization must not lead to a definite design intention but could refer to the process of construction or would rather describe the structure instead of the explicit space it defines.

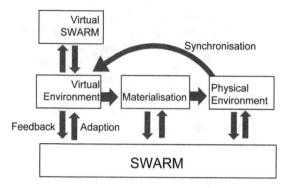

Fig. 6. Swarm processes affecting all aspects of architectural design

In a concluding proposition of Swarm Intelligence in architecture, we see a feedback- and adaption-relation between the physical swarm and the three states of the buildings ontogeny [Fig. 6]. The virtual environment is a simulation environment for a virtual swarm representation. This swarm is potentially making use of evolutionary algorithms to constantly simulate and select different up-to-date options that are then materialized to become part of the architecture in its environment physically. The virtual environment is finally synchronized with any active or passive change of the physical environment to find correlating adaptations. The swarm is therefore part of the virtual environment and part of the physical design but also the agent materializing design. This results in agents that have the following features: (a) ability to sense environment, (b) self-monitoring (c) gather and sending information to a virtual swarm application (d) adaptivity of the agent's state (e) growth by changing the number of agents or amount of material (f) reconfiguration.

However we have to note that applications including virtual swarms and physical swarms in a synchronized set up, could potentially lead to higher resilience, without the features (e) and (f).

References

1. Laureano-Cruces, A., Barcelo, A.A.: Formal verification of multi-agent systems behaviour emerging from cognitive task analysis. J. Exp. Theor. Artif. Intell. **15**(4), 407–431 (2003)
2. Hemberg, M., O'Riley, U., Menges, A., Jonas, K., da Costa Goncalves, M.K., Fuchs, S.: Genr8: architects experience with an emergent design tool. In: Machado, P., Romero, J. (eds.) Artificial Evolution, Chap. 8, pp. 167–188. Springer, Heidelberg (2007)
3. Swarm Tectonics. In: Leach, N., Turnbull, D., Williams, C. (eds.) Digital Tectonics, pp. 70–77. John Wiley & Sons, London (2004)
4. Alexander, C.: Notes on the Synthesis of Form, pp. 28–34. Harvard University Press, Cambridge (1964)
5. Terzidis, K.: Algorithmic design: a paradigm shift in architecture, architecture in the network society. In: 22nd eCAADe Conference Proceedings, Copenhagen, pp. 201–207 (2004)

6. Narahara, T.: Adaptive Growth using robotic fabrication. In: Stouffs, R., Janssen, P., Roudavski, S., Tunçer, B. (eds.) Open Systems Proceedings of the 18th International Conference on Computer-Aided Architectural Design Research in Asia, CAADRIA, pp. 65–74 (2013)
7. Terzidis, K.: Algorithmic Architecture. Architectural Press, Oxford (2006)
8. Camazine, S., Deneubourg, J.L., Franks Nigel, R., Sneyd, J, Theraulaz, G., Bonabeau, E.: Self-Organization in Biological Systems, S. 341–375, S. 405–442 Princeton Studies in Complexity, ISBN 0-691-11624-5 (2001)
9. Murugananthama, A., Zhaoa, Y., Geea, S.B., Qiub, X., Tana, K.C.: Dynamic multiobjective optimization using evolutionary algorithms. In: IES 2013 (2013). http://www.sciencedirect.com/science/article/pii/S1877050913011708
10. Noennig, J.R., Wiesenhütter, S.: Parametric ideation: interactive modeling of cognitive processes. In: Streitz, N., Stephanidis, C. (eds.) DAPI 2013. LNCS, vol. 8028, pp. 225–234. Springer, Heidelberg (2013)
11. Parker, A.C., Zhang, H., Kube, R.C.: Blind bulldozing - multiple robot nest construction. In: Proceedings of the 2003 IEEE/RSJ International Conference on Intelligent Robots and Systems (IROS), Las Vegas, pp. 2010–2015 (2003)
12. Mitchell, W.: The Logic of Architecture, Design, Computation, and Cognition, Cambridge, pp. 64–78 (1990)
13. Tachi, T.: Generalization of rigid foldable quadrilateral mesh origami. In: Domingo, A., Lazaro, C. (eds.) Proceedings of the International Association for Shell and Spatial Structures (IASS) Symposium, Evolution and Trends in Design, Analysis and Construction of Shell and Spatial Structures, Valencia, pp. 2287–2294 (2009)
14. Felbrich, B., Lordick, D., Nönnig, J., Wiesenhütter, S.: Experiments with a folding multi-agent system in the design of triangle mesh structures. In: 16th International Conference on Geometry and Graphics– ICGG 2014 (2014)
15. van Eyck, A.: Aesthetics of number. In: Forum 7/1959, Amsterdam-Hilversum

Shaping Influence and Influencing Shaping: A Computational Red Teaming Trust-Based Swarm Intelligence Model

Jiangjun Tang[1](✉), Eleni Petraki[2], and Hussein Abbass[1]

[1] School of Engineering and Information Technology, University of New South Wales, Canberra, ACT 2600, Australia
{j.tang,h.abbass}@adfa.edu.au
[2] Faculty of Science, Technology, Education, and Mathematics, University of Canberra, Canberra, Australia
Eleni.Petraki@canberra.edu.au

Abstract. Sociotechnical systems are complex systems, where nonlinear interaction among different players can obscure causal relationships. The absence of mechanisms to help us understand how to create a change in the system makes it hard to manage these systems.

Influencing and shaping are social operators acting on sociotechnical systems to design a change. However, the two operators are usually discussed in an ad-hoc manner, without proper guiding models and metrics which assist in adopting these models successfully. Moreover, both social operators rely on accurate understanding of the concept of trust. Without such understanding, neither of these operators can create the required level to create a change in a desirable direction.

In this paper, we define these concepts in a concise manner suitable for modelling the concepts and understanding their dynamics. We then introduce a model for influencing and shaping and use Computational Red Teaming principles to design and demonstrate how this model operates. We validate the results computationally through a simulation environment to show social influencing and shaping in an artificial society.

Keywords: Influence · Shaping · Trust · Boids

1 Introduction

Recently, computational social scientists are attracted to studying means for measuring the concepts of influence and shaping. For influence to work is to exert a form of social power. Servi and Elson [1] introduce a new definition of influence which they apply to online contexts as 'the capacity to shift the patterns of emotion levels expressed by social media users'. They propose that measuring influence entails identifying shifts in users' emotion levels followed by

Portions of this work was funded by the Australian Research Council Discovery Grant number DP140102590.

Y. Tan et al. (Eds.): ICSI 2016, Part I, LNCS 9712, pp. 14–23, 2016.
DOI: 10.1007/978-3-319-41000-5_2

the examination of the extent to which these shifts can be connected with a user. However, if the process of influence creates shifts in patterns of emotions which can be detected in the short-term, can a persistent application of influencing operators create a long-term shift (*i.e.* shaping)?

Shmueli et. al. [2] discuss computational tools to measure processes for shaping and affecting human behaviour in real life scenarios. Trust was identified as a means to influence humans in a social system. Moreover, trust was found to have a significant impact on social persuasion. Trust is a complex psychological and social concept. A review of the concept can be found in [3].

Larson et. al. [4] are among a few to imply a distinction between influence and shaping, whereby shaping is perceived to be a change to the organization or the environment, while influence fosters attitudes, behaviours or decisions of individuals or groups. However, the majority of the literature follows a tendency to assume that social influencing would lead to shaping.

In this paper, we aim to distil subtle differences to distinguish between the two concepts. This distinction is very important for a number of reasons. First, it examines the validity of the implicit assumption that influencing is a sufficient condition for shaping. Second, it is important when studying social sciences using computational models (computational social sciences) to create models that are not ambiguous about the social and psychological phenomena under investigation. Third, it is vital to make it explicit that social influencing and shaping work on different time scales; that is, social influencing is effective in the short run, while shaping requires time and is more effective in the long run.

We will use a computational red teaming (CRT) model, whereby a red agent acts on a blue team to influence, shape and sometimes distract the blue team. The idea of the model should not be seen from a competition or conflict perspective. The model is general, were the red agent can be an agent that promotes a positive attitude within a team (a servant leader) or a social worker correcting the social attitude of a gang.

2 Influence, Shaping, and Trust

Influence will be defined in this paper as: an operation which causes a short-term effect in the attitude or behaviour of an individual, group or an organization. Shaping, on the other hand, is defined as: an operation which causes a long-term effect in the attitude or behaviour of an individual, group or an organization.

We use the more accurate term "effect" rather than the term "change" because sometimes social influence and shaping need to operate to maintain the status quo. If agent A is attempting to influence agent B by changing B's behaviour, agent C can attempt to counteract agent A's influence by influencing B to maintain its behaviour. Therefore, influence does not necessarily require a change to occur.

In a strict mathematical sense, social influence would change the parameters of a model, while social shaping would alter the constraint system.

To illustrate the difference, we will use a model, whereby a group of blue agents attempts to follow a blue leader. A red agent has a self-interest to influence or shape the blue team. All agents are connected through a network. Each agent, excluding the blue leader and the red agent, attempts to align their behaviour with their neighbours (the other agents it is connected to). The blue leader attempts to reach the blue goal (a position in space). When all agents fully trust each other, and in the absence of the red agent's effect, it is expected that the intention of the blue leader will propagate throughout the network. Over time, the blue agents will also move towards the blue goal.

The red agent has a different goal. It aligns with the agents it is connected to, but it also attempts to influence and/or shape them towards its own goal (or away from the blue's goal). Social influence by red is represented through changing red movements; thus, affecting the movements of its neighbours. Social shaping by red is represented through a network re-wiring mechanism. Connections in a network are the constraints on the network's topology. By rewiring the network, the red agent changes the constraints system. We abstract trust to a scale between -1 and 1, whereby "1" implies maximum trust, while "-1" implies maximum distrust. We do not differentiate in this paper between distrust and mistrust. A "0" value is a neutral indicator that is equivalent to not knowing a person.

3   The Model

An agent-based Boids [5] model is proposed in this paper. All agents are randomly initialized with random headings and locations. Agents are connected through a network structure that allows information exchange among the agents. In this setup, the neighborhood is mostly defined by the hamming distance between two agents in the network, while sometimes it will be defined by the proximity of one agent to another in the physical space. This latter definition is the classic and default one used in the Boids model. Three Boids rules: cohesion, alignment, and separation, are still adopted here. However, the first two vectors are sensed by network connections while the separation vector is perceived through the Euclidean distance in the physical space. Each agent has a trust factor value which decides how much this agent trusts the information perceived from others. The first two vectors are scaled using the trust factor before an agent's velocity gets updated. An agent 100 % trusts the cohesion and alignment information from its linked neighbours when it has a trust factor of 1. When the trust factor is -1, the agent totally believe that the information is deliberately altered to the opposite value, and therefore, the agent reverses the information it receives.

In the model, there are three types of agents: blue leader (A_B), red agent (A_R), and blue agent. The blue leader always moves towards a specific location/goal, and attempts to make the other blue agents follow it. The blue agent is an agent that senses its neighbours through network links for both cohesion and alignment but by Euclidean distance for separation, and then makes decisions on its new velocity. The red agent is a special agent in the model who controls the level of noise (η) in the velocity and network connections for influencing and shaping. Many blue agents can exist but there is only a single blue leader and a single red agent.

Agents form a set A and live in a space (S) defined by a given width ($spaceW$) and a given length ($spaceL$). All agents are connected by a random network. To establish network connections, a probability (p) is defined. If we have n agents including one blue leader, one red agent, and $n-2$ blue agents, the network can be denoted as $G(n, p)$. A Goal (G) is a 2-D position that sits at one of the corners of S. Blue leader always aims to move towards G. The area surrounding of G is denoted by δ. Once the blue leader enters this area, the position of G changes. An agent has the following common attributes:

- Position (p), $p \in S$, is a 2-D coordinate.
- Velocity (v) is a 2-D vector representing the agent's movement (heading and speed) in a time unit.
- Cohesion Velocity ($cohesionV$) of an agent is the velocity calculated based on the mass of all agents that are connected to this agent.
- Alignment Velocity ($alignmentV$) of an agent is the velocity calculated based on the average velocity of all agents that are connected to this agent.
- Separation Velocity ($separationV$) of an agent is the velocity that forces this agent to keep a certain small distance from its neighbors and is based on the Euclidean distance.
- Velocity weights:
  - Cohesion weight (w_c): a scaler for the cohesion velocity.
  - Alignment weight (w_a): a scaler for the alignment velocity.
  - Separation weight (w_s): a scaler for the separation velocity.
- Trust factor (τ) defines how much an agent trusts its connected neighbours. It has an impact on both the cohesion velocity and alignment velocity but not on the separation velocity.

All agents except the blue leader attempt moving towards the neighbours' location guided with the cohesion vector. The cohesion vector, $cohesionV_i$, of an agent A_i is:

$$cohesionV_i = \frac{\sum_{j=0}^{|N|} p_j}{|N|} - p_i \tag{1}$$

where, $|N|$ is the cardinality of the neighbourhood N.

The alignment velocity of an agent with its linked neighbours is:

$$alignmentV_i = \frac{\sum_{j=0}^{|N|} v_j}{|N|} - v_i \tag{2}$$

The separation velocity of an agent is calculated using neighbours N_d in the spatial proximity of other agents as follows:

$$separationV_i = -\sum_{j=0}^{|N_d|} (p_j - p_i) \tag{3}$$

The trust factor of a blue agent is updated by the average trust factors of all its connected neighbours (N) as below:

$$\tau_i = 0.5 \times (\tau_i + \frac{\sum_{j=0}^{|N|} \tau_j}{|N|}) \tag{4}$$

The blue leader and red agent's trust factors are not updated.

The velocity of the blue leader always aims at the goal G at each step and it is not affected by any other factor. The velocities at time t of all agents except the blue leader are updated by Eq. 5.

$$v = v + \tau \times (w_c \times cohesionV + w_a \times alignmentV) + w_s \times separationV \tag{5}$$

where, $cohesionV$, $AlignmentV$, and $separationV$ are normalized vectors. The position at time t of each agent can be updated by:

$$p = p + v_t \tag{6}$$

If an agent's new position is outside the bounds of S, the reflection rule is applied. According to Eq. 5, an agent adjusts its own velocity in compliance with both $cohesionV$ and $alignmentV$ when it has a positive trust value, and follows the opposite direction as suggested by $cohesionV$ and $alignmentV$ when its trust factor is negative. If $\tau = 0$, only the separation vector takes effect on this agent so that this agent doesn't anyone.

The red agent introduces heading noise, changes its network structure, or does both at each time step. The heading noise can be propagated to blue agents through the connections of the network to cause a deviation in some blue agents' moving directions. Changes in the network structure may result in long term effects on blue agents.

At each time step, Red agent updates its own velocity $(v_{RedAgent})$ by Eq. 5 and then Eq. 7 uses a normal distribution $(N(0, \eta))$ to generate noise and add it to Red's velocity.

$$v_{RedAgent} = v_{RedAgent} + N(0, \eta) \tag{7}$$

Eq. 6 is used to update the red agent's position.

Furthermore, the red agent has the ability to re-configure network connections by using the noise level η as a probability that governs the eventuality of the following steps:

1. Randomly pick up a blue agent (A_i) who is connected with the red agent.
2. Randomly pick up another blue agent (A_j) who is connected with A_i.
3. Break the connection between A_i and A_j.
4. Connect the red agent with a randomly chosen blue agent A_j.

In this way, the connection between the red agent and blue agents changes but the number of edges of the whole network remains as before. The long term effects of these topological updates are expected because the path along with information propagates changes and some blue agents may not get consistent updates from their neighbours.

The blue leader attempts to lead other blue agents towards a given destination, and the red agent attempts to disorient through influence (deliberate changes in heading) and/or shaping (deliberate re-configuration of network structure). Therefore, the "effect" from our model can be derived as how well the blue agents follow the blue leader given the influence/shaping by the red agent. A straightforward measure of this effect within our model is the average distance between blue agents and the goal when the blue leader reaches the goal. If this distance is small, blue agents followed the blue leader. If it is large, red agent distracted the blue agent.

During a single simulation run, the blue leader is tasked to reach the goal multiple times. Each time it reaches the goal (an iteration), the location of the goal changes. The effect is measured at the end of each iteration. The overall effect of a simulation run is the average of all iterations except the first iteration, which is excluded to eliminate the warm-up period in the system resultant from the random initialisation of agents. In summary, the effect is defined by Eq. 8.

$$\bar{d} = \frac{1}{M} \left(\sum_{m=1}^{M} \frac{1}{n} \sum_{i=1}^{n} d_{m,i} \right) \tag{8}$$

where, M is the number of iterations except the first one, n is the number of blue agents, and $d_{m,i}$ is the distance between agent i and the goal location at the m'th iteration.

4   Experimental Design

Our aim is to evaluate the "effects" of the short term inference and long term shaping caused by the red agent. Two stages are used for the experiments. The first stage focuses on the noise of red agent where the effect from the trust factor is minimised. The second stage investigates both the trust factor and the red agent's noise. The number of blue agents is 25, so there are a total of 27 agents including a blue leader and a red agent. All agents' initial locations are uniformly initialised at random within S, where S is a square space with the dimension of 500×500. All agents' headings are uniformly initialised at random with constant speed of 1. All agents except the blue leader have the same velocity weights: $w_c = 0.4$ for cohesion, $w_a = 0.4$ for alignment, and $w_s = 0.2$ for separation. The initial trust factor values of all blue agents are uniformly assigned at random within the range of $[-1, 1]$. Connections among agents are created by a random network $G(n, 0.1)$, where $n = 27$.

In all experiments, two levels of noise ($\eta^- = 0.1$ and $\eta^+ = 0.9$) are used. In the first stage, to reduce the effect of the trust factor, it is assumed constant with a value of 1 for all agents; that is, all blue agents trust any perceived information, including the information arriving from red. In the second stage, the blue leader has two trust levels: $\tau_B^- = 0.2$ and $\tau_B^+ = 1$, and the red agent has two levels of trust: $\tau_R^- = -0.2$ and $\tau_R^+ = -1$.

Three scenarios are designed for investigating the red agent's impact in our experiments. In Scenario 1, the red agent introduces noise to its heading at

Table 1. Results of red agent's noise impact when $\tau_B = 1$ and $\tau_R = 1$

	R1	R2	R3	R4	R5	R6	R7	R8	R9	R10	Avg	STD	Conf
Scenario1: velocity noise													
$\eta = 0.1$	47.64	46.78	39.26	56.63	47.09	67.29	60.65	38.76	42.99	44.86	49.19	8.90	6.36
$\eta = 0.9$	145.90	155.75	168.04	199.94	171.94	243.61	162.15	144.08	103.82	117.94	161.32	37.65	26.93
e_η	98.27	108.97	128.78	143.31	124.85	176.33	101.50	105.33	60.83	73.08	112.12	31.70	22.68
Scenario 2: network changes													
$\eta = 0.1$	45.71	59.28	47.39	54.31	58.14	69.65	50.27	44.35	43.90	48.83	52.18	7.78	5.56
$\eta = 0.9$	61.23	57.63	56.30	81.25	53.65	74.69	55.76	40.86	47.74	52.03	58.11	11.36	8.13
e_η	15.52	−1.65	8.91	26.94	−4.49	5.04	5.49	−3.49	3.85	3.20	5.93	9.00	6.44
Scenario 3: velocity noise and network changes													
$\eta = 0.1$	45.34	47.09	65.90	54.05	51.93	84.91	54.66	41.11	43.88	52.21	54.11	12.23	8.75
$\eta = 0.9$	213.49	168.69	197.52	188.80	171.62	236.93	174.46	183.98	84.95	122.82	174.32	41.20	29.47
e_η	168.15	121.59	131.62	134.75	119.69	152.02	119.80	142.87	41.07	70.61	120.22	35.90	25.68

each time step thus this noise can immediately affect direct neighbours and can be propagated through the network. In Scenario 2, the red agent changes the network structure at each time step thus shaping the environment of the blue agents. In Scenario 3, the red agent introduces noises to its heading and changes network structures at each time step, so that both influence and shaping take place in our model.

Using 2^k factorial design [6], a total of 2 factor combinations is available at the first stage and 8 at the second stage. Each combination has three individual scenarios to study. Moreover, the randomness exists in the initialisation phase, therefore 10 runs for each factor combination and impact are desired in order to obtain meaningful results. In summary, there are 60 ($3 \times 2 \times 10$) simulation runs in the first stage and 240 ($3 \times 8 \times 10$) runs in the second stage. The results and analysis are provided in the next section.

5   Results and Discussion

The results from the first stage experiments are presented in Table 1. The distance between blue agents and goal location of each run ($\bar{d}$) is listed from the second column to the eleventh column. And the last three columns of the table are the averages of 10 runs, the standard deviations and the confidence intervals that are obtained at $\alpha = 0.05$.

The results show that the more noise the red agent has in its velocity, the more deviation from the goal observed by the blue agents. Changes in the network structure can lower blue agents performance, although the magnitude of this decrease may not be significant. This is expected since the shaping operates work on a smaller timescale than the influence operator. When both influence and shaping work together, the effect is more profound than any of the individual cases in isolation.

Table 2. Results of effects from red agent's noise impact and trust factors. The confidence level is at 0.05.

Effect	R1	R2	R3	R4	R5	R6	R7	R8	R9	R10	Avg	STD	Conf
Scenario 1: velocity noise													
e_{τ_B}	−170.40	−146.54	−83.92	−55.00	−131.83	−6.05	−128.09	−110.66	−184.81	−152.86	−117.02	55.01	39.35
e_{τ_R}	165.32	160.08	95.83	71.21	149.41	8.11	133.20	111.97	167.79	160.23	122.31	51.75	37.02
e_N	1.78	15.84	−9.18	6.22	6.74	3.40	−13.50	0.35	−14.64	−17.58	−2.06	11.04	7.90
Scenario 2: network changes													
e_{τ_B}	−122.99	−165.55	−144.34	−64.94	−168.15	−8.09	−154.27	−170.61	−189.59	−187.66	−137.62	58.31	41.71
e_{τ_R}	142.72	177.17	154.10	77.45	186.62	24.03	164.25	172.21	171.13	172.02	144.17	52.25	37.38
e_N	42.90	1.75	12.81	19.08	8.50	−1.08	−0.29	−12.17	25.41	−15.03	8.19	17.62	12.61
Scenario 3: velocity noise and network changes													
e_{τ_B}	−151.65	−140.59	−132.44	−35.97	−166.98	−7.63	−159.37	−171.89	−194.38	−163.85	−132.47	61.12	43.72
e_{τ_R}	157.63	152.23	147.97	38.06	176.90	16.03	175.96	163.48	171.95	174.82	137.50	59.31	42.43
e_N	2.27	22.60	15.16	14.70	21.23	4.64	5.73	10.22	20.50	8.15	12.52	7.38	5.28

The results from the second stage are summarised in Table 2. Interestingly, the trust factors of the blue leader and red agent become critical but the red agent's noise is not critical. The model responses to the trust factor as expected. When the blue leader has a higher level of trust, all blue agents follow it better (smaller effect values can be observed from e_{τ_B}). On the other hand, the blue agents demonstrate disorder behaviours (larger value of e_{τ_R}) if red agents have small negative trust. These situations are found in all three scenarios and the effects of trust factors are all statistically significant. Although the red agent's noise has some effects on blue agents, it is very little and can be ignored when compared to the effect of the trust factor. Negative trust values taken by the blue agents counteract the influence generated by both blue and red agents.

The red agent's noise can have impact on blue agents' behaviours through short term influence (velocity) and long term shaping (network structures) if the effects of trust are low. When the trust factors are high, the situation changes. Trust has a significant impact on blue agents' behaviours.

Figure 1 illustrates the agents' footprints when the red agent impacts velocity, network structure or both but with minimum trust effects. These footprints are obtained from the first runs of all three scenarios in the first stage and the results are listed in the column "R1" of Table 1.

Figure 1a–c show that the blue leader leads other agents towards the goal well as being demonstrated by a few congested trajectories. When noise increases, blue agents' trajectories are disturbed as shown in Fig. 1d. Figure 1e shows that changes in the network structure seem to not generate much effects on blue agents' behaviours. However, the blue agents behaviours are more random when red affects both velocity and network structure. This manifests disorderliness as more scattered blue agents' footprints can be observed in the last figure.

Figure 2 shows two examples of agents' footprints that are affected by trust with small noise values ($\eta = 0.1$). The footprints presented in Fig. 2a are extracted from the first run of the third scenario in the second stage with $\tau_B = 0.2$ and $\tau_R = -1$. When the red agent's trust is −1, the negative effect on blue agents' trust is continuously broadcasted throughout the network.

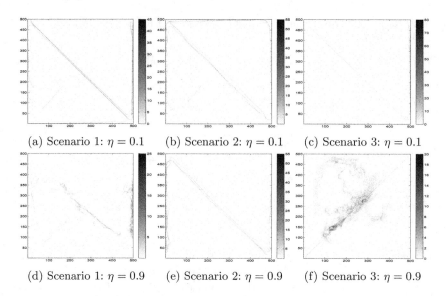

(a) Scenario 1: $\eta = 0.1$ (b) Scenario 2: $\eta = 0.1$ (c) Scenario 3: $\eta = 0.1$

(d) Scenario 1: $\eta = 0.9$ (e) Scenario 2: $\eta = 0.9$ (f) Scenario 3: $\eta = 0.9$

Fig. 1. Agents' footprints under Red agent's noise (η) impacts on velocity and network with minimum trust effects ($\tau_B = 1$ and $\tau_R = 1$).

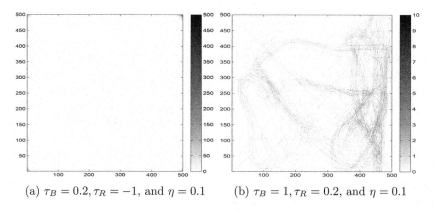

(a) $\tau_B = 0.2, \tau_R = -1$, and $\eta = 0.1$ (b) $\tau_B = 1, \tau_R = 0.2$, and $\eta = 0.1$

Fig. 2. Trust effects on agents behaviours with red agent noise level at 0.1 in scenario 3.

Eventually, all blue agents will have a negative trust value that is close to -1 since the blue leader doesn't have much power ($\tau_B = 0.2$) to compete against the red agent. This results in all blue agents distrusting each other. In this case, the blue agents spread out to the boundaries. However, the reflection rule forces them back into the given space, causing the blue agents to move around the corners after several time steps as shown in Fig. 2a.

The right side of Fig. 2 depicts agents' footprints extracted from the third scenario in the second stage with $\tau_B = 1$, $\tau_R = -0.2$, and $\eta = 0.1$.

Some trajectory patterns can be observed from Fig. 2b. In this case, the blue leader has enough power to beat the red agent in terms of trust. All blue agents will have positive trust that are passed from the blue leader. Although the red agent has influence on their velocity and connections, the blue agents are still capable to follow the blue leader to reach the goal locations (corners) as the trajectory patterns show.

From the above examples and previous results, it can be concluded that trust has a more significant impact on blue agents behaviours than the effect of noise caused by the red agent.

6 Conclusion

In this paper, we presented a CRT trust-based model which is an extension of the classic Boids. The network topologies for situation awareness and a trust factor on perceived information are introduced into our model. They provide the necessary tools to investigate influence and shaping using CRT.

A number of experiments are designed and conducted in order to differentiate the potential impact from influence and shaping on a system. As the results of the first experimental stage suggest, short term influence can have an immediate effect on the system which is easily observed. The long term shaping effects may not be easily observable although it has effect on the system, especially when it interacts with influence. However, trust among agents plays a critical role in the model. Based on our findings in the second experiment, trust dominates the agents' behaviours regardless of noise.

References

1. Servi, L., Elson, S.B.: A mathematical approach to gauging influence by identifying shifts in the emotions of social media users. IEEE Trans. Comput. Soc. Syst. **1**(4), 180–190 (2014)
2. Shmueli, E., Singh, V.K., Lepri, B., Pentland, A.: Sensing, understanding, and shaping social behavior. IEEE Trans. Comput. Soc. Syst. **1**(1), 22–34 (2014)
3. Petraki, E., Abbass, H.: On trust and influence: a computational red teaming game theoreticperspective. In: 2014 Seventh IEEE Symposium on Computational Intelligence for Security and Defense Applications (CISDA), pp. 1–7. IEEE (2014)
4. Larson, E.V., Darilek, R.E., Gibran, D., Nichiporuk, B., Richardson, A., Schwartz, L.H., Thurston, C.Q.: Foundations of effective influence operations: a framework forenhancing army capabilities. Technical report, DTIC Document (2009)
5. Reynolds, C.W.: Flocks, herds and schools: a distributed behavioral model. ACM SIGGRAPH Comput. Graph. **21**, 25–34 (1987). ACM
6. Montgomery, D.C.: Design and Analysis of Experiments. Wiley, Hoboken (2008)

Research Hotspots and Trends in Swarm Intelligence: From 2000 to 2015

Zili Li, Li Zeng[✉], Hua Zhong, and Jinhong Wu

Center for National Security and Strategic Studies (CNSS),
National University of Defense Technology, Changsha 410073, China
{zilili, crack521, wjh}@163.com, gfkj@nudt.edu.cn

Abstract. Swarm Intelligence (SI) is the collective behavior of decentralized, self-organized systems, natural or artificial. This paper was to explore a bibliometric approach to quantitatively assessing current research hotspots and trends on Swarm Intelligence, using the related literature in the Science Citation Index (SCI) database from 2000 to 2015. Articles referring to Swarm Intelligence were concentrated on the analysis of scientific outputs, distribution of countries, institutions, periodicals, subject categories and research performances by individuals. Moreover, innovative methods such as keyword co-citation analysis, semantic clustering and Keyword Frequent Burst Detection were applied to provide a dynamic view of the evolution of swarm intelligence research hotpots and trends from various perspectives which may serve as a potential guide for future research.

Keywords: Swarm intelligence · Bibliometric analysis · Research hotspots

1 Introduction

The term swarm is used for an aggregation of animals such as fish schools, birds flocks and insect colonies such as ant, termites and bee colonies performing collective behavior. The individual agents of a swarm behave without supervision and each of these agents has a stochastic behavior due to her perception in the neighborhood. Local rules, without any relation to the global pattern, and interactions between self-organized agents lead to the emergence of collective intelligence called swarm intelligence [1]. In fact, Swarm Intelligence is a relatively new field of research, which has gained huge popularity in these days. Lots of researchers studied the versatile behavior of different living creatures and especially the social insects (ant, bee, termite, bird or fish) [2–4]. The efforts to mimic such behaviors through computer simulation finally resulted into the fascinating field of SI [5]. In this paper, a bibliometric analysis was performed by investigating annual scientific outputs, distribution of countries, institutions, journals, research performances by individuals and subject categories to offer another perspective on the development of SI research. Moreover, innovative methods such as Keyword co-citation analysis, semantic clustering and Keyword Frequent Burst Detection were applied to provide insights into the global research hotpots and trends from various perspectives which may serve as a potential guide for future research.

© Springer International Publishing Switzerland 2016
Y. Tan et al. (Eds.): ICSI 2016, Part I, LNCS 9712, pp. 24–35, 2016.
DOI: 10.1007/978-3-319-41000-5_3

2   Data and Methods

Data on Swarm Intelligence comes from Science Citation Index (SCI) database. We searched SCI on January 20, 2016 to identify the literature on Swarm Intelligence by entering the keyword "Swarm Intelligence". We restricted our keyword search to Topics (2000–2015) and obtained a total of 3071 papers from SCI database, Full bibliographic records (including their reference lists) of all papers were downloaded. Then we used a bibliometric approach to quantitatively assessing current research hotspots and trends on Swarm Intelligence Results.

3   Result and Discussion

3.1   Characteristics of Article Outputs

Figure 1 shows the number of articles and the prediction of the maturity of the research about swarm intelligence between 2000 and 2015. Black curve stands for the annual number of publications about swarm intelligence. From the curve, we found that a substantial interest in SI research did not emerge until 2002, although a few articles related to SI were published previously. And the highest annual number occurred in the year of 2013 with a number of 417, accounting for 13.58 % of the total number of articles. The red curve stands for the cumulative number of publication in the field of swarm intelligence. According to the theory of technology maturity, the cumulative number of the publication will be presented as an S-curve in general [7]. By using this

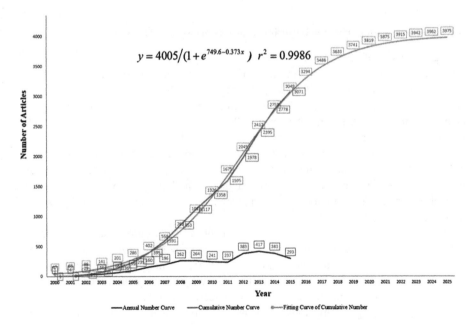

Fig. 1. Variation of article numbers (Color figure online)

theory, the cumulative number of SI articles was accurately approximated by a logistic model $\left(y = 4005/40051 + e^{(749.6-0.373x)}.1 + e^{(749.6-0.373x)}r^2 = 0.9986\right)$, where x and y denote the year and article number, respectively. According to this, we can divide the development of swarm intelligence into four stages: infant period (before 2007), growth period (2007–2012), mature period (2013–2018) and Decline period (after 2018). According to the above stage division, the research of swarm intelligence in 2015 was in the mature period with a maturity of 76.14 %.

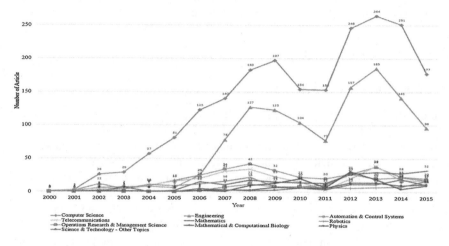

Fig. 2. Distribution of subject categories (Color figure online)

3.2 Subject Categories Distribution and Co-occurring Network

The distribution of the subject categories identified by the Institute for Scientific Information (ISI) was analyzed and the result was displayed in Fig. 3. The total of 3071 articles covered 103 ISI identified subject categories in the SCI databases. The annual articles of the top ten productive subject categories were analyzed. The five most common categories were computer science (2085 articles, 38.87 %), engineering (1157 articles, 21.57 %), automation & control systems (311 articles, 5.80 %), telecommunications (260 articles, 4.85 %), mathematics (212 articles, 3.95 %). We noticed that 60.44 % of all articles were mostly related to computer science and engineering, and the distribution of subject categories also suggested the high priority of automation & control systems, telecommunications, mathematics and robotics issues in SI research (Fig. 2).

We used CiteSpace to visualize a subject categories co-occurring network applying a threshold to the network between centrality in the network of subject categories [8]. Network centrality measures the relative importance of nodes within networks and could be used as an indicator of a subject category's position within the network [9]. We can find that the computer science and engineering took part in more co-occurring relationship, and engineering took the central position in the co-occurring network,

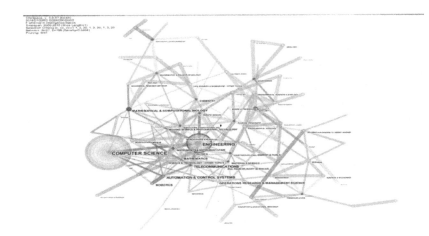

Fig. 3. Subject Categories Co-occurring Network (the thickness of each link represents the intensity of co-occurring, the size of each node represents the number of total articles, and the purple color denotes the high betweenness centrality node)

with mathematical & computational biology, remote sensing, computer science, and telecommunications as close neighbors (Fig. 4).

3.3 Geographic Distribution Map of Countries and International Collaboration

Data on geographic information were generated from author affiliations. Figure 5 shows the geographic distribution of countries (regions) in the field of "intelligence swarm". Overall, The total of 3071 articles covered 50 countries and ten most common countries were CHINA (832 articles, 27.1 %), INDIA (332 articles, 10.8 %), USA (282 articles, 9.2 %), IRAN (131 articles, 4.3 %), BRAZIL (105 articles, 3.4 %), ITALY (95 articles, 3.1 %), ENGLAND (92 articles, 3.0 %), SPAIN(89 articles, 2.9 %), JAPAN (85 articles, 2.8 %), TURKEY (83 articles, 2.7 %).

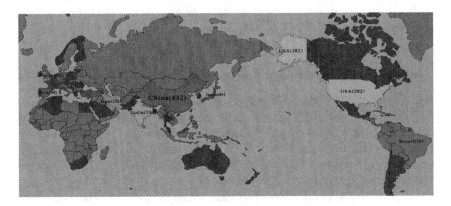

Fig. 4. Geographic distribution map of countries

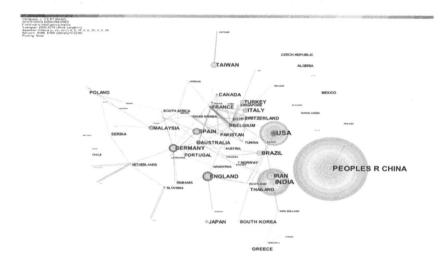

Fig. 5. A network of country of 85 Nodes and 83 Links on SI research publications

Figure 6 depicts a network consisting of 85 nodes and 83 links on behalf of the collaborating countries between 2000 and 2015. As can be seen, the major contribution of the total output mainly came from four countries, namely, CHINA, INDIA, USA and Iran. Clearly, CHINA is the largest contributor publishing 832 papers. In other words, CHINA has a dominant status in the SI publications productions, which produced about which produced about a fourth of world's total during this period. An interesting observation is that there are certain countries which have relatively low frequency but have high value of centrality among all other countries. Australia leads other countries, which are shown as node rings in purple in Fig. 6. This is followed by papers originating from BELGIUM, ENGLAND, GERMANY, FRANCE and so on. In other words, they are pivotal nodes in the network with the highest betweenness centrality. In addition, nine countries are found to have citation bursts: USA (38.3791), SWITZERLAND (5.3012), CANADA (4.2065), FRANCE (4.1281), JAPAN (4.0207), NORWAY (3.6151), MEXICO (3.3748), SAUDI ARABIA (3.1207) and CZECH REPUBLIC (2.9348), suggesting that they have abrupt increases of citations detected, and we listed the details in the Table 1.

3.4 Institutions Distribution and Co-occurring Network

Overall, a total of 1766 research institutes in the world are engaged in intelligence swarm during 2000 and 2015. Figure 7 lists the top ten of them: Chinese Acad Sci (51 articles), Islamic Azad Univ (39 articles), Erciyes Univ (27 articles), Univ Hyderabad (26 articles), Tongji Univ (25 articles), Indian Inst Technol (24 articles), Jadavpur Univ (22 articles), Dalian Univ Technol (22 articles), Univ Libre Bruxelles (21 articles), Univ Extremadura (20 articles).

Figure 8 shows the visualization of the distribution of institutions. In order to show the core institutions of this field, we filter the institutions with only one articles and get

Table 1. Top nine countries with citation bursts

Countries	Year	Strength	Begin	End	2000-2015
USA	2000	38.3791	2001	2006	
SWITZERLAND	2000	5.3012	2002	2005	
CANADA	2000	4.2065	2001	2006	
FRANCE	2000	4.1281	2000	2008	
JAPAN	2000	4.0207	2006	2009	
NORWAY	2000	3.6151	2002	2006	
MEXICO	2000	3.3748	2013	2015	
SAUDI ARABIA	2000	3.1207	2013	2015	
CZECH REPUBLIC	2000	2.9348	2011	2013	

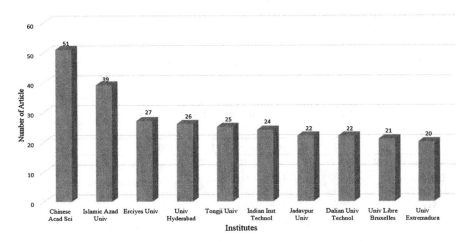

Fig. 6. Distribution of Institutes

a network of institute of 308 nodes and 67 links. Obviously, Chinese Acad Sci in China takes the first place with a frequency of 51 articles. The second place is Islamic Azad Univ with a frequency of 39 articles. Apart from that, there are still other institutions participating in SI research, such as Univ Extremadura, Indian Inst Sci and so on. We also notice that China' institutes such as Chinese Acad, Tongji Univ, Dalian Univ Technol, Harbin Engn Univ, and Peking Univ were more on the top of the list. In addition, ten institutes are found to have citation bursts: Univ Libre Bruxelles (4.8826), Indian Inst Sci (4.4745), Syracuse Univ (4.2705), Univ Missouri (4.1762), Norwegian Univ Sci & Technol (4.1184), N China Elect Power Univ (3.8338), Rochester Inst Technol (3.783), Beijing Inst Technol (3.6897), Univ Delaware (3.6654) and Harbin Engn Univ (3.6473), suggesting that they have abrupt increases of citations detected, and we listed the details in the Table 2.

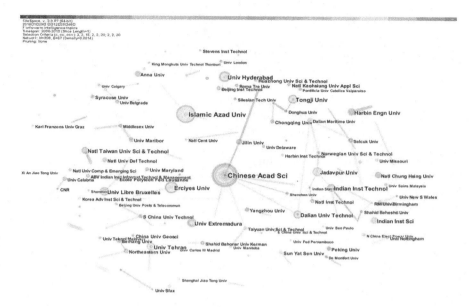

Fig. 7. A network of institute of 308 Nodes and 67 Links on SI research publications

Table 2. Top ten institutes with citation bursts

Countries	Year	Strength	Begin	End	2000-2015
Univ Libre Bruxelles	2000	4. 8826	2006	2009	
Indian Inst Sci	2000	4. 4745	2007	2009	
Syracuse Univ	2000	4. 2705	2004	2009	
Univ Missouri	2000	4. 1762	2004	2007	
Norwegian Univ S&T	2000	4. 1184	2002	2010	
N China Elect Power Univ	2000	3. 8338	2007	2009	
Rochester Inst Technol	2000	3. 783	2003	2005	
Beijing Inst Technol	2000	3. 6897	2006	2008	
Univ Delaware	2000	3. 6654	2004	2008	
Harbin Engn Univ	2000	3. 6473	2004	2008	

3.5   Research Hotspots and Keyword Clustering

Table 3 lists the top 10 high cited articles in the Science Citation Index (SCI) database according our search strategy which may represent the research hotspots in the field of swarm intelligence. Li, Xiaodong (2012) demonstrated a cooperatively co-evolving particle swarms for large scale optimization [10]; Kannan, S(2004) discussed the application of particle swarm optimization technique and its variants to generation expansion planning problem [11]; Yang, Xueming (2007) proposed a modified particle

Table 3. Top ten high cited articles

NO.	Title	First Author	Publication Name	Year	Cited Times
1	Cooperatively Coevolving Particle Swarms for Large Scale Optimization	Li, Xiaodong	IEEE TRANSACTIONS ON EVOLUTIONARY COMPUTATION	2012	97
2	Application of particle swarm optimization technique and its variants to generation expansion planning problem	Kannan, S	ELECTRIC POWER SYSTEMS RESEARCH	2004	95
3	A modified particle swarm optimizer with dynamic adaptation	Yang, Xueming	APPLIED MATHEMATICS AND COMPUTATION	2007	94
4	A powerful and efficient algorithm for numerical function optimization: artificial bee colony (ABC) algorithm	Karaboga, Dervis	JOURNAL OF GLOBAL OPTIMIZATION	2007	911
5	On clarifying misconceptions when comparing variants of the Artificial Bee Colony Algorithm by offering a new implementation	Mernik Marjan	INFORMATION SCIENCES	2015	9
6	Teaching-learning-based optimization with dynamic group strategy for global optimization	Zou, Feng	INFORMATION SCIENCES	2014	9
7	A survey on nature inspired metaheuristic algorithms for partitional clustering	Nanda, Satyasai Jagannath	SWARM AND EVOLUTIONARY COMPUTATION	2014	9
8	Chaotic swarming of particles: A new method for size optimization of truss structures	Kaveh, A.	ADVANCES IN ENGINEERING SOFTWARE	2014	9
9	MRI breast cancer diagnosis hybrid approach using adaptive ant-based segmentation and multilayer perceptron neural networks classifier	Hassanien, Aboul Ella	APPLIED SOFT COMPUTING	2014	9

(Continued)

Table 3. (*Continued*)

NO.	Title	First Author	Publication Name	Year	Cited Times
10	Hybridization strategies for continuous ant colony optimization and particle swarm optimization applied to data clustering	Huang, Cheng Lung	APPLIED SOFT COMPUTING	2013	9

swarm optimizer with dynamic adaptation [12]; Karaboga, Dervis (2007) designed a powerful and efficient algorithm for numerical function optimization: artificial bee colony (ABC) algorithm [13]; Mernik Marjan (2015) clarified the misconceptions when comparing variants of the Artificial Bee Colony Algorithm by offering a new implementation [14]; Zou, Feng(2014) proposed a teaching-learning-based optimization with dynamic group strategy for global optimization [15]; Nanda, Satyasai Jagannath (2014) concluded the nature inspired metaheuristic algorithms for partitional clustering [16]; Kaveh, A designed a new method for size optimization of truss structures named chaotic swarming of particles [17]; Hassanien, Aboul Ella(2014) used adaptive ant-based segmentation and multilayer perceptron neural networks classifier to do a MRI breast cancer diagnosis [18]; Huang, Cheng Lung(2013) proposed

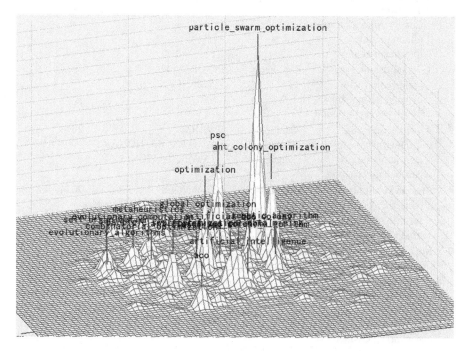

Fig. 8. Keyword clustering and co-occurring network

hybridization strategies for continuous ant colony optimization and particle swarm optimization applied to data clustering [19].

In order to find the research hotspots about swarm intelligence in detail,a keyword clustering method based on word2vec [20] and keyword co-occurring method were used, and Fig. 8 show the result of such method. There are 5674 keyword nodes, 18344 vertex in the network, and the top 10 closed relations of co-occurrence keywords were evolutionary computation-neural network (334), evolutionary computation-genetic algorithm (202), evolutionary computation algorithm (190), swarm intelligence neural network evolutionary (138), computation swarm intelligence (106), evolutionary computation-optimization (92), evolutionary computation-chaos (84), evolutionary computation-abc algorithm (82), particle swarm optimization-evolutionary computation (76), evolutionary computation-robotic arms (70). And the keywords with high betweenness centrality are particle swarm optimization, swarm intelligence, pso, ant colony optimization, optimization, particle swarm optimization, artificial bee colony, artificial bee colony algorithm, genetic algorithm and so on. In addition, twenty keywords are found to have citation bursts: swarm intelligence (14.9816), self-organization (8.2731), collective robotics (6.8131), swarm robotics (6.7234), stigmergy (6.3933), ant colony algorithm (5.5469), ants (5.1945), coordination (4.9626), artificial bee colony (4.6251), clustering (4.2762), combinatorial optimization (4.2314), self-assembly

Table 4. Top twenty keywords with citation bursts

Keywords	Year	Strength	Begin	End	2000-2015
swarm intelligence	2000	14.9816	2001	2005	
self-organization	2000	8.2731	2000	2008	
collective robotics	2000	6.8131	2001	2007	
swarm robotics	2000	6.7234	2003	2007	
ant colony algorithm	2000	5.5469	2004	2008	
ants	2000	5.1945	2002	2007	
coordination	2000	4.9626	2004	2008	
artificial bee colony	2000	4.6251	2013	2015	
clustering	2000	4.2762	2002	2006	
combinatorial optimization	2000	4.2314	2010	2011	
self-assembly	2000	4.1255	2006	2010	
swarm-bot	2000	4.1239	2004	2007	
ad hoc networks	2000	3.7363	2005	2010	
grid	2000	3.5687	2007	2009	
numerical function optimization	2000	3.3481	2013	2015	
spatial organization	2000	3.3128	2008	2009	
agents	2000	3.3004	2000	2007	
multi-agent system	2000	3.2887	2008	2009	

(4.1255), swarm-bot (4.1239), ad hoc networks (3.7363), grid (3.5687), numerical function optimization (3.3481), spatial organization (3.3128), agents (3.3004), multi-agent system (3.2887) and ant colony optimization (aco) (3.2506), suggesting that they have abrupt increases of citations detected, and we listed the details in the Table 4.

4 Conclusion

In this paper, we used a bibliometric method to quantitatively assessing current research hotspots and trends on Swarm Intelligence, using the related literature in the Science Citation Index (SCI) database from 2000 to 2015. Articles referring to Swarm Intelligence were concentrated on the analysis of scientific outputs, distribution of countries, institutions, periodicals, subject categories and research performances by individuals. Moreover, innovative methods such as keyword co-citation analysis, semantic clustering and Keyword Frequent Burst Detection were applied to provide a dynamic view of the evolution of swarm intelligence research hotpots and trends from various perspectives which may serve as a potential guide for future research.

References

1. Beni, G., Wang, J.: Swarm intelligence in cellular robotic systems. In: Proceeding NATO Advanced Workshop on Robots and Biological Systems, Tuscany, Italy, 26–30 June 1989
2. Bonabeau, E., Dorigo, M., Theraulaz, G.: Swarm Intelligence - From Natural to Artificial Systems. Oxford University Press Inc, New York (1999)
3. Poli, R., Kennedy, J., Blackwell, T.: Particle swarm optimization. Swarm Intell. 1(1), 33–57 (2007)
4. Karaboga, D., Akay, B.: A survey: algorithms simulating bee swarm intelligence. Artif. Intell. Rev. 31(1–4), 61–85 (2009)
5. Martens, D., Baesens, B., Fawcett, T.: Editorial survey: swarm intelligence for data mining. Mach. Learn. 82(1), 1–42 (2011)
6. Huang, E.H., Socher, R., Manning, C.D., Ng, A.Y.: Improving word representations via global context and multiple word prototypes. In: Proceedings of the 50th Annual Meeting of the Association for Computational Linguistics: Long Papers, vol. 1, pp. 873–882. Association for Computational Linguistics (2012)
7. Rogosa, D., Brandt, D., Zimowski, M.: A growth curve approach to the measurement of change. Psychol. Bull. 92(3), 726 (1982)
8. Chen, C.: CiteSpace II: Detecting and visualizing emerging trends and transient patterns in scientific literature. J. Am. Soc. Inform. Sci. Technol. 57(3), 359–377 (2006)
9. Freeman, L.C.: Centrality in social networks conceptual clarification. Social Networks 1(3), 215–239 (1979)
10. Li, X., Yao, X.: Cooperatively coevolving particle swarms for large scale optimization. IEEE Trans. Evol. Comput. 16(2), 210–224 (2012)
11. Kannan, S., Slochanal, S.M.R., Subbaraj, P., Padhy, N.P.: Application of particle swarm optimization technique and its variants to generation expansion planning problem. Electr. Power Syst. Res. 70(3), 203–210 (2004)

12. Yang, X., Yuan, J., Yuan, J., Mao, H.: A modified particle swarm optimizer with dynamic adaptation. Appl. Math. Comput. **189**(2), 1205–1213 (2007)
13. Karaboga, D., Basturk, B.: A powerful and efficient algorithm for numerical function optimization: artificial bee colony (ABC) Algorithm. J. Global Optim. **39**(3), 459–471 (2007)
14. Mernik, M., Liu, S.H., Karaboga, D., Črepinšek, M.: On clarifying misconceptions when comparing variants of the Artificial Bee Colony Algorithm by offering a new implementation. Inf. Sci. **291**, 115–127 (2015)
15. Zou, F., Wang, L., Hei, X., Chen, D., Yang, D.: Teaching–learning-based optimization with dynamic group strategy for global optimization. Inf. Sci. **273**, 112–131 (2014)
16. Nanda, S.J., Panda, G.: A survey on nature inspired metaheuristic algorithms for partitional clustering. Swarm Evol. Comput. **16**, 1–18 (2014)
17. Kaveh, A., Sheikholeslami, R., Talatahari, S., Keshvari-Ilkhichi, M.: Chaotic swarming of particles: a new method for size optimization of truss structures. Adv. Eng. Softw. **67**, 136–147 (2014)
18. Hassanien, A.E., Moftah, H.M., Azar, A.T., Shoman, M.: MRI breast cancer diagnosis hybrid approach using adaptive ant-based segmentation and multilayer perceptron neural networks classifier. Appl. Soft Comput. **14**, 62–71 (2014)
19. Huang, C.L., Huang, W.C., Chang, H.Y., Yeh, Y.C., Tsai, C.Y.: Hybridization strategies for continuous ant colony optimization and particle swarm optimization applied to data clustering. Appl. Soft Comput. **13**(9), 3864–3872 (2013)
20. Mikolov, T., Chen, K., Corrado, G., Dean, J.: Efficient estimation of word representations in vector space (2013). arXiv preprint arXiv:1301.3781

Novel Swarm-Based Optimization Algorithms

Duelist Algorithm: An Algorithm Inspired by How Duelist Improve Their Capabilities in a Duel

Totok Ruki Biyanto[1(✉)], Henokh Yernias Fibrianto[1],
Gunawan Nugroho[1], Agus Muhamad Hatta[1], Erny Listijorini[2],
Titik Budiati[3], and Hairul Huda[4]

[1] Engineering Physics Department,
Institut Teknologi Sepuluh Nopember, Surabaya, Indonesia
trb@ep.its.ac.id, joelhenokh@gmail.com
[2] Mechanical Engineering Department,
Univesitas Sultan Ageng Tirtayasa, Cilegon, Indonesia
[3] Food Technology Department, State Polytechnic of Jember,
Jember, Indonesia
[4] Chemical Engineering Department,
Universitas Mulawarman, Samarinda, Indonesia

Abstract. This paper proposes an optimization algorithm based on human fight and learn from each duelist. The proposed algorithm starts with an initial set of duelists. The duel is to determine the winner and loser. The loser learns from the winner, while the winner try their new skill or technique that may improve their fighting capabilities. A few duelists with highest fighting capabilities are called as champion. The champion train a new duelist such as their capabilities. The new duelist will join the tournament as a representative of each champion. All duelist are re-evaluated, and the duelists with worst fighting capabilities is eliminated to maintain the amount of duelists. Several benchmark functions is used in this work. The results shows that Duelist Algorithm outperform other algorithms in several functions.

Keywords: Optimization · Algorithm · Duelist · Fighting

1 Introduction

Optimization is a process to achieve something better. Optimization usually consists of three parts which are optimization algorithm, model and objective function. For example, let there be a problem f(x) as a model then to find optimum value of x which is can be maximum, minimum or at specific value in between is the objective function using an algorithm. Different methods and algorithm have been proposed to solve the optimization problem. One of the most common methods used for optimization is genetic algorithm (GA) which is based on natural selection by evolving a population of candidate solution for defined objective function [1]. On the other hand, a different

© Springer International Publishing Switzerland 2016
Y. Tan et al. (Eds.): ICSI 2016, Part I, LNCS 9712, pp. 39–47, 2016.
DOI: 10.1007/978-3-319-41000-5_4

method for optimization called ant colony optimization is inspired by foraging behavior of real ants [2]. Another type of method is inspired by social behavior of animals which is called as particle swarm optimization (PSO) [3]. There's also a method for optimization which inspired by imperialistic competition called imperialist competitive algorithm (ICA) [4]. All of these mentioned methods are population based algorithm which is mean that there's a set of population and keep improving itself in each iterations [5]. There are other optimization algorithms also commonly used such as predatory search strategy [6], society and civilization optimization [7] and quantum evolutionary algorithm [8]. Nowadays, all this optimization methods are very useful for solving multiple problems such as energy management, scheduling, resource allocation, etc. [9–11].

In this paper, a new algorithm based on genetic algorithm is proposed which is inspired by human fighting and learning capabilities. As an overview, in genetic algorithm there are two ways to develop an individual into a new one. First is crossover where an individual mate with individual to produce a new offspring, this new offspring's genotype are based on their parents. The second one is mutation where an individual mutate into a new one. In duelist algorithm (DA), all the individual in population are called as duelist, all those duelists would fight one by one to determine the champions, winners and losers. The fighting itself just like real life fight where the strongest has possibility as a loser. There is a probability that the weak one would be lucky enough to win. In order to improve each duelist, there are also two ways to evolve. One of them is innovation. Innovation is only applicable to the winner. The other one is called as learning, losers would learn from winners. In GA, both mutation and crossover are seem to be blind in producing any solution to find the best solution. Blind means that each solution or produced individual in genetic algorithm may has not better solution. In fact, it may fall into the worst one. DA tries to minimize this blind effect by giving different treatment on duelists based on their classification. This paper described how duelist algorithm is designed and implemented.

2 Review of a Duel

Duel can be interpreted as a fighting between one or more person(s) with other person (s). Fighting require physical strength, skill and intellectual capability, for example in chess and bridge games. Common type of duel, which include physical strength is boxing, boxing is one of world's most popular sport where two persons need to knock down each of them under certain rules. In every duel, there are consist of the winner and the loser as well as the rules. In a match the probability become the winner depend on strength, skill and luck. After the match, knowing the capabilities of the winner and the loser are very useful. Loser can learn from how the winner, and winner can improve the capability and skill by training or trying something new from the loser. In the proposed algorithm, each duelist do the same to be unbeatable, by upgrading themself whether by learning from their opponent or developing a new technique or skill.

3 Duelist Algorithm

The flowchart of proposed algorithm is shown in Fig. 1. First, population of duelist is registered. Duelists have their properties, which is encoded into binary array. All of the duelists are evaluated to determine their fighting capabilities. The duel schedule is set to each duelist that contain a set of duel participants. In the duel, each duelist would fight one on one with other duelist. This one on one fighting is used rather than gladiator battle to avoid local optimum. Each duel would produce a winner and a loser based on their fighting capabilities and their 'luck'. After the match, the champion is also determined. These champions are the duelist that has the best fighting capabilities.

Then, each winner and loser have opportunity to upgrade their fighting capabilities, meanwhile each champion train the new duelist as such their capabilities. The new duelist will join in the next match. Each loser would learn from their opponents how to be a better duelist by replacing a specific part of their skillset with winner's skillset. On the other hand, winner would try to innovate a new skill by changing their skillset value.

Each duelist fighting capabilities is re-evaluated for the next match. All duelist then re-evaluated through post-qualification and sorted to determine who will be the champions. Since there are new duelists that was trained by champions, all the worst duelists are eliminated to maintain the amount of duelists in the tournament. This process will continue until the tournament is finished. The systematic explanation as follow:

3.1 Registration of Duelist Candidate

Each duelist in a duelist set is registered using binary array. Binary array in duelist algorithm is called as skillset. In a N_{var}-dimensional optimization problem, the duelist would be binary length times N_{var} length array.

3.2 Pre-Qualification

Pre-qualification is a test that given to each duelists to measure or evaluate their fighting capabilities based on their skillset.

3.3 Board of Champions Determination

Board of champions is determined to keep the best duelist in the game. Each champion should trains a new duelist to be as well as himself duel capabilities. This new duelists would replace the champion position in the game and join the next duel.

3.4 Duel Scheduling Between Duelists

The duel schedule between each duelist is set randomly. Each duelist will fight using their fighting capabilities and luck to determine the winner and the loser. The duel is using a simple logic. If duelist A's fighting capabilities plus his luck are higher than duelist B's, then duelist A is the winner and vice versa. Duelist's luck is purely random. The pseudocode to determine the winner and the loser is shown in Algorithm 1.

Algorithm 1. Determination of the winner and the loser.

```
Require : Duelist A and B; Luck_Coefficient
Where : FC = Fighting Capabilities; LC = Luck Coefficient
A(Luck) = A(FC) * (LC + (random(0-1) * LC));
B(Luck) = B(FC) * (LC + (random(0-1) * LC));
If  ((A(FC) + A(Luck)) <= (B(FC) + B(Luck)))
  A(Winner) = 1;
  B(Winner) = 0;
Else
  A(Winner) = 0;
  B(Winner) = 1;

End
```

3.5   Duelist's Improvement

After the match, each duelist are categorized into champion, winner and loser. To improve each duelist fighting capabilities there are three kind of treatment for each categories. First treatment is for losers, each loser is trained by learning from winner. Learning means that loser may copy a part of winner's skillset. The second treatment is for winners, each winner would improve their own capabilities by trying something new. This treatment consist of winner's skillset random manipulation. Finally, each champion would trains a new duelist.

Algorithm 2. Winner and Loser Enhancement.

```
Require : Duelist A and B;
if  A(Winner) = 1;
  for i=1:(skillset_length)
    D = random(0...1);
    If D < Prob_Innovate
      A(skillset) = rand(0...9);
    end
  end
else
  for i=1:(skillset_length)
    E = random(0...1);
    If E < Prob_Learn
      A(skillset) = B(skillset);
    end
  end

end
```

3.6   Elimination

Since there are some new duelists joining the game, there must be an elimination to keep duelists quantity still the same as defined before. Elimination is based on each duelist's dueling capabilities. The duelist with worst dueling capabilities are eliminated.

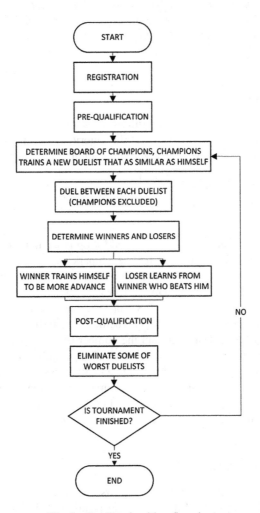

Fig. 1. Duelist algorithm flowchart

4   Experimental Studies

This section discuss about Duelist Algorithm performance using a benchmark for computational speed comparison and 10 other benchmarks for algorithm's robustness comparison. The detail of these functions are shown as follow:

$$f = -\left(\sqrt{x^2 + y^2} * \cos(x - y) * e^{\cos((x*(y+5)/7))}\right) \qquad (1)$$

While 10 other benchmark functions are benchmark function with noise based on real-parameter black box optimization benchmarking [12]. The 10 mathematical optimization problems are f_{110}, f_{111}, f_{113}, f_{115}, f_{116}, f_{119}, f_{120}, f_{121}, f_{122}, f_{123}. The proposed algorithm is compared with a group of commonly used algorithms including Genetic Algorithm [1], Particle Swarm Optimization [3] and Imperialist Competitive Algorithm [4].

4.1 Benchmark Function

Duelist algorithm and several other algorithms are applied on total of 11 benchmark function. The first functions is used to test the algorithm's computational speed and other ten functions are used to test the algorithm's ability in finding the optimum value and robustness.

4.2 Parameter Setting

The proposed Duelist Algorithm has been tested for optimization problems to show the advantages of proposed algorithm. The first benchmark was repeated 10 times and the other ten were repeated 100 times. For the last ten function evaluation, total population and iteration that used in GA and DA is 100 and 500 respectively. Mutation and Crossover probability is set at 0.05 and 0.8. For DA, innovate and learning probability are set as 0.1 and 0.8 respectively with luck coefficient of 0.01. In PSO algorithm, the velocity constants are set 0.4 and 0.6 and the number of swarms at 100. The number colonies in ICA is set at 100, with initial number of imperialists of 8, revolution rate of 0.4 and number of decades of 500. Some change in algorithm's parameter are changed for the first benchmark. The changes in GA are max generation of 200 with mutation probability and crossover probability of 0.5 and 0.8 respectively. In ICA, the decades are shorten into 200 decades. In DA, luck coefficient of 0 and max generations of 200.

4.3 Result of First Evaluation

In this section, each algorithm is tested using first benchmark function. To provide a fair comparison, all algorithm used same initial position or population. The average result of the test are shown as follow:

The experiment shows that DA is able to reach global optimum under lesser number of iterations that GA and PSO (Fig. 2).

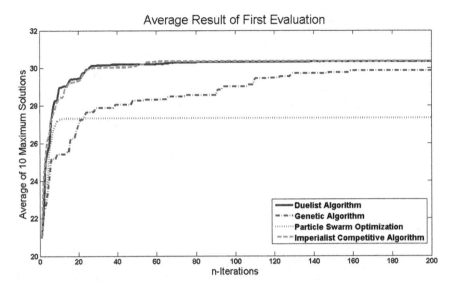

Fig. 2. Comparison result between algorithms (Color figure online)

4.4 Result of Second Evaluation

The statistical results of 100 experiments of ten noisy benchmark functions with twenty dimensions functions are presented in Table 1. The bold font shows the optimum value that achieved for each benchmark function. Based on the table, it can be observed that the proposed algorithm is able to overcome other algorithms. The mean values of proposed algorithms indicate that DA has a good robustness in finding optimum value.

Table 1. Comparison between algorithms for noisy benchmark function

f		GA	PSO	ICA	DA
110	min	614.99719	−21.51084	429.91794	**−125.28074**
	mean	4885.52966	847.83925	2648.06026	**280.93054**
111	min	−134.93092	−134.81238	−132.96055	**−135.01945**
	mean	−91.24741	−87.13642	−69.58788	**−107.01454**
113	min	−79.07112	−82.01971	−80.65960	**−83.57565**
	mean	−59.25298	−73.02693	−69.48691	**−79.75573**
115	min	−52.78967	−52.70470	−27.06354	**−69.60157**
	mean	378.56370	78.94026	219.80703	**45.72266**
116	min	295.49628	13.41322	21.22042	**−70.08705**
	mean	2240.40197	547.91493	685.23940	**−29.25039**
119	min	−57.12290	−57.75113	−57.64251	**−57.89313**
	mean	−56.13720	−57.33738	−56.95080	**−57.86155**

(*Continued*)

Table 1. (*Continued*)

f		GA	PSO	ICA	DA
120	min	−57.89985	−57.89981	−57.89982	**−57.89986**
	mean	−57.88350	−57.87494	−57.87659	**−57.88931**
121	min	−56.01884	−52.66815	−40.04706	**−57.82896**
	mean	−22.50195	64.25804	14.71232	**−52.39053**
122	min	−38.63358	−38.66433	−38.60398	**−38.67088**
	mean	−38.31492	−38.48721	−38.41675	**−38.51987**
123	min	**−38.71998**	−38.71978	−38.71992	−38.71989
	mean	−38.71506	−38.71104	−38.71265	**−38.71602**

5   Conclusion

In this paper, an optimization algorithm based on how duelist improve himself to win a fight is proposed. Each individual in the population is called duelist. Each duelist fight with other duelist to determine who is the winner and the loser. Winner and loser have their own way of improving themselves. The winners are improved by learning theirself. In the other hand, loser improve himself by learning from the winner. After several improvements and duels, some duelists will become the best solution for given problem. The algorithm is tested by using 11 different optimization problems. The result shows that the proposed algorithm is able to surpass the other algorithm and present robust results.

References

1. Melanie, M.: An introduction to genetic algorithms, Cambridge, Massachusetts London, England, Fifth printing, vol. 3 (1999)
2. Dorigo, M., Blum, C.: Ant colony optimization theory: a survey. Theoret. Comput. Sci. **344**, 243–278 (2005)
3. Poli, R., Kennedy, J., Blackwell, T.: Particle swarm optimization. Swarm Intell. **1**, 33–57 (2007)
4. Atashpaz-Gargari, E., Lucas, C.: Imperialist competitive algorithm: an algorithm for optimization inspired by imperialistic competition. In: IEEE Congress on Evolutionary computation, CEC 2007, pp. 4661–4667 (2007)
5. Beasley, J.E., Chu, P.C.: A genetic algorithm for the set covering problem. Eur. J. Oper. Res. **94**, 392–404 (1996)
6. Linhares, A.: Synthesizing a predatory search strategy for VLSI layouts. IEEE Trans. Evol. Comput. **3**, 147–152 (1999)
7. Ray, T., Liew, K.M.: Society and civilization: an optimization algorithm based on the simulation of social behavior. IEEE Trans. Evol. Comput. **7**, 386–396 (2003)
8. Han, K.-H., Kim, J.-H.: Quantum-inspired evolutionary algorithm for a class of combinatorial optimization. IEEE Trans. Evol. Comput. **6**, 580–593 (2002)

9. Hartmann, S.: A competitive genetic algorithm for resource-constrained project scheduling. Naval Res. Logistics (NRL) **45**, 733–750 (1998)
10. Dandy, G.C., Simpson, A.R., Murphy, L.J.: An improved genetic algorithm for pipe network optimization. Water Resour. Res. **32**, 449–458 (1996)
11. Balci, H.H., Valenzuela, J.F.: Scheduling electric power generators using particle swarm optimization combined with the Lagrangian relaxation method. Int. J. Appl. Math. Comput. Sci. **14**, 411–422 (2004)
12. Hansen, N., Auger, A., Finck, S., Ros, R.: Real-parameter black-box optimization benchmarking 2010: experimental setup (2010)

Framework for Robust Optimization Combining Surrogate Model, Memetic Algorithm, and Uncertainty Quantification

Pramudita Satria Palar[1]([⊠]), Yohanes Bimo Dwianto[1],
Lavi Rizki Zuhal[1], and Takeshi Tsuchiya[2]

[1] Bandung Institute of Technology,
Jl. Ganesha No. 10, Bandung, West Java, Indonesia
pramsp@ftmd.itb.ac.id
[2] University of Tokyo, Tokyo 113-8656, Japan

Abstract. In this paper, our main concern is to solve expensive robust optimization with moderate to high dimensionality of the decision variables under the constraint of limited computational budget. For this, we propose a local-surrogate based multi-objective memetic algorithm to solve the optimization problem coupled with uncertainty quantification method to calculate the robustness values. The robust optimization framework was applied to two aerodynamic cases to assess its capability on real world problems. Result on subsonic airfoil shows that the surrogate-based optimizer can produce non-dominated solutions with better quality than the non-surrogate optimizer. It also successfully solved the transonic airfoil optimization problem and found a strong tradeoff between the mean and standard deviation of lift-to-drag ratio.

Keywords: Surrogate model · Multi-objective memetic algorithm · Uncertainty quantification · Robust optimization

1 Introduction

Robust optimization is an optimization framework that incorporates the effect of uncertainty in the real-world problem of design optimization. The main goal of robust optimization is to find optimum robust solution/s whose performances do not change much in the presence of uncertainties [1]. In many practical applications, robust optimization can be conveniently casted into multi-objective optimization problem. Metaheuristic-based optimization method such as evolutionary algorithm (EA) is suitable to solve robust optimization problem due to their capability to find global optimum solution in a single run. Special EAs have been developed to handle problem with uncertainties, such as [2–6].

Due to the expensiveness of robust optimization problem, surrogate model is frequently employed to aid robust optimization. Global surrogate model, which is commonly used in practice, is highly prone to the curse-of-dimensionality. To cope with this, Ong et al. [1] proposed a max-min surrogate-assisted evolutionary

© Springer International Publishing Switzerland 2016
Y. Tan et al. (Eds.): ICSI 2016, Part I, LNCS 9712, pp. 48–55, 2016.
DOI: 10.1007/978-3-319-41000-5_5

algorithm (SAEA) that utilizes local surrogate model for robust design, based on the similar algorithm for deterministic problem [7,8]. The max-min SAEA method is, however, only able to deal with max-min robustness problem and was not specifically designed to solve multi-objective robust optimization problem.

To be able to estimate the robustness properties, uncertainty quantification (UQ) method is needed. Beside of standard Monte Carlo Simulation (MCS), methods such as Bayesian Monte Carlo (BMC) [9] and polynomial chaos expansion (PCE) [10] have been developed to reduce the number of function evaluations. Combination of metaheuristic optimizers and advanced UQ methods have been researched in some previous papers [11,12]. Here, the choice of UQ method is important to reduce computational cost while obtaining great accuracy.

In this paper, we explain the detail of our proposed algorithm that combines multi-objective memetic algorithm, surrogate model, and UQ method. Multi-objective memetic algorithm acts as the optimizer, assisted by local surrogate model for local search. On the other side, UQ serves as a tool to calculate the statistical moments as the objective functions. In this work, we consider only the problem with parametric uncertainties, in which the decision variables are deterministic. The methodology in this paper is specifically designed to handle problem with moderate to relatively high budget of function evaluation, although still limited within EA context.

The rest of the paper is structured as follows: Sect. 2 describes the formulation of robust optimization used in this paper; Sect. 3 presents the detail of the computational methods used for robust optimization; Sect. 4 reports the application and the result of the proposed methodology to two aerodynamic cases; Finally, Sect. 5 concludes the paper with suggestion for the future works.

2 Robust Optimization

Within probability framework, typical goal of robust optimization is, without loss of generality, to minimize mean and standard deviation of the solution. The simplest approach to perform robust optimization is to minimize the weighted sum of mean (μ) and standard deviation (σ) as a function of the decision variable $\boldsymbol{x} = \{x_1, \ldots, x_{n_d}\}$ and random variable $\boldsymbol{\xi} = \{\xi_1, \ldots, \xi_{n_r}\}$:

$$\text{min: } w_m \mu(\boldsymbol{x}, \boldsymbol{\xi}) + w_s \sigma(\boldsymbol{x}, \boldsymbol{\xi}) \tag{1}$$

where w_m and w_s are the weights assigned to the mean and standard deviation, respectively. Another approach is to perform multi objective robust optimization:

$$\text{min: } (\mu(\boldsymbol{x}, \boldsymbol{\xi}), \sigma(\boldsymbol{x}, \boldsymbol{\xi})) \tag{2}$$

We use the multi-objective approach in this paper to alleviate the difficulty in defining the weights. However, there are cases when the designer is more interest in optimizing the deterministic value and its robustness value, as will be shown later in Sect. 4.

3   Computational Methods

The combined framework for robust optimization in this paper consists of local-surrogate assisted multi-objective memetic algorithm and MCS/PCE as the optimizer and UQ tool, respectively. For each individual/solution x in the EA framework, UQ module calculates $\mu(x, \xi)$ and $\sigma(x, \xi)$ as the objective functions using a set of deterministic samples in the probabilistic space Ω with probability density function (PDF) $\rho(\xi)$. The outputs of the UQ module ($\mu(x, \xi)$ and $\sigma(x, \xi)$) are returned to the optimizer as the fitness function values. Constraint in local search can be handled by building an additional surrogate model of the constraint.

3.1   Surrogate Assisted Multi-objective Memetic Algorithm

In this work, we use single-surrogate multi-objective memetic algorithm (SS-MOMA) as the optimizer [8]. The local search is conducted by first building a surrogate model using the individual in action and its nearest neighbors, and then searches this surrogate model using local optimizer. The pseudocode and main loop of SS-MOMA are given in Algorithm 1.

Initialize parent population P_c;
Evaluate initial population;
Start the generation counter $o = 1$;
while *computational budget is not exhausted* **do**
  Perform evolutionary operator and generate offspring population P_o;
  Evaluate the solutions;
  /****Local Search Phase****/;
  for *each individual x in P_o* **do**
    Build the sampling plan by choosing m nearest points to **x** in the database;
    Build surrogate model M for each objective function (achievement scalarizing function) or aggregate function (weighted sum) using this sampling plan;
    Search the surrogate model to find the optimum point **x**$_{\text{opt}}$ of the defined scalarizing function;
    Evaluate **x**$_{\text{opt}}$ with the exact function;
    Enter **x**$_{\text{opt}}$ into the after-local-search solutions A_l;
  end
  Perform non-dominated sorting on P_c, P_o and A_l to create the new parent population P_{c+1};
  Increment the generation counter $o = o + 1$;
end

Algorithm 1. SS-MOMA main loop.

The reason of choosing SS-MOMA as the optimizer is due to its ability to effectively employ local surrogate model to approximate the local landscape of the problem to solve moderate-to-high dimensionality problems. Methods with global surrogate are only suitable for problem with low-dimensionality due to the curse-of-dimensionality problem [7]. Here, we do not distinguish between EA for deterministic or robust optimization since the main difference between them (in our paper) is only on the fitness evaluation. The scalarizing function used in original SS-MOMA is weighted sum (WS), but it has been demonstrated that achievement scalarizing function (ASF) can improve the convergence and diversity performance of SS-MOMA [13].

In this work, we use radial basis function (RBF) and Kriging as the surrogate model for the subsonic and transonic case, respectively. For a detail explanation of RBF and Kriging for surrogate model, readers can refer to [14].

3.2   Uncertainty Quantification

The goal of UQ with probability theory is to calculate the mean and standard deviation of S as a function of random variables $\boldsymbol{\xi}$ with fixed decision variables $\boldsymbol{x}$, and probability density function $\rho(\boldsymbol{\xi})$ in the integration domain $\boldsymbol{\Omega}$. The choice of UQ method mainly depends on the computational cost and the dimensionality of the random variables. The easiest and the most robust UQ method is MCS due to its simplicity and its $1/\sqrt{N_{uq}}$ convergence property. If computational cost is not a big issue in calculating the output uncertainties, MCS is the most suitable method even for problem with low-dimensionality of the random variable. However, MCS is not accurate if the number of deterministic simulation is too low; hence, we should resort to more advanced methods for UQ such as PCE. In engineering simulations, which typically involves computationally expensive partial differential equation (PDE) solver, standard MCS encounters serious difficulty to accurately calculate the output uncertainties. In this paper, we use MCS and PCE for UQ:

Monte Carlo Simulation. MCS works by approximating $\mu(\boldsymbol{x}, \boldsymbol{\xi})$ and $\sigma(\boldsymbol{x}, \boldsymbol{\xi})$ of a function $S(\boldsymbol{x}, \boldsymbol{\xi})$ using N_{uq} samples via the following equations:

$$\mu_S = \frac{1}{N_{uq}} \sum_{i=1}^{N_{uq}} S(\boldsymbol{x}, \boldsymbol{\xi}^{(i)}), \quad \sigma_S^2 = \frac{1}{N_{uq}} \sum_{i=1}^{N_{uq}} (S(\boldsymbol{x}, \boldsymbol{\xi}^{(i)}) - \mu_S)^2 \tag{3}$$

where the samples are generated according to $\rho(\boldsymbol{\xi})$.

Polynomial Chaos Expansion. The concept behind UQ using a PCE is to approximate the functional form between the stochastic response output and each of its random inputs with the following expansion:

$$S(\boldsymbol{\xi}) = \sum_{j=0}^{P} \alpha_j \Psi_j(\boldsymbol{\xi}) \tag{4}$$

where Ψ and α are the product of one-dimensional orthogonal polynomials and PCE coefficients, respectively, and $P + 1$ is the size of polynomial basis. A regression-based with total order expansion that preserves basis polynomials up to a fixed total order specification [15] was used. The mean and standard deviation can then be easily obtained from the PCE coefficients [15].

4 Application to Aerodynamic Optimization

We applied the proposed robust optimization framework to two aerodynamic examples: transonic and subsonic airfoil. Open source computational fluid dynamics (CFD) codes of SU2 [16] and XFOIL [17] were employed to obtain the aerodynamic coefficients for the transonic and subsonic case, respectively.

4.1 Robust Optimization for Maximum Lift Coefficient of Subsonic Airfoil

The first application is the robust optimization of subsonic airfoil to obtain airfoil with optimum maximum lift coefficient ($C_{l_{max}}$) and minimum standard deviation of the lift coefficient (σC_l). The baseline airfoil is NASA SC(2)-0712 airfoil, which was approximated by modified PARSEC with 12 parameter [18] and subjected to $\pm 20\%$ perturbation. This airfoil was optimized at $Re = 1 \times 10^6$ and $M = 0.3$ with uncertain angle of attack [Uniform; $8^0 - 12^0$]. In this case, since $C_{l_{max}}$ on the design condition is the point of interest and not the mean value, it was used as the first objective function. The optimization problem can then be expressed as (minus symbol is to transform maximization into minimization problem):

$$\text{min: } -C_{l_{max}}, \sigma C_l \tag{5}$$

Because XFOIL evaluation is relatively fast, we can use MCS simulation to calculate the statistical moments. In this case, $N_{uq} = 100$ were used for UQ module. The parameters used by SS-MOMA are as follows: population size (N_{pop})=100, initial population(N_{int}) = 100, maximum generation (N_{gen})=100, SBX crossover with probability 0.9, polynomial mutation with probability $1/n_d$, number of neighbors = 150, and number of independent runs = 3. The total CFD call is 2100000 evaluations.

The results, depicted in Fig. 1 (left), show that SS-MOMA was able to found higher quality solutions than standard NSGA-II that used no surrogate. Furthermore, solutions found by SS-MOMA with ASF, in general, have higher convergence property than the SS-MOMA-WS. Depiction of the representative solutions in Fig. 1 (right) shows that, in general, the thickness of the airfoil should be increased to improve the robustness (decreasing σC_l), as can be seen by comparing the results for airfoil A and other airfoils.

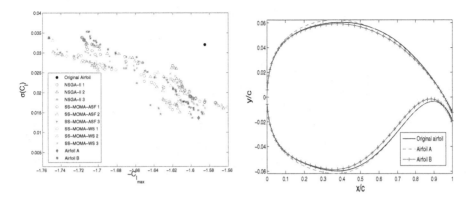

Fig. 1. Non-dominated solutions found by the optimizer (left) and representative solutions (right) on subsonic case (Color figure online)

4.2   Robust Optimization of Transonic Airfoil

The goal of this problem is to find robust optimum transonic airfoil under uncertainties in Mach number and angle of attack with distribution of [Uniform; $0.77 - 0.83$] and [Uniform; $1.5^0 - 2.5^0$], respectively. Optimization of lift-to-drag ratio (L/D), which is the measure of aerodynamic efficiency was considered in this case, and solved by SS-MOMA-ASF. Nine variables PARSEC parameter [19] was first employed to approximate RAE 2822 airfoil (the thickness and ordinate of trailing edge was set to zero). These PARSEC parameters were then subjected to $\pm 20\%$ perturbation and acted as the random variables. The objective functions can be stated as:

$$\text{min:} \ -\mu L/D, \sigma L/D \tag{6}$$

Multi-fidelity version of NIPC [20,21] was used to estimate the statistical moments. The high- and low-fidelity simulations were fully and partially converged simulation, respectively, with the low-fidelity simulation cost a fifth of the high-fidelity one. Our preliminary experiment showed that a low-fidelity 4th-order PCE assisted by 2nd-order correction PCE ($N_{uq} \approx 12.8$) was adequate to obtain accurate value of statistical moments. We then applied SS-MOMA with $N_{int} = 40$, $N_{pop} = 16$, $N_{gen} = 8$, SBX crossover with probability of 0.9, polynomial mutation with probability $1/n_d$, and 150 neighbors to build the local surrogate model to solve the robust optimization problem. The total CFD call is then equivalent to 3778.8 function evaluation.

Result from robust optimization is shown in Fig. 2 (left), where the geometry of representative airfoils is shown in Fig. 2 (right). As it can be seen from Fig. 2 (left), robust SS-MOMA found a strong trade-off between mean and standard deviation of L/D. In other words, airfoil with maximum mean of L/D cannot be achieved without high standard deviation, which means that the solution is unstable. On the other side, airfoils with higher stability (low standard deviation) of L/D have lower mean value.

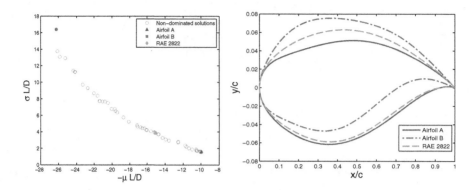

Fig. 2. Non-dominated solutions found by the optimizer (left) and representative solutions (right) on transonic case (Color figure online)

5   Conclusion

Framework for robust optimization that consists of multi-objective memetic algorithm, local surrogate model, and UQ method is presented in this paper. This paper uses multi-objective approach of the performance and robustness due to their typically conflicting nature and specific UQ methods to estimate the value of statistical moments needed for robust optimization. The methodology was demonstrated on two aerodynamic cases of subsonic and transonic airfoil robust optimization. On subsonic case, the SS-MOMA with ASF found higher quality solutions than the standard NSGA-II without surrogate. It also has been successfully deployed and found a strong trade-off between objectives in transonic airfoil case.

Acknowledgements. Part of this work is funded through Riset KK 2016 ITB.

References

1. Ong, Y.S., Nair, P.B., Lum, K.: Max-min surrogate-assisted evolutionary algorithm for robust design. IEEE Trans. Evol. Comput. **10**(4), 392–404 (2006)
2. Tsutsui, S., Ghosh, A.: Genetic algorithms with a robust solution searching scheme. IEEE Trans. Evol. Comput. **1**(3), 201–208 (1997)
3. Jin, Y., Sendhoff, B.: Trade-off between performance and robustness: an evolutionary multiobjective approach. In: Fonseca, C.M., Fleming, P.J., Zitzler, E., Deb, K., Thiele, L. (eds.) EMO 2003. LNCS, vol. 2632, pp. 237–251. Springer, Heidelberg (2003)
4. Maruyama, T., Igarashi, H.: An effective robust optimization based on genetic algorithm. IEEE Trans. Magn. **44**(6), 990–993 (2008)
5. Tsai, J.T., Liu, T.K., Chou, J.H.: Hybrid Taguchi-genetic algorithm for global numerical optimization. IEEE Trans. Evol. Comput. **8**(4), 365–377 (2004)

6. Shimoyama, K., Oyama, A., Fujii, K.: Development of multi-objective six sigma approach for robust design optimization. J. Aerosp. Comput. Inf. Commun. **5**(8), 215–233 (2008)
7. Ong, Y.S., Nair, P.B., Keane, A.J.: Evolutionary optimization of computationally expensive problems via surrogate modeling. AIAA J. **41**(4), 687–696 (2003)
8. Lim, D., Jin, Y., Ong, Y.S., Sendhoff, B.: Generalizing surrogate-assisted evolutionary computation. IEEE Trans. Evol. Comput. **14**(3), 329–355 (2010)
9. Rasmussen, C.E., Ghahramani, Z.: Bayesian Monte Carlo. In: Becker, S., Thrun, S., Obermayer, K. (eds.) Advances in Neural Information Processing Systems, vol. 15, pp. 489–496. The MIT Press, Cambridge (2003)
10. Xiu, D., Karniadakis, G.E.: Modeling uncertainty in flow simulations via generalized polynomial chaos. J. Comput. Phys. **187**(1), 137–167 (2003)
11. Dodson, M., Parks, G.T.: Robust aerodynamic design optimization using polynomial chaos. J. Aircr. **46**(2), 635–646 (2009)
12. Ho, S., Yang, S.: A fast robust optimization methodology based on polynomial chaos and evolutionary algorithm for inverse problems. IEEE Trans. Magn. **48**(2), 259–262 (2012)
13. Palar, P.S., Tsuchiya, T., Parks, G.T.: A comparative study of local search within a surrogate-assisted multi-objective memetic algorithm framework for expensive problems. Appl. Soft Comput. **43**, 1–19 (2016)
14. Forrester, A., Sobester, A., Keane, A.: Engineering Design via Surrogate Modelling: A Practical Guide. Wiley, London (2008)
15. Eldred, M., Burkardt, J.: Comparison of non-intrusive polynomial chaos and stochastic collocation methods for uncertainty quantification. AIAA Pap. **976**(2009), 1–20 (2009)
16. Palacios, F., Colonno, M.R., Aranake, A.C., Campos, A., Copeland, S.R., Economon, T.D., Lonkar, A.K., Lukaczyk, T.W., Taylor, T.W., Alonso, J.J.: Stanford university unstructured (su2): an open-source integrated computational environment for multi-physics simulation and design. AIAA Pap. **287** (2013)
17. Drela, M.: XFOIL: an analysis and design system for low reynolds number airfoils. In: Mueller, T.J. (ed.) Low Reynolds Number Aerodynamics. Lecture Notes in Engineering, vol. 54, pp. 1–12. Springer, Berlin (1989)
18. Yotsuya, T., Kanazaki, M., Matsushima, K.: Design performance investigation of modified parsec airfoil representation using genetic algorithm
19. Sobieczky, H.: Parametric airfoils and wings. In: Fujii, K., Dulikravich, G.S. (eds.) Recent Development of Aerodynamic Design Methodologies. Notes on Numerical Fluid Mechanics (NNFM), vol. 65, pp. 71–87. Springer, Berlin (1999)
20. Ng, L.W.T., Eldred, M.S.: Multifidelity uncertainty quantification using nonintrusive polynomial chaos and stochastic collocation. In: Proceedings of the 53rd AIAA/ASME/ASCE/AHS/ASC Structures, Structural Dynamics and Materials Conference, Number AIAA 2012–1852, Honolulu, HI, USA. AIAA (2012)
21. Palar, P.S., Tsuchiya, T., Parks, G.T.: Multi-fidelity non-intrusive polynomial chaos based on regression. Comput. Methods Appl. Mech. Eng. **305**, 579–606 (2016)

Autonomous Search in Constraint Satisfaction via Black Hole: A Performance Evaluation Using Different Choice Functions

Ricardo Soto[1]([✉]), Broderick Crawford[1], Rodrigo Olivares[1,2],
Stefanie Niklander[3,4,5], and Eduardo Olguín[6]

[1] Pontificia Universidad Católica de Valparaíso, Valparaíso, Chile
{ricardo.soto,broderick.crawford}@pucv.cl, rodrigo.olivares@uv.cl
[2] Universidad de Valparaíso, Valparaíso, Chile
[3] Universidad Adolfo Ibañez, Viña del Mar, Chile
stefanie.niklander@uai.cl
[4] Universidad Autónoma de Chile, Santiago, Chile
[5] Universidad Científica del Sur, Lima, Peru
[6] Universidad San Sebastián, Santiago, Chile
eduardo.olguin@uss.cl

Abstract. Autonomous Search is a modern technique aimed at introducing self-adjusting features to problem-solvers. In the context of constraint satisfaction, the idea is to let the solver engine to autonomously replace its solving strategies by more promising ones when poor performances are identified. The replacement is controlled by a choice function, which takes decisions based on information collected during solving time. However, the design of choice functions can be done in very different ways, leading of course to very different resolution processes. In this paper, we present a performance evaluation of 16 rigorously designed choice functions. Our goal is to provide new and interesting knowledge about the behavior of such functions in autonomous search architectures. To this end, we employ a set of well-known benchmarks that share general features that may be present on most constraint satisfaction and optimization problems. We believe this information will be useful in order to design better autonomous search systems for constraint satisfaction.

Keywords: Autonomous Search · Constraint programming · Constraint satisfaction · Optimization · Choice functions

1 Introduction

Autonomous Search (AS) is a modern technique addressed to the incorporation of self-tuning/self-adjusting features to problem-solvers [4]. Different examples have been reported ranging from approximate methods to exact methods such as constraint satisfaction [1]. Particularly in constraint satisfaction, the idea is to let the solver to autonomously control their solving strategies. A solving strategy is composed of two strategies: enumeration and propagation. The

© Springer International Publishing Switzerland 2016
Y. Tan et al. (Eds.): ICSI 2016, Part I, LNCS 9712, pp. 56–65, 2016.
DOI: 10.1007/978-3-319-41000-5_6

enumeration strategy is responsible for controlling the assignment of values to variables in order to generate the potential solutions of the problem, while propagation attempts the delete the values leading to unfeasible solutions. Several autonomous search solvers [2,3,8] hold a portfolio of strategies and interleave them during solving time with the aim of keeping acting always the best one for each part of the search space. In this way, when poor-performing strategies are detected, they are straightforwardly replaced by more promising ones. This replacement is governed by a choice function, which decides based on relevant performance information collected during solving time, which strategies must operate next. However, the design of choice functions can be done in very different ways, leading of course to very different resolution processes.

In this paper, we present a performance evaluation of 16 rigorously designed choice functions. We incorporate those choice functions to an AS framework based on Black Hole optimization [6,7], which is able to operate with 24 solving strategies. We solve a set of well-known benchmarks that collect general features that may be present on most constraint satisfaction and optimization problems (N-Queens problem, the Sudoku puzzle, the Magic & Latin Square, the Knight Tour and Quasigroup problem). Our goal is to provide new and interesting knowhow about the behavior of such functions. We believe that this information will be useful in order to design better AS systems for efficiently solving constraint satisfaction problems. To the best of our knowledge, this matter has not been explored yet in the literature.

This paper is organized as follows: In Sect. 2, basic notions of constraint solving are introduced. In Sect. 3, the AS framework employed is described. Finally, experimentation results and conclusions are presented in Sects. 4 and 5, respectively.

2 Constraint Solving

As previously mentioned, in a CP context, problems are formulated as CSPs. Formally, a CSP $\mathcal{P}$ is defined by a triple $\mathcal{P} = \langle \mathcal{X}, \mathcal{D}, \mathcal{C} \rangle$ where:

- $\mathcal{X}$ is an n-tuple of variables $\mathcal{X} = \langle x_1, x_2, \ldots, x_n \rangle$.
- $\mathcal{D}$ is a corresponding n-tuple of domains $\mathcal{D} = \langle D_1, D_2, \ldots, D_n \rangle$ such that $x_i \in D_i$, and D_i is a set of values, for $i = 1, \ldots, n$.
- $\mathcal{C}$ is an m-tuple of constraints $\mathcal{C} = \langle C_1, C_2, \ldots, C_m \rangle$, and a constraint C_j is defined as a subset of the Cartesian product of domains $D_{j_1} \times \cdots \times D_{j_{n_j}}$, for $j = 1, \ldots, m$.

A solution to a CSP is an assignment $\{x_1 \to a_1, \ldots, x_n \to a_n\}$ such that $a_i \in D_i$ for $i = 1, \ldots, n$ and $(a_{j_1}, \ldots, a_{j_{n_j}}) \in C_j$, for $j = 1, \ldots, m$.

3 Autonomous Search

Autonomous Search (AS) [4,5] appeared as a new technique that enables the problem solver to control and adapt its own parameters and heuristics during

Table 1. Indicators used during the search process.

Name	Description
VFP	Number of variables fixed by propagation
VFE	Number of variables fixed by enumeration
Step	Number of steps or decision points (n increments each time a variable is fixed during enumeration)
$T_n(S_j)$	Number of steps since the last time that an enumeration strategy S_j was used until step n^{th}
SB	Number of Shallow Backtracks
B	Number of Backtracks
d_{max}	Maximum depth in the search tree
MDV	Represents a Variation of the Maximum Depth. It is calculated as: $CurrentDepth_{Maximum} - PreviousDepth_{Maximum}$
DV	Calculated as: $CurrentDepth - PreviousDepth$. A positive value means that the current node is deeper than the one explored at the previous step
SSR	Search Space Reduction. It is calculated as: $(PreviousSearchSpace - SearchSpace) \,/\, PreviousSearchSpace$
TR	The solving process alternates enumerations and backtracks on a few variables without succeeding in having a strong orientation. It is calculated as: $d_{t-1} - VFP_{t-1}$

solving in order to be more efficient without the knowledge of an expert user. The goal of AS is to give more capabilities to the solver in order to let them manage, control, and adapt their search (based on some metrics and automated self-tuning) to be more efficient. In this this way, the user does not require any expert knowledge for efficiently solving problems. Under this paradigm, the self-adaptive property of AS uses several metrics which allow to analyze the performance exhibits in order to improve the process in search tree. The metrics used in this work are detailed in Table 1.

3.1 Black Hole Optimization

Black hole optimization is a population-based method inspired on the black hole phenomenon [6,7]. A black hole is a region of space that has so much mass concentrated in it that there is no way for a nearby object to escape its gravitational pull. Anything falling into a black hole, including light, can not escape.

Similar to other population-based algorithms, the black hole algorithm starts with an initial population of potential solutions to an optimization problem and an objective function that is calculated for them. At each iteration of the black hole algorithm, the best candidate is selected to be the black hole, which then starts pulling other candidates around it, called stars. If a star gets too close to the black hole, it will be swallowed and it is gone forever. In such a case, a new star (potential solution) is randomly generated and placed in the search space and starts a new search.

The black hole has the ability to absorb the stars that surround it. After initializing the black hole and stars, the black hole begins by absorbing the stars around it and all the stars start moving towards the black hole. The absorption of stars by the black hole is formulated as follows:

$$x_i^d(t + 1) = x_i^d(t) + r[x_{BH}^d - x_i^d(t)], \ \forall \ i \in \{1, \ldots, N\} \tag{1}$$

where $x_i^d(t)$ and $x_i^d(t + 1)$ are the locations of the ith-star at iterations t and $t + 1$, respectively. x_{BH}^d is the location of the black hole in the search space. r is a random number in the interval $[0, 1]$. N is the number of stars (potential solutions).

While moving towards the black hole, a star may reach a location with lower cost than the black hole. In such a case, the black hole moves to the location of that star and vice versa. Then the algorithm will continue with the black hole in the new location and then stars start moving towards this new location.

In addition, there is the probability of crossing the event horizon during moving stars towards the black hole. Every star (potential solution) that crosses the event horizon of the black hole will be swallowed by the black hole. Every time a candidate (star) dies - it is sucked in by the black hole - another potential solution (star) is born and distributed randomly in the search space and starts a new search. This is done to keep the number of potential solutions constant. The next iteration takes place after all the stars have been moved.

The radius of the event horizon in the black hole algorithm is calculated using the following equation:

$$R = \frac{f_{BH}}{\sum_{i=1}^{N} f_i} \tag{2}$$

where f_{BH} is the fitness value of the black hole and f_i is the fitness value of the ith-star. N is the number of stars (potential solutions). When the distance between a potential solution and the black hole (best candidate) is less than R, that candidate is collapsed and a new candidate is created and distributed randomly in the search space.

Algorithm 1 details the main steps in the black hole. First step consist in to initialize a population of stars with random locations in the search space. Then, while a termination criterion (a maximum number of iterations or a sufficiently good fitness) is met, each fitness of a potential solution is evaluated. Then, the best star is selected that has the best fitness value as the black hole x_{BH} and the location of each star is changed according to Eq. 1.

If a star reaches a location with lower cost than the black hole, exchange their locations and if a star crosses the event horizon of the black hole (calculated via Eq. 2), replace it with a new star in a random location in the search space.

Algorithm 1. Black Hole Optimization

1: Initialize a population of stars with random locations in the search space;
2: **while** $t < T$ **do**
3:   **for all** x_i **do**
4:     Evaluate the objective function in f_i;
5:     Select the best star that has the best fitness value as the black hole x_{BH};
6:     Change the location of each star according to Eq. 1
7:     **if** $f_i < f_{BH}$ **then**
8:       Exchange their locations;
9:     **end if**
10:    **if** $R < x_i$ **then**
11:      Replace it with a new star in a random location in the search space;
12:    **end if**
13: **end for**
14: **end while**
15: Postprocess results and visualization;

---

4   Experimental Results

In this section we present the case studies with the corresponding experimental results. We consider seven well-known problems widely employed in the literature: N-queens with N = $\{10, 12, 15, 20, 50, 75\}$, Sudoku puzzle $\{1, 2, 5, 7, 9\}$,

Table 2. CFs employed for the experiments.

Choice functions	
CF_1:	$w_1B + w_2Step + w_3T_n(S_j)$
CF_2:	$w_1SB + w_2MDV + w_3DV$
CF_3:	$w_1B + w_2Step + w_3MDV + w_4DV$
CF_4:	$w_1VF + w_2dmax + w_3DB$
CF_5:	$w_1VFP + w_2SSR - w_3B$
CF_6:	$w_1VFP + w_2SSR - w_3SB$
CF_7:	$w_1VFP + w_2SSR - w_3SB - w_4B$
CF_8:	$w_1VFE + w_2SSR - w_3B$
CF_9:	$w_1VFE + w_2SSR - w_3SB$
CF_{10}:	$w_1VFE + w_2SSR - w_3SB - w_4B$
CF_{11}:	$w_1VFP + w_2VFE - w_3SSR - w_4B$
CF_{12}:	$w_1VFP + w_2VFE - w_3SSR - w_4SB$
CF_{13}:	$w_1VFP + w_2VFE - w_3SSR - w_4SB - w_5B$
CF_{14}:	$w_1VFP + w_2SSR - w_3TR$
CF_{15}:	$w_1VFE + w_2SSR - w_3TR$
CF_{16}:	$w_1VFP + w_2VFE - w_3SSR - w_4TR$

Table 3. Portfolio of the enumerations strategies used

Variable ordering	Heuristic description
Input Order	the first entry in the list is chosen
First Fail	the entry with the smallest domain size is chosen
Anti First Fail	the entry with the largest domain size is chosen
Occurrence	the entry with the largest number of attached constraints is chosen
Smallest	the entry with the smallest value in the domain is chosen
Largest	the entry with the largest value in the domain is chosen
Most Constrained	the entry with the smallest domain size is chosen
Max Regret	the entry with the largest difference between the smallest
	and second smallest value in the domain is chosen
Value ordering	**Heuristic description**
Min	Values are tried in increasing order.
Mid	Values are tried beginning from the middle of the domain.
Max	Values are tried in decreasing order.
Enumeration strategies S_j	
$S_1 = Input\ Order + Min$	$S_{13} = Smallest + Mid$
$S_2 = First\ Fail + Min$	$S_{14} = Largest + Mid$
$S_3 = Anti\ First\ Fail + Min$	$S_{15} = Most\ Constrained + Mid$
$S_4 = Occurrence + Min$	$S_{16} = Max\ Regret + Mid$
$S_5 = Smallest + Min$	$S_{17} = Input\ Order + Max$
$S_6 = Largest + Min$	$S_{18} = First\ Fail + Max$
$S_7 = Most\ Constrained + Min$	$S_{19} = Anti\ First\ Fail + Max$
$S_8 = Max\ Regret + Min$	$S_{20} = Occurrence + Max$
$S_9 = Input\ Order + Mid$	$S_{21} = Smallest + Max$
$S_{10} = First\ Fail + Mid$	$S_{22} = Largest + Max$
$S_{11} = Anti\ FirstFail + Mid$	$S_{23} = Most\ Constrained + Max$
$S_{12} = Occurrence + Mid$	$S_{24} = Max\ Regret + Max$

Knight's Tour with N=$\{5, 6\}$, Magic Square with N=$\{3, 4, 5\}$, Latin Square with N=$\{4, 5, 6\}$ and Quasigroup with N=$\{5, 6, 7\}$.

The adaptive enumeration component has been implemented on the Ecl^ips^e Constraint logic Programming Solver v6.10, and the black hole optimizer has been developed in Java. The experiments have been launched on a 3.30 GHz Intel Core $i3 - 2120$ with 4 GB RAM running Windows 7 Professional 32 bits. The instances are solved to a maximum number of 65535 steps as equally done in previous work [2]. If no solution is found at this point the problem is set to *t.o.* (time-out).

Table 2 illustrates the choice functions that we used for the experiments. These CFs were the best performing ones after the corresponding training phase of the algorithm. Finally, we have studied the choice function impact on 8 variable selection heuristics and 3 value selection heuristics, which when combined, provide a portfolio of 24 strategies (see Table 3). Results are analyzed using the number of backtracks and solving time (for space reason in sec.) as performance indicators to evaluate yield of search process. For this evaluation, we used the relative percentage deviation (RPD). RPD value quantifies the deviation of the

Table 4. Backtracks required for all instances.

CF_i	NQ								SK							KT				MS					LS					QG						
	10	12	15	20	50	75	Σ	%	1	2	5	7	9	Σ	%	5	6	Σ	%	3	4	5	Σ	%	4	5	6	Σ	%	1	3	5	6	7	Σ	%
CF_1	3	5	0	6	7	15	36	0.24	0	1	78	89	0	168	0.71	878	10229	11107	0.11	0	0	35	35	1.33	0	0	0	0	0	0	0	0	1	1	2	2
CF_2	2	1	1	9	8	17	38	0.31	0	1	56	76	0	133	0.36	781	12071	12852	0.29	0	1	79	80	4.33	0	0	0	0	0	0	0	0	0	0	0	0
CF_3	2	3	1	9	11	31	57	0.97	0	2	117	87	0	206	1.1	771	12981	13752	0.38	0	0	98	98	5.53	0	0	0	0	0	1	0	0	0	0	1	1
CF_4	3	1	0	8	45	75	132	3.55	0	1	87	112	0	200	1.04	675	12129	12804	0.28	0	2	178	180	11	0	0	0	0	0	0	1	0	0	0	1	2
CF_5	2	1	1	9	12	27	52	0.79	0	2	81	74	0	157	0.60	598	14157	14755	0.48	0	0	141	141	8.4	0	0	0	0	0	0	1	0	0	1	2	3
CF_6	1	3	0	7	0	39	50	0.72	0	1	57	48	0	106	0.08	679	11542	12221	0.223	0	0	73	73	3.87	0	0	0	0	0	1	0	0	0	0	1	1
CF_7	2	2	1	10	2	22	39	0.34	0	1	31	97	0	129	0.32	981	13501	14482	0.45	0	1	14	15	0	0	0	0	0	0	0	0	0	0	0	0	0
CF_8	2	4	2	11	1	17	37	0.28	0	1	28	111	0	140	0.43	762	10857	11619	0.16	0	2	87	89	4.63	0	0	0	0	0	1	0	0	0	0	1	1
CF_9	5	1	0	11	1	41	59	1.03	0	2	173	117	0	292	1.98	812	13092	13904	0.39	0	1	148	149	8.375	0	0	0	0	0	0	1	0	0	0	1	1
CF_{10}	2	3	1	13	3	34	56	0.93	0	3	68	64	0	135	0.38	801	10777	11578	0.16	0	0	108	108	5.81	0	0	0	0	0	0	0	0	0	0	0	0
CF_{11}	1	2	0	8	2	67	80	1.76	0	2	37	59	0	98	0	912	11101	12013	0.2	0	0	187	187	10.75	0	0	0	0	0	1	0	0	0	1	2	2
CF_{12}	3	2	2	12	2	29	49	0.69	0	1	197	79	0	277	1.32	479	9501	9980	0	0	1	15	16	0.0625	0	0	0	0	0	0	1	0	0	0	1	1
CF_{13}	4	3	2	11	1	32	53	0.83	0	0	71	87	0	158	0.44	387	13001	13388	0.28	0	0	101	101	1.51	0	0	0	0	0	0	0	0	0	0	0	0
CF_{14}	2	3	1	12	0	12	29	0	0	0	97	76	0	173	0.55	327	15112	15439	0.44	0	0	165	165	2.63	0	0	0	0	0	1	1	0	1	0	3	3
CF_{15}	1	1	1	11	2	23	39	0.29	0	3	84	49	0	136	0.28	561	11755	12316	0.19	0	3	54	57	0.74	0	0	0	0	0	0	1	0	0	0	1	1
CF_{16}	0	1	1	9	5	19	35	0.17	0	1	121	61	0	183	0.46	301	12574	12875	0.22	0	2	121	123	0.88	0	0	0	0	0	0	1	0	0	0	1	1

Table 5. Solving time for all instances in seconds

CF_i	NQ						Σ	%	SK					Σ	%	KT		Σ	%	MS			Σ	%	LS			Σ	%	QG					Σ	%
	10	12	15	20	50	75			1	2	5	7	9			5	6			3	4	5			4	5	6			1	3	5	6	7		
CF_1	0.6	0.7	1	1.5	8	19.6	31.4	*0.01	0.6	0.5	0.5	0.6	0.5	2.7	0.04	3.2	4.5	7.7	0.03	0.6	0.9	0.7	2.2	0.05	0.4	0.4	0.5	1.3	0.3	0.3	0.2	0.7	0.6	0.5	2.3	0.15
CF_2	0.6	0.8	1.1	1.5	7.9	19.7	31.6	0.01	0.5	0.6	0.6	0.6	0.5	2.7	0.04	3.3	4.4	7.7	0.03	0.7	0.8	0.7	2.2	0.0	0.3	0.3	0.4	1	0	0.2	0.2	0.6	0.6	0.4	2	0
CF_3	0.6	0.8	1	1.6	7.8	19.7	31.5	*0.01	0.6	0.5	0.5	0.6	0.5	2.7	0.03	3.2	4.5	7.7	0.03	0.7	0.9	0.7	2.3	0.1	0.3	0.3	0.5	1.1	0.1	0.2	0.2	0.6	0.6	0.5	2.1	0.05
CF_4	0.6	0.7	0.9	1.6	8.3	19.7	31.8	0.01	0.7	0.6	0.6	0.6	0.5	3	0.15	3.3	4.5	7.8	0.04	0.7	1	0.8	2.5	0.19	0.3	0.4	0.4	1.1	0.1	0.3	0.2	0.7	0.7	0.5	2.4	0.2
CF_5	0.6	0.8	1.1	1.6	8.1	21	33.2	0.06	0.5	0.6	0.6	0.6	0.4	2.6	0	3.1	4.6	7.7	0.03	0.6	0.8	0.8	2.2	0.05	0.4	0.5	0.4	1.3	0.3	0.3	0.2	0.6	0.7	0.4	2.2	0.1
CF_6	0.7	0.7	1	1.5	7.8	19.7	31.4	0	0.6	0.6	0.5	0.6	0.4	2.7	0.04	3.2	4.7	7.9	0.05	0.5	0.8	0.8	2.1	*0.01	0.4	0.4	0.5	1.3	0.3	0.2	0.2	0.6	0.7	0.4	2.1	0.05
CF_7	0.6	0.7	1	1.5	7.9	19.7	31.4	0	0.6	0.6	0.7	0.5	0.6	3	0.15	3.3	4.5	7.8	0.04	0.6	0.9	0.7	2.2	0.05	0.4	0.3	0.5	1.2	0.2	0.3	0.2	0.6	0.6	0.4	2.1	0.05
CF_8	0.7	0.8	0.9	1.6	8.2	19.7	31.9	0.02	0.5	0.5	0.6	0.6	0.5	2.7	0.04	3.2	4.5	7.7	0.03	0.6	0.9	0.7	2.2	0.05	0.3	0.3	0.5	1.1	0.1	0.3	0.2	0.7	0.7	0.5	2.4	0.2
CF_9	0.6	0.8	1	1.5	7.7	19.9	31.5	*0.01	0.7	0.6	0.6	0.6	0.5	3	0.15	3.3	4.5	7.8	0.04	0.6	0.8	0.8	2.2	0.05	0.4	0.5	0.4	1.3	0.3	0.2	0.2	0.7	0.7	0.5	2.3	0.15
CF_{10}	0.6	0.8	1	1.5	8.1	19.8	31.8	0.01	0.6	0.6	0.5	0.6	0.4	2.7	0.038	3.1	4.4	7.5	0	0.5	0.8	0.8	2.1	*0.01	0.3	0.4	0.5	1.2	0.2	0.2	0.2	0.6	0.6	0.4	2.1	0.05
CF_{11}	0.5	0.6	1	1.4	8	20.8	32.3	0.03	0.5	0.5	0.5	0.6	0.5	2.7	0.04	3.1	4.6	7.7	0.03	0.6	1.1	0.7	2.4	0.14	0.2	0.4	0.4	1	0	0.2	0.2	0.6	0.6	0.5	2.1	0.05
CF_{12}	0.6	0.7	0.9	1.6	7.8	20.7	32.3	0.03	0.6	0.5	0.5	0.5	0.6	2.7	0.04	3.2	4.5	7.7	0.03	0.5	1	0.9	2.4	0.14	0.2	0.4	0.5	1.1	0.1	0.2	0.2	0.6	0.7	0.4	2.1	0.05
CF_{13}	0.5	0.8	1	1.5	8	19.7	31.5	*0.01	0.5	0.6	0.6	0.6	0.4	2.6	0	3.2	4.4	7.6	0.01	0.5	1	0.8	2.3	0.1	0.3	0.4	0.4	1.1	0.1	0.3	0.2	0.7	0.7	0.4	2.3	0.15
CF_{14}	0.6	0.7	0.9	1.6	8	19.6	31.4	*0.01	0.6	0.7	0.5	0.5	0.6	2.9	0.12	3.2	4.7	7.9	0.05	0.6	0.8	0.7	2.1	0	0.3	0.4	0.4	1.1	0.1	0.2	0.2	0.7	0.6	0.5	2.2	0.1
CF_{15}	0.6	0.6	0.9	1.5	8	19.8	31.4	0	0.6	0.5	0.6	0.6	0.5	2.8	0.08	3.2	4.5	7.7	0.03	0.6	0.8	0.7	2.1	0	0.4	0.4	0.4	1.2	0.2	0.3	0.2	0.7	0.6	0.4	2.2	0.1
CF_{16}	0.6	0.6	1.1	1.5	8	20.3	32.1	0.02	0.5	0.5	0.6	0.6	0.4	2.6	0	3.2	4.6	7.8	0.04	0.7	0.7	0.8	2.2	0.05	0.3	0.5	0.4	1.2	0.05	0.2	0.2	0.7	0.6	0.4	2.1	0.05

*: RPD is lees than 0.01

accumulated value $\sum_i^n CF_j$ from $min(\sum_i^n CF_j)$, where n corresponds to the number of instances, and it is calculated as follows:

$$RPD = \left(\frac{\sum_i^n CF_j - min(\sum_i^n CF_j)}{min(\sum_i^n CF_j)} \right) \times 100, \ \forall j \in \{1, \ldots, 16\} \qquad (3)$$

Table 4 illustrates the performance in terms of called backtrack and Table 5 exposes the efficiency in terms of the solving time required to find a solution for each choice function and the proposed optimizer. Results show the high-impact to use a choice function to correctly select the enumeration strategy during the search process. For instance, considering the total number of backtracks required for solving the CSPs, the best choice function is CF_{12}, where this selection method required only 738 backtracks. Now, analyzing the choice functions in terms of the solving time required to find a solution, we can observe that the best choice function is CF_{15} required 49783 ms, fallowed from close by CF_5, CF_2 and CF_7, respectively.

5 Conclusions

Autonomous search appeared as a new technique that enables the problem solver to control and adapt its own parameters and heuristics during solving in order to be more efficient without the knowledge of an expert user. In this paper, we have focused in the impact to use choice functions for correct selection the strategy and improving as a consequence the performance of the solver during the resolution process. To this end, we have exposed a performance evaluation of 16 different carefully constructed choice functions tuned by the black hole algorithm. The experimental results have demonstrated which in terms of the backtracks, the CF_{12} is better than others choice functions, and CF_{15} converges faster than its closest competitors: CF_5, CF_2 and CF_7, respectively. As future work, we plan to design of a similar adaptive framework for interleaving different domain filtering techniques. Additionally, we propose to add a new performance indicator to validate the achieved results in terms of backtracks and solving time.

Acknowledgments. Ricardo Soto is supported by Grant CONICYT / FONDECYT / REGULAR / 1160455, Broderick Crawford is supported by Grant CONICYT / FONDECYT / REGULAR / 1140897 and Rodrigo Olivares is supported by Postgraduate Grant Pontificia Universidad Católica de Valparaíso 2016.

References

1. Crawford, B., Soto, R., Castro, C., Monfroy, E.: A hyperheuristic approach for dynamic enumeration strategy selection in constraint satisfaction. In: Ferrández, J.M., Álvarez Sánchez, J.R., de la Paz, F., Toledo, F.J. (eds.) IWINAC 2011, Part II. LNCS, vol. 6687, pp. 295–304. Springer, Heidelberg (2011)

2. Crawford, B., Castro, C., Monfroy, E., Soto, R., Palma, W., Paredes, F.: Dynamic selection of enumeration strategies for solving constraint satisfaction problems. Rom. J. Inf. Sci. Tech. **15**, 106–128 (2012)
3. Crawford, B., Soto, R., Castro, C., Monfroy, E., Paredes, F.: An extensible autonomous search framework for constraint programming. Int. J. Phys. Sci. **6**(14), 3369–3376 (2011)
4. Hamadi, Y., Monfroy, E., Saubion, F.: What is autonomous search? In: van Hentenryck, P., Milano, M. (eds.) Hybrid Optimization: The Ten Years of CPAIOR. Springer, Heidelberg (2011)
5. Hamadi, Y., Monfroy, E., Saubion, F. (eds.): Autonomous Search. Springer Science + Business Media, Heidelberg (2012)
6. Hatamlou, A.: Black hole: a new heuristic optimization approach for data clustering. Inf. Sci. **222**, 175–184 (2013)
7. Kumar, S., Datta, D., Singh, S.: Black hole algorithm and its applications. In: Azar, A.T., Vaidyanathan, S. (eds.) Computational Intelligence Applications in Modeling and Control. Studies in Computational Intelligence, vol. 575, pp. 147–170. Springer International Publishing, Heidelberg (2015)
8. Soto, R., Crawford, B., Misra, S., Palma, W., Monfroy, E., Castro, C., Paredes, F.: Choice functions for autonomous search in constraint programming: Ga vs pso. Tech. Gaz. **20**(4), 621–629 (2013)

Scatter Search for Homology Modeling

Mouses Stamboulian and Nashat Mansour[(⊠)]

Department of Computer Science and Mathematics,
Lebanese American University, Beirut, Lebanon
{mouses.stamboulian,nmansour}@lau.edu.lb

Abstract. Homology modeling is an effective technique in protein structure prediction (PSP). However this technique suffers from poor initial target-template alignments. To improve homology based PSP, we propose a scatter search (SS) metaheuristic algorithm. SS is an evolutionary approach that is based on a population of candidate solutions. These candidates undergo evolutionary operations that combine search intensification and diversification over a number of iterations. The metaheuristic optimizes the initial poor alignments and uses fitness functions. We assess our algorithm on a number of proteins whose structures are present in the Protein Data Bank and which have been used in previous literature. Results obtained by our SS algorithm are compared with other approaches. The 3D models predicted by our algorithm show improved root mean standard deviations with respect to the native structures.

Keywords: Comparative modeling · Homology modeling · Protein structure prediction · Scatter search · Metaheuristics

1 Introduction

Proteins are considered to be large biological molecules that are composed of a specific sequence of amino acids (AA). A protein is distinguished from another by the sequence of AAs. These sequences of AAs fold and take three dimensional shapes (tertiary structure) forming complex structures of proteins. The function of a protein is decided by the structure of the protein and the way it folds after it gets transcribed [1]. Many diseases in humans result from misfolded proteins; examples are Cystic Fibrosis and Parkinson's disease [2].

The protein Structure Prediction (PSP) problem refers to determining the 3D conformation of proteins, given the initial sequence of AAs. Knowing their structure allows designing drugs and implementing personalized medicine. Next generation sequencing technologies have resulted in a huge explosion in genomic data and in the number of protein sequences. This has led to a growing gap between protein sequences and the structures discovered so far, thus encouraging the development of faster and more effective methods. Accurate wet lab methods exist for the PSP. These techniques are very time consuming and could still be error prone. Therefore, computational methods for PSP constitute a significant alternative.

Computational determination of a protein's 3D conformation is an intractable problem [3]. Hence, heuristic approaches are needed and have been developed to tackle

© Springer International Publishing Switzerland 2016
Y. Tan et al. (Eds.): ICSI 2016, Part I, LNCS 9712, pp. 66–73, 2016.
DOI: 10.1007/978-3-319-41000-5_7

this problem. Computational methods could be categorized into three groups. Ab-Initio methods predict protein structures using only their sequence of AAs, guided by energy functions [4]. Protein Threading, or fold recognition, assumes a limited number of distinct protein folds and uses fold libraries to map protein folds to sequences [5]. Homology modeling is highly based on previous knowledge. It determines the structure of a target protein by using template protein structures [6]. Homology modeling predicts accurate models, provided that suitable template structures exist for the prediction. The rationale for this approach is that protein structures are more conserved than their respective sequences during evolution [7]. Homology protein modeling takes advantage of this fact and aims to predict the structure of a certain target sequence using 3D structures of known homologous proteins. Whenever the sequence similarity falls below 25 %, homology modeling suffers from serious misalignments, resulting in poor comparative models. Three heuristic approaches have been reported for homology modeling: the Genetic Algorithm (GA) [8], Tabu Search (TS), and Particle Swarm Optimization (PSO) [9].

In this paper we focus on homology modeling for predicting 3D protein structures. We use, for the first time, the Scatter Search (SS) metaheuristic to explore the search space for new and better target-template alignments. SS is used within the framework of homology modeling by satisfying spatial restraints [10]. Experiments are conducted using our proposed approach and the results are compared with results from previous literature.

2   SS on Homology Modeling

2.1   Preliminaries

In general, the steps for homology modeling start with a search for appropriate template (s) (given the target sequence), aligning the target with the discovered templates, proposing a 3D structure for the target sequence based on this alignment, and finally evaluating the predicted model's accuracy. To perform the first step, online search tools are used to search for templates in databases of known protein structures. Popular searching tools are BLAST [11] and FASTA [12]. After choosing appropriate template structures, the next step is to perform pairwise or multiple sequence alignment between the target and the template(s) [13]. After aligning the sequences, the actual building of the model is done. One common technique used to build a model is 'modeling by satisfying spatial restraints'. Based on the target-template alignment, the latter technique constrains the possible structure for the target protein based on the restraints extracted from the template structures taking into account their sequence similarity [6]. Stereochemical restraints are added to the extracted constraints, such as the lengths of the bonds, their angles and molecular mechanics. Information from the two sources are combined together into an objective function. The most probable comparative model is suggested by optimizing this objective function.

The fitness functions used to guide the SS metaheuristic is the DOPE score [14] and MODELLER [10] is employed to develop 3D models. DOPE stands for Discrete

Optimized Protein Energy. It aims to obtain a global optimum value of a scoring system, which would be a direct indication of a native structure for that particular protein sequence. This is achieved through using Joint Probability Density Function of the positions, in 3D space, for all the atoms present in the protein molecule. Details are found in [14].

MODELLER is an automated comparative modeling tool for protein structure prediction, whose aim is to find the most likely 3D conformation for a given target sequence of amino acids. This process is initiated through the alignment of the target sequence with at least one or more template sequence(s) of known structures. The 3D comparative model of a sequence X of unknown structure is predicted by comparing it with structure(s) of one or more close homologues. If there are more than one template structure, then these structures are first compared with each other and spatial features get extracted from them. After that, the extracted features are sent to the target sequence, and hence a group of spatial restraints about the structure to-be- predicted is obtained. The final predicted 3D model is optimized by maximizing the spatial restraint satisfaction as much as possible. Details are found in [10].

Scatter Search (SS) is a population based evolutionary search strategy and has been adapted for solving a number of intractable optimization problems [15]. Starting with an initial random or controlled-random population, SS maintains a small population set, the reference-set; then, the algorithm combines' solutions and updates the reference-set. The basic steps representing the template for Scatter Search are the following:

- *Diversification-Generation Method:* an initial population of candidate solutions are created completely randomly or with controlled sampling;
- *Solution-Improvement-Method;*
- *Reference-Set-Update-Method:* a small group of candidate solutions are maintained in the 'reference set'. The best-quality and most-diverse solutions obtained after the improvement phase get admitted to the reference set, hence allowing SS to combine both properties of diversification and intensification;
- *Subset-generation-method:* subsets of candidate solutions of defined sizes are created;
- *Solution-Combination-Method:* the subsets of candidate solutions get combined to create new solutions.

In the next sections, we explain how the SS metaheuristic was adapted for the PSP problem.

2.2   Solution Representation

Candidate solutions are represented by objects that refer to the target protein. For practical purposes, they also contain the sequence for the template proteins. These sequences are represented by arrays composed of single letter amino acid representations and gaps, which are manipulated by the SS algorithm.

2.3 Diversification Generation Method (DGM)

This method creates an initial random population of candidate solutions, and enables SS to explore wider ranges of the solution space. This method takes an input which is an initial target-template alignment. The initial alignment is generated using a dynamic programing algorithm for local sequence alignment, with affine gap penalties. Based on this initial alignment, randomly generated alignments are created whose number is specified by *PSIZE* (population size), which is set to 100. In generating the initial population, the following rules are upheld: The length of the initial seed target-template alignment is respected; the number of gaps within the sequence alignments (both in the template and target) remain the same; the order of residues in which they appear in the initial seed alignment is respected. Restricting the process to these rules ensures the generation of feasible solutions. The DGM is called once at the beginning of the algorithm.

2.4 Solution Improvement Method (SIM)

This method aims to improve the quality of the solutions produced by DGM or by the solution combination method. For this purpose, we employ hill climbing for locally improve the solutions. The improvement is done in two phases. First, the method takes the target-template alignment and, using the template sequence, it reshuffles the gap positions randomly. This is done by choosing the gap position and its length randomly. This process is repeated 5 times. Each time the alignment is altered, the respective comparative model is calculated by the program MODELLER and the obtained model gets assessed by the DOPE function. Whenever the score of the new model is better than the old one, the new alignment replaces the old one. If by the end of the specified number of attempts no improvements are made, the original model is kept. The same process is repeated in the second phase using the target sequence instead of the template sequence.

2.5 Reference Set Creation Method (RSCM)

In this method, the initial reference set (RSet) is created. RSet is divided into two sets, the high quality solution set, *HQRefSet,* and the diverse solution set, *DivRefSet.* Usually these two subsets contain equal number of solutions ($|RSet|$ = 20, $|HQRefSet|$ = 10 and $|DivRefSet|$ = 10). To create the *RSet* for the first time, the population resulting from the SIM method is sorted from best to worst solution. After sorting, the first $|HQRefSet|$ solutions are chosen to form the *HQRefSet.* Concerning the *DivRefSet,* the set of most diverse solutions are admitted to this set. The diversity of a candidate solution is characterized as the Levenshtein distance (LD) between the sequence of the considered candidate solution in the population and the solutions already present in the *RSet.* For each candidate solution in the population, the LD between that solution and each of the solutions present in the *RSet* is calculated and the minimum of these distances is recorded for each solution. Then, the solutions with the maximum minimum-distance are added to the *DivRefSet.*

2.6   Subset Generation Method and Solution Combination Method (SCM)

This Subset Generation Method generates subsets using the *RSet*. For computational reasons, we limit this method in enumerating all the subsets of size two.

The SCM combines the information from the candidate solutions present in each subset and gives rise to new solutions. Since we restricted subsets of size two, a crossover operator is used, which crosses over information between two solution points and gives rise to a new solution carrying information from both initial solutions. The two target and template sequences from the two parents grouped together by the SGM undergo crossover separately.

First, the two target sequences from the parent alignments are extracted. The sequential gap indices for the two sequences are extracted into vectors. A location for crossover is chosen among the gap indices where the vectors differ. The vector which ends up with the smaller index gets combined with the vector starting with the larger index. The resulting combined vector specifies the locations of the gaps in the child target sequence. This entire procedure is repeated for the template sequence as well. The respective homology model is built afterwards by MODELELR using the new target-template sequences.

2.7   Reference Set Update Method (RSUM)

This method updates the reference set by using the population resulting from SCM. New solutions are admitted to the *RSET* and replace an existing solution in either of the two cases:

- If the considered solution's fitness score is better than that of the worst Solution in the *HQRefSet,* then it replaces it, and the updated *HQRefSet* is sorted again from best to worst.
- If the solution considered is more diverse than the least diverse Solution in the *DivRefSet.*

This entire process is repeated until all of the candidate solutions generated by the SCM are considered.

3   Experimental Results and Discussion

3.1   Experimental Procedure

The proposed approach is tested by predicting the 3D conformations of proteins of known structures that are stored in PDB. We compare our algorithm to three other approaches (GA; TS; PSO) proposed in [8, 9] that fall under the same framework of satisfying spatial restraints and using the same proteins. A total of eight target-template protein pairs were used. In addition to this, we compare our approach to a pure ab-initio approach [16] and a fragment based approach [17]. This is done by predicting the structures of three proteins that were also used in their experiments.

The results were assessed by calculating the root mean standard deviation (RMSD) between the predicted structures and the native ones found in the protein data bank (PDB). The RMSD values were calculated using PyMOL [18], which is also used to visualize the predicted structures. The algorithm starts with 100 initial random population of solutions. A reference set of size 20 was maintained throughout its execution. Each solution was allowed five iterations of improvement in the solution improvement phase. The termination criterion for the algorithm was that the reference set did not get updated.

3.2 Comparison with GA, TS and PSO

We compared the effectiveness of SS with that of genetic algorithm (GA), tabu search (TS), and particle swarm optimization (PSO). The comparison is based on the best RMSD results of the algorithms. TS and PSO were only tested on two protein pairs, hence the rest of the experiments were carried out by comparing our algorithm to the GA. The results are summarized in Table 1. RMSD values for the first two cases, using 2CCY-1BBH and 2RHE-3HLA target-template protein pairs, indicate that Tabu-Search and PSO perform poorly compared to GA and SS. Similar results were obtained while modeling the protein 3HLA. TS and PSO resulted in poor RMSD values, 15.209 and 15.24 respectively, whereas GA and SS maintained much lower RMSDs, 7.579 and 5.791. Furthermore, for both proteins, SS led to lower RMSD values than those of GA.

For the remaining proteins, SS produced better RMSD values than those of GA for three cases out of the total six and a similar RMSD for one, while it failed on the remaining two. The total count shows that SS yielded competitive results in comparison with GA, TS, and PSO.

Table 1. RMSD values for 8 protein.

Template	Target	Target length(AA)	% identity	% coverage	GA(Å)	TS(Å)	PSO(Å)	SS(Å)
2CCY (5:A-128:A)	1BBH (5:A-131:A)	126	21.3%	97.0%	2.362	3.048	3.762	1.752
2RHE (3-108)	3HLA (4:B-98:B)	94	2.4%	96.0%	7.579	15.209	15.24	5.791
1BOV (2:A-69:A)	1LTS (17:D-102:D)	85	4.4%	83.5%	8.579	N/A	N/A	7.84
1PAZ (3-93)	1AAJ (21-105)	84	27.5%	84.7%	2.1	N/A	N/A	1.339
1EGO (1-84)	1ABA (1-87)	86	16.9%	100.0%	4.1	N/A	N/A	4.164
2RHE (8-112)	3CD4 (2-100)	98	21.7%	100.0%	6.5	N/A	N/A	3.938
1FLB (15-86)	1HOM (7-57)	50	17.6%	75.0%	1.1	N/A	N/A	5.02
9RNT (2-104)	2SAR (7:A-91:A)	84	13.1%	88.5%	4.8	N/A	N/A	5.472

3.3 Comparison with Ab-Initio Method and Fragment Based Assembly Method

Further experiments were conducted by comparing our algorithm to a pure Ab-initio based method proposed in [16], and to a fragment based assembly method proposed in

[17]. The three methods compared in this section are based on entirely different techniques, making the comparison difficult. The performance of these techniques is directly related to the amount of information known about the predicted proteins. We ran our algorithm on the three proteins that were tested in the other two methods. The results are summarized in Table 2. To make this comparison as fair as possible, we chose the worst hits returned by BLAST as template proteins, provided they are evolutionary related to the target at least. Clearly, the results that the homology-based PSP is a better choice made provided that enough information exists. This is supported by the RMSD results for predicting 1CRN and 1UTG. To make things even more difficult, the template chosen to predict the structure of 1ROP only covers 57 % of the target protein sequence. This means that no spatial restraints could be extracted to constrain the target structure's prediction for the rest of the 43 %. Despite this, our approach returned an RMSD value of 7.033 Å, which is lower than 12.14 Å resulting by the pure ab-initio approach, and somewhat worse than the 5.43 Å returned by the fragment based approach.

Table 2. Comparison with ab-initio method and fragment based modelling.

Template	Target	% identity	% coverage	Mansour et al. (Å)	Fragment based SS (Å)	SS(Å)
1EDO(2-44)	1CRN(1-47)	51.0%	93.0%	9.01	8.05	0.952
1UTR(23:A-90:A)	1UTG(1-70)	57.0%	97.0%	14.78	12.34	1.71
4LCT(271:A-304:A)	1ROP(1-63)	42.0%	57.0%	12.14	5.43	7.033

4   Conclusion

We have presented a homology based protein modeling using SS to predict the 3D folds by satisfying spatial restraints. The heuristic was guided by assessing the resulting structures using two scoring functions, GA341 and DOPE. Our algorithm was evaluated by running it on 11 target-template protein pairs and the results were assessed by measuring the RMSD errors between the native and predicted structures. Comparisons were made between our results and those produced by 3 algorithms from previous literature. Out of 8 protein pairs, our approach resulted in lower RMSD values in 5 cases. Furthermore, we compared our approach to a pure ab-initio and a fragment-based assembly methods, by predicting 3 protein structures. Our algorithm was superior in terms of RMSD values for the first two cases, and returned comparable values in the third case.

References

1. Skolnick, J., Fetrow, J.: From genes to protein structure and function: novel applications of computational approaches in the genomic era. Trends Biotechnol. **18**(1), 34–39 (2000)
2. Welch, W.: Role of quality control pathways in human diseases involving protein misfolding. Semin. Cell Dev. Biol. **15**(1), 31–38 (2004)

3. Guyeux, C., Côté, N., Bahi, J., Bienia, W.: Is protein folding problem really a NP-Complete one? First investigations. J. Bioinf. Comput. Biol. **12**(01), 1350017 (2014). 24 pages
4. Abbass, J., Nebel, J.C., Mansour, N.: Ab Initio protein structure prediction: methods and challenges. In: Elloumi, M., Zomaya, A.Y. (eds.) Biological Knowledge Discovery Handbook: Preprocessing, Mining and Postprocessing of Biological Data. IEEE-Wiley, New Jersey (2014)
5. Jones, D.: GenTHREADER: an efficient and reliable protein fold recognition method for genomic sequences. J. Mol. Biol. **287**(4), 797–815 (1999)
6. Kopp, J., Schwede, T.: Automated protein structure homology modeling: a progress report. Pharmacogenomics **5**(4), 405–416 (2004)
7. Chothia, C., Lesk, A.: The relation between the divergence of sequence and structure in proteins. EMBO **5**(4), 823–826 (1986)
8. John, B.: Comparative protein structure modeling by iterative alignment, model building and model assessment. Nucleic Acids Res. **31**(14), 3982–3992 (2003)
9. Doong, S.: Protein homology modeling with heuristic search for sequence alignment. In: 40th Annual Hawaii International Conference on System Sciences, p. 128, Waikoloa (2007)
10. Šali, A., Blundell, T.: Comparative protein modelling by satisfaction of spatial restraints. J. Mol. Biol. **234**(3), 779–815 (1993)
11. Altschul, S.: Basic local alignment search tool. J. Mol. Biol. **215**(3), 403–410 (1990)
12. Pearson, W.: Empirical statistical estimates for sequence similarity searches. J. Mol. Biol. **276**(1), 71–84 (1998)
13. Mishra, S., Saxena, A., Sangwan, R.: Fundamentals of homology modeling steps and comparison among important bioinformatics tools: an overview. Sci. Int. **1**(7), 237–252 (2013)
14. Shen, M., Sali, A.: Statistical potential for assessment and prediction of protein structures. Protein Sci. **15**(11), 2507–2524 (2006)
15. Marti, R., Laguna, M.: Scatter Search: Basic Design and Advanced Strategies. Int. Artif., vol. 7, no. 19 (2003)
16. Mansour, N., Ghalayini, I., Rizk, S., El-Sibai, M.: Evolutionary algorithm for predicting all-atom protein structure. In: Proceedings of the ISCA 3rd International Conference on Bioinformatics and Computational Biology, pp. 7–12, New Orleans (2015)
17. Mansour, N., Terzian, M.: Fragment-based computational protein structure prediction. In: The Eighth International Conference on Advanced Engineering Computing and Applications in Sciences, pp. 108–112 (2015)
18. The PyMOL Molecular Graphics System. Schrödinger, LLC

Cuckoo Search Algorithm Inspired
by Artificial Bee Colony and Its Application

Yin Gao, Xiujuan Lei[✉], and Cai Dai

School of Computer Science, Shaanxi Normal University Xi'an,
Shaanxi 710119, China
xjlei168@163.com

Abstract. Cuckoo search algorithm with advanced levy flight strategy, can greatly improve algorithm's searching ability and increase the diversity of population. But it also has some problems. We improve them in this paper. First, in order to address the randomness of levy flight fluctuating significantly in the later and its poor convergence performance, we combine artificial bee colony algorithm with cuckoo search algorithm since artificial bee colony algorithm considers the group learning and cognitive ability, individuals learn from each other in the iterative process, which improves the local search ability of the later, and can find the optimal solution more quickly. Second, we use mutation operation to create the worst nest's position so as to increase the diversity of the population. Then put forward the ABC-M-CS algorithm and use the thought of K-means to cluster UCI data. The experimental results on UCI data sets indicate that ABC-M-CS algorithm has the fastest convergence speed, highest accuracy and stability.

Keywords: Cuckoo search algorithm · Artificial bee colony algorithm · Mutation operation · Clustering

1 Introduction

Clustering is a common method of data analysis and data mining. Clustering analysis derives from machine learning, pattern recognition, and data mining, *etc*. When clustering, the data objects are divided into several clusters and the objects in the same cluster have high similarity. Traditional clustering methods are probably divided into the following categories [1]: the method based on hierarchy [2], the method based on density [3–5], the method based on grid and the method based on model [6]. Heuristic intelligent algorithm is one of the research hotspots in the field of computer algorithm design and analysis, which attracted a group of scholars and researchers with its simplicity, distribution, high robustness, and scalability [7]. In recent years, many researchers have used swarm intelligence optimization algorithm for clustering, *e.g.*, T. Niknam proposed a new hybrid evolutionary algorithm to solve nonlinear partition clustering problem, which could find better cluster partition and has the higher robustness [8]; Simulated annealing algorithm and PSO algorithm were applied to cluster respectively [9, 10]; P.S Shelokar put forward an ant colony approach for clustering that the computational simulations revealed very encouraging results in terms of the quality of solution found, the average

© Springer International Publishing Switzerland 2016
Y. Tan et al. (Eds.): ICSI 2016, Part I, LNCS 9712, pp. 74–85, 2016.
DOI: 10.1007/978-3-319-41000-5_8

number of function evaluations and the processing time required [11]; Khaled S. Al-Sultan came up with a new algorithm based on Tabu Search(TS) algorithm for clustering, the algorithm performance was better than that of K-means and simulated annealing algorithm [12]; Jie Zhao *et al.* proposed a new algorithm, which improved cuckoo search algorithm inspired by particle swarm optimization (PSO) algorithm [13].

Cuckoo search (CS) algorithm is a new metaheuristic optimization algorithm developed by Yang and Deb in 2009 [14]. CS algorithm is based on the obligate brood parasitic behavior of some cuckoo species in combination with the Levy flight behavior of birds to solve the optimization problem effectively [15, 16]. Studies have shown that the CS optimization algorithm, with the outstanding search capabilities, effective random search path and few parameters, has been applied to many research areas [17, 18]. But CS algorithm has disadvantage of high randomness and fluctuation in the later stage. To address this problem, we propose a new CS algorithm which combined artificial colony algorithm named as ABC-M-CS and carry it into clustering. The proposed algorithm uses the UCI data sets for simulation, the experimental results reveal very encouraging results in terms of convergence performance and clustering effect.

2 Description of Basic Cuckoo Search Algorithm

In nature, the cuckoo looks for suitable birds' nest randomly, for simplicity in describing the CS algorithm, we now use the following three idealized rules [14]:

(1) Cuckoo lays an egg at a time, and dumps its egg in randomly chosen nest;
(2) The best nests with high quality of eggs will carry over to the next generations;
(3) The number of available host nests is fixed, foreign eggs are detected by host bird with a probability, marked as *pa* and *pa* $\in [0, 1]$.

Under these assumptions, the cuckoos update their own position to find new solutions by levy flight. The cuckoo randomly chooses the nest position to lay egg as follow:

$$x_i^{(t+1)} = x_i^{(t)} + \alpha * Levy(\beta), \ (i = 1, 2, \ldots n). \tag{1}$$

Where $x_i^{(t)}$ denotes the nest position of the ith in the tth generation; α is a random step and in most cases, it can be set as a certain constant 1; the product $*$ means entrywise multiplications; $Levy(\beta)$ is a random search path and random walk which conforms to levy distribution [15, 16, 19]:

$$Levy \sim \mu = t^{-1-\beta}, (0 < \beta \leq 2). \tag{2}$$

The step of Levy flight can be obtained by the following formulas. The μ and v are normal distribution. Type of Γ is the standard gamma function:

$$s = \frac{\mu}{|v|^{1/\beta}} \tag{3}$$

$$\mu \sim N\left(0, \sigma_\mu^2\right) v \sim N(0, \sigma_v^2) \tag{4}$$

$$\sigma_\mu = \left\{ \frac{\Gamma(1+\beta)\sin(\pi\beta/2)}{\Gamma[(1/\beta)/2]\beta 2^{(\beta-1)/2}} \right\}^{1/\beta}, \ \sigma_v = 1 \tag{5}$$

The worst of the bird's nest are detected by host bird with a probability $pa \in [0,1]$ and then the new solution is created according to the random walk:

$$x_{worst}^{(t+1)} = x_{worst}^{(t)} + \alpha * rand(). \tag{6}$$

3 Cuckoo Search Algorithm Inspired by Artificial Bee Colony Algorithm

3.1 ABC-M-CS Algorithm

Basic CS algorithm increases species diversity and all the values can be searched in the solution space due to the use of the levy flight, which greatly enhances the search ability and search optimization ability. But its randomness of levy flight influences the convergence and stability of the algorithm, and meanwhile the algorithm is lack of vigor and the search speed is slow. While the artificial bee colony (ABC) algorithm [20] has fast convergence speed, with the increase of the number of iterations, search space of the bees gradually decreases, which is helpful to improve the precision of the algorithm in the process of location update.

Inspired by the way that bees look for solutions of nectar sources, we improve the performance of CS algorithm. In the CS algorithm, cuckoos search the target using levy flight method, which can be regarded as the ability of 'self-learning', then 'social-learning' ability of the cuckoo increases when we combine the position updating formula of ABC algorithm, meanwhile, the searching capability of the CS algorithm improves. In this paper, we apply ABC-M-CS to clustering, the new algorithm has good stability and effectiveness, moreover, the search speed has great improved. Cuckoo's new position updates as follows:

$$x_i^{(t+1)} = x_i^{(t)} + \alpha * Levy(\beta) + R * (randnest - x_i^{(t)}), \ (i = 1, 2, \ldots, n). \tag{7}$$

The formula consists of three parts $x_i^{(t)}$ is the last generation of the cuckoo's nest solution; $Levy(\beta)$ is the cuckoo's nest location updated by Levy flight $Levy(\beta)$; and $R * (randnest - x_i^{(t)})$ means the new solution of sources searching introduced from ABC algorithm. Where $x_i^{(t)}$ denotes the position of the ith nest in the tth generation; α is a random step and in most cases, it can be set as a certain constant 1; *randnest* is the random nest of all; the product * means entrywise multiplications; $Levy(\beta)$ is a random search path and r represents uniform random number between [0,1].

The worst of the bird's nest are detected by host bird with a probability $pa \in [0, 1]$, and then the new solution is created according to the mutation operation:

$$nest_i(t) = nest_a(t) + F * (nest_b(t) - nest_c(t)). \tag{8}$$

where we choose three nests $nest_a(t)$, $nest_b(t)$, $nest_c(t)$ except the worst nest, and $a, b, c \in \{1, 2, \ldots, n\}$, n is the number of cuckoo. F is the zoom factor. $F \in [0, 2]$.

3.2   Clustering Criterion Function

Clustering criterion function is used to evaluate the quality of the algorithm, we define clustering criterion function as follows [2, 13]:

$$Jc = \sum_{k=1}^{n} \sum_{x_i \in c_k} d(x_i, z_k). \tag{9}$$

where n is the number of sample; Z_k represents the kth clustering center; $d(x_i, z_k)d(x_i, z_k)$ is the Euclidean distance from x_i to the corresponding clustering center. The smaller the clustering function value is, the shorter distance in internal classes and the better clustering effect could be.

3.3   Procedure of ABC-M-CS for Clustering

Step1: Generate initial parameters. The number of the bird's nest m, on behalf of the clustering center; the number of cuckoo n, in order to choose the best bird's nest; the iteration number *iter*; maximum iterations number *maxiter*; the number of class c; the random walk α the objective function Jc, here we denote as $f(x)$, $x = (x_1, x_2, \ldots, x_n)$. First, we generated the initial position of nests x_i, $i = (1, 2, \ldots, m)$ and calculate Euclidean distance between each point and cluster center, then clustering on the thought of K-means.

Step2: Calculate fitness of every bird's nest according to formula (9) and compare the fitness of each, so as to obtain locations of optimal bird's nest and the worst bird's nest.

Step3: In addition to the optimal bird's nest, the other cuckoos' positions update to find new bird's nest by the formula (7), then clustering and calculate the fitness value.

Step4: Compare the updated fitness value with prior one, retain the better bird's nest. The worst of the bird's nest is found in a probability, marked as *pa*. Generate a random number R which donates that foreign eggs are detected by host bird. If $R > pa$, produce the new nest according to the formula (8) and calculate the fitness of new bird's nest, then compare the new fitness with the current optimal objective function value of the bird's nest, and retain an optimal solution.

Step5: Perform the above steps, output the optimal solution until it meets the end of conditions, otherwise, go to **Step2**.

4   Experimental Results and Evaluation

4.1   The Experimental Data and Environment

In order to test the effectiveness and accuracy of the proposed ABC-M-CS algorithm, we simulated on the UCI data set, such as the Iris, Glass, Wine, Cancer, Sonar, Breast data. The characteristics of the data set are showed in Table 1.

Table 1. The properties of data sets

Data set	Classes	Dim	Size of data set(the number of each class)
Iris	3	4	150(50,50,50)
Glass	6	9	214(29,76,70,17,13,9)
Wine	3	13	178(59,71,48)
Cancer	2	9	683(444,239)
Sonar	2	60	208(97,111)
Breast	2	9	277(81,196)

4.2   Parameter of ABC-M-CS

4.2.1   Parameter Analysis

In order to describe the influence of pa and α on the result, we take different values for experiments. We test them on Iris data set and the results are shown in Fig. 1:

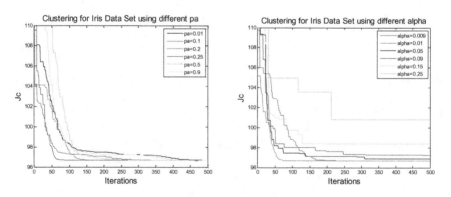

Fig. 1. Parameter analysis results (Color figure online)

It can be seen from Fig. 1 that the result is not stable and the convergence speed is slow when pa are set 0.01, 0.1, 0.2, 0.5 and 0.9, respectively. When pa are 0.5 and 0.2, algorithm is converged around 270 generations and 150 generations, respectively;

when pa is 0.01 or 0.9, algorithm is converged after 400 generations and the algorithm fluctuates obviously; while pa is 0.25, the algorithm is converged only at about 80 generations and performs better. So we choose $pa = 0.25$. From the different values of α, we can see that when α are taken 0.009, 0.05, 0.09, 0.15 and 0.25, the algorithm's convergence speed and accuracy are poor. While taken $\alpha = 0.01$, the algorithm has the best convergence and it's the most stable of all. We can see from the experimental results that a reasonable value is critical because the values of α have great influence on the performance of the algorithm. So we choose $\alpha = 0.01$.

4.2.2   Set the Parameters' Setting

The parameters of HBMO, SA, ACO, GA, TS algorithms were set in [8, 27], and the parameters of PSO: $C_1 = C_2 = 2$, $W_{min} = 0.4$, $W_{max} = 0.9$, $interation = 1000$ (Table 2).

Table 2. The parameters' setting of ABC-M-CS

Parameter	Description	Values settings
n	Number of cuckoo	20
pa	The worst of the bird's nest is found in a probability	0.25
α	Step control volume	0.01

4.3   Analysis of the Convergence

To validate the convergence of ABC-M-CS, we compare it with K-means, PSO and the basic CS algorithm on the Iris data set and Cancer data set, the results are shown in Fig. 2:

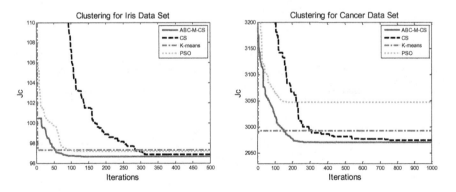

Fig. 2. Convergence curves in Iris data set and Cancer data set (Color figure online)

We can see clearly from the Fig. 2 (Jc denotes the fitness value) that K-means and PSO converges faster, but has lower convergence precision than CS and ABC-M-CS, and both of them are easy to fall into local optimum, and have the premature convergence phenomenon [9]; compared with ABC-M-CS, CS has slow convergence speed and obvious fluctuation. On the whole, ABC-M-CS has the fastest convergence rate, its convergence precision is the highest and performance is the best of all.

4.4   Analysis of Clustering Results

In order to evaluate the clustering results more intuitively and clearly, we firstly retreat the clustering result of ABC-M-CS algorithm through the principal component analysis (PCA) [21]. And then simulate the experiment on the Iris, Glass, Wine, Cancer, Sonar, Breast data set. The experimental results are shown in Fig. 3:

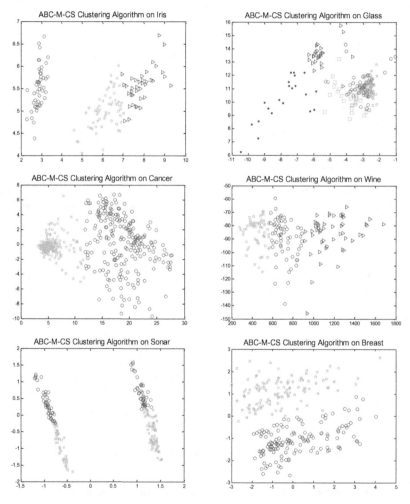

Fig. 3. The clustering results of ABC-M-CS in each data set

It is pointed out that ABC-M-CS algorithm performs better on these data sets. The algorithm can separate each cluster clearly except Glass and Sonar data sets. The clustering result is poor on Glass and Sonar data sets due to their high dimensions and the number of classes. High dimensional data set has many irrelevant attributes and the distance between data is almost equal, the algorithm has not do the dimension reduction, so its clustering effect doesn't improved obviously.

Experiment of ABC-M-CS clustering algorithm carries out thirty times on the six data sets and compares with K-means [22–25], Simulated Annealing (SA) [9, 22, 24, 25], Genetic Algorithm (GA) [10, 11, 22, 24, 25], Ant Colony Optimization (ACO) [11, 22, 24, 25], PSO, CS, TS [8, 22], Honey Bee Mating Optimization (HBMO) [8–10, 22], Improved Cuckoo Search (ICS) [13], then we list the best value, worst value, mean value and accuracy of different algorithms on different data sets. Among them, some scholars introduced the clustering process and experiment results through K-means, ACO, PSO, SA, GA, TS, HBMO and CS algorithm on different data sets and illustrates the advantages and disadvantages of the algorithms [8–12, 22–27]. The comp Tables 3, 4, 5, 6, 7, 8:

Table 3. The comparison of iris data set

Algorithm	Best Jc	Worst Jc	Mean Jc	Precision
K-means [23]	97.32	/	102.57	83.95 %
ACO [8, 25]	97.10	97.81	97.17	/
SA [8, 22]	97.46	102.01	99.96	/
GA [8, 22]	113.99	139.78	125.20	/
TS [8, 22]	97.37	98.57	97.87	/
HBMO [8, 22]	96.75	97.76	96.95	/
PSO	97.15	102.88	101.82	83.51 %
ICS [13]	96.66	/	96.68	90.40 %
CS	96.90	97.31	97.21	86.82 %
ABC-M-CS	**96.65**	**96.68**	**96.66**	**93.01 %**

Table 4. The comparison of glass data set

Algorithm	Best Jc	Worst Jc	Mean Jc	Precision
K-means [23]	213.42	/	241.03	51.70 %
ACO [25]	269.72	280.08	273.46	/
SA [22]	275.16	287.18	282.19	/
GA [22]	278.37	286.77	282.32	/
TS [8, 22]	279.87	286.47	283.79	/
HBMO [8, 22]	245.73	249.54	247.71	
PSO	228.28	255.12	239.36	43.35 %
ICS [13]	210.95	/	213.84	56.07 %
CS	213.31	217.84	215.97	51.50 %
ABC-M-CS	**210.23**	**212.56**	**211.59**	**57.40 %**

Table 5. The comparison of wine data set

Algorithm	Best Jc	Worst Jc	Mean Jc	Precision
K-means [23]	16555.68	/	17662.73	66.62 %
ACO [8, 25]	16530.53	16530.53	16530.53	/
SA [8, 22]	16473.48	18083.25	17521.09	/
GA [8, 22]	16530.53	16530.53	16530.53	/
TS [8]	16666.23	16837.54	16785.46	/
HBMO [8]	16357.28	16357.28	16357.28	/
PSO	16307.16	16579.05	16320.67	71.67 %
ICS [13]	16295.67	/	16302.40	72.36 %
CS	16315.00	16477.28	16370.85	70.30 %
ABC-M-CS	**16292.21**	**16292.21**	**16292.21**	**77.80 %**

Table 6. The comparison of cancer data set

Algorithm	Best Jc	Worst Jc	Mean Jc	Precision
K-means [22]	2988.43	2999.19	2988.99	/
ACO [8, 25]	2970.49	3242.01	3046.06	/
SA [8, 22]	2993.45	3421.95	3239.17	/
GA [8, 22]	2999.32	3427.43	3249.46	/
TS [8, 22]	2982.84	3434.16	3251.37	/
HBMO [8, 22]	2989.94	3210.78	3112.42	/
PSO	2977.80	3069.10	2980.30	80.82 %
CS	2971.30	2977.63	2972.80	86.80 %
ABC-M-CS	**2967.10**	**2971.12**	**2968.33**	**89.52 %**

Table 7. The comparison of sonar data set

Algorithm	Best Jc	Worst Jc	Mean Jc	Precision
K-means [23]	234.77	/	235.06	55.05 %
PSO	239.87	262.81	250.70	53.43 %
ICS [13]	232.20	/	238.58	55.77 %
CS	240.55	259.44	246.10	54.30 %
ABC-M-CS	**231.50**	**235.06**	**233.99**	**58.56 %**

Table 8. The comparison of breast data set

Algorithm	Best Jc	Worst Jc	Mean Jc	Precision
PSO	709.83	725.84	710.45	63.00 %
CS	703.90	710.30	705.91	65.90 %
ABC-M-CS	**700.52**	**702.57**	**701.89**	**71.80 %**

It is observed that the best value, mean value and worst value of ABC-M-CS algorithm on the six data sets respectively have small change, which indicates that the algorithm is stable. ABC-M-CS algorithm has a high accuracy when compares with K-means, ACO, PSO, SA, GA, TS, HBMO, ICS and CS algorithm. K-means algorithm is not stable and has low accuracy because it is sensitive to the initial clustering center; global optimization ability of ACO, PSO, CS, GA, TS, HBMO and SA are strong, but their accuracy are lower than ABC-M-CS algorithm; the ICS algorithm combining with PSO, has a fast convergence speed, while it is easy to fall into local optimum, so the effect is not very ideal. At the same time, we can see from the tables that different algorithms have different effects on different data sets, *i.e.*, the clustering effect relates to the data type and the data dimension, so we should considered all factors synthetically such as the accuracy, clustering criterion function and data types.

5　Conclusion

In this work, we propose ABC-M-CS algorithm and apply it to cluster, then simulate the experiments on the Iris, Glass, Wine, Cancer, Sonar, and Breast data set, respectively. The experimental results show that ABC-M-CS algorithm obviously overcomes the slow convergence speed and instability of CS algorithm. Moreover, the proposed algorithm improves the searching precision compared with K-means and PSO, meanwhile its Best *Jc*, Worst *Jc* and Mean *Jc* are also better than the other clustering algorithms we involved above. It is clearly seen that the new algorithm has the strong optimization ability and stable performance. Clustering results are outstanding on the Iris, Wine, Breast, Cancer data set, but the results are not very ideal on Glass, Sonar data set because of their high dimensions and the number of classes. In future, we will continue to make further improvement on the high dimensional data.

Acknowledgments. This paper is supported by the National Natural Science Foundation of China (61502290, 61401263), Industrial Research Project of Science and Technology in Shaanxi Province (2015GY016), the Fundamental Research Funds for the Central Universities, Shaanxi Normal University (GK201501008), and China Postdoctoral Science Foundation (2015M582606).

References

1. Li, R.-Y.: Clustering Analysis Algorithm in Data Mining Research and Application. Xi'an university of electronic science and technology, Xi'an (2012)
2. Serna, A.: Implementation of hierarchical clustering methods. J. Comput. Phy. **129**(1), 30–40 (1996)
3. Ester, M., Kriegel, H.-P., Sander, J., Xu, X.: A density-based algorithm for discovering clusters in large spatial databases with noise. In: National Conferences on Aritificial Intelligence 1998–1999, pp. 226–231 (1998)

4. Ankerst, M., Breunig, M.M., Kriegel, H.-P., Sander, J.: OPTICS: ordering points to identify the clustering structure. In: Proceedings of the 1999 ACM SIGMOD International Conference on Management of data, pp. 49–60 (1999)
5. Guha, S., Rastogi, R., Shim, K.: ROCK: a robust clustering algorithm for categorical attributes Inform. Inf. Syst. **25**(5), 345–366 (2000)
6. Han, J., Kamber, M.: Data Mining Concepts and Techniques. Morgan Kaufmann Publishers, San Francisco (2001)
7. Lei, X.: Swarm Intelligent Optimization Algorithms and their Application. Science Press (in Chinese), Beijing (2012)
8. Niknam, T., Amiri, B.: An efficient hybrid approach based on PSO, ACO and k-means for cluster analysis. Appl. Soft Comput. **10**(1), 183–197 (2010)
9. Selim, S.Z., Alsultan, K.: A simulated annealing algorithm for the clustering problem. Pattern Recogn. J. Pattern Recogn. Soc. **24**(10), 1003–1008 (1991)
10. Van Der Merwe, D.W., Engelbrecht, A.P.: Data clustering using particle swarm optimization. In: Proceeding of 2003 Congress on Evolutionary Computation(CEC 2003), pp. 215–220 (2003)
11. Shelokar, P.S., Jayaraman, V.K., Kulkarni, B.D.: An ant colony approach for clustering. Anal. Chim. Acta **509**(2), 187–195 (2004)
12. Khaled, S.: Al-Sultan.: A tabu search approach to the clustering problem. Pattern Recogn. J. Pattern Recogn. Soc. **28**(9), 1443–1451 (1995)
13. Zhao, J., Lei, X., Wu, Z., Tan, Y.: Clustering using improved cuckoo search algorithm. In: Tan, Y., Shi, Y., Coello, C.A. (eds.) ICSI 2014, Part I. LNCS, vol. 8794, pp. 479–488. Springer, Heidelberg (2014)
14. Yang X.S., Deb, S.: Cuckoo Search via Levy flights. In: World Congress on Nature & Biologically Inspired Computing, pp. 210–214. IEEE Publication, USA (2009)
15. Ouaarab, A., Ahiod, B., Yang, X.-S.: Random-key cuckoo search for the travelling salesman problem. Soft Comput. **19**(4), 1099–1106 (2015)
16. Yang, X.S., Deb, S.: Engineer optimization by cuckoo search. Int. J. Math. Modeling Num. Optimization **1**(4), 330–343 (2010)
17. Afzalan, E., Joorabian, M.: An improved cuckoo search algorithm for power economic load dispatch. Int. Trans. Electr. Energy Syst. **25**(6), 958–975 (2015)
18. Wang, L.-J., Yin, Y.-L., Zhong, Y.-W.: Cuckoo Search Algorithm with Dimension by Dimension Improvement. J. Softw. **11** (2013)
19. Reynolds, A.M., Rhodes, C.J.: The levy flight paradigm: random search patterns and mechanism. Ecology **90**, 877–887 (2009)
20. Karaboga, D., Basturk, B.: A powerful and efficient algorithm for numerical function optimization: artificial bee colony (ABC) algorithm. J. Global Optim. **39**(3), 459–471 (2007)
21. Zhang, X.-G.: Pattern recognition, 3rd edn. Tsinghua university press, Beijing (1999)
22. Krishnasamy, G., Kulkarni, A.J., Paramesran, R.: A hybrid approach for data clustering based on modified cohort intelligence and K-means. Expert Syst. Appl. **41**(13), 6009–6016 (2014)
23. Haasanzadeh T, Meybodi M R.: A new hybrid approach for data clustering using firefly algorithm and K-means. In: International Symposium on Artificial Intelligence and Signal Processing 16th, pp. 007–011 (2012)
24. Cowgill, M.C., Harvey, R.J., Watson, L.T.: A genetic algorithm approach to cluster analysis. Comput. Math Appl. **37**(7), 99–108 (1999)
25. Hatamlou, A., Abdullah, S., Nezamabadi-pour, H.: A combined approach for clustering based on K-means and gravitational search algorithms. Swarm Evol. Comput. **6**, 47–52 (2012)

26. Kao, Y.-T., Zahara, E., Kao, I.-W.: A hybridized approach to data clustering. Expert Syst. Appl. **34**(3), 1754–1762 (2007)
27. Niknam, T., Fard, E.T., Pourjafarian, N., Rousta, A.: An efficient hybrid algorithm based on modified imperialist competitive algorithm and K-means for data clustering. Eng. Appl. Artif. Intell. **24**(2), 306–317 (2010)

An Ideal Fine-Grained GAC Algorithm for Table Constraints

Limeng Qiao, Zhenhui Xu, Jin Dong, Yuan Shao, Xin Tong,
and Zhanshan Li$^{(\boxtimes)}$

College of Computer Science and Technology,
Jilin University, Changchun, China
{qiaolm2113,xuzh2113,shaoyuan2113}@mails.jlu.edu.cn,
dongjin0712@gmail.com, tongxin.jlu@gmail.com,
lizs@jlu.edu.cn

Abstract. AC5TC is a generic value-based domain-consistency framework for GAC propagation algorithms, which stores the information of removed values with a propagation queue. Three efficient implementations of AC5TC algorithm have been proposed in [1]. One of these algorithms (called AC5TC-Tr) has an optimal time complexity theoretically, but its space complexity is inefficient because of the dynamic data structure. This paper proposes a novel algorithm based on AC5TC framework, called AC5TC-Alter, which leverages a more efficient search strategy to establish GAC. AC5TC-Alter accesses both allowed table and valid table alternately during the process of supports seeking, also our method is scalable since it does not need data structure to maintain the validity of tuples. The experimental results show that AC5TC-Alter outperforms most baseline algorithms on time complexity and AC5TC-Tr on space complexity.

Keywords: Constraint satisfaction problem · Table constraints · AC5TC · Value-based · Search strategy · AC5TC-Alter

1 Introduction

Many important problems in real life such as computer version, resource allocation and scheduling, can be modeled as constraint satisfaction problem (CSP). Also, there are a lot of filtering algorithms proposed to solve CSP. As the most important part of constraint programming, arc consistency (AC) has been implemented by many different methods such as: AC3 [2], AC4 [3], AC5 [4], AC6 [5], AC7, AC8 [6], and AC2001 [7]. According to [8], we know that AC5 algorithm has the best performance (i.e. both time complexity and space complexity) among all these algorithms.

In this paper, we propose a new method which maintains GAC (i.e. extension of AC on non-binary constraints) to improve the performance of AC5TC algorithms, i.e. AC5 algorithm for table constraint. What's more, a table constraint is defined by a set of $r - ary$ tuples and a CSP is solved *iff* each constraint in this problem could find a support, i.e. a tuple that is both allowed and valid. As a result, traditional AC5TC

© Springer International Publishing Switzerland 2016
Y. Tan et al. (Eds.): ICSI 2016, Part I, LNCS 9712, pp. 86–94, 2016.
DOI: 10.1007/978-3-319-41000-5_9

algorithms iterate over the list of valid tuples or the list of allowed tuples to find supports. The new algorithm that we proposed combines AC5TC algorithm with a more efficient search strategy which visits both valid tuples and allowed tuples alternately during the process of supports searching [9].

The rest of this paper is organized in the following way. In the next section, we recall the basic definitions of CSP and fix the notations used in this paper. Section 3 describes AC5TC framework and introduces three implementations of AC5TC, i.e. AC5TC-Bool, AC5TC-Recomp and AC5TC-Tr. Then, for the most important one, we will state the new algorithm of AC5TC which employs a more efficient search strategy in Sect. 4. By the end of this paper we will analyze our experimental results and come to conclusion.

2 Preliminary Knowledge

A constraint satisfaction problem is described by a triple $(X, D(X), C)$, which contains a set of n variables $X = \{x_1, \cdots, x_n\}$, a set of domains $D(X) = \{D(x_1), \cdots, D(x_n)\}$ where $D(x_i)$ is consist of all possible values of x_i, and a set of constraints $C = \{c_1, \cdots, c_n\}$ with that each constraint is composed of a plenty of $r-ary$ tuples. A solution of a CSP is an assignment of all variables such that each constraint in this problem could find a support.

2.1 Generalized Arc Consistency

Generalized arc consistency is the mainstream method of constraints propagation. In many applications, AC is then said to be Generalized (GAC) when non-binary constraints naturally arise [9]. Now assuming that we have a constraint network (CN) $N = (X, D, C)$, and there are a constraint $c \in C$ and a variable $x_i \in X(c)$ in it. Then a value $v_i \in D(x_i)$ is compatible with c *iff* there exists a tuple σ with $v_i = \sigma(x_i)$ which satisfies the constraint c, a domain D is compatible with c *iff* all values in D are satisfied with the constraint c, a constraint networks N has GAC *iff* all domains in CN are compatible with constraints in CSP.

2.2 Table Constraints

Given a set of tuples T of arity r, a table constraint c over T holds if $(x_1, \cdots x_r) \in T[1]$, i.e. c is described by a set of $r-ary$ tuples. For a table constraint, we say that a tuple σ is allowed when it belong to the table and the allowed table is consist of all these allowed tuples. Besides, we define that a tuple σ is valid *iff* $\sigma \in \prod_{Y=var(c)} dom(Y)$, i.e. each $r-ary$ constraint has $var(c) = \{x_1 \cdots \cdots x_r\}$. Likewise, the valid table is consist of all these valid tuples.

2.3   Related Work

A lot of research effort has been spent on arc consistency and table constraints. The existing propagators can be mainly classified into two categories: coarse-grained algorithms (i.e. constraint-based) and fine-grained algorithms (i.e. value-based). And all these algorithms ensure that each value in the domain of each variable is supported by some value in the domain of a variable by which it is constrained. The following will give a brief description of these arc consistency algorithms.

AC3. AC3 [2] revises the domains repeatedly in order to remove unsupported values and keeps all the constraints which do not guarantee AC in a propagation queue to avoid useless calls of method *REVISE*. However, AC3 still performs many ineffective checks.

AC4. AC4 [3] confirms the existence of a supports by storing all supports for each value in accessorial data structures, but not identifying it throughout search. However, the huge data structures makes its space complexity is inefficient.

AC5. AC5 [4] can be implemented on AC3 or AC4. This propagator maintains the index of its first current support in the table and the index of the next tuple sharing the same value for this variable for each value of each variable.

AC6. AC6 [5] ensures that a value has one support at least but not counts all supports for a constraint. That is to say, AC6 maintains a data structure lighter than AC4.

AC8. AC8 [6] is like AC3 but propagations are made over values. Besides, AC8 records the references of the variables in a list and maintains their status in an array.

AC2001. AC2001 [7] follows the same framework with AC3, but it stores the smallest support for each value on each constraint.

Hence we can see that there are a lot of improvements have been achieved by: changing the way of propagation from arcs to values, using new structures to store more information which is more useful, storing the first support found or the smallest support found to improve the performance of algorithms, etc. [10]

3   AC5TC-Framework and Three Implementations

3.1   AC5TC: A Generic Arc Consistency Framework

AC5TC is a generic value-based framework for GAC propagation algorithms. Here, value-based means that the algorithms store the information of the removed values with a propagation queue at all times. In addition to maintaining a queue, the concrete propagators must implement the methods *post* and *valRemove* [1]. The former one initializes the specific structures for the propagation and makes the first pruning. And the latter one (i.e. which is the central method of AC5TC) reflects the deletion of value a from $dom(x)$ and removes non-GAC values which are no longer supported in the constraint c. Besides, the propagation queue implements the method *enqueue* to put the necessary information in the queue when some values become invalid during searching supports.

3.2 Special Data Structure of AC5TC Algorithm

Each value-based algorithms needs to maintain a data structure called $FS(x, a)$ to record the index of first support for pair (x, a), and to speed up the table traversal, most of them also need a second data structure named $next(x, i)$ to link together the tuples with the same value for variable x. The constraint table of CSP is static, but the $FS(x, a)$ must be dynamic to record the index of support for the constraint c during seeking supports. However, depending on the structure of $next(x, i)$ is static or dynamic, we will get different implementations of AC5TC.

3.3 Three Implementations of AC5TC Algorithm

AC5TC-Bool. As an implementation of AC5TC, AC5TC-Bool maintains a bool array *isQValid* for each table in addition. This approach uses *isQValid[i]* to identify the validity of tuple $\sigma_{c,i}$ in the table. Obviously, the array must be trailed at all times when we traverse entire table during the propagation.

AC5TC-Recomp. This variation of AC5TC doesn't need any data structure like *FS* or *next* to record the validity of tuples, but a function *isValidTC* which tests the tuple's validity is necessary. The function *isValidTC* will be called once when the propagator wants to know the validity of a tuple $\sigma_{c,i}$.

AC5TC-Tr. This implementation of AC5TC is optimal theoretically by changing the *next* structure dynamically. The idea is to make the validity information of tuples included in next structure. This approach maintains that $next(x, i)$ is always equal to the smallest tuple index $j > i$ with the same value for variable x which is valid. The idea which can be implemented easily by a doubly-linked list speeds up the algorithm and avoids lots of unnecessary checks.

Performance Comparison of these Three Algorithms. Given a table constraint where r is the arity of the constraint, t is the number of the table and d is the size of the largest domain, a domain consistency algorithm will check the validity of each tuple in the table and remove some or even all the values from the domains. Thus it has a complexity $\Omega(r \cdot t + r \cdot d)$ in the worst case. AC5TC-Tr which maintains the *next* structure to seek supports directly has an optimal time complexity of $O(r \cdot t + r \cdot d)$ per table constraint. However, the other two algorithms have the complexity of $O(r \cdot t + r \cdot d)$ because the validity of tuples must be obtained by additional calculation.

4 AC5TC-Alter: With a New Search Strategy

All implementations of AC5TC that we describe in Sect. 3 visit both lists of allowed tuples and valid tuples during the process of seeking supports. Now, we propose a novel algorithm called AC5TC-Alter to establish GAC by leveraging a more efficient search strategy which avoids to access lots of irrelevant tuples in constraint table. The core methods of AC5TC-Alter are *seekSupport* (see Algorithm 1, i.e. which returns the index of support for pair (x, a) in constraint c and *setNextValid* (see Algorithm 2), i.e.

which computes the smallest valid tuple σ built from c such that $\sigma(x) = a$ or T if it does not exist.

Initially, the call of *seekSupport* uses three data structures to store some necessary information: θ is a sub-table with $value(x) = a|x \in var(c)$ of constraint c, I is an array which stores the index of elements in θ, and σ computes the first valid tuple for constraint c as initial condition. The loop will finish when all valid tuples which should be considered have been tested (line 9). Each execution of loop body will compute a valid tuple. Traditional approaches iterate over the valid table or allowed table sequentially to find supports. As a result, its time complexity is linear $O(t)$ because each valid tuple needs to be considered. Now considering that all tuples in allowed table and valid table are organized by lexicographic ordering, so our approach uses method *binarySearch* to find the smallest allowed tuple σ' whose validity must be checked (line 10). If σ' is equal to σ (i.e. method *compare* return zero), the function will return the index of σ' in allowed table because the tuple σ is both allowed and valid (line 12). And an invalid value will be returned if σ' is equal to T which means we can't find a support in the table (line 15). Otherwise, the algorithm will use the method *setNextValid* to compute the smallest valid tuple σ which maintains $\sigma' < \sigma$ and $\sigma[x] = a$(line 16). This function, described by *setNextValid*, uses a bool variable *flag* to record whether the function finds the target tuple σ or not. Then, to find a tuple strictly greater than σ, we have to iteratively look for next value following $\sigma.value(i)$ in $dom(y)$(line 11). The function will return a target tuple σ or T(i.e. means can't find a valid tuple anymore).

The method *post* which achieves domain consistency has a complexity $O(r \cdot t + r \cdot d)$ per table constraint. Then, we consider the time complexity of the method *valRemove*. In the best cases, the number of searches will be reduced from $O(t)$ to $O(\log t)$ by utilizing the method *binarySearch*. However, it is impossible to get optimal result every time which means we may still access all tuples in the table for unsuccessful search during the process of looking supports. Hence, the time complexity of the AC5TC-Alter is in the range of $O(r \cdot r \cdot \log t + r \cdot d)$ to $O(r \cdot r \cdot t + r \cdot d)$ which is strictly better than AC5TC-Bool and AC5TC-Recomp.

5   Experimental Results

In order to compare the performance of different approaches accurately, all proposed algorithms have been implemented with a same AC5 framework by Java. And we have tested plenty of instances used as benchmarks for some competitions of CSP solver. (see http://www.cril.univ-artois.fr/~lecoutre/benchmarks.html). These instances are from different backgrounds and distinguish between structured instances and random instances (the more structured an instance is, the lower its entropy is). Besides, only positive table constraints are selected to test. During our experiments, performances have been measured in terms of the number of tuples visited by AC5TC (when looking for supports) and the mean propagation time of CPU in seconds during search. What's more, the results of Tables 1 and 2 were conducted on an Intel(R) Core(TM) i5-4200U CPU @ 1.60 GHz 2.30 GHz using Windows 10.

Algorithm 1. Function *seekSupport* for AC5TC-Alter.

```
Program seekSupport (int)
    var    c: Constraint;
           x: Variable;
           a: Value;
    begin
      Θ := setSubtable(c,x,a);
      I := computeIndex(c.allowedtable,Θ);
      σ := setFirstValid(c,x,a);
      while(σ is valid) do
        σ' = binarySearch(Θ,I,σ);
        if(compare(Θ,I,σ,σ')  == 0)
          return I[σ'];
        else
          if(σ' >= I.length)
            return T;
          σ = setNextValid(c,x,σ,σ',I);
      return T;
    end
```

Algorithm 2. Function *setNextValid* for AC5TC-Alter.

```
Program setNextValid (Tuple)
    var    c: Constraint;
           x: Variable;
           σ: Tuple;
    begin
      for i ranging from 1 to σ.arity do
        if(c.variable[i] != x)
          y = c.variable[i];
          if(flag == false)
            if(σ.value[i] < c.get(res).value[i])
              σ.value[i] = dom(y).next;
              if(failed to find a correct value)
                return T;
              flag = true;
          else σ.value[i] = dom(y).first;
      return σ;
    end
```

First, we have tested three series of CSP instances included the *Geom*, *RandReg-ular* and *TSP* problems. Table 1 gives the results on these structured instances. Although AC5TC-Tr has the best time complexity theoretically, its dynamic data structure makes the performance of the algorithm decreased. From Table 1, we know that AC5TC-Tr is even overflow (i.e. represented by asterisk) on some *TSP* instances. In fact, AC5TC-Alter makes it possible to greatly reduce the number of visited tuples but the reduction of propagation time is almost the same as AC5TC-Tr or even faster in most instances.

Next, we have tested other four groups of instances included the *QCP, BQWH, aim* and *pret* problems and give our experimental results in Table 2. And the *aim* and *pret* problem belong to the classes of BOOL instances (i.e. instances only involving Boolean variables). Obviously, from Table 2, AC5TC-Alter largely reduces the number of tuples visited but the reduction of propagation time is a little limited. What's more, our new algorithm has bad performance on some *pret* instances.

Finally, we have tested random instances belong to classes of the form < k, n, d, e, t > where denote the arity of the constraints, number of variables, uniform domain size, number of constraints and the constraint tightness. From Table 3 (i.e. the results obtained on an Intel(R) Core(TM) i7-4700MQ CPU @ 1.60 GHz 2.40 GHz using Windows 10), we conclude that our new algorithm does not have advantage on propagation time for random instance. Indeed, each random instance's entropy is high, which means that the method *binarysearch* will be aborted more times in the process of seeking supports. Not surprisingly, with the increase of the constraint tightness (i.e. the number of allowed tuple is lower), the reduction of the number of visit tuples is getting smaller and smaller. That is to say, the more structured an instance is, the better performance our algorithm will has.

Table 1. Results obtained from the series of *Geom, RandRegular* and *TSP* instances. AC5TC framework is maintained during search according to 4 different schemas.

Instances		Bool	Tr	Recomp	Alter
Geo-50-20-11-sat	cpu	32.3	31.5	33.3	31.6
(r = 2)	visit	52.2 M	66.2 M	14.3 M	4.7 M
Geo-50-20-85-sat	cpu	182	163	161	157
(r = 2)	visit	265 M	372 M	63.6 M	18.2 M
RandRegular-1-unsat	cpu	93	89.8	95.6	92.5
(r = 3)	visit	142 M	299 M	85 M	59 M
RandRegular-12-sat	cpu	11.3	11.1	11.8	13.3
(r = 3)	visit	21.1 M	39.8 M	13.0 M	9.2 M
tsp-20-190-sat	cpu	31.1	*	26.7	25.6
(r = 3)	visit	77.4 M	*	23.9 M	7.0 M
tsp-20-901-sat	cpu	151	*	133	119
(r = 3)	visit	498 M	*	153 M	39 M
tsp-25-38-sat	cpu	1123	*	925	877
(r = 3)	visit	2736 M	*	748 M	186 M
tsp-25-66-sat	cpu	10	*	7.8	7.5
(r = 3)	visit	24 M	*	6.9 M	1.7 M

Table 2. Results obtained from the series of *QCP*, *BQWH*, *aim* and *pret* instances. AC5TC framework is maintained during search according to 4 different schemas.

Instances		Bool	Tr	Recomp	Alter
qcp-10-67-3-sat	cpu	0.08	0.13	0.09	0.08
(r = 2)	visit	64 k	121 k	27 k	11 k
qcp-15-120-0-sat	cpu	233	*	225	200
(r = 2)	visit	236 M	*	68.8 M	14.7 M
bqwh-15-106-18-sat	cpu	6.3	8	6.5	6.7
(r = 2)	visit	4.1 M	10.9 M	2.7 M	2.3 M
bqwh-18-141-72-sat	cpu	48.2	67.3	53.9	51.1
(r = 2)	visit	26.5 M	66.6 M	17.6 M	15.3 M
aim-50-1-6-sat	cpu	0.22	0.29	0.21	0.46
(r = 3)	visit	245 k	713 k	189 k	142 k
aim-100-1-6-sat	cpu	111	103	95.5	107
(r = 3)	visit	109 M	370 M	86.7 M	63.5 M
pret-60-40-unsat	cpu	24.9	25.9	28	30.7
(r = 3)	visit	26.7 M	77.6 M	25.4 M	39.2 M

Table 3. Results obtained from the series 7 random instances with different constraint tightness. AC5TC framework is maintained during search according to 4 different schemas.

Instances		Bool	Tr	Recomp	Alter
< 2, 40, 11, 414, 0.2 > - unsat	cpu	485	585	425	434
	visit	1043 M	1954 M	296 M	123 M
< 2, 40, 16, 250, 0.35 > - unsat	cpu	279	265	224	256
	visit	953 M	922 M	257 M	104 M
< 2, 40, 25, 180, 0.5 > - sat	cpu	342	*	218	236
	visit	985 M	*	282 M	108 M
< 2, 40, 40, 135, 0.65 > - sat	cpu	37.4	*	27.1	28.4
	visit	124 M	*	38.7 M	15.2 M
< 2, 40, 80, 103, 0.8 > - sat	cpu	174	*	156	165
	visit	622 M	*	208 M	88.2 M
< 2, 40, 180, 84, 0.9 > - sat	cpu	145	*	167	176
	visit	341 M	*	162 M	60.0 M

6   Conclusion

In this paper, a new generic arc consistency algorithm AC5TC-Alter using AC5TC framework has been proposed. The new algorithm not only uses a new search strategy to speed up AC5TC, but also doesn't need any data structure to store the validity information in constraint tables. Considering that allowed table is organized by lexicographic order in CSP, the approach seeks supports for each constraint by jumping over the sequences of irrelevant tuples in the table. Experimental results show that AC5TC-Alter will greatly reduce the number of visit tuples for most instances. And, the propagation time of AC5TC-Alter is less than AC5TC-Tr sometimes. Above all, we conclude that AC5TC-Alter is an ideal fine-grained algorithm for table constraints which is better on performance and more stable than the other three algorithms.

Acknowledgments. The authors would like to express sincere thanks to our instructor, Professor Li because of his careful guidance and valuable suggestions in the process of topic selection and paper writing. Besides, our research is supported by the National Undergraduate Training Programs for Innovation and Entrepreneurship as a national project. We are grateful for this opportunity.

References

1. Mairy, J.-B., Van Hentenryck, P., Deville, Y.: An optimal filtering algorithm for table constraints. In: Milano, M. (ed.) CP 2012. LNCS, vol. 7514, pp. 496–511. Springer, Heidelberg (2012)
2. Mackworth, A.K.: Consistency in networks of relations. Artif. Intell. **8**, 99–118 (1977)
3. Mohr, R., Henderson, T.C.: Arc and path consistency revisited. Artif. Intell. **28**, 225–233 (1986)
4. Van Hentenryck, P., Deville, Y., Teng, C.M.: A generic arc-consistency algorithm and its specializations. Artif. Intell. **57**, 291–321 (1992)
5. Bessière, C.: Arc-consistency and arc-consistency again. Artif. Intell. **65**(1), 179–190 (1994)
6. Chmeiss, A., Jégou, P.: Efficient path-consistency propagation. Int. J. Artif. Intell. Tools **7**(2), 79–89 (1998)
7. Zhang, Y., Yap, R.: Making ac-3 an optimal algorithm. In: IJCAI 2001, pp. 316–321, Seatle (2001)
8. Dib, M., Abdallah, R., Crminada, A.: Arc-consistency in constraint satisfaction problems: a survey. In: Second International Conference on Computational Intelligence, Modelling and Simulation, pp. 291–296 (2010)
9. Lecoutre, C., Szymanek, R.: Generalized arc constraint for positive table constraints. Proc. CP **06**, 284–298 (2006)
10. Arangu, M., salido, M.A., Barber, F.: A filtering technique to achieve 2-consistency in constraint satisfaction problem. Int. J. Innovative Comput. Inf. Control **8**(6), 3891–3906 (2012)

Particle Filter Optimization: A Brief Introduction

Bin Liu[1,2]([✉]), Shi Cheng[3], and Yuhui Shi[4]

[1] School of Computer Science and Technology,
Nanjing University of Posts and Telecommunications, Nanjing 210023, China
bins@ieee.org
[2] State Key Laboratory for Novel Software Technology, Nanjing University,
Nanjing 210023, China
[3] School of Computer Science, Shaanxi Normal University,
Xi'an 710062, China
cheng@snnu.edu.cn
[4] Department of Electrical and Electronic Engineering,
Xi'an Jiaotong-Liverpool University, Suzhou 215123, China
Yuhui.Shi@xjtlu.edu.cn

Abstract. In this paper, we provide a brief introduction to particle filter optimization (PFO). The particle filter (PF) theory has revolutionized probabilistic state filtering for dynamic systems, while the PFO algorithms, which are developed within the PF framework, have not attracted enough attention from the community of optimization. The purpose of this paper is threefold. First, it aims to provide a succinct introduction of the PF theory which forms the theoretical foundation for all PFO algorithms. Second, it reviews PFO algorithms under the umbrella of the PF theory. Lastly, it discusses promising research directions on the interface of PF methods and swarm intelligence techniques.

Keywords: Particle filter · Optimization · Swarm intelligence

1 Introduction

Particle filters (PFs), also known as Sequential Monte Carlo methods [1], consist of a set of generic type Monte Carlo sampling algorithms to solve the state filtering problem [2,3]. The objective of state filtering is to compute the posterior distributions of the states of the dynamic system, given some noisy and/or partial observations. The PF theory provides a well-established methodology for generating samples from the required distribution without requiring assumptions about the state-space model or the state distributions [4–6]. Under mild assumptions, the convergence of the PF methodology has been proved [7,8]. With the notable exception of linear-Gaussian signal-observation models, the PF theory has become the dominated approach to solve the state filtering problem in dynamic systems [9–11]. In the last two decades, applications of the PF theory have expanded to diverse fields such as object tracking [2], navigation and

© Springer International Publishing Switzerland 2016
Y. Tan et al. (Eds.): ICSI 2016, Part I, LNCS 9712, pp. 95–104, 2016.
DOI: 10.1007/978-3-319-41000-5_10

guidance [12,13], robotics [14], fault diagnosis [15], signal processing in digital communication systems [16], financial time series [17], biological engineering [18] and optimization [19,20].

As a generic type mutation-selection sampling approach [21], the PF theory based algorithm uses a set of particles (also called samples) to represent the posterior distribution given some noisy and/or partial observations. The working mechanism of the PF algorithm allows the state-space model to be nonlinear and the initial state and noise distributions to take any form required. Provided that the particles are properly placed, weighted and propagated, the posterior distributions can be estimated sequentially over time. Despite many advantages, the PF theory is usually faced with a fatal problem, termed sample impoverishment, due to the suboptimal sampling mechanism used [1,22]. The population based swarm intelligence techniques consist in one branch of approaches to reduce the effect of sample impoverishment. The basic principle is to treat the particles and the likelihood function involved in the PF theory as the individuals and the fitness function involved in swarm intelligence methods, respectively. Based on the above principle, a number of swarm intelligence algorithms, e.g., the particle swarm optimization (PSO) [23,24], the genetic algorithm [25,26], artificial fish swarm [27], have been used to improve the PF methods in reducing the effect of sample impoverishment.

Recently a class of optimization algorithms, termed particle filter optimization (PFO), has been proposed [28–31]. Although the PF theory has revolutionized probabilistic state filtering for dynamic systems, the PFO algorithms have not attracted enough attention from the community of optimization. The focus of this paper just lies in the introduction of the PF theory and the PFO algorithms.

The remainder of the paper is organized as follows. In Sect. 2, we present the basic components of the PF theory. In Sect. 3, we make a brief overview on the PFO algorithms under the umbrella of the PF theory. In Sect. 4, we compare two versions of PFO algorithms with a benchmark PSO algorithm in performance. Lastly in Sect. 5, we conclude the paper and make several remarks on the interface between the PF theory and the swarm intelligence techniques.

2   Particle Filter Theory

Here we give a brief overview on the PF theory, which forms the theoretical foundation for the PFO algorithms discussed in Sect. 3. Refer to [3,32] on further details on the PF theory.

The PF theory addresses the problem how to sample from a set of target probabilistic density functions (pdf) $\{\pi_k\}_{k\in[1,...,K]}$ in a sequential manner, namely first sample from π_1, then from π_2 and so on. In the context of sequential Bayesian state inference, π_k could be the posterior distribution of the state x_k given the data collected until the kth time step, i.e., $\pi_k = p(x_k|y_{1:k})$, where y denotes the collected data and $y_{1:k} \triangleq \{y_1,\ldots,y_k\}$. At a given time k, the basic idea is to obtain a large collection of N weighted random particles

$\{w_k^i, x_k^i\}_{i=1:N}, w_k^i > 0, \sum_{i=1}^{N} w_k^i = 1$, whose empirical distribution converges asymptotically ($N \to \infty$) to π_k, i.e., for any π_k-integrable function ϕ ($\phi(x) \in \mathbb{R}$):

$$\sum_{i=1}^{N} w_k^i \phi(x_k^i) \to \mathbb{E}_{\pi_k}(\phi) \qquad \text{almost surely,} \tag{1}$$

where '$\to$' denotes 'converges to' and

$$\mathbb{E}_{\pi_k}(\phi) = \int_X \phi(x)\pi_k(x)dx. \tag{2}$$

In what follows, any operation involving the superscript i must be understood as performed for $i = 1 : N$, where N is the total number of particles. Assume that we have at hand a weighted sample set $\{w_k^i, x_k^i\}_{i=1:N}$, which satisfies Eq. (1), a combination of sequential importance sampling (SIS) and resampling ideas [33] is used in the PF theory to carry forward these particles over time, yielding another weighted sample set $\{w_{k+1}^i, x_{k+1}^i\}_{i=1:N}$, which can provide a Monte Carlo approximation to π_{k+1} as follows,

$$\pi_{k+1} \simeq \sum_{i=1}^{N} w_{k+1}^i \delta(x_{k+1} - x_{k+1}^i), \tag{3}$$

where δ denotes the delta-mass function.

According to the principle of importance sampling [33], we first draw random particles x_{k+1}^i from a proposal pdf $q(x_{k+1}|y_{1:k+1})$, then the weights in Eq. (3) are defined to be

$$w_{k+1}^i \propto \frac{\pi_{k+1}(x_{k+1}^i)}{q(x_{k+1}^i|y_{1:k+1})}. \tag{4}$$

To obtain a sequential solution, the proposal pdf is chosen to factorize such that [3]

$$q(x_{k+1}|y_{1:k+1}) = q(x_{k+1}|x_k, y_{1:k+1})q(x_k|y_{1:k}). \tag{5}$$

Note that $\pi_{k+1}(x_{k+1})$ can be derived based on the Bayesian theorem as follows

$$\begin{aligned} \pi_{k+1}(x_{k+1}) &= \frac{p(y_{k+1}|x_{k+1})p(x_{k+1}|x_k)}{p(y_{k+1}|y_{1:k})}\pi_k(x_k) \\ &\propto p(y_{k+1}|x_{k+1})p(x_{k+1}|x_k)\pi_k(x_k), \end{aligned} \tag{6}$$

where $p(x_{k+1}|x_k)$ is determined by the evolution law of the state sequence $x_1, x_2, \ldots$, and $p(y_{k+1}|x_{k+1})$ denotes the likelihood function. By substituting Eqs. (5) and (6) into (4), we have

$$\begin{aligned} w_{k+1}^i &\propto \frac{p(y_{k+1}|x_{k+1}^i)p(x_{k+1}^i|x_k^i)\pi_k(x_k^i)}{q(x_{k+1}^i|x_k^i, y_{1:k+1})q(x_k^i|y_{1:k})} \\ &= w_k^i \frac{p(y_{k+1}|x_{k+1}^i)p(x_{k+1}^i|x_k^i)}{q(x_{k+1}^i|x_k^i, y_{1:k+1})}. \end{aligned} \tag{7}$$

The SIS algorithm thus consists of recursive propagation of the weights and support points as each measurement is received sequentially. A common problem with the SIS algorithm is the degeneracy phenomenon, which means that after a few iterations, all but one particles will have negligible weights. This degeneracy indicates that a large computational effort is devoted to updating particles whose contribution to the approximation to π_k is almost zero. In conventional PF algorithms, a resampling step is used as a practical way to reduce the effect of degeneracy. The basic idea of resampling is to eliminate small weight particles and to replicate large weight particles. The resampling step involves generating a new set x_k^{i*} by resampling (with replacement) N times from an approximate discrete representation of $p(x_k|y_{1:k})$ given by

$$p(x_k|y_{1:k}) \approx \sum_{i=1}^{N} w_k^i \delta(x_k - x_k^i) \tag{8}$$

so that $\Pr(x_k^{i*} = x_k^j) = w_k^j$. Note that after the resampling step, the particles with bigger weights will be replicated into multiple copies, while particles with much smaller weights will be dropped out from the resulting particle set. So the resampling procedure can be regarded as a selection method in terms of replicating bigger weight particles and deleting smaller weight ones. Refer to [1,34] for more details on different approaches to implement the resampling step.

3   Particle Filter Optimization Algorithms

In this section, we give a brief overview on PFO algorithms, which are grounded in the PF theory presented in Sect. 2. The basic principle of PFO algorithms consists of transforming the objective function to be a target pdf and then performing PF to simulate the target pdf. The hope is that the optimum of the objective function can be assessed from the resulting samples.

Here we focus on the maximization problem

$$\max_{x \in X} f(x) \tag{9}$$

where f denotes the objective function, X denotes the nonempty solution space defined in $\mathbb{R}^d$, d is the dimension of x, f $(f(X) \in \mathbb{R})$ is a continuous real-valued function that is bounded on X. We denote the maximal function value as f^*, i.e., there exists an x^* such that $f(x) \leq f^* \triangleq f(x^*), \forall x \in X$.

Based on the PF theory as presented in Sect. 2, the working mechanism of PFO algorithms consists of sampling from a sequence of target distributions $\{\pi_k\}_{k=1:K}$ sequentially, whereas the purpose of sampling is different. In conventional PF theory, the purpose of sampling is to obtain a numerical approximation to the stochastic integrals (such as the one in Eq. (2)) required for state inference, which is otherwise intractable in computation. For PFO algorithms, the purpose of sampling is to find the optimum, namely x^*.

A basic PFO algorithm consists of K sequentially processed iterations. Specifically, the kth $(k > 1)$ iteration consists of the following steps. First, generate a particle set, $\{x_k^i\}_{i=1:N}$, from the proposal pdf $q(\cdot|\{x_{k-1}^i\}_{i=1:N})$, where $\{x_{k-1}^i\}_{i=1:N}$ is the outputted particle set of the $(k-1)$th iteration. Then, weight the particles by $w_k^i \propto \pi_k(x_k^i)/q(x_k^i)$, $\sum_{i=1}^{N} w_k^i = 1$. Finally, perform resampling, obtaining an equally weighted particle set $\{x_k^{i*}\}_{i=1:N}$ and then set $x_k^i = x_k^{i*}$ and $w_k^i = 1/N$.

The PFO algorithms can differentiate from each other by the definitions of the target pdf π_k and of the proposal pdf q. A specific definition of π_k and q determines how the objective function is implanted in the sampling process and how the random samples (i.e., candidate solutions) are generated, respectively.

3.1   On the Definition of the Target Pdf

Now we focus on the different definitions of π_k in the context of PFO. A simple definition of the target pdf is shown to be [32]

$$\pi_k(x) \propto (f(x))^{\phi(k)}, \tag{10}$$

where $0 < \phi(k)$ and $\phi(k) < \phi(k+1)$, $\forall k \in [1, \ldots, K]$. The underlying idea is to make the initial target pdf π_1 (usually corresponding to a $\phi(1) < 1$) easy to sample from and the final target π_K concentrate around a very proximal area of the ideal optimum, i.e., x^*.

In [28,35], the target pdf π_k is defined to be

$$\pi_k(x) = \frac{\varphi(f(x) - y_k)\pi_{k-1}(x)}{\int \varphi(f(x) - y_k)\pi_{k-1}(x)dx}, \text{for}\quad k > 1, \tag{11}$$

where $\varphi(\cdot)$ denotes a positive and strictly increasing pdf that is continuous on its support $[0, \infty)$; y_k is the maximum objective function value that has been found at the kth iteration, taken here as a noisy observation of the optimal function value $f(x^*)$. Eq. (11) implies that π_k is tuned towards the more promising area where $f(x)$ is greater than y_k, since $\varphi(f(x) - y_k)$ is positive if $f(x) \geq y_k$ and is zero otherwise. Note that such a definition of the target pdf can theoretically lend the estimate of $f(x^*)$ to monotonically increasing and asymptotically converge to the true optimal function value [28]. A similar definition of the target pdf can be found in [20], and it is shown that PFO algorithms derived based on such kind of target pdf definitions are equivalent to the cross entropy methods [36].

The Boltzmann distribution is another choice in defining π_t for PFO algorithms [29,31]. Specifically, the target pdf at the kth iteration is defined to be

$$\pi_k(x) = \frac{1}{Z_k} \exp\left(\frac{f(x)}{T_k}\right), \tag{12}$$

where $Z_k = \int \exp(f(x)/T_k)dx$ is the normalization constant, and T_k is referred to as the annealing temperature at the kth iteration. It is proved that the temperature T_k does not have to be monotonically decreasing, as long as $T_k \to 0$

and the absolute change $\triangle k = \frac{1}{T_k} - \frac{1}{T_{k-1}}$ is monotonically decreasing to 0 as $k \to \infty$ [31]. So the Boltzmann distribution based definition allows more flexible temperature cooling schedule.

3.2   On the Definition of the Proposal Pdf

One common practice in defining the proposal pdf q for PFO algorithms is to use a symmetric such as normal distribution [31]. Specifically, it generates the value of x_k^i from a proposal

$$q(\cdot|\{x_{k-1}^j\}_{j=1:N}) = \mathcal{N}(x_{k-1}^i, \Sigma), \tag{13}$$

where $\mathcal{N}(x, \Sigma)$ denotes a normal distribution with mean x and covariance matrix Σ. Note that such a definition just corresponds to the local random walk proposal, which has been widely used in the Metropolis-Hastings algorithm [37].

The mainstream approach to define $q(x_k^i|\{x_{k-1}^j\}_{j=1:N})$ consists of representing it by a parametric model. The resulting PFO algorithm was therefore termed the model based optimization approach [35]. The optional model forms include but not limited to Gaussian, mixture of Gaussian pdfs and mixture of Student's t pdfs [38]. Note that the underlying idea of seeking a model representation of the proposal to generate new particles (i.e., the candidate solutions in the context of optimization) also appeared in the cross entropy methods [39], the adaptive importance sampling methods [40,41] and the estimation of distribution algorithms (EDA) [42].

3.3   On Practical Issues in Implementing PFO Algorithms

Here we present several practical issues that have been left out but are important in designing a specific PFO algorithm.

First, the temperature cooling schedule needs to be determined. The most simple and straightforward way is to set T_k as a constant $\forall k \in [1, \ldots, K]$, such as in [19,20,28,35]. In addition, an empirically selected monotonically decreasing (corresponding to Eq. (12)) or increasing (corresponding to Eq. (10)) cooling schedule is usually the choice. Recently some efforts have been made in designing an adaptive annealing temperature schedule based on information gathered during randomized exploration [38,43]. Despite this progress, how to derive a theoretically optimal as well as practical cooling schedule remains an open problem.

Second, the form of the model in defining the proposal pdf needs to be fixed up. The mixture model has shown to be a flexible and powerful choice, especially for multimodal objective functions [38]. Given a specific form of the model for representing the proposal, a corresponding parameter estimation approach needs to be selected for determining the values of the model's parameters.

In addition, the sample size N, the resampling strategy and the number of iterations can be specified empirically based on the amount of available resources in computation, memory and time.

4   Performance Evaluation

Here we focus on an optimization task in searching the global maximum of the minus Michalewicz function defined as below

$$f(\mathbf{x}) = \sum_{i=1}^{d} \sin(x_i) \left(\sin \left(\frac{ix_i^2}{\pi} \right) \right)^{2m}. \tag{14}$$

This function is multi-modal, having $d!$ local maxima. The parameter m defines the steepness of this function. A larger m leads to a more difficult search. We use the recommended value $m = 10$ here. We evaluate on the hypercube $x_i \in [0, \pi]$, for all $i = 1, \ldots, d$. The global maximum $f(\mathbf{x}^\star)$ in 5 and 10 dimensional cases are 4.687658 and 9.66015, respectively.

We use the Trelea type vectorized PSO [44] as the benchmark algorithm for performance comparison. A basic version of the PFO algorithm (BPFO) [20], and one of the most advanced PFO algorithms, termed posterior exploration based Sequential Monte Carlo (PE-SMC) [38], are involved in this comparison. The key difference between BPFO and PE-SMC lies in the definitions of the target and the proposal pdfs. Refer to [20,38] for details on the BPFO and PE-SMC algorithms, respectively. Following [44], we set the swarm size of the PSO algorithm to be 50, and the maximum number of iterations to be 10000. For BPFO and PE-SMC, the particle size is set to be 2000 and 5000 for the 5D and 10D cases, respectively. The termination condition is that in the last 10 iterations, no better solution is found. Each algorithm is ran 100 times independently for each case and the convergence results are listed in Table 1. Boldface in the table indicates the best result obtained. As is shown, PE-SMC beats PSO, while PSO is preferable to BPFO, for both cases. Refer to [38] for more results on performance comparisons between the PSO and PFO algorithms.

Table 1. Convergence results of the involved algorithms

d	Goal($f(\mathbf{x}^\star)$)	PSO	BPFO	PE-SMC
5	4.687658	$4.6624 \pm 4.01 \times 10^{-2}$	4.2103 ± 0.1648	$\mathbf{4.6875} \pm 2.81 \times 10^{-4}$
10	9.66015	9.4562 ± 0.16	5.9357 ± 0.2852	$\mathbf{9.6596} \pm 1.77 \times 10^{-2}$

5   Concluding Remarks

In this paper, we provided a very brief introduction to the PF theory and the PFO algorithms. It is shown that the success of PFO algorithms depends on appropriate designs of the target and proposal pdfs. In what follows, let us make several remarks on the relationships between the PFO algorithms and the swarm intelligence methods. Our intention is to provide some insight into the behaviour of the PFO algorithm and its relationship with some existing swarm intelligence algorithms.

It is shown that, PFO algorithms equipped with annealed target pdfs, as defined in Eqs. (10) and (12), find a straightforward relationship with the simulated annealing (SA) method [45]. The SA is a serially processed algorithm relying on a Markov Chain Monte Carlo sampling mechanism, while it is natural to find parallel implementations for the PFO algorithms. In addition, the PFO algorithm can provide a forward looking mechanism for automatically determining an appropriate annealing temperature value for the following iteration based on the information collected at the current iteration [38]. If the target pdf in the PFO algorithm keeps invariant per iteration, namely $\pi_k(x) = \pi_{k-1}(x)$, $\forall k \in [2, \ldots, K]$, the PFO algorithm just degenerates into the estimation of distribution algorithm in concept, although their sampling and selection mechanisms look different in form. Actually the PFO algorithm can be regarded as a generic type swarm intelligence technique, in which the candidate solutions are generated, weighted and propagated based on probability theorem, other than meta-heuristics. Inherited from the PF theory, the PFO algorithm could own satisfactory theoretical properties in e.g., convergence and global searching, while it consumes much computational resources in always maintaining a strict probabilistic interpretation of the landscape of the evaluated solutions in each iteration. We argue that a promising future direction lies in developing novel approaches that can make best use of the advantages and bypass the disadvantages of the PFO and the meta-heuristics based algorithms.

Acknowledgments. This work was partly supported by the National Natural Science Foundation (NSF) of China under grant Nos. 61571238, 61302158 and 61571434, the NSF of Jiangsu province under grant No. BK20130869, China Postdoctoral Science Foundation under grant No.2015M580455, China Postdoctoral International Academic Exchange Program and the Scientific Research Foundation of Nanjing University of Posts and Telecommunications under grant No. NY213030.

References

1. Liu, J.S., Chen, R.: Sequential Monte Carlo methods for dynamic systems. J. Am. Stat. Assoc. **93**(443), 1032–1044 (1998)
2. Gordon, N.J., Salmond, D.J., Smith, A.F.: Novel approach to nonlinear/non-Gaussian Bayesian state estimation. IEE Proc. F (Radar Sig. Process.) **140**(2), 107–113 (1993)
3. Arulampalam, M.S., Maskell, S., Gordon, N., Clapp, T.: A tutorial on particle filters for online nonlinear/non-Gaussian Bayesian tracking. IEEE Trans. Sig. Process. **50**(2), 174–188 (2002)
4. Del Moral, P.: Feynman-Kac Formulae. Springer, Heidelberg (2004)
5. Del Moral, P.: Measure-valued processes and interacting particle systems. Application to nonlinear filtering problems. Ann. Appl. Probab. **8**, 438–495 (1998)
6. Del Moral, P.: Non-linear filtering: interacting particle resolution. Markov process. Relat. Fields **2**(4), 555–581 (1996)
7. Crisan, D., Doucet, A.: A survey of convergence results on particle filtering methods for practitioners. IEEE Trans. Sig. Process. **50**(3), 736–746 (2002)

8. Hu, X.L., Schön, T.B., Ljung, L.: A basic convergence result for particle filtering. IEEE Trans. Sig. Process. **56**(4), 1337–1348 (2008)
9. Daum, F.: Nonlinear filters: beyond the Kalman filter. IEEE Aerosp. Electron. Syst. Mag. **20**(8), 57–69 (2005)
10. Ristic, B., Arulampalam, S., Gordon, N.: Beyond the Kalman filter. IEEE Aerosp. Electron. Syst. Mag. **19**(7), 37–38 (2004)
11. Chen, Z.: Bayesian filtering: From Kalman filters to particle filters, and beyond. Statistics **182**(1), 1–69 (2003)
12. Gustafsson, F., Gunnarsson, F., Bergman, N., Forssell, U., Jansson, J., Karlsson, R., Nordlund, P.: Particle filters for positioning, navigation, and tracking. IEEE Trans. Sig. Process. **50**(2), 425–437 (2002)
13. Gordon, N., Salmond, D., Ewing, C.: Bayesian state estimation for tracking and guidance using the bootstrap filter. J. Guidance Control Dyn. **18**(6), 1434–1443 (1995)
14. Pozna, C., Precup, R.E., Földesi, P.: A novel pose estimation algorithm for robotic navigation. Robot. Autonom. Syst. **63**, 10–21 (2015)
15. De Freitas, N.: Rao-Blackwellised particle filtering for fault diagnosis. In: Proceedings of IEEE Aerospace Conference, vol. 4, pp. 4–1767. IEEE (2002)
16. Clapp, T.C., Godsill, S.J.: Fixed-lag blind equalization and sequence estimation in digital communications systems using sequential importance sampling. In: Proceedings of IEEE International Conference on Acoustics, Speech, and Signal Processing (ICASSP), vol. 5, pp. 2495–2498 (1999)
17. Fearnhead, P.: Using random Quasi-monte-carlo within particle filters, with application to financial time series. J. Comput. Graph. Stat. **14**(4), 751–769 (2005)
18. DiMaio, F., Kondrashov, D.A., Bitto, E., Soni, A., Bingman, C.A., Phillips, G.N., Shavlik, J.W.: Creating protein models from electron-density maps using particle-filtering methods. Bioinformatics **23**(21), 2851–2858 (2007)
19. Chen, X., Zhou, E.: Population model-based optimization with Sequential Monte Carlo. In: Proceedings of 2013 Winter Simulation Conference, pp. 1004–1015 (2013)
20. Zhou, E., Fu, M.C., Marcus, S., et al.: Particle filtering framework for a class of randomized optimization algorithms. IEEE Trans. Autom. Control **59**(4), 1025–1030 (2014)
21. Whiteley, N., Johansen, A.M.: Recent developments in auxiliary particle filtering. In: Cemgil, B., Chiappa, S. (eds.) Inference and Learning in Dynamic Models, vol. 38, pp. 39–47. Cambridge University Press, Cambridge (2010)
22. Doucet, A., Godsill, S., Andrieu, C.: On sequential Monte Carlo sampling methods for Bayesian filtering. Stat. Comput. **10**(3), 197–208 (2000)
23. Zhang, M., Xin, M., Yang, J.: Adaptive multi-cue based particle swarm optimization guided particle filter tracking in infrared videos. Neurocomputing **122**, 163–171 (2013)
24. Tong, G., Fang, Z., Xu, X.: A particle swarm optimized particle filter for nonlinear system state estimation. In: IEEE Congress on Evolutionary Computation (CEC), pp. 438–442 (2006)
25. Kwok, N.M., Fang, G., Zhou, W.: Evolutionary particle filter: re-sampling from the genetic algorithm perspective. In: Proceedings of IEEE/RSJ International Conference on Intelligent Robots and Systems, pp. 2935–2940 (2005)
26. Han, H., Ding, Y.S., Hao, K.R., Liang, X.: An evolutionary particle filter with the immune genetic algorithm for intelligent video target tracking. Comput. Math. Appl. **62**(7), 2685–2695 (2011)

27. Xiaolong, L., Jinfu, F., Qian, L., Taorong, L., Bingjie, L.: A swarm intelligence optimization for particle filter. In: Proceedings of 7th World Congress on Intelligent Control and Automation, pp. 1986–1991 (2008)

28. Zhou, E., Fu, M.C., Marcus, S.I.: A particle filtering framework for randomized optimization algorithms. In: Proceedings of 40th Conference on Winter Simulation, pp. 647–654 (2008)

29. Ji, C., Zhang, Y., Tong, M., Yang, S.: Particle filter with swarm move for optimization. In: Rudolph, G., Jansen, T., Lucas, S., Poloni, C., Beume, N. (eds.) PPSN 2008. LNCS, vol. 5199, pp. 909–918. Springer, Heidelberg (2008)

30. Medina, J.C., Taflanidis, A.A.: Adaptive importance sampling for optimization under uncertainty problems. Comput. Methods Appl. Mech. Eng. **279**, 133–162 (2014)

31. Zhou, E., Chen, X.: Sequential monte carlo simulated annealing. J. Global Optim. **55**(1), 101–124 (2013)

32. Del Moral, P., Doucet, A., Jasra, A.: Sequential Monte Carlo samplers. J. Roy. Stat. Soc.: Ser. B (Stat. Methodol.) **68**(3), 411–436 (2006)

33. Liu, J.S., Chen, R., Logvinenko, T.: A theoretical framework for sequential importance sampling with resampling. In: Doucet, A., de Freitas, N., Gordon, N. (eds.) Sequential Monte Carlo methods in Practice. SEIS, pp. 225–246. Springer, New York (2001)

34. Kitagawa, G.: Monte Carlo filter and smoother for non-Gaussian nonlinear state space models. J. Comput. Graph. Stat. **5**(1), 1–25 (1996)

35. Hu, J., Wang, Y., Zhou, E., Fu, M.C., Marcus, S.I.: A survey of some model-based methods for global optimization. Optimization, Control, and Applications of Stochastic Systems. SCFA, pp. 157–179. Springer, Heidelberg (2012)

36. De Boer, P.T., Kroese, D.P., Mannor, S., Rubinstein, R.Y.: A tutorial on the cross-entropy method. Ann. Oper. Res. **134**(1), 19–67 (2005)

37. Hastings, W.K.: Monte Carlo sampling methods using Markov chains and their applications. Biometrika **57**(1), 97–109 (1970)

38. Liu, B.: Posterior exploration based sequential Monte Carlo for global optimization. arXiv preprint (2015). http://arxiv.org/abs/1509.08870

39. Rubinstein, R.: The cross-entropy method for combinatorial and continuous optimization. Methodology Comput. Appl. Probab. **1**(2), 127–190 (1999)

40. Oh, M.S., Berger, J.O.: Integration of multimodal functions by Monte Carlo importance sampling. J. Am. Stat. Assoc. **88**(422), 450–456 (1993)

41. Cappé, O., Douc, R., Guillin, A., Marin, J.M., Robert, C.P.: Adaptive importance sampling in general mixture classes. Stat. Comput. **18**(4), 447–459 (2008)

42. Hauschild, M., Pelikan, M.: An introduction and survey of estimation of distribution algorithms. Swarm Evol. Comput. **1**(3), 111–128 (2011)

43. Yang, C., Kumar, M.: An information guided framework for simulated annealing. J. Global Optim. **62**(1), 131–154 (2015)

44. Trelea, I.C.: The particle swarm optimization algorithm: convergence analysis and parameter selection. Inf. Process. Lett. **85**(6), 317–325 (2003)

45. Kirkpatrick, S., Gelatt, C.D., Vecchi, M.P., et al.: Optimization by simulated annealing. Science **220**(4598), 671–680 (1983)

Immunological Approach for Data Parameterization in Curve Fitting of Noisy Points with Smooth Local-Support Splines

Andrés Iglesias[1,2(✉)], Akemi Gálvez[1,2], and Andreina Avila[1]

[1] Department of Applied Mathematics and Computational Sciences,
University of Cantabria, Avda. de Los Castros, s/n, 39005 Santander, Spain
[2] Department of Information Science, Faculty of Sciences, Toho University,
2-2-1 Miyama, 274-8510 Funabashi, Japan
iglesias@unican.es
http://personales.unican.es/iglesias

Abstract. This paper addresses the problem of computing the parameterization of a smooth local-support spline curve for data fitting of noisy points by using an immunological approach. Given an initial set (not necessarily optimal) of breakpoints, our method applies a popular artificial immune systems technique called clonal selection algorithm to perform curve parameterization. The resulting optimal data parameters are then used for further refinement of the breakpoints via the deBoor method. In this way, the original non-convex optimization method is transformed into a convex one, subsequently solved by least-squares singular value decomposition. The method is applied to two illustrative examples (human hand and ski goggles, each comprised of three curves) of two-dimensional sets of noisy data points with very good experimental results.

1 Introduction

Fitting curves to data points is a classical problem in many fields, such as image reconstruction [23], geometric modeling and processing [1], and computer-aided design and manufacturing (CAD/CAM) [24,29]. In general, the fitting curves are assumed to be parametric, very often in the form of piecewise spline curves. However, the use of spline curves is still challenging because they typically depend on many different continuous variables in an intertwined and highly nonlinear way [2,26]. This implies that this problem cannot be partitioned into independent sub-problems for the different sets of variables. Furthermore, it is not possible in general to compute all these parameters analytically [2,8].

A suitable approach to overcome this limitation is to formulate this curve fitting problem as a continuous nonlinear optimization problem [8]. In this case, the optimization problem involves three sets of variables: data parameters, breakpoints, and spline coefficients. However, the traditional mathematical optimization methods have failed to solve the general problem for parametric splines.

© Springer International Publishing Switzerland 2016
Y. Tan et al. (Eds.): ICSI 2016, Part I, LNCS 9712, pp. 105–115, 2016.
DOI: 10.1007/978-3-319-41000-5_11

Therefore, in this paper we follow a different approach. We start with an initial (not necessarily optimal) vector of breakpoints; then, we perform curve parameterization to obtain the optimal data parameters. This parameterization is then used for further refinement of the breakpoints via the deBoor method [2]. As a result, we obtain a convex optimization problem for the spline coefficients, subsequently solved by least-squares singular value decomposition. In other words, the original curve fitting problem is transformed into that of data parameterization.

Unfortunately, classical mathematical optimization techniques also fail to solve this parameterization problem. Several alternative methods have been proposed in the literature to tackle this issue, such as neural networks, machine learning, and so on. Although they perform properly for some particular instances, the general case still remains elusive. Recently, the methods based on biological systems (the so-called *bio-inspired computation*) are receiving increasing attention owing to their good behavior for complex optimization problems involving ambiguous and noisy data and with little (or none) information about the problem. However, there are still few works reported in the literature on this subject. Recent schemes are described for particle swarm optimization [10–12,15], genetic algorithms [19,30], artificial immune systems [16,17], simulated annealing [22], firefly algorithm [14], estimation of distribution algorithms [31] and hybrid approaches [13,18,27]. However, most of these methods are designed for simpler cases, such as Bézier curves (that do not have breakpoints) or explicit functions (that do not require any parameterization). Therefore, their solutions are not actually applicable to this case. In this paper, we solve it by applying an artificial immune systems method called clonal selection algorithm.

The structure of this paper is as follows: in Sect. 2 we describe the artificial immune systems, with focus on the clonal selection algorithm, the optimization method used in this paper. Then, Sect. 3 describes the basic concepts about parametric spline curves along with the curve fitting problem to be solved. Section 4 describes our method to solve this problem. Finally, two illustrative examples along with some computational issues are reported in Sect. 5.

2 Artificial Immune Systems

Biological systems have recently inspired new computational algorithms for optimization. Amongst them, the *Artificial Immune Systems* (AIS) are receiving attention because of their ability to solve complex optimization problems. By AIS we refer to a powerful methodology based on the human immune system [3,7]. However, the immune system is very complicated and no single AIS encompasses all its features. Instead, there are several models in AIS, each focused on one or a few of those features [21,28]. Relevant examples include negative selection [9], artificial immune network [25], dendritic cells [20] and clonal selection [5]. In this paper we focus on the clonal selection theory [6].

2.1   The Clonal Selection Algorithm

The *Clonal Selection Algorithm* (CSA) is an AIS method based on the clonal selection principle, a widely accepted theory used to explain the basic features of an adaptive immune response to an pathogenic microorganisms called *antigens* (Ag) [4–6]. When exposed to an Ag, the immune system responds by producing *antibodies* (Ab). They are molecules whose aim is to recognize and bind to Ag's with a certain degree of specificity called the *affinity* of the couple Ag-Ab. Under the clonal selection theory, only those cells that recognize the antigens are selected to proliferate. The selected cells are subject to an *affinity maturation* process, which improves their affinity to the selective Ag's over the time.

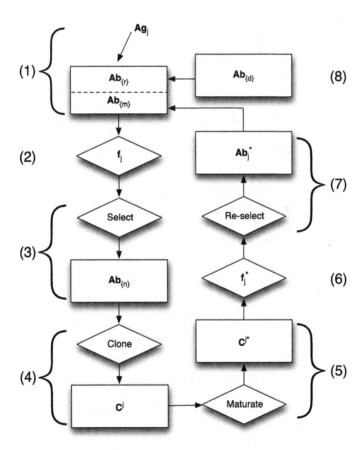

Fig. 1. Flow chart of the clonal selection algorithm [4,6].

In the immune system, the learning process involves raising the relative population size and affinity of those lymphocytes valuable in recognizing a given Ag. A clone will be created temporarily and those progenies with low affinity will be discarded. Typically, an organism would be expected to encounter a given Ag repeatedly during its lifetime. The effectiveness of the immune response to secondary encounters is enhanced considerably by the presence of memory cells associated with the first infection, capable of producing high-affinity Ab's just after subsequent encounters.

Ab's in a memory response have a higher affinity than those of the early primary response, a process called *maturation* of the immune response. Then, random changes (*mutation*) are introduced into the genes responsible for the Ag-Ab interactions, increasing occasionally the affinity of the Ab. A rapid accumulation of mutations is necessary for a fast maturation of the immune response. When a B cell recognizes an antigen, it is stimulated to proliferate)at an extremely high rate (*somatic hypermutation*), while low-affinity cells may be further mutated and die if they do not improve their antigenic affinity.

Figure 1 shows the flowchart of the original clonal selection algorithm [4,6]. The algorithm considers two repertoires: a set of antigens $Ag_{\{M\}}$ and a set of antibodies $Ab_{\{N\}}$. For the sake of clarity, cardinality is indicated by the subindexes within brackets. The latter is further divided into two subsets: memory Ab repertoire $Ab_{\{m\}}$ and remaining Ab repertoire $Ab_{\{r\}}$, such that $m + r = N$. We also keep track of two other sets: the set $Ab_{\{n\}}$ of the n Ab's with the highest affinities to a given Ag, and the set $Ab_{\{d\}}$ of the d new Ab's that will replace the low-affinity Ab's from $Ab_{\{r\}}$. The algorithm can be summarized as follows:

1. Random choice of an antigen Ag_j. It is presented to all antibodies of $Ab_{\{N\}}$.
2. Compute the vector affinity $\mathbf{F} = (F_1, F_2, \ldots, F_N)$ where $F_i = Af(Ab_i, Ag_j)$.
3. Select the n highest affinity components of $\mathbf{F}$ to generate $Ab_{\{n\}}$.
4. Elements of $Ab_{\{n\}}$ will be cloned adaptively. The number of clones is pro-
   portional to the affinity according to: $N_c = \sum_{j=1}^{n} round\left(\dfrac{\omega_j.N_{ab}}{j}\right)$ where N_c
   represents the number of clones, ω_j is a positive multiplying factor, N_{ab} is the total number of Ab's and $round(.)$ rounds toward the closest integer.
5. The clones resulting from the previous step are subjected to somatic hypermutation. The affinity maturation rate is inversely proportional to the antigenic affinity: the higher the affinity, the smaller the maturation rate.
6. Compute the vector affinity of Ag_j with respect to the new matured clones.
7. From this set of matured clones, select the one with the highest affinity to be candidate to enter into the set $Ab_{\{n\}}$. If $Af(Ab_k, Ag_j) > Af(Ab_l, Ag_j)$ for a given $Ab_l \in Ab_{\{n\}}$, then Ab_k will replace Ab_l.
8. Replace the d Ab's with lowest affinity in $Ab_{\{r\}}$ by new individuals in $Ab_{\{r\}}$.

Each execution of steps 1-8 for all given Ag_j, $(j = 1, \ldots, M)$ is called a *generation*. The algorithm is repeated for a certain number of generations, N_{gen}. This algorithm has proved to be very well suited for optimization problems, having been successfully applied to problems such as character recognition and multimodal optimization with very good performance.

3   The Problem

In this section we introduce some basic concepts on parametric spline functions (see [2, 26] for further details). Then, we describe our optimization problem. Note that in this paper vectors are denoted in bold.

3.1   Parametric Spline Curves

Let $\boldsymbol{\Phi}(\tau) = (\phi^1(\tau), \dots, \phi^\varsigma(\tau))$ be a parametric function defined on a finite interval $[\alpha, \beta]$. Consider now a strictly increasing sequence of real numbers $\xi_0 = \alpha < \xi_1 < \dots \xi_v < \xi_{v+1} = \beta$ called breakpoints. The function $\boldsymbol{\Phi}(\tau)$ is a *parametric polynomial spline of degree* $\eta \geq 0$ *with breakpoints* $\bar{\xi} = \{\xi_k\}_k$ if the following two conditions are fulfilled for $i = 0, \dots, v$, and $j = 1, \dots, \varsigma$:

(1) $\phi^j(\tau)$ is a polynomial spline of degree up to η on each interval $[\xi_i, \xi_{i+1}]$, and
(2) $\phi^j(\tau)$ and its derivatives up to order $\eta - 1$ are continuous on $[\xi_i, \xi_{i+1}]$.

The support of a function is the subset of the function domain where the function does not vanish. Since each component $\phi^j(\tau)$ is only defined on the interval $[\xi_i, \xi_{i+1}]$, $\boldsymbol{\Phi}(\tau)$ is a local-support function. That means that any perturbation of the function $\phi^j(\tau)$ on the interval $[\xi_i, \xi_{i+1}]$ only affects to this part of the curve, while the other pieces remain unaltered, a very valuable feature for interactive design in CAD/CAM industry and many other real-world applications.

Of course, different basis functions can be used for polynomial splines. In this paper, we will focus particularly on the B-spline basis functions. Note, however that our method does not preclude any other basis functions to be used instead. The B-spline basis functions of degree ν defined on $[\xi_i, \xi_{i+1}]$ can be computed according to the Cox-de-Boor recursive formula [2]:

$$\psi_{i,\nu+1}(\tau, \bar{\xi}) = \varphi_{i,\nu}^+(\tau, \bar{\xi})\psi_{i,\nu}(\tau, \bar{\xi}) + \varphi_{i+1,\nu}^-(\tau, \bar{\xi})\psi_{i+1,\nu}(\tau, \bar{\xi}) \qquad (1)$$

for $i = 0, \dots, v - \nu$ and $\nu > 1$, where $\varphi_{i,\nu}^+(\tau, \bar{\xi}) = \dfrac{\tau - \xi_i}{\xi_{i+\nu} - \xi_i}$, $\varphi_{i,\nu}^-(\tau, \bar{\xi}) = \dfrac{\xi_{i+\nu} - \tau}{\xi_{i+\nu} - \xi_i}$, and $\psi_{i,1}(\tau, \bar{\xi})$ is the unit function with support on the interval $[\xi_i, \xi_{i+1})$. The dimension of the vector space of functions satisfying conditions (1) and (2) is $v + \eta + 1$. The given vector of breakpoints $\bar{\xi}$ yields $v - \eta + 1$ linearly independent basis functions of degree η. The remaining 2η basis functions are obtained by introducing the boundary breakpoints $\xi_{-\eta} = \xi_{-\eta+1} = \dots = \xi_{-1} = \xi_0 = \alpha$ and $\xi_{v+1} = \xi_{v+2} = \dots = \xi_{v+\eta+1} = \beta$. With this choice of boundary breakpoints all basis functions vanish outside the interval domain $[\alpha, \beta]$. Every parametric spline curve $\boldsymbol{\Phi}(\tau, \bar{\xi})$ of degree η is represented by:

$$\boldsymbol{\Phi}(\tau, \bar{\xi}) = \sum_{i=-\eta}^{v} \boldsymbol{\Xi}_i \, \psi_{i,\eta+1}(\tau, \bar{\xi}) \qquad (2)$$

where $\{\boldsymbol{\Xi}_i\}$ are the spline coefficients of the curve, and $\psi_{i,\eta+1}(\tau, \bar{\xi})$ are the basis functions defined above.

3.2   The Optimization Problem

Let us suppose that we are provided with a set of measured data points $\{\Theta_k\}_{k=1,\ldots,\rho}$. The goal consists of obtaining a parametric spline curve $\Phi(\tau, \bar{\xi})$ of degree η fitting the $\{\Theta_k\}_k$. Due to the conditions on the boundary breakpoints, we can take $\Phi(\tau_1, \bar{\xi}) = \Theta_1$, and $\Phi(\tau_\rho, \bar{\xi}) = \Theta_\rho$, and approximate the remaining parameters:

$$\Theta_k \approx \Phi(\tau_k, \bar{\xi}) = \sum_{j=-\eta}^{v} \Xi_j \psi_{j,\eta+1}(\tau_k, \bar{\xi}) \qquad (k = 2, \ldots, \rho - 1) \qquad (3)$$

Eq. (3) can be written in matrix notation as:

$$\Theta = \Psi.\Xi \qquad (4)$$

where $\Theta = (\Theta_2, \ldots, \Theta_{\rho-1})^T$, $\Xi = (\Xi_{-\eta}, \ldots, \Xi_v)^T$, $(.)^T$ represents the transpose of a vector or matrix, and $\Psi = \left(\{\psi_{j,\eta+1}(\tau_k, \bar{\xi})\}_{k=2,\ldots,\rho-1\,;\,j=-\eta,\ldots,v}\right)$ is the matrix of sampled basis functions. The dimension of the search space in Eq. (4) is: $\zeta(v+\eta-1)+v+\rho-2$, which could be of hundreds or thousands of variables for non-trivial shapes. From this observation, it becomes clear that the system (4) is overdetermined. This implies that the matrix of basis functions is not invertible and hence no direct solution can be obtained. Therefore, we consider the least squares approximation of (3), defined as the minimization problem given by:

$$\underset{\substack{\{\xi_i\}_i \\ \{\Xi_j\}_j \\ \{\tau_k\}_k}}{\text{minimize}} \left(\sum_{k=2}^{\rho-1} \left\| \Theta_k - \sum_{j=-\eta}^{v} \Xi_j \psi_{j,\eta+1}(\tau_k, \bar{\xi}) \right\|_{\ell_2}\right) \qquad (5)$$

where ℓ_2 represents the Euclidean norm. Note that the parameters and breakpoints are related by Eq. (1), thus leading to a high-dimensional continuous nonlinear optimization problem. Assuming that a suitable data parameterization can be obtained, we have to solve a nonlinear continuous optimization problem involving both the spline coefficients and the breakpoints as free variables of the problem. Unfortunately, this approach makes the optimization problem non-convex, because $\Phi(\tau, \bar{\xi})$ is a non-convex function of the breakpoints [2,8]. To overcome this problem, we firstly give a rough estimation of the breakpoints before the optimization process is executed and then refine them for better fitting. With this strategy, the resulting problem is convex, so a global optimum can eventually be found. In order to apply this strategy, we need to obtain a suitable parameterization of data points. We solve this problem by applying the clonal selection algorithm described in Sect. 3, as explained in next paragraphs.

4   Our Method

The CSA described above is only suitable for supervised problems. To overcome this limitation, some modifications are required. On one hand, there is no need

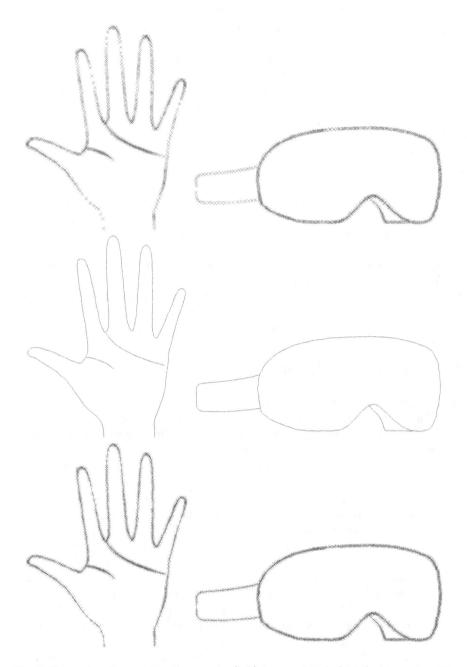

Fig. 2. Examples discussed in this paper: (left) human hand; (right) ski goggles. The figure shows the input sets of noisy data points (top), the reconstructed curves (middle) and the combination of both pictures for better visualization (bottom).

to maintain a separate subset of memory $Ab_{\{m\}}$, since no specific Ag_j has to be recognized. Instead, the whole population of antibodies will compose the memory set. In our representation, the antibodies are encoded as real-valued vectors initialized with uniformly distributed random values on the interval $(0,1)^{\rho-2}$ and then sorted to reflect the structure of data points. Another modifications are that several Ab's with high affinity are selected in step 7 of the algorithm, rather than just the best one, and that all Ab's in the population can be selected for cloning in step 3, so there is no need to maintain set $Ab_{\{n\}}$; in that case, the affinity proportionate cloning is no longer necessary. All antibodies can be cloned at the same rate (i.e. the number of clones generated for each antibody will be the same). Our mutation mechanism is a single-point operator that introduces uniform random perturbations on one randomly chosen component τ_k of the antibody according to the rule $\chi \to \chi + \Delta(\sigma - \gamma)$ where $\sigma \in U(0,1)$ and $\Delta \in (\gamma\chi, \gamma - \gamma\chi)$. Finally, before searching for the solution of the problem, some control parameters should be set up. They are: the total number of antibodies, N_{ab}, number of antibodies in Ab_d, number of antibodies to be cloned, n, number of clones N_c, and the number of generations, N_{gen}. All these parameters are selected empirically. To this aim, we perform several executions for different values of the parameters and then select the best ones for our final experiments. Then, the CSA is executed for the given number of generations. The antibody with the best affinity value is selected as the best solution of the problem.

5   Experimental Results

To evaluate the performance of our method, it has been applied to several examples of noisy data points. For the sake of limitations of space, only two of them are described here. Their input sets of noisy data points are displayed as red $\times$ symbols in Fig. 2(top). They correspond to an organic shape (a human hand, comprised of three open curves with 500, 30, and 70 data points, respectively) and a synthetic real-word shape (a ski goggles model, containing two open curves with 40 and 50 data points, and a closed curve with 500 data points), depicted on left and right, respectively. We applied our method to these examples with the following parameter values: $N_{ab} = 100$, $d = 10$, $n = 10$, $N_{gen} = 10$, and $\gamma = 0.5$. The number of clones was taken proportionally to the affinity measure. In particular, we select $\omega_1 = 0.6$, $\omega_2 = 0.4$, $\omega_3 = 0.2$, $\omega_k = 0.1$, $(k = 4, \ldots, n)$ for N_c, so the total number of clones becomes $N_c = 190$. These clones undergo somatic hypermutation at a rate that is inversely proportional to the affinity.

The corresponding fitting curves are displayed as blue solid lines in Fig. 2(middle). As the reader can see, the method performs very well and we have been able to capture the underlying shape of data points with very high accuracy. This fact is even better perceived in Fig. 2(bottom), where the data points and the fitting curves are combined into a single picture for better visualization. Note the very good matching between the input data points and their fitting curves. These visual observations are confirmed by our numerical results. We obtained a RMSE (root-mean square error) of 9.785E-5, 2.674E-5, and 9.745E-5, for the three curves of the human hand, and 1.982E-3, 1.573E-3, and 3.428E-3, for the three curves in the ski goggles example (affected by noise of higher intensity than the first example). These values have been obtained as the mean value of 30 independent executions to prevent any spurious effect due to randomness. These are very remarkable results, because our data points are substantially affected by measurement noise. This fact is particularly visible in some parts of Fig. 2(middle), where the fitting curve shows small oscillations caused by the noise, such as, for instance, the vertical part of the rightmost region of the ski goggles, or in the vertical parts of index and ring fingers of the human hand. This shows that our method performs very well even in cases of noisy data points, being able to capture even the subtlest details of the input data. As a consequence, it can be used in noisy environments such as those in many real-world settings. In other words, it can be applied to real-world problems without any further pre-/post-processing.

All computations in this paper have been performed on a 2.4 GHz Intel Core 2 Duo processor with 4 GB. of RAM. The source code has been implemented by the authors in the native programming language of the popular scientific program *Matlab*, version 2013b. Regarding the computation times, in our trials we found our method to be very affordable in terms of CPU times: all simulations in this paper took less than 10 seconds to be obtained.

Acknowledgements. This research has been kindly supported by the Computer Science National Program of the Spanish Ministry of Economy and Competitiveness, Project Ref. #TIN2012-30768, Toho University, and the University of Cantabria.

References

1. Barnhill, R.E.: Geometric Processing for Design and Manufacturing. SIAM, Philadelphia (1992)
2. de Boor, C.A.: Practical Guie to Splines. Springer, Heidelberg (2001)
3. De Castro, L.N., Timmis, J.: Artificial Immune Systems: A New Computational Intelligence Approach. Springer, London (2002)
4. De Castro, L.N., Von Zuben, F.J.: Artificial Immune Systems: Part I - Basic Theory and Applications, Technical report-RT DCA 01/99 (1999)
5. De Castro, L.N., Von Zuben, F.J.: The clonal selection algorithm with engineering applications. In: Proceedings of GECCO00, Workshop on Artificial Immune Systems and their Applications, Las Vegas, USA, pp. 36–37 (2000)
6. De Castro, L.N., Von Zuben, F.J.: Learning and optimization using the clonal selection principle. IEEE Trans. Evol. Comput. **6**(3), 239–251 (2002)
7. Dasgupta, D. (ed.): Artificial Immune Systems and Their Applications. Springer, Berlin (1999)
8. Dierckx, P.: Curve and Surface Fitting with Splines. Oxford University Press, Oxford (1993)
9. Forrest, S., Perelson, A.S., Allen, L., Cherukuri, R.: Self-nonself discrimination in a computer. In: Proceedings of IEEE Symposium on Research in Security and Privacy. Los Alamitos, CA, pp. 202–212 (1994)

10. Gálvez, A., Cobo, A., Puig-Pey, J., Iglesias, A.: Particle swarm optimization for Bézier surface reconstruction. In: Bubak, M., van Albada, G.D., Dongarra, J., Sloot, P.M.A. (eds.) ICCS 2008, Part II. LNCS, vol. 5102, pp. 116–125. Springer, Heidelberg (2008)
11. Gálvez, A., Iglesias, A.: Efficient particle swarm optimization approach for data fitting with free knot B-splines. Comput. Aided Des. **43**(12), 1683–1692 (2011)
12. Gálvez, A., Iglesias, A.: Particle swarm optimization for non-uniform rational B-spline surface reconstruction from clouds of 3D data points. Inf. Sci. **192**(1), 174–192 (2012)
13. Gálvez, A., Iglesias, A.: A new iterative mutually-coupled hybrid GA-PSO approach for curve fitting in manufacturing. Appl. Soft Comput. **13**(3), 1491–1504 (2013)
14. Gálvez, A., Iglesias, A.: Firefly algorithm for polynomial Bézier surface parameterization. J. Appl. Math. 9p. (2013). Article ID 237984
15. Gálvez, A., Iglesias, A.: Particle-based meta-model for continuous breakpoint optimization in smooth local-support curve fitting. Appl. Math. Comput. **275**, 195–212 (2016)
16. Gálvez, A., Iglesias, A., Avila, A.: Hybridizing mesh adaptive search algorithm and artificial immune systems for discrete rational Bézier curve approximation. Vis.Comput. **32**, 393–402 (2016)
17. Gálvez, A., Iglesias, A., Avila, A., Otero, C., Arias, R., Manchado, C.: Elitist clonal selection algorithm for optimal choice of free knots in B-spline data fitting. Appl. Soft Comput. **26**, 90–106 (2015)
18. Gálvez, A., Iglesias, A., Cobo, A., Puig-Pey, J., Espinola, J.: Bézier curve and surface fitting of 3D point clouds through genetic algorithms, functional networks and least-squares approximation. In: Gervasi, O., Gavrilova, M.L. (eds.) ICCSA 2007, Part II. LNCS, vol. 4706, pp. 680–693. Springer, Heidelberg (2007)
19. Gálvez, A., Iglesias, A., Puig-Pey, J.: Iterative two-step genetic-algorithm-based method for efficient polynomial B-spline surface reconstruction. Inf. Sci. **182**(1), 56–76 (2012)
20. Greensmith, J., Aickelin, U.: Artificial dendritic cells: multi-faceted perspectives. In: Bargiela, A., Pedrycz, W. (eds.) Human-Centric Information Processing Through Granular Modelling. SCI, vol. 182, pp. 375–395. Springer, Heidelberg (2009)
21. Hart, E., Timmis, J.: Application areas of AIS: past, present and future. Appl. Soft Comput. **8**(1), 191–201 (2008)
22. Iglesias, A., Gálvez, A., Loucera, C.: Two simulated annealing optimization schemas for rational Bézier curve fitting in the presence of noise. Math. Prob. Eng. 17p. (2016). Article ID 8241275
23. Jacobson, T.J., Murphy, M.J.: Optimized knot placement for B-splines in deformable image registration. Med. Phys. **38**(8), 4579–4592 (2011)
24. Patrikalakis, N.M., Maekawa, T.: Shape Interrogation for Computer Aided Design and Manufacturing. Springer, Heidelberg (2002)
25. Jerne, N.K.: Toward a network theory of the immune system. Ann. Immunol. **125**, 373–389 (1974)
26. Piegl, L., Tiller, W.: The NURBS Book. Springer Verlag, Berlin/Heidelberg (1997)
27. Sarfraz, M., Raza, S.A.: Capturing outline of fonts using genetic algorithms and splines. In: Proceedings of Fifth International Conference on Information Visualization IV 2001, pp. 738–743. IEEE Computer Society Press (2001)
28. Timmis, J., Hone, A., Stibor, T., Clark, E.: Theoretical advances in artificial immune systems. Theoret. Comput. Sci. **403**(1), 11–32 (2008)

29. Varady, T., Martin, R.R., Cox, J.: Reverse engineering of geometric models - an introduction. Comput. Aided Des. **29**(4), 255–268 (1997)
30. Yoshimoto, F., Harada, T., Yoshimoto, Y.: Data fitting with a spline using a real-coded algorithm. Comput. Aided Des. **35**, 751–760 (2003)
31. Zhao, X., Zhang, C., Yang, B., Li, P.: Adaptive knot placement using a GMM-based continuous optimization algorithm in B-spline curve approximation. Comput. Aided Des. **43**, 598–604 (2011)

Swarming Behaviour

Quantifying Swarming Behaviour

John Harvey$^{(\boxtimes)}$, Kathryn Merrick, and Hussein Abbass

School of Engineering and Information Technology,
The University of New South Wales, ADFA, Canberra, ACT 2600, Australia
{j.harvey,k.merrick,h.abbass}@adfa.edu.au

Abstract. Swarming behaviour has been the subject of extensive objective analysis since Reynolds introduced the first-computer based 'boids' model in 1987. The current study extends that work by applying a range of measures — comprising existing and novel 'group' and 'order' measures and measures originally applied to chaotic systems — to a simplified version of Reynolds' original boids model. Classifier models are then developed to identify preferred measures and combinations of measures that can accurately identify and quantify swarming behaviour. Novel combinations of existing measures show promise in providing a single measure to identify and quantify swarming. The results also suggest that there may be degrees of 'swarminess' rather than a simple swarming/not-swarming situation. Better understanding of swarming systems, including identifying measures that can define the parameter space in which swarming occurs and quantifying the resulting swarming dynamics, is potentially useful in a range of applications from targeted use of swarm intelligence systems to developing realistic graphics for visual effects. The predictive models developed in this study can be used as a basis for improving the predictability and tuning the behaviour of swarming systems.

Keywords: Swarming · Chaotic systems · Classifier models

1 Introduction

Swarming behaviour has been the subject of extensive objective analysis since Reynolds introduced the first-computer based 'boids' model in 1987 [9]. Vicsek and Zafeiris [13] provide a good summary of the broad field of collective motion, of which swarming is a subset, as well as mathematical methods and models for simulating swarming behaviour. Vicsek et al. emphasise the importance of developing simple models from which understanding of complex systems in the real-world can be developed. Numerous such models have been developed to help explain real-world swarming behaviour, for example by Mecholsky et al. [7], Vicsek et al. [12], and Gazi and Passino [4].

The key identifying characteristics of swarming behaviour as displayed by the boids model are that particles are 'Grouped'; that is, particles tend to form clusters or groups, and are 'Not ordered'; that is, particles are continuously moving

Y. Tan et al. (Eds.): ICSI 2016, Part I, LNCS 9712, pp. 119–130, 2016.
DOI: 10.1007/978-3-319-41000-5_12

and not uniformly ordered/aligned in direction of motion, i.e. there is diversity of directions of motion, even within a group/cluster. Two simple objective measures based on these characteristics — 'groups' and 'order' — have been used to identify the presence of swarming behaviour, for example by Vicsek et al. [12] and Ferrante et al. [2,3].

The current study applies a range of objective measures to identify objective measures and/or combinations of measures that can accurately identify the presence of swarming behaviour and quantify that behaviour for the simplified boids model. Such measures would be useful in a range of applications from targeted use of swarm intelligence systems to developing realistic graphics for visual effects.

Harvey et al. [6] developed a simplified version of Reynolds' original boids model — comprising the interaction of attract, align and repel forces — to conduct testing of the applicability of objective measures associated with chaotic dynamics to a swarming system. (A detailed description of the simplified boids model is provided in [6]). The eight possible combinations of the presence or absence of the attract, align and repel forces in the simplified boids model leads to eight 'rule-sets'.

Three of these rule-sets — 'Attract', 'Attract + Repel' and 'Attract + Repel + Align' — were shown to produce swarming behaviour, based on existing measures of 'group' and 'order' [6]. These results are consistent with the results of previous studies using different swarm models [9–11]. Harvey et al. [6] refer to these three rule-sets as the 'swarming cases' and the remaining five rule-sets as the 'non-swarming cases'.

For the current study the same simplified boids model was used but the parameter space was expanded to include three values of the maximum particle velocity (V_{max}), four values of the attract and repel forces (controlled by $1/F_{FT}$ and $1/F_{KA}$ respectively), and three values of the align force (controlled by $1/F_{RV}$), leading to 144 test cases, as shown in Table 1.

This article is organised as follows. In Sect. 2 results from application of the objective measures to the test cases are presented and discussed. In Sect. 3 classifier models to identify the presence of swarming behaviour and predict human perception of swarming are developed based on the results from Sect. 2. Conclusions are drawn in Sect. 4, along with a discussion on future work.

2   Results of Application of Objective Measures

Quantitative measures used in this study that may be useful in identifying and quantifying swarming behaviour, classified into those that provide a measure of 'group' and those that provide a measure of 'order', comprise:

- **'Group' measures**: $Group_{objective}$ and $Coverage_{pos}$
- **'Order' measures**: $Disorder$, $Coverage_{dir}$, $ApEn$, $CoDi$, LLE and Div

- $Group_{objective}$: The 'group' measure used to identify the presence/number of clusters among the population of particles follows the hierarchical clustering methodology used by Ferrante et al. [3].

Table 1. Experiment control parameters. Values used in original study by [6] are underlined.

Parameter	Description	Value
N	Total number of particles	100
R_{max}	Region size	500
V_{max}	Maximum velocity	1, 2, 4
T	Total iterations	1500
R_c	Attraction range	100
R_a	Alignment range	100
R_s	Repulsion range	10
$1/F_{FT}$	Flock together factor	0, 0.005, 0.01, 0.02
$1/F_{RV}$	Relative velocity factor	0, 0.0625, 0.125
$1/F_{KA}$	Keep away factor	0, 0.50, 1, 2

– *Disorder*: The *Disorder* measure used in the study is based on the use of an *order* measure applied by Ferrante et al. [3] and originally developed by Vicsek et al. [12]. The *Disorder* measure is calculated using the absolute value of the average normalised velocity of the particles.
– *Coverage*$_{pos}$: *Coverage*$_{pos}$ is based on dividing the model area into equally sized 'bins', equal in number to the total number of particles. Each 'bin' is then examined to see if one or more particles are located in it.
– *Coverage*$_{dir}$: *Coverage*$_{dir}$ is based on dividing the total 2π radians of possible particle directions (ignoring magnitude of velocity) into N direction bins.
– Information generation rate — approximate entropy: For the current study, information generation rate is calculated using an approximate entropy measure ($ApEn$), based on work by Pincus [8].
– System complexity — correlation dimension: For the current study system complexity was assessed using a measure of correlation dimension ($CoDi$) originally developed by Grassberger and Procaccia [5] and as implemented by Ding et al. [1]. The correlation integral, $C(\varepsilon)$, is defined as the probability that a pair of points chosen randomly on the attractor is separated by a distance less than ε on the attractor.
– Sensitivity to initial conditions - *LLE*: LLE of a dynamic system can be calculated based on the time series of any observable variable of the system. Calculation is achieved through reconstruction of the phase space of the system. *LLE* can be estimated as the mean divergence rate of the nearest neighbours.
– Sensitivity to initial conditions — *Div*: Harvey et al. [6] introduced a novel 'divergence' (*Div*) measure to identify sensitivity to initial conditions. *Div* is calculated by displacing a single boid one unit in a random direction from its original position and measuring the divergence from its original path is then measured over a fixed time period, in this case 500 time units.

The previous objective measures are applied to the time series — comprising position and velocity information for each particle for each time step — for each of the 144 test cases.

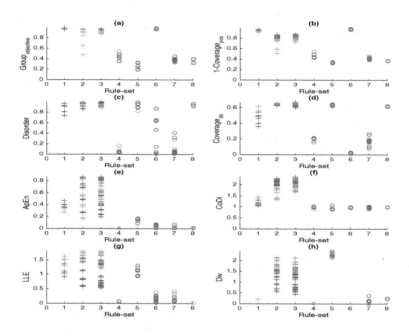

Fig. 1. Results from application of the eight individual measures at $t = 1500$ for 10 seed values for each of the 144 cases. Results are categorised by rule-set and V_{max}. X-axis shows rule-sets which comprise: 1-Attract, 2-Attract + Repel, 3-Attract + Align + Repel, 4-Align, 5-Repel, 6-Attract + Align, 7-Repel + Align, 8-No Rules. Y-axis shows value for each objective measure. The swarming cases comprise the first three rule-sets, shown as a '+'. Non-swarming cases are shown as a 'o'. Blue-$V_{max} = 1$, Red-$V_{max} = 2$, Green-$V_{max} = 4$. (Color figure online)

Figure 1(a), (b) show the results of application of the two objective 'group' measures — $Group_{objective}$ and $1 - Coverage_{pos}$ — to the time series. Both measures show that all but two instances of the swarming cases (where V_{max} is high for the Attract + Repel rule-set) have very high levels of the 'group' measure, consistent with the grouped nature of swarming behaviour. All non-swarming cases show much lower values of both objective group measures with the exception of the Attract + Align case which shows very high values of the 'group' measures. The overlap in 'group' measures for swarming and non-swarming cases shows that the presence of grouping by itself is not sufficient to identify swarming behaviour.

Figure 1(c)–(h) show the results of application of the six objective 'order' measures to the time series from the 144 test cases. In general the swarming cases tend to have a medium to high level of the respective 'order' measure (in this case, shown as the level of 'Disorder'), consistent with swarming behaviour being 'not-ordered'. The six non-swarming cases generally have a low value for the 'order' measures, exceptions being:

- all instances of the Repel rule-set have a high value of the 'order' measure for the $Disorder$, $Coverage_{dir}$, LLE and Div objective measures.
- all instances of the No Rules rule-set have a high value for the $Disorder$ and $Coverage_{dir}$ measure.

For each of the 'order' measures there is at least some overlap between the swarming and non-swarming cases. No individual 'order' measure, therefore, is able to differentiate between swarming and non-swarming behaviour. Differences in results between different 'order' measures, however, may mean that combinations of 'order' measures may be able to provide differentiation between swarming and non-swarming behaviour and potentially provide differentiation between the individual rule-sets that make up the swarming cases. For example, while the LLE measure shows a high value for all the swarming cases, the Div measure has a very low value for the Attract rule-set. Further, the four chaos measures show a high degree of sensitivity to V_{max} for the three swarming cases (with the exception of Div for the Attract rule-set) which suggests these measures may be useful in quantifying swarming behaviour.

Application of the eight objective measures to the 144 time series generated from the simplified boids model show that neither individual 'group' nor individual 'order' measures can unambiguously differentiate between swarming and non swarming behaviour. For the two 'group' measures there is little difference between the two results. For the six 'order' measures there is considerable difference between the results, with each showing different sensitivity to the individual rule-sets. Classifier models developed based on combinations of measures, therefore, may be able to accurately differentiate between swarming and non swarming cases and potentially between different types of swarming behaviour. The range of values for the 'order' measures for the swarming cases also suggests there may in fact be degrees of 'swarminess' which can be quantified by the individual measures or combinations of the measures.

3   Classifier Models

In this section results from the previous section are used to develop classifier models to identify and swarming behaviour from time series generated by the simplified boids model. The J48 classification tree algorithm was used in unsupervised learning mode with a 10-fold cross-validation and 66 % split between training and test sets in all cases. The classification accuracy reported for all models is based on model cross-validation results. Classification trees shown are based on using all data as the training set.

3.1   Classifier Models Based on Model Input Parameters

The first classifier model developed was based on the model input parameters that control the strength of the attract, align and repel forces ($1/F_{FT}, 1/F_{RV}$, and $1/F_{KA}$ respectively), and maximum velocity (V_{max}) as the independent variables, and the objectively defined swarming/non-swarming cases as the dependent variable. The resulting model is shown below.

attract ≤ 0 : not-swarming (36/0)

attract > 0

— repel ≤ 0

— — align ≤ 0: swarming (9/0) *(Attract)*

— — align > 0: not-swarming (18/0)

— repel > 0: swarming (81/0) *(Attract + Repel, Attract + Align + Repel)*

As expected, because the swarm cases were defined by the input parameters, the classifier model identifies the three rule-sets that make up the swarming cases with 100 % accuracy, there are no classification errors. All other cases lead to a not-swarming result. The classifier model also shows that the swarming/not-swarming classification is not sensitive to V_{max}.

3.2   Classifier Models Based on Time Series from the Test Cases

Where input control parameters for the swarm model are not available, identification of swarming behaviour will need to be based on objective measures calculated from the time series data generated by the boids model.

Classification Based on Individual 'Group' and 'Order' Measures. Individual 'group' and 'order' measures (independent variable) were first tested against the objectively identified swarm cases (dependent variable) to determine their suitability to identify swarming behaviour. Classification accuracy and complexity results are shown in Table 2.

Table 2. Accuracy and complexity of classifier models using individual 'group' and 'order' objective measures based on objectively identified swarming cases.

Measure	Accuracy (%)/Leaves	Measure	Accuracy (%)/Leaves
$Group_{objective}$	93.1/5	$ApEn$	100/4
$1\text{-}Coverage_{pos}$	100.0/5	$CoDi$	100/6
$Disorder$	100/24	LLE	100/16
$Coverage_{dir}$	92.1/5	Div	100/15

As Table 2 shows, only six of the eight objective measures can classify the swarm cases with 100 % accuracy. The complexity of these models varies from four to 24 leaves. Examination of the classifier models that achieve 100 % accuracy, even for the model that requires only four leaves ($ApEn$), shows that classification accuracy is achieved by the inclusion of 'bands' (often very narrow bands) of alternating swarming and non-swarming behaviour. As there is no reason to believe swarming behaviour occurs in such bands for any of the objective measures, all these models are assessed to be overfitting the data. Classifier models developed using individual objective measures, therefore, are assessed to be unsuitable for accurately identifying or predicting swarming behaviour.

Classification Based on Paired 'Group' and 'Order' Measures. Based on the key features of swarming behaviour, i.e. particles are grouped and not ordered, paired 'group' and 'order' measures were tested to see whether they could be used to develop classify models that can accurately identify swarming behaviour. Accuracy and complexity results for the 12 possible paired 'group'/'order' measures, using objectively identified swarm cases as the dependent variable, are shown in Table 3. Results are also shown in Fig. 2, as a scatter plot of eight rule-sets plotted using the respective 'group' and 'order' measure.

Table 3. Accuracy and complexity of classifier models using paired 'group' and 'order' objective measures. '*' shows minimum complexity models.

Measure 1	Measure 2	Accuracy (%)/Leaves	Measure 1	Measure 2	Accuracy (%)/Leaves
$Group_{objective}$	$Disorder$	100/5	$1\text{-}Coverage_{pos}$	$Coverage_{dir}$	100/3*
$Group_{objective}$	$Coverage_{dir}$	100/3*	$1\text{-}Coverage_{pos}$	$Disorder$	100/4
$Group_{objective}$	$ApEn$	100/4	$1\text{-}Coverage_{pos}$	$ApEn$	100/4
$Group_{objective}$	$CoDi$	100/4	$1\text{-}Coverage_{pos}$	$CoDi$	100/4
$Group_{objective}$	LLE	100/3*	$1\text{-}Coverage_{pos}$	LLE	100/3*
$Group_{objective}$	Div	100/7	$1\text{-}Coverage_{pos}$	Div	100/5

General observations from Table 3 are as follows:

- All classifier models use both measures, which means paired measures are always better than any single objective measure.
- All paired measures identify the objectively defined swarming cases with 100 % accuracy but the complexity of the classifier model varies from three to seven leaves.
- Four models require only three leaves to achieve 100 % accuracy for the swarming cases.

Theoretically, the simplest model that can classify the swarming/non-swarming cases using two variables with 100 % accuracy requires three leaves. Additional leaves beyond three indicate there are 'bands' of swarming/non-swarming instances within at least one of the variables. As there is no reason to believe the swarming/non-swarming instances occur in such bands for any of the objective measures, models that require more than three leaves are likely to be trying to overfit the data to maximise classification accuracy. For

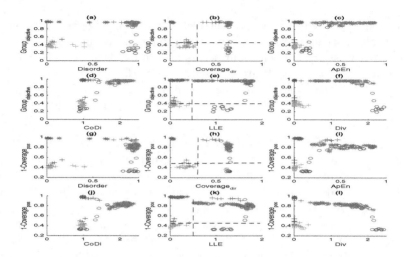

Fig. 2. Correlation between one objective 'group' measure and one objective 'order' measure. (a) $Group_{objective}$ and (b) $Group_{objective}$ and, (c) $Group_{objective}$ and 'not ordered', (d) $Group_{objective}$ and, (e) $Group_{objective}$ and, (f) $Group_{objective}$ and, (g) $1 - Coverage_{pos}$ and, (h) $1 - Coverage_{pos}$ and, (i) $1 - Coverage_{pos}$ and $ApEn$, (j) $1 - Coverage_{pos}$ and $CoDi$, (k) $1 - Coverage_{pos}$ and LLE, (l) $1 - Coverage_{pos}$ and Div. Swarming cases are shown in red. Cut-off values for the four models requiring only three classification leaves — (b), (e), (h), and (k) — are shown. (Color figure online)

the current study four models that achieve 100 % classification accuracy and only require only three leaves. These models are: $Group_{objective}$ & $Coverage_{dir}$, $Group_{objective}$ & LLE, $Coverage_{pos}$ & $Coverage_{dir}$ and $Coverage_{pos}$ & LLE. These models, and the classification thresholds they use, are shown in Fig. 2(b), (e), (h) and (k). As the figures show, all four pairs of measures provide clear separation between the swarming and non-swarming cases.

Classification Based on Other Pairs of Objective Measures. For completeness, the remaining pairs of objective measures — comprising 17 paired 'order' measures and one paired 'group' measure — were also examined for ability to identify and quantify swarming behaviour. The results for the 16 paired measures that resulted in 100 % classification accuracy are shown in Table 4.

Analysis of the results in Table 4 shows:

- With one exception, all classifier models use both measures, which means paired measures are better than any single objective measure. The exception is the $CoDi$ & LLE pair where the LLE measure is not used.
- All paired measures identify the objectively defined swarming cases with 100 % accuracy but the complexity of the classifier model varies from three to seven leaves.

Table 4. Accuracy and complexity of classifier models using pairs of objective measures — excluding paired 'group' and 'order' measures. '*' indicates measures that result in 100 % classification accuracy with only three leaves in the model.

Measure 1	Measure 2	Accuracy (%)/Leaves	Measure 1	Measure 2	Accuracy (%)/Leaves
$ApEn$	$CoDi$	100/3*	$Disorder$	LLE	100/11
$ApEn$	LLE	100/3*	$Disorder$	Div	100/9
$ApEn$	Div	100/4	$Disorder$	$Coverage_{dir}$	100/19
$CoDi$	$LLE(1)$	100/4	$Coverage_{dir}$	$ApEn$	100/3*
$CoDi$	Div	100/4	$Coverage_{dir}$	$CoDi$	100/6
LLE	Div	100/4	$Coverage_{dir}$	LLE	100/11
$Disorder$	$ApEn$	100/4	$Coverage_{dir}$	Div	100/9
$Disorder$	$CoDi$	100/5	$Group_{objective}$	$Coverage_{pos}$	100/5

- Two models — $ApEn$ & LLE and $Coverage_{dir}$ & $ApEn$ — require only three leaves to achieve 100 % classification accuracy for the swarming cases and classification accuracy equivalent to results from the inputs-based model.

The existence of paired 'order' measures that accurately classify swarming behaviour was not expected — a 'group' measure was also expected to be required. The outcome is, however, along the lines of the chaos composite measure proposed by Harvey et al. [6] which was based on the combination of three chaos measures without any grouping measure.

Classifier Models Using Composite Measure Based on Product of Objective Measures. Following the approach by Harvey et al. [6], composite measures were calculated based on the product of pairs of objective measures. The basis of this approach is that all measures (when 1-$Coverage_{pos}$ is used) are structured such that the magnitude of the measure is low for non-swarming behaviour and high for swarming behaviour. The product of the two will magnify this effect, the product only being high if both are high or low if both are low. For the $CoDi$ measure, the result was adjusted to $CoDi - 1$, because, as explained in [6], for chaotic motion $CoDi$ must be greater than 1. One additional composite measure was added to the paired measures, one combining all three chaos measures, i.e. $ApEn * (CoDi - 1) * LLE$, a modified version of the one used in [6].

An advantage of the composite measures over the paired measures is that they provide a single objective measure that can be used to quantify the dynamics of the behaviour. Testing the 28 possible composite paired measures (independent variable) showed that 10 of them achieved a classification accuracy of 100 % using two leaves in the classifier tree — the minimum number for a single measure — as shown in Table 5. These 10 composite measures, therefore, provide clear differentiation between swarming and non-swarming behaviour and establish a single metric to quantify swarming, i.e. the degree of 'swarminess' present.

Table 5. Classifier models using a composite measure based on the product of two objective measures achieving 100 % classifier accuracy with two leaves.

Measures	Accuracy (%)/Leaves	Min	Max	Threshold
$Group_{objective} * Coverage_{dir}$	100/2	0.01	0.65	0.25
$Group_{objective} * ApEn$	100/2	0.00	0.84	0.07
$Group_{objective} * CoDi$	100/2	0.00	1.31	0.03
$Group_{objective} * LLE$	100/2	0.00	1.75	0.38
$Coverage_{pos} * Coverage_{dir}$	100/2	0.01	0.57	0.23
$Coverage_{pos} * ApEn$	100/2	0.00	0.71	0.07
$Coverage_{pos} * CoDi$	100/2	0.00	1.14	0.03
$Coverage_{pos} * LLE$	100/2	0.00	1.52	0.44
$ApEn * CoDi$	100/2	0.00	0.13	0.01
$ApEn * CoDi * LLE$	100/2	0.00	1.66	0.01

The first eight of these composite measures, comprising the combination of a 'group' measure and an 'order' measure, is consistent with the key characteristics of swarming behaviour. Interestingly, while both the existing $Group_{objective}$ measure and novel $Coverage_{pos}$ are included in the successful composite measures, the existing 'order' measure (adjusted to the $Disorder$ measure) is not included, only the novel $Coverage_{dir}$ measure is included.

This indicates that existing measures of 'group' and 'order' (instantiated here as $Group_{objective}$ and $Disorder$ respectively) cannot be successfully combined to create a single measure to identify or quantify swarming behaviour.

The last two successful composite measures, $ApEn * CoDi$ and $ApEn * CoDi * LLE$, do not include a measure of 'group' and are more interesting as they represents a significantly different approach to identifying swarming behaviour to the current approach. Of interest the $Chaos_{composite}$ measure used in [6] required three leaves to achieve 100 % accuracy, indicating that the novel Div measure introduced in that study is not as useful for identifying swarming behaviour as the LLE measure.

All 10 composite measures were subsequently tested against time series generated by a random walk model to determine whether the measures could discriminate swarming behaviour from random walk behaviour. Of the 10 composite measures in Table 5, only $Group_{objective} * Coverage_{dir}$ was successful for all instances of V_{max}. $Group_{objective} * CoDi$ and $Coverage_{pos} * CoDi$ were successful but only for the case where $V_{max} = 1$.

Analysis Based on Using the Full Range of Measures and Overall Results. Classification based on the full range of objective measures was also analysed. A large number of models based on combinations of more than two measures were able to achieve the same accuracy as combinations of two measures but none were able to improve on them. Given the computational overhead of using additional measures, models comprising more than two measures were not preferred.

4   Conclusion

In this paper, we used eight objective measures — comprising existing and novel 'group' and 'order' measures and measures originally applied to chaotic systems — and combinations of those measures, to develop classifier models to identify and quantify swarming behaviour based on time series generated by a simplified version of Reynolds' original boids model. Application of the eight objective measures showed that no individual measure was able to accurately identify swarming cases.

Sixteen classifier models were developed — comprising four paired 'group'/'order' measures, two paired 'order' measures and 10 composite measures (based on the products of two measures) — which provided 100 % classification accuracy of the time series generated by the simplified boids model without over-fitting the data. Interestingly the composite measures do not always comprise a 'group' and an 'order' measure, as expected from the two key features of swarming behaviour. In many cases the combination of two 'order' measures were able to provide 100 % classification accuracy. The 10 successful composite measures have the advantage of providing a single objective measure that can be used to map the parameter space where swarming occurs, based on a single threshold level. The range in values for the single measure also suggests the existence of degrees of 'swarminess' rather than a simple swarming/not-swarming situation. The same composite measure may therefore be able to be used to quantify the degree of 'swarminess' present. Analysis using the same measures shows that one of the composite measures is also able to successfully discriminate between swarming behaviour and random walk behaviour.

References

1. Ding, M., Grebogi, C., Ott, E., Sauer, T., Yorke, J.A.: Plateau onset for correlation dimension: when does it occur? Phys. Rev. Lett. **70**(25), 3872 (1993)
2. Ferrante, E., Turgut, A.E., Mathews, N., Birattari, M., Dorigo, M.: Flocking in stationary and non-stationary environments: a novel communication strategy for heading alignment. In: Schaefer, R., Cotta, C., Kołodziej, J., Rudolph, G. (eds.) PPSN XI. LNCS, vol. 6239, pp. 331–340. Springer, Heidelberg (2010)
3. Ferrante, E., Turgut, A.E., Stranieri, A., Pinciroli, C., Birattari, M., Dorigo, M.: A self-adaptive communication strategy for flocking in stationary and non-stationary environments. Nat. Comput. **13**(2), 225–245 (2014). 1567-7818
4. Gazi, V., Passino, K.M.: Stability analysis of social foraging swarms. IEEE Trans. Syst. Man Cybern. Part B Cybern. **34**(1), 539–557 (2004)
5. Grassberger, P., Procaccia, I.: Measuring the strangeness of strange attractors. Physica D **9**, 189–208 (1983)
6. Harvey, J., Merrick, K., Abbass, H.A.: Application of chaos measures to a simplified boids flocking model. Swarm Intell. **9**(1), 23–41 (2015). 1935-3812

7. Mecholsky, N.A., Ott, E., Antonsen, T.M., Guzdar, P.: Continuum modeling of the equilibrium and stability of animal flocks. Phys. D Nonlinear Phenom. **241**(5), 472–480 (2012)
8. Pincus, S.M.: Approximate entropy as a measure of system complexity. Proc. Nat. Acad. Sci. **88**(6), 2297–2301 (1991)
9. Reynolds, C.W.: Flocks, herds and schools: a distributed behavioral model. In: ACM SIGGRAPH Computer Graphics, vol. 21, pp. 25–34. ACM (1987)
10. Strömbom, D.: Collective motion from local attraction. J. Theor. Biol. **283**(1), 145–151 (2011)
11. Strömbom, D.: Attraction based models of collective motion (2013)
12. Vicsek, T., Czirók, A., Ben-Jacob, E., Cohen, I., Shochet, O.: Novel type of phase transition in a system of self-driven particles. Phys. Rev. Lett. **75**(6), 1226 (1995)
13. Vicsek, T., Zafeiris, A.: Collective motion. Phys. Rep. **517**(3), 71–140 (2012)

A Simulation Study on Collective Motion of Fish Schools

Fatih Cemal Can$^{(\boxtimes)}$ and Hayrettin Şen

Department of Mechatronics Engineering, Izmir Katip Celebi University,
Izmir, Turkey
{fatihcemal.can,hayrettin.sen}@ikc.edu.tr

Abstract. This research presents a computer simulation that illustrates collective motion of fish. The simulation is developed by using individual-based model. Four modes are considered during the simulation. These modes are introduced as searching, swarm, feeding and escape modes. The simulation consists of three different sizes of fish. The medium fish try to hunt down the small ones while the big fish follows the medium ones. Feeds always have escape mode when they are very close to medium fish. Originality of the paper comes from the investigation of collective motion of fish with respect to these four modes. Also it is possible to add new individuals and moving feeds while the simulation is running. Polarization and expanse changes of swarm is also analyzed for two modes.

Keywords: Swarm simulation · Swarm behavior · Fish school

1 Introduction

Collective motion is created by individuals that are moving with respect to some pair-wise interactions. This motion is exhibited by some animals such as fish, bees and ants in nature. The pair-wise interactions between individuals are proposed by simple mathematical model based on three rules namely attraction, parallel orientation and repulsion fields. Fish tries to move toward their neighbors in attraction field. Parallel orientation field lets them remain close to their neighbors. Repulsion field avoids collision with each other during collective motion.

One of the earliest swarm simulations was created by Craig Reynolds [1] in 1987. The collective motion of fish schools was investigated by Inada [2]. The effect of variation of preferred direction was analyzed with his model. His simulation consists of three rules namely attraction, parallel orientation and repulsion. Strömbom [3] proposed a collective motion model including as a single rule attraction. Furthermore, the simulations of collective motion are proposed by using elastic springs between nearby individuals [4, 5].

Oboshi [6] carried out one computer simulation of prey-predator system. He observed the behavior of a fish swarm escaping from a predator. Two new methods are presented for direction sensing of a robot swarm in order to perform some applications that include landmine detection and firefighting by Venayagamoorthy [7]. The first method indicates an embedded fuzzy logic approach in the particle swarm optimization

© Springer International Publishing Switzerland 2016
Y. Tan et al. (Eds.): ICSI 2016, Part I, LNCS 9712, pp. 131–141, 2016.
DOI: 10.1007/978-3-319-41000-5_13

algorithm. The second one presents a swarm of fuzzy logic controllers. Castro [8] improved a tool that has strategies for a hunting game between predators and prey by using particle swarm optimization.

Atyabi et al. [11] designed a robotic swarm which is navigated by a simulation that has two phases, training and testing. The training phase is the participation of agents in survivor rescuing missions as a team. In test phase, performance of agents is improved by using obtained knowledge in training phase. Swarm robots are controlled by using wireless sensory network and multi mobile robot approach in the study of Lee and Shen [10]. In their study six and twelve individuals were used to simulate swarm behaviors. Their simulation results of swarm groups are shown by using figures.

In Table 1, the comparison of the previous simulations are listed. In this research, we can add more than 500 individuals to the simulation. However, time interval is delayed due to the large number of individuals. The improvement of our simulation is explained in next section. Then mathematical model of swarm behavior is given in the Sect. 3.

Table 1. Comparison of the previous simulations

Research	N (Number of individuals)	Swarm modeling rules
Huth and Wissel [9]	8	Attraction, parallel orientation and repulsion
Inada [2]	50	Attraction, parallel orientation and repulsion
Oboshi et al. [6]	100	Avoidance, parallel orientation, attraction and searching
Strömbom [3]	80	Attraction
Lee and Shen [10]	6 and 12	Separation, alignment, cohesion

2   Collective Motion in the Previous Simulation and the Recent One

The recent simulation is extensively developed based on the previous one [12]. Currently more than two individuals can be inserted in the same simulation. Simply by clicking on the simulation area, one individual is added to the simulation. The feeds are programmed to be movable in this simulation. The number of feeds changes during the simulation. If the feeds are so close to the predators, either the feeds may be eaten or the feeds gain speed in order to increase escaping possibility.

Our Simulation program includes two forms (or graphical user interface) as shown in Figs. 1 and 2. User can open and close these forms at any time in simulation because of the fact that main program for these two forms are completely same. However, as seen from the figures, graphical methods for describing individuals are different. Therefore, two forms are working simultaneously by means of using same coordinates

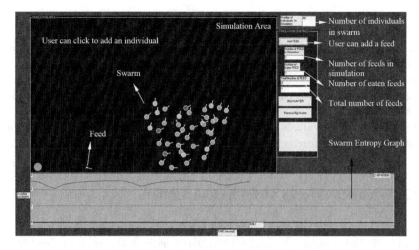

Fig. 1. The first form of the simulation

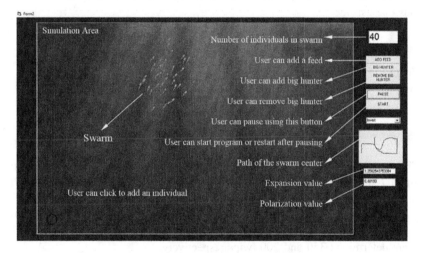

Fig. 2. The second form of the simulation

and orientations for individuals. In the second form, five buttons are used to control simulations such that user can add, remove feed and big hunter as well as start and pause simulation. All variables such as coordinates, orientations of individuals, polarization, and expansion values for each time step are saved in five *output.txt* after closing the simulation. Hence, we can use simulation data in other programs (Matlab, Mathematica and Excel).

Some assumptions are considered when the simulation algorithm is created. These assumptions are as follows;

1. The number of individuals is not set to a constant number.
2. A limited 2-D world is defined for movement of fishes.
3. The motion of individuals is provided by two parameters: velocity and direction.
4. The position and orientation of each individual is computed in every 20 ms. The decision of movement is also accomplished in the same time interval.
5. The motion of each fish can be influenced by every swarm individual. Euclidean distance between individuals is calculated for the decision of individuals.
6. The motion of individuals is affected by external influences. For example, any number of movable feeds can be added and this affects the motion of individuals.

3   Mathematical Model of Collective Motion

A very well-known mathematical model is selected to determine positions of individuals [2, 9]. The starting position $(x_j(0), y_j(0))$ of individuals are determined by the user.

Time is increased at every time step (20 ms). At the beginning of the simulation, time has no value. The new time is calculated by addition of the time step to the previous one as,

$$t = t + \Delta t \tag{1}$$

Where t is time and Δt is time step.

$$x_j(t) = x_j(t - \Delta t) + \Delta t\, v_j(t) \cos \beta_j(t) \tag{2}$$

$$y_j(t) = y_j(t - \Delta t) + \Delta t\, v_j(t) \sin \beta_j(t) \tag{3}$$

Where $x_j(t)$ and $y_j(t)$, $v_j(t)$ and $\beta_j(t)$ describe position, velocity and orientation of individual j at time t, respectively. The orientation was determined randomly. The velocity was defined 10 pixel per second.

The radii of attraction (R_a), parallel orientation (R_p) and repulsion (R_r) fields were chosen to be 2500, 1800 and 500 twips respectively. Body length of the individual and feed were determined 350 and 200 twips respectively.

Euclidean distance $R_{jk}(t)$ between j^{th} and k^{th} individuals @ time t can be calculated as follows;

$$R_{jk}(t) = \sqrt{\left(x_j(t) - x_k(t)\right)^2 + \left(y_j(t) - y_k(t)\right)^2} \tag{4}$$

If $R_{jk}(t) < R_a$ then firstly $\beta_{jk}(t)$ which is angle difference between j^{th} and k^{th} individuals @ time t, is calculated in Eq. (5);

$$\beta_{jk}(t) = \beta_j(t) - \beta_k(t) \tag{5}$$

The angles of the individuals are determined by comparing $\beta_{jk}(t)$ as following three cases;

- Case 1: If $\beta_{jk}(t) < 0$ and $abs(\beta_{jk}(t)) < 90$, then angles of individuals are not changed.
- Case 2: If $\beta_{jk}(t) < 0$ or $\beta_{jk}(t) > 0$ and $abs(\beta_{jk}(t)) > 90$, then angles are calculated using following equations;

$$\beta_j(t) = Mod\left[ArcTan\left[\frac{\left(y_k(t) - y_j(t)\right)}{\left(x_k(t) - x_j(t)\right)}\right], 2\pi\right] - \left(\frac{\left(R_a - R_{jk}(t)\right)}{R_{jk}(t)}\right)\pi \tag{6}$$

$$\beta_k(t) = Mod\left[ArcTan\left[\frac{\left(y_j(t) - y_k(t)\right)}{\left(x_j(t) - x_k(t)\right)}\right], 2\pi\right] - \left(\frac{\left(R_a - R_{jk}(t)\right)}{R_{jk}(t)}\right)\pi \tag{7}$$

- Case 3: If $\beta_{jk}(t) > 0$ and $abs(\beta_{jk}(t)) < 90$, then angles of the individuals are determined by following equations;

$$\beta_j(t) = \beta_j(t) - \frac{\beta_{jk}(t)}{2} \tag{8}$$

$$\beta_k(t) = \beta_k(t) - \frac{\beta_{jk}(t)}{2} \tag{9}$$

If $R_{jk}(t) < R_p$ then mathematical model is going to be as follows;

$$\beta_j(t) = \beta_j(t) + 2\cos(B\pi) \tag{10}$$

$$\beta_k(t) = \beta_k(t) + 2\cos(B\pi) \tag{11}$$

Where B is a random number.

If $R_{jk}(t) < R_r$ then mathematical model is going to be as follows;

$$\beta_j(t) = \beta_j(t) - \left(\frac{600}{R_{jk}(t)}\right)C \tag{12}$$

$$\beta_k(t) = \beta_k(t) - \left(\frac{600}{R_{jk}(t)}\right)C \tag{13}$$

Where C is a positive random number. In equations, 600 was found by testing the simulation.

If $R_{jk}(t) > R_a$, then angles of individuals are going to be as follows;

$$\beta_j(t) = \beta_j(t) + 4\cos(A\pi) \tag{14}$$

$$\beta_j(t) = \beta_j(t) + 5\cos(C\pi) \tag{15}$$

Where A and C are random numbers.

In the simulation, hunting behavior of medium fish was mathematically modeled as follows,

$$R_{f,j}(t) = \sqrt{\left(x_j(t) - x_{f,ii}(t)\right)^2 + \left(y_j(t) - y_{f,ii}(t)\right)^2} \tag{16}$$

$$P_{ii}(t) = \frac{1}{\beta^\alpha} e^{\frac{R_{f,j}(t)}{1000\beta}} \tag{17}$$

$$S_{t,ii}(t) = \frac{1 - P_{ii}(t)^q}{1 - q} \tag{18}$$

Where $R_{f,j}(t)$ is Euclidean distance between individual and feed. $P_{ii}(t)$ is the probability of hunt. If the $P_{ii}(t) > 0.86$ then the fish eat the feed (Fig. 3). $S_{t,ii}(t)$ is Tssalis entropy between predator (fish) and prey (feed). Here q is a real parameter, which ranges from 0 to 1. As q approaches 1, Tsallis entropy term becomes the Boltzman-Gibbs entropy term which is known as classical entropy term. In our study, real parameter was taken as 0.6 for simulation, $j = 1, 2, ., N, ii = 1, 2, .., N_f$. Where N and Nf are number of the fish and number of the feeds, respectively.

$$S_{tmin,k}(t) = Min\left(S_{t,ii}(t)\right) \tag{19}$$

Where k is index of feed that gives minimum entropy value obtained from all entropy values for j^{th} individual. Therefore, individual is directed to the feed which has minimum entropy value, by using Eq. (20).

$$\beta_j(t) = ArcTan\left[\frac{\left(x_j(t) - x_{f,k}(t)\right)}{\left(y_j(t) - y_{f,k}(t)\right)}\right]\frac{180}{\pi} + 180 \tag{20}$$

Where $\beta_j(t)$ is orientation, $x_j(t)$ and $y_j(t)$ is position of jth individual at time t. $x_{f,k}(t)$ and $y_{f,k}(t)$ is position of kth feed in simulation.

3.1   Characterization of Swarm Simulation

Polarization p, is defined as the arithmetic average of angle differences between the orientation of each fish and the movement direction of swarm [9].

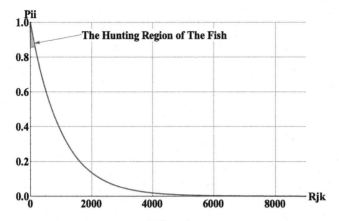

Fig. 3. Probability of hunt

$$\text{Pol}_{av}(t) = \left[\sum_k^n \beta_k(t)\right]/n \tag{21}$$

$$\text{Pol}(t) = \left[\sum_{k=1}^n \text{Pol}_{av}(t) - \beta_k(t)\right] \tag{22}$$

Where $Pol_{av}(t)$ is the average of polarization and $Pol(t)$ is the polarization of swarm.

Expanse a, is defined as the arithmetic average of distances between each individual of swarm and the center of fish group or swarm [9].

$$X_{av}(t) = \left[\sum_{k=1}^n X_k(t)\right]/n \tag{23}$$

$$Y_{av}(t) = \left[\sum_{k=1}^n Y_k(t)\right]/n \tag{24}$$

$$a(t) = \left[\sum_{k=1}^n \sqrt{(X_{av}(t) - X_k(t))^2 - (Y_{av}(t) - Y_k(t))^2}\right]/n \tag{25}$$

Where $X_{av}(t)$ is the average value of horizontal positions of individuals, $Y_{av}(t)$ is the average value of vertical positions of individuals and $a(t)$ is the expanse of swarm.

The simulation was represented by color figures (Figs. 1 and 2). But collective motion of fish cannot be seen from these figures while time is changing. Therefore, a clear representation of collective motion fishes is needed to be represented on the paper. Grayscale representation can be used for describing motion of fish in time. The motion of individuals can be determined by contrast of the arrows (Fig. 4). The contrast of

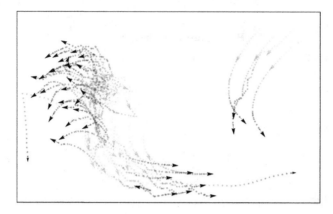

Fig. 4. Grayscale representation of individuals

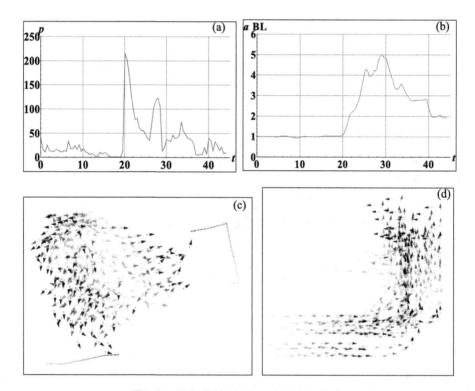

Fig. 5. 80 individuals try to hunt two feeds

arrows gradually increases from the beginning to the end of the motion. This illustration will be used for observing polarization and expanse values of swarm in a certain time interval.

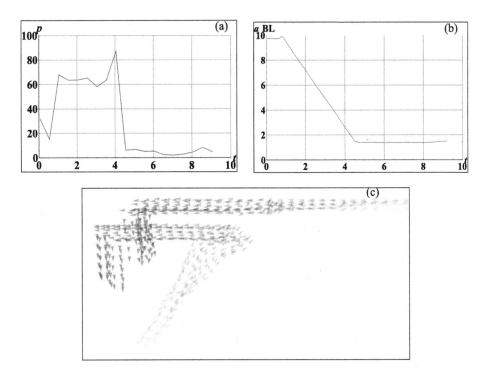

Fig. 6. Two groups (10 individuals) of swarm merge in to one swarm

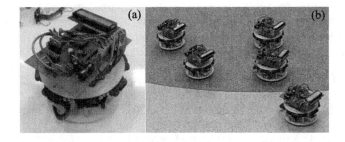

Fig. 7. One robot (a) and Swarm Robots (b)

The simulation was started including eighty individuals. These eighty individuals were in the swarm mode until the 20th second. Therefor polarization and expanse values are very low as shown in the Fig. 5(a) and (b). After the 20th second, the polarization and expanse values rapidly increase due to the adding two feeds to the simulation. The behavior of the individuals between 23rd and 25th seconds are shown in the Fig. 5(c) by using grayscale representation.

The polarization and expanse values decrease after the 28th second, since the feeds were eaten by the fish until that time. The behavior of the fish between 30th and 34th seconds are shown in the Fig. 5(d).

The simulation was paused for adding two different swarm groups having ten individuals on two distinct locations of simulation area in order to observe meeting of these groups. The expansion and polarization values considerably decrease after 4, 5th seconds as shown in the Fig. 6(a) and (b). Therefore two groups were combined into one swarm group (Fig. 6(c)).

4 Conclusion

In this study, a computer simulation that illustrates collective motion of fish was investigated. The previous simulation [12] was enhanced well enough. The behavior model of individuals was explained mathematically for feeding and swarm modes. The definitions of polarization and expanse were given and the graphics of polarization and expanse were plotted for these modes.

Future work plans include the application of this algorithm on 2D mobile swarm robots. Interactions between robots will be provided by using same pair-wise interactions of fishes. The swarm robots have been already designed and manufactured (Fig. 7(a) and (b)). The coding robots and experiment studies are still in progress.

References

1. Reynolds, C.W.: Flocks, herds and schools: a distributed behavioral model. In: ACM SIGGRAPH Computer Graphics. ACM (1987)
2. Inada, Y.: Steering mechanism of fish schools. Complex. Int. **8**, 1–9 (2001)
3. Strömbom, D.: Collective motion from local attraction. J. Theor. Biol. **283**(1), 145–151 (2011)
4. Triandaf, I., Schwartz, I.B.: A collective motion algorithm for tracking time-dependent boundaries. Math. Comput. Simul. **70**(4), 187–202 (2005)
5. Varghese, B., McKee, G.: A mathematical model, implementation and study of a swarm system. Robot. Auton. Syst. **58**(3), 287–294 (2010)
6. Oboshi, T., et al.: A simulation study on the form of fish schooling for escape from predator. FORMA-TOKYO- **18**(2), 119–131 (2003)
7. Venayagamoorthy, G.K., Grant, L.L., Doctor, S.: Collective robotic search using hybrid techniques: fuzzy logic and swarm intelligence inspired by nature. Eng. Appl. Artif. Intell. **22**(3), 431–441 (2009)
8. Castro, E.G., Tsuzuki, M.S.G.: Swarm intelligence applied in synthesis of hunting strategies in a three-dimensional environment. Expert Syst. Appl. **34**(3), 1995–2003 (2008)
9. Huth, A., Wissel, C.: The simulation of the movement of fish schools. J. Theor. Biol. **156**(3), 365–385 (1992)
10. Li, W., Shen, W.: Swarm behavior control of mobile multi-robots with wireless sensor networks. J. Netw. Comput. Appl. **34**(4), 1398–1407 (2011)

11. Atyabi, A., Phon-Amnuaisuk, S., Ho, C.K.: Navigating a robotic swarm in an uncharted 2D landscape. Appl. Soft Comput. **10**(1), 149–169 (2010)
12. Can, F.C., et al.: Characterization of swarm behavior through pair-wise interactions by Tsallis entropy. In: ICAI (2005)

Swarmscape: A Synergistic Approach Combining Swarm Simulations, Body Movement and Volumetric Projections to Generate Immersive Interactive Environments

Nimish Biloria[(✉)] and Jia-Rey Chang

HyperBody, Faculty of Architecture and the Built Environent, TU Delft,
Julianalaan 134, 2628 BL Delft, The Netherlands
{N.M.Biloria,J.R.Chang}@tudelft.nl

Abstract. The paper illustrates a real-time interactive, fully immersive spatial installation titled Ambiguous Topology. The installation creatively combines dynamic movement of the human body and swarm intelligence driven generative geometry production techniques to drive volumetric projection systems. Speed, frequency and intensity of movement of human body are used as parameters for activating a swarm of volumetrically projected digital particles/agents in space in real-time. Evolving 3D topological nuances within which the participant navigates and in doing so triggers further evolution of these immersive topologies are thus materialized.

Keywords: Swarm simulation · Dynamic geometry · Volumetric projection · Interactive and pro-active behavior · Bi-directional real-time data exchange

1 Introduction

Ambiguous Topology is a fully immersive interactive installation which blends the domains of human behavior, art, architecture and science. The installation is developed as a part of a European Union Culture Grant: METABODY. The installation, in real-time, tracks dynamic human behavior associated with the movement of the body, and communicates this data to customized swarm algorithms for generating emergent geometric topologies. This geometric data is subsequently transcribed into volumetric light projections using specialized hardware and custom scripting routines to materialize an immersive swarm-scape.

Translation of swarm computing based geometric outputs, which, are otherwise visualized within the constraints of the computer screen are materialized in real-world spatial settings using volumetric projections. As opposed to other performance art driven investigations into swarm computing [1–3] which employ planar projections, the volumetric projection aspect allows for a fully immersive corporeal experience of continuously transforming geometric topologies.

© Springer International Publishing Switzerland 2016
Y. Tan et al. (Eds.): ICSI 2016, Part I, LNCS 9712, pp. 142–153, 2016.
DOI: 10.1007/978-3-319-41000-5_14

1.1   Swarm: Behavioral Premise

A behavioral interaction process is initiated by analyzing and harnessing the manner in which emergent collective behavior of autonomous computational agents are generated at a global scale, as a result of their decentralized, local interactions. These local interactions predominantly stem from a very basic level of intelligence that is ascribed to the constituting agent population in the form of separation, alignment and cohesion values. In accordance with Reynolds [4, 5], separation implies avoiding crowding next to each other, alignment implies steering towards the average direction of the neighboring flocks, and cohesion implies driving the agent's movement towards the average position of the local agents.

The core behavioral premise, constitutes of a swarm of dormant virtual particles/agents encoded with these rule sets in three dimensional space. 640 of these dormant virtual agents volumetrically fill up the designated physical interaction space. The presence and movement of physical bodies, within this space is tracked in real-time, and serves as an external force, which disturbs the otherwise dormant swarm. This causes the agents to displace in the digital space in accordance with the aforementioned rule sets, thus entailing a change in their location, velocity and acceleration parameters. Each displacement amongst the agents of the swarm results in the generation, dissipation and transfer of virtual energy per agent. Eventually, the generated field of energy dies out (over time when there is no human presence detected to disturb the dormant field) and allows the agent systems to return back to a state of dormancy. Each agent in real-time transmits data pertaining to its current state as regards energy, position, velocity and acceleration in space. These values form the basis for an algorithmic structure, responsible for establishing geometric linkage between the agents in the form of line segments and NURB geometries. These agents in the digital space (in the swarm simulations) are projected as light beams via the volumetric projection system deployed as an output medium for the installation. Aspects of width, intensity and the color of the resulting light beams are directly impacted by the position, energy level and proximity status of the agents.

A threshold concerning total energy gain of the entire swarm is set within the swarm algorithm, which, once attained, assigns full autonomy to the light projections. This implies a shift in the engagement and perception of users, wherein, the earlier triggering of light beams, which, is directly associated with individual body movement is reversed such that the light projections now acquire the leading role, resulting in the individual, to gradually follow its cues. A gradual shift from 'Interactive to Pro-active' behavior is thus inculcated. The state of each agent's propulsion and energy levels are furthermore communicated via color changes within the projections. Aggressive colors, such as red and yellow indicate high value of flux compared to blue and green, which express passive and stable agent movement.

1.2   Technical Implementation

The agent-based simulation is created using an open-source programming language; Processing, developed by Ben Fry and Casey Reas in 2001 [6]. It is mainly designed for graphic designers, architects and interactive artists to develop a range of 2-Dimensional and 3-Dimensional visual graphics, Max/MSP and Open Sound Control (OSC). In terms of hardware, Microsoft Kinect is used for motion tracking and is associated with SimpleOpenNi (a motion tracking library of Processing) and 4 HD projectors. All computational processes are simulated in a 3d environment in Processing based on the aforementioned agent rule sets, which, are directly interfaced with the skeleton tracking data obtained from Kinect.

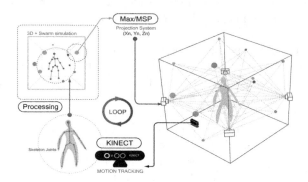

Fig. 1. Diagram showing the interactive loop of data streams.

The data communication process involves Processing simulation instantaneously transmitting the required data (co-ordinates and color coding per agent, geometric output) to custom patches set up in Max/MSP through an OSC (Open Sound Control) protocol. This assures real-time data synchronization of the X-Y-Z coordinates of each swarm agent between Processing and Max/MSP. Max/MSP patches hereafter communicate with the render mode of the HD projectors (Fig. 1).

2 Setting up the Narrative

After meticulous development and user testing, at the Protospace Lab, Hyperbody, TU Delft, Ambiguous Topology, was successfully set-up as a real-time immersive public installation at the Media Lab, Prado, Madrid, Spain. The site allocated for the installation allowed, an effective interaction zone (the convergence point of the four projectors) as 6 meters in width (X-direction), 5 m in length (Y-direction) and 5 m in Height (Z- direction). 640 agents/particles are embedded virtually in the space and wait to be triggered by the bodies of participants. Seven narrative modes are developed and organized in a fluent sequence in order to facilitate a holistic experience to the participants in a limited period of time. These modes are sequenced in the following order: Rain Mode, Follow Mode, Spike Mode, Disturb Mode, Attract Mode, Nurbs Mode, Rain-Up Mode. These modes are described in the following sections in conjunction with the participant's experiences.

2.1   Rain Mode

The Rain Mode is the first mode to be triggered by the presence of people (tracked by Kinect) within the allocated installation space. A high velocity (velocity) downpour of 640 agents/particles (P1) constituting the installation akin to heavy rainfall is immediately set in motion. The agents gradually reduce (*decreasingVelFactor) their speed of falling and completely cease to do so in certain locations in space. This is accompanied by a change in the color gradient of the agents (from magenta to dark blue), indicating the change in the velocity: from high velocity to a calm state. Observed participant reactions to this narrative involved slow movements and subtle gestures akin to holding your palm out to feel the rain-like droplets of agents (Fig. 2).

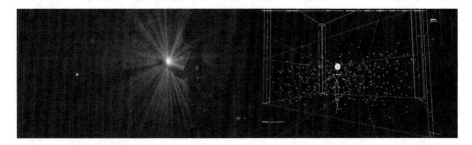

Fig. 2. Volumetric projection system and its corresponding 3 Dimensional real-time simulation showing the rain narrative.

Algorithm 1: Rain Mode. 0 sec. < T1 < 5 sec.
```
-----------------------------------------------------------
Input: set of 640 random spatial points, P1={x1, y1, z1};
P1 are generated randomly.
   const decreasingVelFactor = 0.98;
   const maxTime01 = 5;
   var time;
   var velocity;
       begin
          time:= 0;
           velocity:= -1200;
            repeat
                velocity := velocity * decreasingVelFactor;

                if value of velocity <=0.05
                     then velocity = 0;
                end if
                        P1 := P1 + velocity;
                        Q1 := P1;
            until time = maxTime01
end.
```
Output: set of 640 spatial points, Q1 = {x2, y2, z2}.
The initial locations of the input spatial points are randomized in a 5m width by 6m length and 5m height cubic space.
The unit associated with time is 1 second.

2.2   Follow Mode

This is the first instance that participants provide an impulse to the agents (P2). Each movement of the participants (User1 and User2), creates a flux in the agent field within which they are immersed. The swarm logic further entails that the agent propels its movement to the nearest neighbors and thus a ripple is sent through the virtual field as an emergent global outcome. It was observed that over time, the swarm of agents in space tends to follow the average direction of movement of the participants (torsoVelocitySum) (if they move in the same direction). However, if two participants (torsoVelocityUser1 and torsoVelocityUser2) attempt to move in opposite directions, the swarm tends to remain stable (Fig. 3). Furthermore, differential agent velocities of the entire swarm are the result of the participant's movement velocity and thus tend to speed up or slow down, with high velocity states depicted as magenta and low velocity state as blue. This mode subtly engages the participant via responsive interaction and provokes physical movement of the participants.

Fig. 3. 3D simulation showing how the participants interact with the agents: while the participant is moving toward one direction, the cloud of agents would follow the same direction and modify their color according to their respective velocity.

Algorithm 2: Follow Mode. 5 sec. < T2 < 45 sec.

```
-------------------------------------------------------------------
Input: set of 640 spatial points, P2 = Q1 = {x2, y2, z2};
Q1 is the result from Algorithm 1.
   const decreasingVelFactor = 0.98;
   const maxTime02 = 45;
   var time;
   var torsoVelocityUser1;
   var torsoVelocityUser2;
   var torsoVelocitySum;
     begin
      time:= 5;
      repeat
        torsoVelocitySum := torsoVelocityUser1+torsoVelocityUser2;
        torsoVelocitySum := torsoVelocitySum*decreasingVelFactor;

        if value of torsoVelocitySum <=0.05 then torsoVelocity = 0;
        end if
                      P2 := P2 + torsoVelSum;
                      Q2 := P2;
        until time = maxTime02
end.
-------------------------------------------------------------------
Output: set of 640 spatial points, Q2 = {x3, y3, z3}.
torsoVelocityUser1 and torsoVelocityUser2 represent the tracked torso
velocity of both users.
The unit associated with time is 1 second.
```

2.3   Spike Mode

The Spike Mode, introduces geometric connections; In this narrative, along with all the existing colored agents (P3), pure white lines are exhibited (line-connections). These

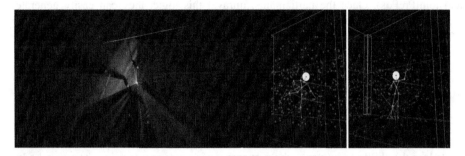

Fig. 4. White line connections visualized via volumetric projection and associated 3D simulations showing how joint tracking generate line-based geometric connections.

```
Algorithm 3: Follow Mode + Spike Mode. 45 sec. < T3 < 95 sec.
-------------------------------------------------------------------------
Input: set of 640 spatial points, P3 = Q2 = {x3, y3, z3};
Q2 is the result from Algorithm 2.
   const decreasingVelFactor = 0.98;
   const maxTime03 = 95;
   var time;
   var torsoVelocityUser1;
   var torsoVelocityUser2;
   var torsoVelocitySum;
   var jointCurPos;
   var jointPrePos;
     begin
       time:= 45;
       repeat
           torsoVelocitySum := torsoVelocityUser1+torsoVelocityUser2;
           torsoVelocitySum := torsoVelocitySum*decreasingVelFactor;

           if value of torsoVelocitySum <=0.05 then torsoVelocity = 0;
           end if
                        P2 := P2 + jointTorsoVelocitySum;
                        Q2 := P2;
           if(distance between jointCurrentPos and P3 < 500)
             then make line-connections between jointCurPos and jointPrePos;
           end if
       until time = maxTime03
     end.
-------------------------------------------------------------------------
Output: set of 640 spatial points, Q3 = {x4, y4, z4}.
Output: Several line-connections.
Output: torsoVelocitySum as each particle's velocity.
Sets of velocity as torsoVelocitysum= tV={tV1, tV2, tV3};
jointCurPos are the selected joints' locations in current frame,
jointPrePos are the selected joints' locations in previous frame.
Both User1 and User2's left and right hands and feet joints are
considered as reference joints in the algorithm; 8 in total.
The unit associated with time is 1 second.
```

lines are directly connected to the distances between the nearest agents of the swarm and are correlated with the participant's hand and feet joints tracked with Kinect. While gesturing with one's hands and feet, any two agents falling within this waving path establish a connection depicted by a white line to be drawn between them (jointCurPos and jointPrePos). Once the participants, unravel this logic, they can easily generate these flashing lines and start manipulating (Fig. 4). Some characteristics of the Follow Mode, such as the panning effect (torsoVelocitySum) and color gradations are retained in this narrative.

2.4   Disturb Mode

The Disturb Mode involves a shift from responsive to pro-active interaction. Participants lose the ability to influence the movement of swarm agents using their own body movements. Additionally, all the agents (P4), as autonomous entities start losing their energy, turn transparent and become almost invisible in space. In reality, once the agents loose their momentum, they become imperceptible and acquire a state of readiness for new stimulation from the participants. By touching, pushing, swinging the invisible agents, the participants actually feed the agents with energy and trigger their movement again. Each participant's hands and knees, now, become activating nodes, which, in turn influence the agents, based on the impetus produced by the movement of these joints. The faster the participants move (jointHandAcceleration), the larger the area of influence over the agents is. This directly impacts the agent's velocity and energy levels (Fig. 5). The swarm logic behind the scenes, implies that active agents seek to influence other passive neighbors thus triggering non-linear associations. It was observed that the participants tend to become keen and keep trying different body postures and movements to gradually set the dormant swarm in action once more.

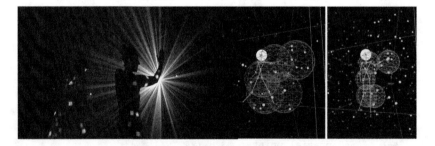

Fig. 5. Volumetric projections and 3D simulations showing how the participant influence the agent's behavior with their hands and knees. The sphere shapes in the simulation represent the dynamic area of influence exerted by the hand and knee joint movement.

```
Algorithm 4: Disturb Mode. 95 sec. < T4 < 140 sec.
----------------------------------------------------------------------
Input: set of 640 spatial points, P4 = Q3 = {x4, y4, z4};
Q3 is the result from Algorithm 3.
Input: velocity of each particle, originVel = tV={tV1, tV2, tV3};
tV is the result from Algorithm 3.
   const decreasingVelFactor = 0.98;
   const maxTime04 = 95;
const divisionFactor = 8^2;
var time;
var velocity;
var jointHandPos;
var jointHandAcceleration;
 begin
  time:= 95;
  velocity = originVel;
  repeat
     if(distance between jointHandPos & P4)
     <=(200*jointHandAcceleration/divisionFactor);
       then velocity = velocity + jointHandAcceleration;
     end if

     velocity := velocity * decreasingVelFactor;

     if value of velocity <=0.05 then velocity = 0;
     end if

             P4 = P4 + velocity;
             Q4 = P4;

   until time = maxTime04
end.
----------------------------------------------------------------------
Output: set of 640 spatial points, Q4 = {x5, y5, z5}.
/Velocity as velocity of each particle starts its initial value remained
from Algorithm 3.
Joints of jointHandPos and jointHandAcceleration refers to user1 and
user2's right and left hands; 4 in total.
The unit associated with time is 1 second.
```

2.5 Attract Mode

The Attract Mode involves the swarm of agents (P5) to suddenly and aggressively move rapidly towards the participant (attVelocity). This is also accompanied by the agent's switching their color to an aggressive red and yellow gradient. In this mode, the agent simulations are programmed to be attracted towards the participant's hands and feet (joint1–6) in order to create virtual polygonal geometries in space. Over a period of time, these virtual polygons unknowingly produced by the participants also appear in white along with other colored agents thus distinguishing the polygonal geometries the participants generate. Once the participants become aware of this game-play, they instinctively start attracting the agents via producing strange but interesting movements, such as changing moving direction rapidly, jumping up and down radically, and curling or stretching bodies oddly (Fig. 6).

Algorithm 5: Attract Mode. 140 sec. < T5 < 220 sec.

```
Input: set of 640 spatial points, P5 = Q4 = {x5, y5, z5};
Q4 is the result from Algorithm 4.
   const num = 100;
   const decreasingVelFactor = 0.98;
   const maxVelocity = 300;
   const maxTime05 = 220;
   var time;
   var attVelocity;
   var joint1, joint2, joint3, joint4, joint5, join6;
   begin
     time:= 140;
     repeat
       if ((numbers of particles' velocity>maxVelocity)> num)
         then
           particles will be attracted to follow the polyline as a
           path constructed by joint1, joint2, joint3, joint4, joint5,
           and joint6;
       end if

       velocity := velocity * decreasingVelFactor;

       if value of attVelocity <=0.05 then velocity = 0;
       end if
                 P5 := P5 + attVelocity;
                 Q5 := P5;
     until time = maxTime05
end.
```

```
Output: set of 640 spatial points, Q5 = {x6, y6, z6}.
joint 1~3: User 1's left hand, right hand and right foot.
joint 4~6: User 2's left hand, right hand and right foot.
The unit associated with time is 1 second.
```

Fig. 6. Volumetric projection and simulation with the participant interacting with the colored agents while white polygonal geometry connects with the position of the hand and feet joints.

2.6   Nurbs Mode

In the Nurbs Mode, the participants can push, wave, and touch the agents (P6). In addition to this, a continuous transforming nurbs (spline-line) is materialized based on the agent aggregation based density in space. On an average, ten locations coinciding with ten densest locations of the agents in space are selected as control points to construct the nurbs.

Since the agent densities can be impacted directly by the participant's movement in space (jointHandAcceleration), the nurbs geometry fluidly morphs from one shape to another (Fig. 7). The volumetric projections acquire a conical shape. Participant

Fig. 7. Volumetric projections illustrating the ambience of the Nurbs mode and 3D simulations showing the generation of Nurbs in space. The bigger green dots represent the densest spots of agent aggregation, which serve as vertices input for constructing the spline.

```
Algorithm 6: Nurbs Mode + Disturb Mode. 220 sec. < T6 < 265 sec.
-------------------------------------------------------------------------
Input: set of 640 spatial points, P6 = Q5 = {x6, y6, z6};
Q5 is the result from Algorithm 5.
   const decreasingVelFactor = 0.98;
   const maxTime06 = 265;
   const divisionFactor = 8^2;
   var time;
   var velocity;
   var jointHandPos;
   var jointHandAcceleration;
    begin
      time:=220;
        velocity = originVel;
          repeat
            if(distance between jointHandPos & P6)
            <=(200*jointHandAcceleration/divisionFactor);
               then velocity = velocity + jointHandAcceleration;
            end if

            velocity := velocity * decreasingVelFactor;

            if value of velocity <=0.05 then velocity = 0;
            end if
                    P6 := P6 + velocity;
                    Q6 := P6;
            10 densest spots of distributed particles will be the
            reference points to construct a continuous nurbs.
                    N1 := {xN0, yN0, zN0};
          until time = maxTime06
end.
-------------------------------------------------------------------------
Output: set of 640 spatial points, Q6 = {x7, y7, z7}.
Output: set of 10 spatial points, N1 = {xN0, yN0, zN0}.
Output: a dynamic continuous nurbs constructed by N1.
Joints of jointHandPos and jointHandAcceleration referencing both user1
and user2's right and left hands; 4 in total.
Velocity as velocity of each particle starts its initial value remained
from Algorithm 5.
The unit associated with time is 1 second.
```

reactions depicting the desire to touch the light based surface of the nurbs, inserting one's head inside the light cone, or moving their bodies along with the curvatures of the nurbs (Fig. 7) were observed. After few minutes of this relatively contemplative mode, the "Attract Mode" (Algorithm 7) is introduced back to simultaneously operate with the "Nurbs Mode".

```
Algorithm 7: Nurbs Mode + Attract Mode. 265 sec. < T7 < 310 sec.
---------------------------------------------------------------------
Input: set of 640 spatial points, P7 = Q6 = {x7, y7, z7};
Q6 is the result from Algorithm 6.
Input: set of 10 spatial points, NR1 = N1 = {xN0, yN0, zN0};
N1 is the result from Algorithm 6.
  const num = 100;
  const maxVelocity = 300;
  const maxTime07 = 310;
  var time;
  var attVelocity;
  var joint1, joint2, joint3, joint4, joint5, join6;
  begin
   time:=265;
   repeat
       if ((numbers of particles' velocity > maxVelocity) > num)
          then
             particles will be attracted to follow the polyline as a
             path constructed by joint1, joint2, joint3, joint4, joint5,
             and joint6;
       end if
                   P7 := P7+ attVelocity;
                   Q7 := P7;

       10 densest spots of distributed particles will be the
       reference points to construct a continuous nurbs.
                   N2 := NR1;
     until time = maxTime07
end.
---------------------------------------------------------------------
Output: set of 640 spatial points, Q7 = {x8, y8, z8}.
Output: set of 10 spatial points, N2 = {xN1, yN1, zN1}.
Output: a relatively lively dynamic continuous nurbs constructed by N2.
The unit associated with time is 1 second.
```

2.7   Rain-Up Mode

Before the "Rain-up Mode" (Algorithm 8), the "Follow Mode" is exhibited again to gently inform the participants that the experiential installation is nearly towards the end. After a few minutes of "Follow Mode", the participants entirely lose control of the swarm agents and instead witness the agents flying upwards to the sky (velocity). All agents (P9) propel up with high velocity and gradually cease movement. In terms of color, all the agents start with magenta representing higher speed and gradually become dark blue corresponding to the velocity, which each agent embodies. Towards the end, all the agents lose their momentum, turn transparent and tend to fully disappear. The entire space returns back into an entirely dark state, awaiting the next group of participants to engage with it.

Algorithm 8: Rain-up Mode. 312 sec. < T9 < 322 sec.

```
-----------------------------------------------------------------------
Input: set of 640 spatial points, P9 = {x9, y9, z9};
Q8 is the result from Algorithm 8.
  const decreasingVelFactor = 0.98;
  const maxTime09 = 322;
  var time;
  var velocity;

  begin
    time:= 312;
    velocity:= 1200;
      repeat
        velocity := velocity * decreasingVelFactor;

        if value of velocity <=0.05 then velocity = 0;
        end if
                P9 := P9 + velocity;
                Q9 := P9;
      until time = maxTime09;
end.
-----------------------------------------------------------------------
Output: set of spatial 640 points, Q9 = {x10, y10, z10}.
The unit associated with time is 1 second.
```

3 Conclusion

Ambiguous Topology, a swarm based ambient landscape (Swarmscape) is an immersive real-time interactive spatial installation. The installation thrives on a synergistic merger between the domains of swarm intelligence, human psychology and volumetric projections, resulting in real-time generation of emergent computational and human interactions. The Swarmscape is theorized, experimented with and analyzed in real-life physical settings and has proven to be an apt experiment to physically materialize emergent geometric outputs driven by swarm intelligence. The experiment falls in the category of 2d and 3d virtual swarm system which is deployed as an innovative socio technical system for enhancing human machine interaction.

References

1. Bisig, D., Palacio, P.: Phantom limb - hybrid embodiments for dance. In: Proceedings of the 17th Generative Art Conference, Rome, Italy, pp. 92–107 (2014)
2. Unemi, T., Bisig, D.: Visual deformation by swarm - a technique for virtual liquidization of objects. In: Proceedings of the 17th Generative Art Conference, Rome, Italy, pp. 348–356 (2014)
3. Boyd, J.E., Hushlak, G., Jacob, C.J.: SwarmArt: interactive art from swarm intelligence. In: Proceedings of the 12th Annual ACM International Conference on Multimedia. ACM (2004)
4. Reynolds, C.W.: Steering behaviors for autonomous characters. In: Proceedings of Game Developers Conference, 1999 (San Jose, California), pp. 763–782. Miller Freeman Game Group, San Francisco, California (1999)
5. Reynolds, C.W.: Flocks, herds and schools: a distributed behavioral model. Comput. Graph. 21(4), 25–34 (1987). ACM SIGGRAPH
6. Reas, C., Fry, B.: Processing: A Programming Handbook for Visual Designers and Artists. The MIT Press, London (2014)

Fundamental Diagrams of Single-File Pedestrian Flow for Different Age Groups

Shuchao Cao[1,2], Jun Zhang[2(✉)], Daniel Salden[2], and Jian Ma[3]

[1] State Key Laboratory of Fire Science,
University of Science and Technology of China,
Hefei 230027, China
[2] Institute for Advanced Simulation,
Forschungszentrum Jülich GmbH,
52425 Jülich, Germany
ju.zhang@fz-juelich.de
[3] School of Transportation and Logistics,
Southwest Jiaotong University, Chengdu 610031, China

Abstract. In this paper properties of pedestrian movement are investigated with series of single-file experiments by considering the age composition of the crowd. Pedestrian trajectories with different age groups (young students group, old people group and mixed group) are extracted from the software *PeTrack*. It is found that the free velocity and maximum specific flow of young student group are the largest among three groups due to different mobility between young students and old people. More interestingly, the maximum specific flow of mixed group is smaller than that of old group, which indicates the jam occurs more easily in crowd composed of people with different movement abilities than that with homogeneous composition. At last, a nondimensional method considering pedestrian free velocity and body size is used to scale the fundamental diagrams for different age groups. The study is helpful to understand evolution of pedestrian dynamics with different ages.

Keywords: Pedestrian dynamics · Traffic and crowd · Single-file experiment · Fundamental diagram

1 Introduction

In recent decades, the safety of pedestrians is paid more and more attentions and the research on pedestrian dynamics has become a hot issue. It is important to understand the evolution of pedestrian dynamics, especially for the design of pedestrian facilities with mass crowds. However, the pedestrian system is very complex, since people with different body sizes, ages and educational backgrounds interact with each other in this system, which can cause large discrepancies in the fundamental diagrams. Meanwhile, some interesting phenomena such as self-organization [1], herding behavior [2], lane formation [3], stop-and-go [4] etc. are also found from pedestrian experiments and field observations.

© Springer International Publishing Switzerland 2016
Y. Tan et al. (Eds.): ICSI 2016, Part I, LNCS 9712, pp. 154–161, 2016.
DOI: 10.1007/978-3-319-41000-5_15

Up to now, the researches on pedestrian dynamics mainly focus on people with young or middle age. However, the aging population is growing so fast in recent decades. The relative weight of the population shifts from younger to elderly in the most of the region on the planet. Current aging rate in the developed countries varies from 12 % to 13 % and is expected to increase up to 21–37 % in 2050. The increase of aging rate will lead to mobility problems especially in large cities [5]. Aging population and more people with mobility impairments are bringing new challenges to the management of routine and people movement. Therefore, to understand the basic characteristics of pedestrian dynamics regarding to the situation with cohabitation of younger and elderly generation, the movement properties of pedestrian crowds with different age compositions should be studied carefully. Some experiments with different age groups have been done in the past. Kholshchevnikov et al. [6] analyzed the evacuation of children from pre-school educational institutions. A theoretical analysis was done for establishing the relationships between speed and density of flows of children of different age groups during their upward and downward movement on stairs, through door openings, along horizontal routes at a normal walking speed, and while running. Cueta and Gwynne [7] collected movement data of 4 to 16 years old children from evacuation drills, including pre-evacuation times, travel speeds, route use and evacuation arrival times. Kuligowski et al. [8] presented data on movement speeds of occupants with various types of mobility impairments evacuating from two residential facilities for older adults to better understand pedestrian behavior during evacuation with and without assistance in stairwell. Shimura et al. [5] conducted experiment with mixed younger and elder pedestrians. It was found that the emergent spatial formation was characterized by the initial formation, walking speed and overtaking behavior. The pedestrians with slower speed were considered as moving obstacles to others acting as a transient bottleneck during overtaking process.

However, the current study on pedestrian dynamics with old people is very little. In Ref. [5], the elder pedestrians are not the real old people, just some students intend to move slowly and pretend to be old people. Meanwhile, the current experiments were carried out under different conditions. In this case, it is difficult to compare the movement properties of pedestrians under various age ranges. Therefore, we conducted a series of single-file pedestrian flow experiments in this paper, which involves purely longitudinal interactions among pedestrians and avoids lateral ones. Actually, the single-file experiment has been carried out by some researchers. Seyfried et al. [9] investigated the relation between the velocity and the inverse of density through the single-file experiment. It was found there was an unexpected conformance between the fundamental diagrams in the single-file experiment and other experiments for movement in a plane. Fang and Song [10] found both the pedestrian step length and frequency decreased with the increasing global pedestrian density by the controlled single-file experiment. Furthermore, there was a linear relationship between the step frequency and the distance headway. Jelić et al. [11] conducted experiment inside a ring formed by inner and outer round walls to study the properties of pedestrians moving in line. The results showed that three distinct regimes existed between velocity and headway relation, and the transitions between these regimes occurred at spatial headway of about 1.1 m and 3 m respectively.

The rest of this paper is organized as follows. In Sect. 2, the details of experiment setup are described. In Sect. 3, the experiment results are discussed. The fundamental diagrams are compared among the three different groups, and a nondimensional method is used to unify the density-velocity and density-flow relations. In Sect. 4, we close the paper by summarizing our findings and discussing our future research area.

2 Experiment Setup

The experiment was carried out in 2015 in Tianshui Health School, China. Figure 1 shows the geometry of experiment scenario. The whole length of the corridor is 25.7 m, including two straight parts and two semicircles. The width of the corridor is 0.8 m and overtaking is not allowed in the experiment. There are 80 young students (average age is about 17) and 47 aged adults (average age is about 52) participate the experiments. The whole experiment includes mainly three parts: (1) the young student group experiment, (2) the old group experiment and (3) the mixed group experiment (the mixed ratio of young student and old people is 1:1). Different numbers of pedestrians are adopted to obtain data under different density level in each run which lasts about three minutes. The whole process of the experiment is recorded by the camera fixed on the fourth floor of teaching building. At last, pedestrian trajectories are extracted by using the software *PeTrack* [12, 13]. We display some trajectories with different number of participants in Fig. 2. It shows that the more pedestrians are in the corridor, the more lateral oscillations are observed in the trajectories. It becomes more difficult for pedestrians to move forward in the corridor under high densities, which causes lateral movement.

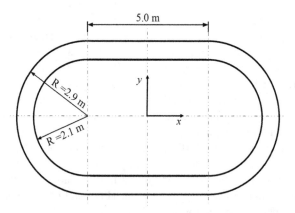

Fig. 1. The sketch of experiment scenario.

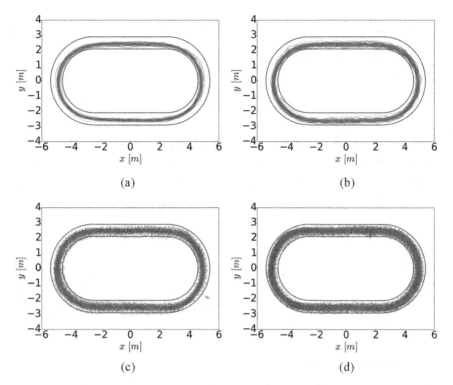

Fig. 2. Trajectories of pedestrians in different runs of experiment. (a) Number of young students $N_Y = 5$, (b) number of old people $N_O = 21$, (c) number of mixed people $N_M = 60$, (d) number of young students $N_Y = 66$.

3   Result and Analysis

For the simplification of analysis, we transfer the trajectory data from 2D coordinate system to 1D. Each pedestrian is regarded as a new one when he/she passes the position $(-2.5$ m, -2.5 m$)$. With such transformation, the maximum of one's coordinate is $x = 25.7$ m in the new 1D coordinate system.

3.1   Method Analysis

Microscopically, an individual density can be defined based on headway or Voronoi tessellation [14]. For any pedestrian i at time t, the one dimensional Voronoi distance $d_{v,i}(t)$ is the half distance between the centers of his follower and predecessor. Due to the high precision and low fluctuation of Voronoi method as described in Ref. [14], the individual density $\rho_i(t)$ is calculated based on the Voronoi distance in this paper:

$$\rho_i(t) = 1/d_{v,i}(t) \tag{1}$$

The individual instantaneous velocity $v_i(t)$ is calculated as:

$$v_i(t) = \frac{x_i(t + \Delta t/2) - x_i(t - \Delta t/2)}{\Delta t} \tag{2}$$

where $x_i(t)$ is the coordinate of pedestrian i in the 1D coordination system at time t. Δt is the time interval and $\Delta t = 0.4$ s is used in the paper.

Macroscopically, the average density $\rho(t)$ and velocity $v(t)$ in a measurement length l_m at time t are calculated based on Voronoi method [15]:

$$\rho(x, t) = \rho_i(t) \text{ and } v(x, t) = v_i(t) \quad \text{if } x \in l_m \tag{3}$$

$$\rho(t) = \frac{\int \rho(x, t)dx}{l_m} \tag{4}$$

$$v(t) = \frac{\int v(x, t)dx}{l_m} \tag{5}$$

3.2 Fundamental Diagram

The fundamental diagram [11, 14, 15], the relationship between density and velocity or between density and flow, is a basic relation in pedestrian dynamics. In this section, we investigate the fundamental diagrams of single-file flow for different age groups.

As shown in Fig. 3(a), pedestrian velocity decreases with the increase of density for all three groups. Under the low density situation for $\rho < 0.5$ m^{-1}, pedestrians move with their preferred velocity. The free velocity is about 1.23 ± 0.22 m/s for young students group, 0.95 ± 0.16 m/s for old group and 1.05 ± 0.16 m/s for mixed group respectively. Obviously, the free velocity of young students is the highest and the old group has the smallest free velocity. However, in the medium density for density ρ between 0.5 m^{-1} and 1.2 m^{-1}, the velocity of mixed group is higher than that of old group, which indicates that inhomogeneous crowd with different age components makes pedestrian movement more complex and difficult compared to homogenous crowd. At high densities, congestion occurs and pedestrian velocity is nearly 0. The crowd goes into stagnate phase. There is no enough space for people to move forward and they need to stop and wait.

The density-flow relation is displayed in Fig. 3(b). It can be divided into three stages. For $\rho < 0.9$ m^{-1}, pedestrian specific flow increases with the increment of pedestrian density and the flow reaches to the highest value (1.3 s^{-1}, 0.9 s^{-1} and 0.7 s^{-1} for young, old and mixed group respectively) around $\rho = 0.9$ m^{-1} for all groups. Due to the different mobility of young student and old people, the maximum specific flow of young student group is the largest among three groups. It is also found that the maximum specific flow of the mixed group is smaller than that of the old group, which demonstrates again that the movement of pedestrians with distinct age components is more complicated than that with similar age. When the density ρ is higher than 0.9 m^{-1}, pedestrian jam appears and the flow starts to decrease. However, the decline rates of the flow with the increasing density become different from 1.6 m^{-1} for the young student group and from 1.4 m^{-1} for the mixed group. When the density is higher than 1.6 m^{-1} (1.4 m^{-1}), stop-and-go waves occur frequently and dominate pedestrian motion in the young student group (old group). Meanwhile, the decline rate becomes smaller under this condition.

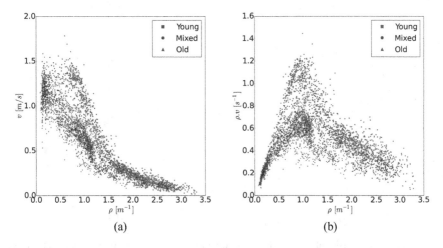

(a)   (b)

Fig. 3. The fundamental diagrams obtained from Voronoi methods. The red, blue and green points stand young student, old and mixed group respectively. (a) Density-velocity relation. (b) Density-flow relation. (Color figure online)

From Fig. 3, we can see that the fundamental diagrams for different age groups are not comparable. In this case, we try to use a nondimensional method [15] to scale the fundamental diagram in the following analysis. We use v / v_0 and ρ. L_0 to replace the original v and ρ in the fundamental diagram, where v_0 represents the free velocity of pedestrian in the experiment and is set as $v_0 = 1.23$ m/s, 0.95 m/s and 1.05 m/s for young student group, old group and mixed group respectively. L_0 stands the body size of pedestrian. Before experiment, the chest circumferences of all the test persons are measured. For simplicity in the calculation and analysis, we assume that the shape of pedestrian body's projection on the ground is a circle and we use the diameter of the circle as L_0. Based on such assumption, we get the average body size $L_0 = 0.30$ m and $L_0 = 0.34$ m for young students and old people. For the mixed group, we use the mean value $L_0 = 0.32$ m of young student and old people (the mixed ratio is 1:1). The nondimensionalization result is shown in Fig. 4, the difference of these three fundamental diagrams seems smaller in both density-velocity and density-flow relations. They agree well with each other when ρ. $L_0 < 0.15$ (absolute free flow state) and ρ. $L_0 > 0.45$ (congested states with stop-and-go waves). However, obvious differences can still be observed under medium density situations especially for the mixed group. It seems that the movement properties of the mixed groups are quite complex and are not able to be obtained by simply scaling only with free velocity and body dimensions. Regarding to the differences of the movement characteristics between young student and old people, more empirical data of old group are needed. In a word, this nondimensional method can reduce the difference of the fundamental diagrams for different age groups. However, some distinct is still observed and more investigation is needed to validate this method in the future.

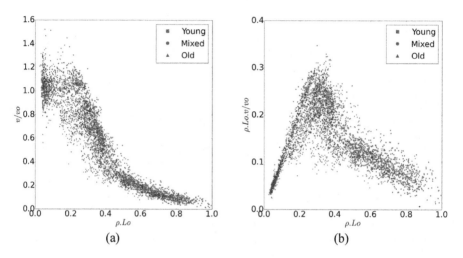

Fig. 4. The nondimensionalized fundamental diagram by using the free velocity (1.23 m/s, 0.95 m/s and 1.05 m/s for young students, old group and mixed group respectively) and body size (0.30 m, 0.34 m and 0.32 m for young students, old group and mixed group respectively). (a) Density-velocity relation. (b) Density-flow relation. (Color figure online)

4   Conclusion

In this paper, series of single-file experiments are conducted to investigate the effect of different age groups on pedestrian dynamics. Pedestrian trajectories are extracted through video tracking. From the analysis of fundamental diagrams for different age groups, we find that when the density is very low, pedestrian can move with their preferred velocity. With the increase of pedestrian density, the velocity decreases monotonously, however, the specific flow increases firstly, then reaches to the highest value around 0.9 m^{-1}, and reduces to nearly 0 at large densities. Among these three groups, the free velocity and maximum specific flow of young student group are the largest. Furthermore, the velocity and specific flow of mixed group are smaller than those of old group in the medium density, which indicates the congestion occurs more easily in the crowd with different movement motilities than that in the homogeneous crowd. To reduce the difference of fundamental diagrams for various age groups, a nondimensional method considering pedestrian free velocity and body size is adopted. The scaling result declares that they agree well with each other when $\rho.L_0 < 0.15$

(absolute free flow state) and ρ. $L_0 > 0.45$ (congested states with stop-and-go waves). But the difference among the three groups is still large in the medium density. Therefore, we should be careful to use this method and more experiment data are also needed to validate it in the future. We believe this study can help to understand the evolution of pedestrian dynamics with different age groups, especially under the aging population background.

Acknowledgments. The foundation supports from the State Key Laboratory of Fire Science in University of Science and Technology of China (HZ2015-KF11).

References

1. Helbing, D., Molnar, P., Farkas, I.J., Bolay, K.: Self-organizing pedestrian movement. Environ. Plann. B **28**, 361–384 (2001)
2. Burger, M., Markowich, P., Pietschmann, J.F.: Continuous limit of a crowd motion and herding model: analysis and numerical simulations. Kinet. Relat. Models **4**, 1025–1047 (2011)
3. Song, W.G., Lv, W., Fang, Z.M.: Experiment and modeling of microscopic movement characteristic of pedestrians. Procedia Eng. **62**, 56–70 (2013)
4. Portz, A., Seyfried, A.: Analyzing stop-and-go waves by experiment and modeling. In: Peacock, R.D., Kuligowski, E.D., Averill, J.D. (eds.) Pedestrian and Evacuation Dynamics, pp. 577–586. Springer, New York (2011)
5. Shimura, K., Ohtsuka, K., Vizzari, G., Nishinari, K., Bandini, S.: Mobility analysis of the aged pedestrians by experiment and simulation. Pattern Recogn. Lett. **44**, 58–63 (2014)
6. Kholshchevnikov, V.V., Samoshin, D.A., Parfyonenko, A.P., Belosokhov, I.P.: Study of children evacuation from pre-school education institutions. Fire Mater. **36**, 349–366 (2012)
7. Cuesta, A., Gwynne, S.: The collection and compilation of school evacuation data for model use. Saf. Sci. **84**, 24–36 (2016)
8. Kuligowski, E., Peacock, R., Wiess, E., Hoskins, B.: Stair evacuation of people with mobility impairments. Fire Mater. **39**, 371–384 (2015)
9. Seyfried, A., Steffen, B., Klingsch, W., Boltes, M.: The fundamental diagram of pedestrian movement revisited. J. Stat. Mech.: Theory Exp. **10**, P10002 (2005)
10. Fang, Z.M., Song, W.G.: A continuous distance model (CDM) for the single-file pedestrian movement considering step frequency and length. Phys. A **391**, 307–316 (2012)
11. Jelić, A., Appert, R.C., Lemercier, S., Pettré, J.: Properties of pedestrians walking in line: Fundamental diagrams. Phys. Rev. E **85**, 036111 (2012)
12. Boltes, M., Seyfried, A., Steffen, B., Schadschneider, A.: Automatic extraction of pedestrian trajectories from video recordings. In: Klingsch, W.W.F., Rogsch, C., Schadschneider, A., Schreckenberg, M. (eds.) Pedestrian and Evacuation Dynamics 2008, pp. 43–54. Springer, Heidelberg (2010)
13. Boltes, M., Seyfried, A.: Collecting pedestrian trajectories. Neurocomputing **100**, 127–133 (2013)
14. Steffen, B., Seyfried, A.: Methods for measuring pedestrian density, flow, speed and direction with minimal scatter. Phys. A **389**, 1902–1910 (2010)
15. Zhang, J., Mehner, W., Holl, S., Boltes, M., Andresen, E., Schadschneider, A., Seyfried, A.: Universal flow-density relation of single-file bicycle, pedestrian and car motion. Phys. Lett. A **378**, 3274–3277 (2014)

Some Swarm Intelligence Algorithms and Their Applications

A Discrete Monarch Butterfly Optimization for Chinese TSP Problem

Gai-Ge Wang[1,2(✉)], Guo-Sheng Hao[1], Shi Cheng[3], and Quande Qin[4]

[1] School of Computer Science and Technology,
Jiangsu Normal University, Xuzhou, Jiangsu, China
gaigewang@163.com, gaigewang@gmail.com,
guoshenghaoxz@tom.com

[2] Department of Electrical and Computer Engineering, University of Alberta,
9107-116 Street, Edmonton, AB T6G 2V4, Canada

[3] School of Computer Science,
Shaanxi Normal University, Xi'an, China
cheng@snnu.edu.cn

[4] Department of Management Science,
Shenzhen University, Shenzhen, China

Abstract. Recently, Wang *et al.* proposed a new kind of metaheuristic algorithm, called Monarch Butterfly Optimization (MBO), for global continuous optimization tasks. It has experimentally proven that it has better performance than some other heuristic search strategies. On the other hand, travelling salesman problem (TSP) is one of the most representative NP-hard problems that are hard to be solved by traditional methods. It has been widely studied and solved by several metaheuristic algorithms. In this paper, MBO is discretized, and then a discrete MBO (DMBO), and firstly used to solve Chinese TSP (CTSP). In the basic MBO, Wang et al. had made little effort to fine-tune the parameters. In our present work, the parametric study for one of the most parameter, butterfly adjusting rate (BAR), is also provided. The best-selected BAR is inserted into the DMBO method and then solve CTSP problem. By comparing with three other algorithms, experimental results presented clearly demonstrates DMBO as an attractive addition to the portfolio of swarm intelligence techniques.

Keywords: Travelling salesman problem · Monarch butterfly optimization · Butterfly adjusting rate · Discrete optimization

1 Introduction

As one of the most representative NP-hard problems, TSP has attracted wide attention from various fields since it is firstly formulated in 1930s. In general, TSP is not just a problem, it is a class of problem, because many different real-world problems can be modelled into TSP problem. Accordingly, these problems can be solved by the methods that can solve TSP problem. However, as an NP-hard problem, TSP is difficult to be dealt with by traditional methods within given condition. Modern metaheuristic algorithms can solve the TSP problem well.

© Springer International Publishing Switzerland 2016
Y. Tan et al. (Eds.): ICSI 2016, Part I, LNCS 9712, pp. 165–173, 2016.
DOI: 10.1007/978-3-319-41000-5_16

By idealizing the rule of the biological, physical and chemical phenomenon in nature, a huge number of metaheuristic algorithms has been put forward and successfully utilized to address TSP, which are differential evolution (DE) [1–3], cuckoo search (CS) [4–9], particle swarm optimization (PSO) [10–14], biogeography-based optimization (BBO) [15–17], harmony search (HS) [18–21], gravitational search algorithm (GSA) [22, 23], fireworks algorithm (FWA) [24], earthworm optimization algorithm (EWA) [25], elephant herding optimization (EHO) [26, 27], water wave optimization [28], ant lion optimizer (ALO) [29], multi-verse optimizer (MVO) [30], firefly algorithm (FA) [31, 32], ant colony optimization (ACO) [33], bat algorithm (BA) [34–37], grey wolf optimizer (GWO) [38], and krill herd (KH) [39–41].

Inspired by the migration behavior of monarch butterflies in North American, a new kind of metaheuristic algorithm, called Monarch Butterfly Optimization (MBO), is proposed by Wang et al. for global continuous optimization problem [42–44]. By comparing with five state-of-the-art metaheuristic algorithms, it has experimentally proven that MBO has better performance on most benchmark cases. In our current work, a new version of MBO method, called discrete monarch butterfly optimization (DMBO), is firstly proposed by discretizing the basic MBO method. Subsequently, DMBO is then used to solve Chinese TSP (CTSP). In addition, differentiating with the basic MBO method, one of the most parameter, butterfly adjusting rate (BAR), is fully investigated via an array of experiments. The best-selected BAR is used to guide DMBO method with the aim of searching for the shortest path for CTSP. By comparing with three other metaheuristic algorithms, experimental results presented clearly show that DMBO is well capable of finding the shortest paths among four methods in most cases.

The organization of this paper is outlined here. Sections 2 and 3 provide the Chinese TSP and the basic (continuous) MBO, respectively. Subsequently, discrete MBO method is presented in Sect. 4, and is followed by several simulation results, comparing DMBO with other optimization methods for CTSP as presented in Sect. 5. Additionally, the parametric study for butterfly adjusting rate (BAR), is provided in this section. Section 6 concludes our work.

2   Chinese TSP

As a typical NP-complete problem, the travelling salesman problem (TSP) can be simply defined as: Given a list of cities and the distances between each pair of cities, what is the shortest possible route that visits each city exactly once and returns to the origin city? Usually, the distance between each of n cities has been given. A traveling salesman starts to visit each city once and only once from a city, and finally returns to the starting city. How to arrange this traveling is to make the shortest route. In short, TSP is to find a shortest path traversal among n cities, or search for a permutation $\pi(X) = \{V_1, V_2, \cdots, V_n\}$ in a natural subset $X = \{1, 2, \cdots, n\}$ (the elements of X represent the n city's number). That is, we minimize the total distance in Eq. (1).

$$T_d = \sum_{i=1}^{n-1} d(V_i, V_{i+1}) + d(V_n, V_1) \tag{1}$$

where $d(V_i, V_{i+1})$ is the distance between city V_i and city V_{i+1}.

In the present work, MBO is applied to solve the Chinese TSP (CTSP) problem. There are 31 main cities in China, and the distance between those has been given.

3 Monarch Butterfly Optimization

By idealizing the migration behaviour of monarch butterflies, Wang *et al.* proposed a new swarm intelligence algorithm, called MBO [42].

In MBO, the number of monarch butterflies in Land 1 and Land 2 is NP_1 and NP_2. NP is population size; p is the ratio of monarch butterflies in Land 1. Subpopulation 1 and Subpopulation 2 are composed of monarch butterflies in Land 1 and Land 2, respectively. Accordingly, migration operator can be given as

$$x_{i,k}^{t+1} = x_{r_1,k}^t \tag{2}$$

where $x_{i,k}^{t+1}$ indicates the kth element of x_i at generation $t + 1$. Similarly, $x_{r_1,k}^t$ indicates the kth element of x_{r_1}. t is the current generation number. Butterfly r_1 is randomly selected from Subpopulation 1. When $r \le p$, $x_{i,k}^{t+1}$ is generated by Eq. (2). On the contrast, if $r > p$, $x_{i,k}^{t+1}$ is updated as follows:

$$x_{i,k}^{t+1} = x_{r_2,k}^t \tag{3}$$

where $x_{r_2,k}^t$ indicates the kth element of x_{r_2} [42].

For all the elements in butterfly j, if $rand \le p$, it can be updated as

$$x_{j,k}^{t+1} = x_{best,k}^t \tag{4}$$

where $x_{j,k}^{t+1}$ indicates the kth element of x_j at generation $t + 1$. Similarly, $x_{best,k}^t$ indicates the kth element of x_{best} with the fittest solution in butterfly population. On the contrast, if $rand > p$, it can be updated as

$$x_{j,k}^{t+1} = x_{r_3,k}^t \tag{5}$$

where $x_{r_3,k}^t$ indicates the kth element of x_{r_3}. More MBO can be found in [42].

4 Discrete Monarch Butterfly Optimization

The basic MBO is only used to solve the continuous optimization problems, therefore it cannot be used to address discrete optimization problem directly. In our current work, the basic MBO method is discretized, and a discrete version of MBO method is then proposed. So, TSP can be solved by discrete MBO (see Algorithm 1).

According to Algorithm 1, firstly, all the parameters are initialized, and then the initial population are generated. Subsequently, the positions of all butterflies are updated step by step until certain conditions are satisfied. After updating all the butterflies, the whole population is discretized in many different ways. In our current work, *round* function is used to discretize the whole population.

Algorithm 1 Discrete MBO

Begin

  Step 1: Initialization. Initializing all the parameters is the same with MBO.

  Step 2: While $t<MaxGen$ **do**

      Divide butterflies into two subpopulations;

      for $i=$ 1 to NP_1 **do**

         Generate new Subpopulation 1 by Section 3.1.

      end for i

      for $j=$ 1 to NP_2 **do**

         Generate new Subpopulation 2 by Section 3.2.

      end for j

      Combine the two new subpopulations;

      Discretize the whole population;

      $t=t+1$.

  Step 3: end while

End.

---

5 Simulation Results

In this section, CTSP is dealt with by SCS method. In order to obtain fair results, all the implementations are conducted under the same conditions as shown in [45]. The parameters used in MBO, ACO, BBO and DE are set the same as [42, 46].

Here, seventy independent implementations are carried out with aim of get the representative statistical results. In the following experiments, the shortest path is highlighted in bold font. The path is found by using the following parameters: population size = 100, maximum generation = 100, butterfly adjusting rate = 5/12. In the following, the parametric study of butterfly adjusting rate will be given.

TSP can be solved by DMBO method. DMBO will find the shortest path among these cities (see Figs. 1 and 2). Firstly, the initial paths are randomly generated as follows:

19— > 31— > 24— > 29— > 23— > 12— > 30— > 8— > 26— > 10— > 25
— > 13— > 15— > 1— > 22— > 5— > 17— > 20— > 2— > 18— > 14— > 16
— > 9— > 3— > 4— > 11— > 27— > 28— > 21— > 7— > 6— > 19

Its initial total distance is 42649 km (see Fig. 1).

DMBO has the ability of guiding the butterflies find much shorter path. The final path is shown as follows:

15— > 14— > 29— > 1— > 31— > 30— > 27— > 28— > 26— > 21— > 22
— > 3— > 18— > 24— > 5— > 2— > 8— > 10— > 9— > 16— > 23— > 11
— > 25— > 20— > 19— > 17— > 4— > 7— > 6— > 13— > 12— > 15

Its distance is 18762 km that is very approaching to the known shortest distance 15378 km (see Fig. 2).

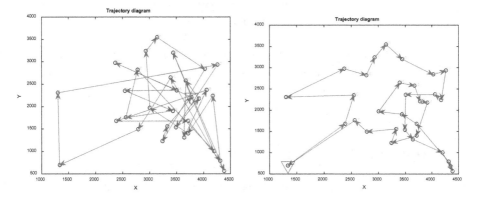

Fig. 1. Randomly generated path.   **Fig. 2.** Final trajectory obtained by DMBO.

In this section, DMBO is compared with that of the other methods on CTSP. In order to get a fair comparison, all the methods are implemented in the same limited conditions. The other parameters are similar to those described in the above Section 70 trials are conducted in order to get more accurate statistical results (see Table 1).

From Table 1, for $NP = MaxGen = 50$, DMBO performs the best for the worst paths, while ACO and BBO performs the best on mean, worst, Std (standard deviation) and best paths. For $NP = MaxGen = 100$, DMBO has the best and shortest paths, while ACO has the best mean, worst, Std paths. With the increment of population size and maximum generations, when $NP = MaxGen = 150$ and $NP = MaxGen = 200$, DMBO can find the shortest paths on best, mean and worst paths, while ACO still has the smallest Std. Table 1 indicates that DMBO can find much shorter paths with the increment of population size and maximum generations, and it can successfully find the best solution on most cases. On contrast, other methods have little improvements with the increment of population size and maximum generations. However, ACO has smaller Std than DMBO on all the cases. How to reduce the Std of DMBO is worthy of further study.

Table 1. Optimization results obtained by four methods with different population size and maximum generations.

		ACO	BBO	DE	DMBO
50	Best	20568	**18873**	27189	19811
	Mean	**22017**	23894	29910	23123
	Worst	**22308**	22992	28729	**22308**
	Std	**219**	1674	1102	1927
100	Best	20740	19486	25798	**18425**
	Mean	**21399**	22170	28321	21478
	Worst	**21581**	23497	27729	22751
	Std	**215**	1366	786	1673
150	Best	20416	18886	24644	**17643**
	Mean	21067	21474	27036	**20557**
	Worst	**21562**	**21562**	28372	**21562**
	Std	**269**	1344	964	1509
200	Best	20302	18279	23934	**17133**
	Mean	20799	20821	26418	**19669**
	Worst	**21533**	**21533**	26424	**21533**
	Std	**280**	1293	783	1305

6 Discussion and Conclusions

As one of the most representative NP-hard problem, travelling salesman problem (TSP) is scarcely possible to be addressed by traditional methods when nodes are big enough. Modern metaheuristic algorithms are used to solve TSP. With the aim of addressing TSP more efficiently and effectively, in our present work, the basic MBO method is discretized and then a discrete version of MBO method is proposed. After that, DMBO is subsequently used to solve Chinese TSP (CTSP). In addition, in order to make DMBO method search for much shorter paths, the influence of butterfly adjusting rate (BAR) on DMBO method is fully investigated. Through this intensive parametric study with different population and maximum generations, the best BAR is selected. The best-selected BAR is used to guide DMBO method search. By comparing with three other metaheuristic algorithms, experimental results presented clearly shows that DMBO with best BAR is well capable of finding the shortest paths among four methods in most cases.

Acknowledgements. This work was supported by Jiangsu Province Science Foundation for Youths (No. BK20150239), National Natural Science Foundation of China (No. 61503165) and Jiangsu Provincial Natural Science Foundation (No. BK20131130).

References

1. Storn, R., Price, K.: Differential evolution-a simple and efficient heuristic for global optimization over continuous spaces. J. Glob. Optim. **11**, 341–359 (1997)
2. Wang, G.-G., Gandomi, A.H., Alavi, A.H., Hao, G.-S.: Hybrid krill herd algorithm with differential evolution for global numerical optimization. Neural Comput. Appl. **25**, 297–308 (2014)
3. Teoh, B.E., Ponnambalam, S.G., Kanagaraj, G.: Differential evolution algorithm with local search for capacitated vehicle routing problem. Int. J. Bio-Inspired Comput. **7**, 321–342 (2015)
4. Yang, X.S., Deb, S.: Cuckoo search via Lévy flights. In: Abraham, A., Carvalho, A., Herrera, F., Pai, V. (eds.) Proceeding of World Congress on Nature & Biologically Inspired Computing (NaBIC 2009), pp. 210–214. IEEE Publications, Coimbatore (2009)
5. Li, X., Yin, M.: Modified cuckoo search algorithm with self adaptive parameter method. Inf. Sci. **298**, 80–97 (2015)
6. Wang, G.-G., Deb, S., Gandomi, A.H., Zhang, Z., Alavi, A.H.: Chaotic cuckoo search. Soft Comput. (2015)
7. Wang, G.-G., Gandomi, A.H., Yang, X.-S., Alavi, A.H.: A new hybrid method based on krill herd and cuckoo search for global optimization tasks. Int. J. Bio-Inspired Comput. (2012)
8. Wang, G.-G., Gandomi, A.H., Zhao, X., Chu, H.E.: Hybridizing harmony search algorithm with cuckoo search for global numerical optimization. Soft. Comput. **20**, 273–285 (2016)
9. Wang, G., Guo, L., Duan, H., Liu, L., Wang, H., Wang, J.: A hybrid meta-heuristic DE/CS algorithm for UCAV path planning. J. Inf. Comput. Sci. **9**, 4811–4818 (2012)
10. Kennedy, J., Eberhart, R.: Particle swarm optimization. In: Proceeding of the IEEE International Conference on Neural Networks, vol. 4, pp. 1942–1948. IEEE, Perth, Australia (1995)
11. Shieh, H.-L., Kuo, C.-C., Chiang, C.-M.: Modified particle swarm optimization algorithm with simulated annealing behavior and its numerical verification. Appl. Math. Comput. **218**, 4365–4383 (2011)
12. Mirjalili, S., Lewis, A.: S-shaped versus V-shaped transfer functions for binary Particle Swarm Optimization. Swarm Evol. Comput. **9**, 1–14 (2013)
13. Wang, G.-G., Gandomi, A.H., Yang, X.-S., Alavi, A.H.: A novel improved accelerated particle swarm optimization algorithm for global numerical optimization. Eng. Comput. **31**, 1198–1220 (2014)
14. Grillo, H., Peidro, D., Alemany, M., Mula, J.: Application of particle swarm optimisation with backward calculation to solve a fuzzy multi–objective supply chain master planning model. Int. J. Bio-Inspired Comput. **7**, 157–169 (2015)
15. Simon, D.: Biogeography-based optimization. IEEE Trans. Evol. Comput. **12**, 702–713 (2008)
16. Zheng, Y.-J., Ling, H.-F., Xue, J.-Y.: Ecogeography-based optimization: enhancing biogeography-based optimization with ecogeographic barriers and differentiations. Comput. Oper. Res. **50**, 115–127 (2014)
17. Duan, H., Zhao, W., Wang, G., Feng, X.: Test-sheet composition using analytic hierarchy process and hybrid metaheuristic algorithm TS/BBO. Math. Probl. Eng. **2012**, 1–22 (2012)
18. Geem, Z.W., Kim, J.H., Loganathan, G.V.: A new heuristic optimization algorithm: harmony search. Simulation **76**, 60–68 (2001)

19. Gholizadeh, S., Barzegar, A.: Shape optimization of structures for frequency constraints by sequential harmony search algorithm. Eng. Optimization **45**(6), 1–20 (2012)
20. Wang, G., Guo, L., Duan, H., Wang, H., Liu, L., Shao, M.: Hybridizing harmony search with biogeography based optimization for global numerical optimization. J. Comput. Theor. Nanosci. **10**, 2318–2328 (2013)
21. Pandi, M., Premalatha, K.: Clustering microarray gene expression data using enhanced harmony search. Int. J. Bio-Inspired Comput. **7**, 296–306 (2015)
22. Rashedi, E., Nezamabadi-pour, H., Saryazdi, S.: GSA: a gravitational search algorithm. Inf. Sci. **179**, 2232–2248 (2009)
23. Yin, M., Hu, Y., Yang, F., Li, X., Gu, W.: A novel hybrid K-harmonic means and gravitational search algorithm approach for clustering. Expert Syst. Appl. **38**, 9319–9324 (2011)
24. Tan, Y., Zhu, Y.: Fireworks algorithm for optimization. In: Tan, Y., Shi, Y., Tan, K.C. (eds.) ICSI 2010, Part I. LNCS, vol. 6145, pp. 355–364. Springer, Heidelberg (2010)
25. Wang, G.-G., Deb, S., Coelho, L.D.S.: Earthworm optimization algorithm: a bio-inspired metaheuristic algorithm for global optimization problems. Int. J. Bio-Inspired Comput. (2015)
26. Wang, G.-G., Deb, S., Coelho, L.D.S.: Elephant herding optimization. In: 2015 3rd International Symposium on Computational and Business Intelligence (ISCBI 2015), pp. 1–5. IEEE, Bali, Indonesia (2015)
27. Wang, G.-G., Deb, S., Gao, X.-Z., Coelho, L.D.S.: A new metaheuristic optimization algorithm motivated by elephant herding behavior. Int. J. Bio-Inspired Comput. (2016)
28. Zheng, Y.-J.: Water wave optimization: a new nature-inspired metaheuristic. Comput. Oper. Res. **55**, 1–11 (2015)
29. Mirjalili, S.: The ant lion optimizer. Adv. Eng. Softw. **83**, 80–98 (2015)
30. Mirjalili, S., Mirjalili, S.M., Hatamlou, A.: Multi-verse optimizer: a nature-inspired algorithm for global optimization. Neural Comput. Appl. **27**, 495–513 (2016)
31. Yang, X.S.: Firefly algorithm, stochastic test functions and design optimisation. Int. J. Bio-Inspired Comput. **2**, 78–84 (2010)
32. Guo, L., Wang, G.-G., Wang, H., Wang, D.: An effective hybrid firefly algorithm with harmony search for global numerical optimization. Sci. World J. **2013**, 1–10 (2013)
33. Dorigo, M., Maniezzo, V., Colorni, A.: Ant system: optimization by a colony of cooperating agents. IEEE Trans. Syst. Man Cybern. B Cybern. **26**, 29–41 (1996)
34. Yang, X.S.: Nature-inspired Metaheuristic Algorithms. Luniver Press, Frome (2010)
35. Wang, G., Guo, L.: A novel hybrid bat algorithm with harmony search for global numerical optimization. J. Appl. Math. **2013**, 1–21 (2013)
36. Wang, G.-G., Chu, H.E., Mirjalili, S.: Three-dimensional path planning for UCAV using an improved bat algorithm. Aerosp. Sci. Technol. **49**, 231–238 (2016)
37. Xue, F., Cai, Y., Cao, Y., Cui, Z., Li, F.: Optimal parameter settings for bat algorithm. Int. J. Bio-Inspired Comput. **7**, 125–128 (2015)
38. Mirjalili, S., Mirjalili, S.M., Lewis, A.: Grey wolf optimizer. Adv. Eng. Softw. **69**, 46–61 (2014)
39. Gandomi, A.H., Alavi, A.H.: Krill herd: a new bio-inspired optimization algorithm. Commun. Nonlinear Sci. Numer. Simul. **17**, 4831–4845 (2012)
40. Wang, G.-G., Gandomi, A.H., Alavi, A.H.: Stud krill herd algorithm. Neurocomputing **128**, 363–370 (2014)
41. Wang, G.-G., Gandomi, A.H., Alavi, A.H.: An effective krill herd algorithm with migration operator in biogeography-based optimization. Appl. Math. Model. **38**, 2454–2462 (2014)

42. Wang, G.-G., Deb, S., Cui, Z.: Monarch butterfly optimization. Neural Comput. Appl. (2015)
43. Wang, G.-G., Zhao, X., Deb, S.: A novel monarch butterfly optimization with greedy strategy and self-adaptive crossover operator. In: 2015 2nd International. Conference on Soft Computing & Machine Intelligence (ISCMI 2015), pp. 45–50. IEEE, Hong Kong (2015)
44. Feng, Y., Wang, G.-G., Deb, S., Lu, M., Zhao, X.: Solving 0-1 knapsack problem by a novel binary monarch butterfly optimization. Neural Comput. Appl. (2015)
45. Wang, G., Guo, L., Wang, H., Duan, H., Liu, L., Li, J.: Incorporating mutation scheme into krill herd algorithm for global numerical optimization. Neural Comput. Appl. 24, 853–871 (2014)
46. Wang, G.-G., Guo, L., Gandomi, A.H., Hao, G.-S., Wang, H.: Chaotic krill herd algorithm. Inf. Sci. 274, 17–34 (2014)

Truss Structure Optimization Using Co-variance Based Artificial Bee Colony Algorithm

Shashank Gupta[(✉)], Divya Kumar, and K.K. Mishra

Computer Science and Engineering Department,
Motilal Nehru National Institute of Technology Allahabad, Allahabad, India
shanku.cs@gmail.com, {divyak,kkm}@mnnit.ac.in

Abstract. To minimize the weight or volume of the truss structures while satisfying all design constraints, is one of the main goal of truss optimization problems. In this paper we have proposed a novel co-variance based Artificial Bee Colony (ABC) algorithm for this structural optimization task. To gauge the effectiveness of our proposed algorithm we have tested it on well known bench-marked structural optimization problems. We have also compared the performance of our algorithm with some of the other well established structural optimization algorithms. The result shows that the proposed algorithm has outperformed other algorithms in evolving of optimal design of space structures. The present approach not only achieves the high quality desired optimal solutions but also takes fewer iterations to produce these solutions.

Keywords: Structural optimization problem · Truss design · Swarm intelligence · Artificial Bee Colony (ABC) · Co-variance

1 Introduction

Structural design optimization is one of the most challenging task in the field of structural or civil engineering and truss weight minimization is one of the most crucial sub-problem of this field. This problem has a great significance since minimization of the weight leads to minimization of required steel material which therefore bring down the time and cost budgeting of development of truss structures. Many mathematical models have been proposed to solve this optimization problem such as linear programming, non linear programming, gradient based methods, dynamic programming and others [15,25,43]. All the suggested approaches have some limitations because finding gradient in higher dimensions is in-feasible. Secondly truss optimization is a proven NP-Complete problem [34]. A lot of meta-heuristics have been proposed for solving this discrete optimization task in the last decade [33,35] as the growing power of meta-heuristic computation has attracted the attention of researchers at a large.

Evolutionary Algorithms (EA) are the powerful stochastic search procedures that have been tremendously applied for optimization tasks [3,28,29]. Evolutionary Algorithm are the population based search techniques which are inspired by

© Springer International Publishing Switzerland 2016
Y. Tan et al. (Eds.): ICSI 2016, Part I, LNCS 9712, pp. 174–183, 2016.
DOI: 10.1007/978-3-319-41000-5_17

the natural biological evolution of the species and Darwin's theory of natural selection [6]. The four well known classes of EA's are: (i) Genetic Algorithms [12], (ii) Genetic Programming [27], (iii) Evolutionary Strategies [39] and (iv) Evolutionary Programming [9]. Evolutionary computation offers multiple advantages to the researches which includes their simplicity, flexibility and robustness [1]. Thus EA's can be very well deployed for NP-Complete problems which do not have concrete heuristics. Swarm Intelligence (SI) [24] is the another class of nature inspired algorithms. These algorithm are based on the allied behavior of natural swarms which work in groups. These swarms acts and achieve well in coordination with the whole of their team without any centralized control [8]. Some of the well known SI based algorithms are: (i) Artificial Bee Colony (ABC) [19] (ii) Particle Swarm Optimization (PSO) [23] (iii) Ant Colony Optimization (ACO) [7].

EAs and SIs have been well applied to solve difficult optimization problems in varied fields of engineering, medical or finance. EAs and SIs have also been predominately used for the optimization of truss structures in the literature. Some of the notable works on truss optimization using Genetic Algorithms are described in [5,11,26,37,38]. PSO is also used for truss optimization in [13,22,35]. ACO based approaches are used in [4,21] and authors in [14,40,41] have used ABC algorithm for structural optimization.

In this research article we have proposed a novel co-variance [36] based ABC algorithm and have deployed this algorithm for truss optimization. The rest of the paper is organized as follows: In the next section we have detailed the truss optimization problem then we have described our proposed co-variance based ABC algorithm. Afterwords the effectiveness of our algorithm is detailed and ascertained on two $3D$ bench-mark problems of 25 bars and 72 bars each. The paper culminates with conclusions in the end.

2   Problem Description

In real world the of each structural problem (construction problem) has one main objective which is to minimize the weight of structure without violating the design constraints [2,16]. Trusses are most famous and accepted structural design procedure which is used to design pillars, bridges, roofs, towers, air-crafts, and ships etc. Trusses are defined as set of joints(nodes) and members(edges), the relation between joints and members is expressed as:

$$m \geq 2j - r \tag{1}$$

where m, j, and r are respectively, number of members, number of joints, and number of reactions. All joints should be in equilibrium condition(Sum of all the forces and torques should be zero).

$$\sum F^i = 0 \;\Rightarrow\; \sum F_x^i = 0, \sum F_y^i = 0, \sum F_z^i = 0 \tag{2}$$

$$\sum M^i = 0 \;\Rightarrow\; \sum M_{xy}^i = 0, \sum M_{yz}^i = 0, \sum M_{zx}^i = 0 \tag{3}$$

where $\forall i \in \{1, \ldots, j\}$

The main objective of structure optimization is to minimize the weight of structure, and structure is designed via members and joints so total weight of structure is equal to the sum of weights of all the members. Weight of each member depends upon the volume of member and density of material used, material and length of each member is fixed so the weight will be the function of cross-section area vector $\overrightarrow{A} = \{A_1, \ldots, A_m\}$. The cross-section area vector is selected between $\overrightarrow{A^l} = \{A_1^l, \ldots, A_m^l\}$ and $\overrightarrow{A^u} = \{A_1^u, \ldots, A_m^u\}$. So our objective function can be written as:

$$W(\overrightarrow{A}) = \sum_{k=1}^{m} A_k L_k \rho_k \qquad (4)$$

where A_k, L_k, and ρ_k are respectively area of cross-section, length, and density of materiel of kth member. We minimize the Eq. 4 by satisfying the following constraints:

$$s_{k,l}(\overrightarrow{A}) = \frac{\sigma_{k,l}}{\sigma_{k,allowed}} - 1 \leq 0, \ k = 1, 2, \ldots, m \qquad (5)$$

$$b_{k,l}(\overrightarrow{A}) = \frac{\lambda_{k,l}}{\lambda_{k,allowed}} - 1 \leq 0, \ k = 1, 2, \ldots, m \qquad (6)$$

$$d_{n,l}(\overrightarrow{A}) = \frac{u_{n,l}}{u_{n,allowed}} - 1 \leq 0, \ n = 1, 2, \ldots, m_m \qquad (7)$$

where $s_{k,l}(\overrightarrow{A})$, $b_{k,l}(\overrightarrow{A})$ and $d_{n,l}(\overrightarrow{A})$ are respectively, the member stress ratio, member buckling ratio and nodal displacement ratio constraint function. $\sigma_{k,l}$ and $\lambda_{k,l}$ are the stress and slenderness ratio of kth member due to the loading condition l, respectively; $\sigma_{k,allowed}$ and $\lambda_{k,allowed}$ are the allowed(ultimate) stress and slenderness ratio of kth member. $u_{n,l}$ and $u_{n,allowed}$ are the nodal displacement and allowable displacement of nth degree of freedom due to the loading condition l, respectively.

3   Proposed Solution Methodology

Artificial Bee Colony (ABC), proposed by Karaboga [17] in 2005, is based on the foraging behavior of honey bee swarms. ABC is extensively used to solve constrained and unconstrained optimization problems with proven results [18,20]. ABC algorithm consists of three sets of food searching swarms which are (i) Employed Bees, (ii) Onlooker Bees and (iii) Scout Bees. The basic structure of ABC algorithm is explained in Algorithm 1. We have combined concepts of statistical co-variance matrices [31,36] conjoined with the onlooker bee phase of ABC algorithm. The reason behind using co-variance matrix in onlooker bee phase is to generates new point around the weighted mean value of good points. With the help of co-variance matrix points can be generated near to mean with high probability and far with low probability, this approach increases the convergence rate of proposed algorithm. The detailed algorithm of the proposed strategy is described in Algorithm 1.

Algorithm 1. Basic Structure of ABC Algorithm

Step 1: Initialize Population
Step 2: while not happy repeat step 2.1 through 2.3
    Step 2.1: Employed Bee Phase
        //searching new points in whole search space
    Step 2.2: Onlooker Bee Phase
        //according to the information shared by employed bees,
        //exploit food sources based on their nectar amount.
    Step 2.3: Scout Bees Phase
        //searching new points instead of the points which can not be further evolved
Step 3: end while

---

4 Tests and Results

Suggested algorithm is tested on computer having intel core $i7$, 3.40 GHz processor, 4 GB RAM, and Windows 8 operating system. The parameters of co-variance based ABC Algorithm 2 are: $Runtime = 10$, $NP = 200$, $FoodNumber = NP/2$, $\mu = NP/4$, $Limit = 300$, and $MaximumCycle = 10000$. The Algorithm 2 is applied on two classical $3D$ bench-mark problems of 25 bars and 72 bars each. The 25 bar problem, first reported by Fox et al. in [10], is shown in Fig. 1. The 72 bar problem, first reported by Venkayya et al. in [42], is shown in Fig. 2. The optimal cross sectional area and the subsequent calculated weight for the two problems (25 bars and 72 bars) is shown in Table 1 with material having $Young's Modulus = 10^7 psi$ and $Density = 0.1 lb/in^3$. Design constraints for 25-Bar truss is described in [10] and design constraint for 72-Bar truss is described in [42]. For estimating the performance of our proposed algorithm we have compared the results with the four previously tested well known algorithms, which are (i) Lee and Geem [32], (ii) Li et al. [33], (iii) Lamberti [30], and (iv) Sonmez [40].

From the results it has been observed that a comparable weight has been achieved for 25 bars problem and best weight has been obtained for 72 bars problem. A performance parameter that can be significantly outlined is the **number of iterations** in which the best result is achieved. Our best results i.e. 545.16271 lb for 25 bars and 363.82375 lb for 72 bar is achieved within 1000 **iterations**, while most of the algorithms available in the literature took nearly $1000 \times problem\, dimension$ iterations. The computational cost of any algorithm is measured on basis of **number of function evaluation**, and function evaluation involved in each iteration is equal to the population size(food size) which is taken as 200 in our approach so the total number of function evaluation becomes

---

Algorithm 2. Main body of Co-variance based ABC

Inputs:	Description
	Dimensions Size of Problem: D,
	Total Number of Forager(Population Size): NP,
	Lower Bound of Problem: $lower$,
	Upper Bound of Problem: $upper$,
	Co-variance matrix $C = I$
Parameters	Description
NP:	denotes the population size which is divided in two parts half for exploration and half for exploitation. Each solution $\vec{x_i} = \{x_{i,1}, x_{i,2}, \ldots, x_{i,D}\}$ for $i = 1, 2, \ldots, NP/2$, which is a row vector, is of D-dimension.
C:	denotes the co-variance matrix of $D \times D$ of best μ points.
μ:	denotes the points with good fitness.
w_i:	denotes the weight of each best particle and $\sum_{i=1}^{\mu} w_i = 1$.
σ:	denotes the step-size.
$x_{i,j}$:	denotes the jth dimension of ith solution.
$x_{i:NP/2}$:	denotes the ith best solution in $NP/2$ solutions
$N(0, I)$:	normally distributed row vector with row matrix of $0's$ as mean and identity matrix as standard deviation.
$rand(0, m)$:	denotes any random number between 0 and m.
$n(0, sd)$:	is normally distributed number with 0 mean and sd standard deviation which is linearly dependent on number of cycles.
cf:	is convergence factor taken as 1.5.
$f(x)$:	Fitness function, equation (4).
$c(x)$:	Constraint functions, equation (5), (6) and (7).

Step 1: Randomly Initialize Food sources $x_i = lower + rand(0, 1) \times (upper - lower)$

Step 2: Initialize the $C = I$ and $cycle = 1$

Step 3: Evaluate foods and constrained condition $f(x)$ and $c(x)$

Step 4: while cycle<maximum number of cycles repeat step 4.1 through 4.8

   Step 4.1: Employed Bee Phase:
      for $\forall i \in \{1, \ldots, NP/2\}$ and for $\forall j \in \{1, \ldots, D\}$
      $x_{i,j} = x_{i,j} + n(0, sd) \times (x_{i,j} - x_{i,neighbour}) + rand(0, cf) \times (x_{best,j} - x_{i,j})$

   Step 4.2: Evaluate $f(x)$ and $c(x)$ for each bee

   Step 4.3: Calculate the weighted mean of μ best points $mean = \sum_{i=1}^{\mu} w_i x_{i:NP/2}$

   Step 4.4: Calculate Probability of each food source for selection purpose $p_i = $
   $$\begin{cases} 0.5 + \left(\frac{fitness_i}{\sum_{j=1}^{NP/2} fitness_j} \right) \times 0.5, & if feasible \\ \left(1 - \frac{fitness_i}{\sum_{j=1}^{NP/2} fitness_j} \right) \times 0.5, & otherwise \end{cases}$$

   Step 4.5: Onlooker Bees Phase:
      for $\forall i \in \{1, \ldots, NP/2\}$
      $x_i = mean + \sigma \times N(0, I) \times C^{1/2}$

   Step 4.6: Scout Bees Phase:
      for all bees b who are now abundant
      $x_b = lower + rand(0, 1) \times (upper - lower)$

   Step 4.7: Memorize the best solution x_{best} found so far

   Step 4.8: Update the value of C and cycle=cycle+1

Step 5: end while

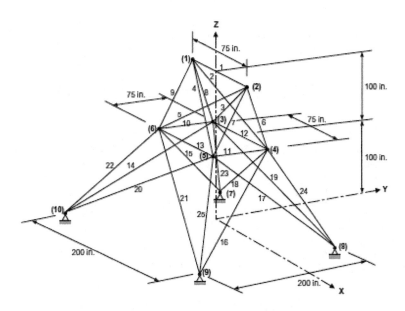

Fig. 1. 25 bar truss structure benchmark

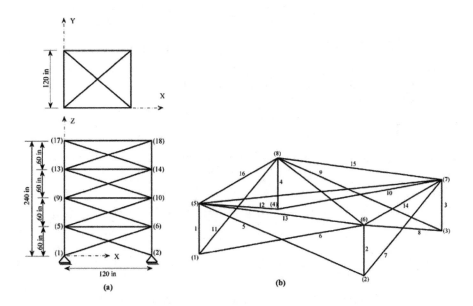

Fig. 2. 72-Bar truss structure

Table 1. 25-Bar and 72-Bar Truss optimization result

	Variables					
	Element no.	Lee and Geem [32]	Li et al. [33]	Lamberti [30]	ABC [40]	Co-variance base ABC
25-Bar	A_1	0.047	0.01	0.01	0.011	**0.010000**
	$A_2 - A_5$	2.022	1.97	1.987	1.979	**1.98705**
	$A_6 - A_9$	2.95	3.016	2.994	3.003	**2.993471**
	$A_{10} - A_{11}$	0.01	0.010	0.010	0.010	**0.010000**
	$A_{12} - A_{13}$	0.014	0.010	0.010	0.010	**0.010000**
	$A_{14} - A_{17}$	0.688	0.694	0.694	0.69	**0.683958**
	$A_{18} - A_{21}$	1.657	1.681	1.681	1.679	**1.67689**
	$A_{22} - A_{25}$	2.663	2.643	2.643	2.652	**2.662148**
	Weight (lb)	544.38	545.19	545.161	545.193	**545.16271**
	Violation	0.0122	None	None	None	None
72-Bar	$A_1 - A_4$	1.963	1.907	0.1665	0.1675	**0.166481**
	$A_5 - A_{12}$	0.481	0.524	0.5363	0.5346	**0.536215**
	$A_{13} - A_{16}$	0.01	0.01	0.446	0.4443	**0.445806**
	$A_{17} - A_{18}$	0.011	0.01	0.5761	0.5803	**0.575948**
	$A_{19} - A_{22}$	1.233	1.288	0.5207	0.5208	**0.521012**
	$A_{23} - A_{30}$	0.506	0.523	0.518	0.5178	**0.518128**
	$A_{31} - A_{34}$	0.011	0.01	0.01	0.01	**0.010000**
	$A_{35} - A_{36}$	0.012	0.01	0.1141	0.1048	**0.113973**
	$A_{37} - A_{40}$	0.538	0.544	1.2903	1.2968	**1.290164**
	$A_{41} - A_{48}$	0.533	0.528	0.517	0.5191	**0.516987**
	$A_{49} - A_{52}$	0.01	0.019	0.01	0.01	**0.010000**
	$A_{53} - A_{54}$	0.167	0.02	0.01	0.0101	**0.010000**
	$A_{55} - A_{58}$	0.161	0.176	1.8866	1.8907	**1.887474**
	$A_{59} - A_{66}$	0.542	0.535	0.5169	0.5166	**0.516844**
	$A_{67} - A_{70}$	0.478	0.426	0.01	0.01	**0.010000**
	$A_{71} - A_{72}$	0.551	0.612	0.01	0.01	**0.010000**
	Weight (lb)	364.33	364.86	363.803	363.8392	**363.82375**
	Violation	12.06	13.701	0.00004	None	None

200000, which is also very less in comparison of other algorithms. The convergence rate of co-variance based ABC is shown in Fig. 3.

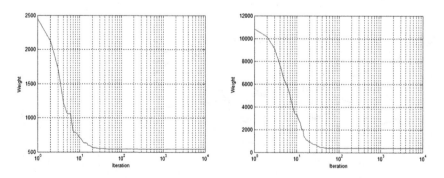

Fig. 3. Convergence rate of Co-variance based ABC for 25-Bar truss(left) and 72-Bar truss(right)

5   Results and Conclusions

Artificial Bee Colony (ABC) algorithm is a well known swarm intelligence based optimization technique. In the present research article we have proposed a novel co-variance based ABC algorithm for space structure optimization. The effectiveness of the proposed approach is tested on two classical benchmark problems of truss optimization of 25 bars and 72 bars. From the results it can be armed that the proposed approach out performs various other approaches proposed in the past. We have achieved significantly better results that too in fewer iteration only. From the results it may be well concluded that co-variance based ABC algorithm can remarkably change the present structural design practices pertaining to truss optimization.

References

1. Abraham, A., Nedjah, N., de Macedo, L.: Mourelle: evolutionary computation: from genetic algorithms to genetic programming. In: Abraham, A., Nedjah, N., de Macedo, L. (eds.) Genetic Systems Programming. ACI, vol. 13, pp. 1–20. Springer, Heidelberg (2006)
2. Adeli, H., Kamal, O.: Efficient optimization of space trusses. Comput. Struct. **24**(3), 501–511 (1986)
3. Bäck, T., Schwefel, H.-P.: An overview of evolutionary algorithms for parameter optimization. Evol. Comput. **1**(1), 1–23 (1993)
4. Camp, C.V., Bichon, B.J.: Design of space trusses using ant colony optimization. J. Struct. Eng. **130**(5), 741–751 (2004)
5. Coello, C.A., Christiansen, A.D.: Multiobjective optimization of trusses using genetic algorithms. Comput. Struct. **75**(6), 647–660 (2000)
6. Darwin, C., Bynum, W.F.: The origin of species by means of natural selection: or, the preservation of favored races in the struggle for life (2009)
7. Dorigo, M., Birattari, M., Stützle, T.: Ant colony optimization. IEEE Comput. Intell. Mag. **1**(4), 28–39 (2006)
8. Eberhart, R.C., Shi, Y., Kennedy, J.: Swarm Intelligence. Elsevier, London (2001)
9. Fogel, L.J.: Evolutionary programming in perspective: the top-down view. In: Computational Intelligence: Imitating Life, vol. 1 (1994)
10. Fox, R.L., Schmit Jr., L.A.: Advances in the integrated approach to structural synthesis. J. Spacecraft Rockets **3**(6), 858–866 (1966)
11. Galante, M.: Genetic algorithms as an approach to optimize real-world trusses. Int. J. Numer. Meth. Eng. **39**(3), 361–382 (1996)
12. Goldberg, D.E.: Genetic Algorithms. Pearson Education India, New Delhi (2006)
13. Herbert Martins Gomes: Truss optimization with dynamic constraints using a particle swarm algorithm. Expert Syst. Appl. **38**(1), 957–968 (2011)
14. Hadidi, A., Azad, S.K., Azad, S.K.: Structural optimization using Artificial Bee Colony algorithm. In: 2nd International Conference on Engineering Optimization (2010)
15. Haftka, R.T., Gürdal, Z.: Elements of Structural Optimization, vol. 11. Springer, Heidelberg (2012)
16. Hansen, S.R., Vanderplaats, G.N.: Approximation method for configuration optimization of trusses. AIAA J. **28**(1), 161–168 (1990)

17. Karaboga, D.: An idea based on honey bee swarm for numerical optimization. Technical report-tr06, Erciyes University, Engineering Faculty, Computer Engineering Department (2005)
18. Karaboga, D., Basturk, B.: Artificial Bee Colony (ABC) optimization algorithm for solving constrained optimization problems. In: Melin, P., Castillo, O., Aguilar, L.T., Kacprzyk, J., Pedrycz, W. (eds.) IFSA 2007. LNCS (LNAI), vol. 4529, pp. 789–798. Springer, Heidelberg (2007)
19. Karaboga, D., Basturk, B.: A powerful and efficient algorithm for numerical function optimization: Artificial Bee Colony (ABC) algorithm. J. Global Optim. **39**(3), 459–471 (2007)
20. Karaboga, D., Basturk, B.: On the performance of Artificial Bee Colony (ABC) algorithm. Appl. Soft Comput. **8**(1), 687–697 (2008)
21. Kaveh, A., Farhmand Azar, B., Talatahari, S.: Ant colony optimization for design of space trusses. Int. J. Space Struct. **23**(3), 167–181 (2008)
22. Kaveh, A., Talatahari, S.: Particle swarm optimizer, ant colony strategy and harmony search scheme hybridized for optimization of truss structures. Comput. Struct. **87**(5), 267–283 (2009)
23. Kennedy, J.: Particle swarm optimization. In: Gass, S.I., Fu, M.C. (eds.) Encyclopedia of Machine Learning, pp. 760–766. Springer, Heidelberg (2011)
24. Kennedy, J., Kennedy, J.F., Eberhart, R.C., Shi, Y.: Swarm Intelligence. Morgan Kaufmann, San Francisco (2001)
25. Kirsch, U.: Structural Optimization: Fundamentals and Applications. Springer, Heidelberg (2012)
26. Koumousis, V.K., Georgiou, P.G.: Genetic algorithms in discrete optimization of steel truss roofs. J. Comput. Civ. Eng. **8**(3), 309–325 (1994)
27. Koza, J.R.: Genetic Programming: On the Programming of Computers by Means of Natural Selection, vol. 1. MIT Press, Cambridge (1992)
28. Koziel, S., Michalewicz, Z.: Evolutionary algorithms, homomorphous mappings, and constrained parameter optimization. Evol. Comput. **7**(1), 19–44 (1999)
29. Lagaros, N.D., Papadrakakis, M., Kokossalakis, G.: Structural optimization using evolutionary algorithms. Comput. Struct. **80**(7), 571–589 (2002)
30. Lamberti, L.: An efficient simulated annealing algorithm for design optimization of truss structures. Comput. Struct. **86**(19), 1936–1953 (2008)
31. Ledoit, O., Wolf, M.: A well-conditioned estimator for large-dimensional covariance matrices. J. Multivar. Anal. **88**(2), 365–411 (2004)
32. Lee, K.S., Geem, Z.W.: A new structural optimization method based on the harmony search algorithm. Comput. Struct. **82**(9), 781–798 (2004)
33. Li, L.J., Huang, Z.B., Liu, F., Wu, Q.H.: A heuristic particle swarm optimizer for optimization of pin connected structures. Comput. Struct. **85**(7), 340–349 (2007)
34. Overbay, S., Ganzerli, S., De Palma, P., Brown, A., Stackle, P.: Trusses, NP-completeness, and genetic algorithms. In: Proceedings of the 17th Analysis and Computation Specialty Conference (2006)
35. Perez, R.E., Behdinan, K.: Particle swarm approach for structural design optimization. Comput. Struct. **85**(19), 1579–1588 (2007)
36. Pinheiro, J.C., Bates, D.M.: Unconstrained parametrizations for variance-covariance matrices. Stat. Comput. **6**(3), 289–296 (1996)
37. Rajan, S.D.: Sizing, shape, and topology design optimization of trusses using genetic algorithm. J. Struct. Eng. **121**(10), 1480–1487 (1995)
38. Rajeev, S., Krishnamoorthy, C.S.: Discrete optimization of structures using genetic algorithms. J. Struct. Eng. **118**(5), 1233–1250 (1992)

39. Rechenberg, I.: Evolution strategy. In: Computational Intelligence: Imitating Life, vol. 1 (1994)
40. Sonmez, M.: Artificial Bee Colony algorithm for optimization of truss structures. Appl. Soft Comput. **11**(2), 2406–2418 (2011)
41. Sonmez, M.: Discrete optimum design of truss structures using Artificial Bee Colony algorithm. Struct. Multi. Optim. **43**(1), 85–97 (2011)
42. Venkayya, V.B., Khot, N.S., Reddy, V.S.: Optimization of structures based on the study of energy distribution. Technical report, DTIC Document (1968)
43. Yates, D.F., Templeman, A.B., Boffey, T.B.: The complexity of procedures for determining minimum weight trusses with discrete member sizes. Int. J. Solids Struct. **18**(6), 487–495 (1982)

Solving Manufacturing Cell Design Problems by Using a Bat Algorithm Approach

Ricardo Soto[1,2,3], Broderick Crawford[1,4,5], Andrés Alarcón[1], Carolina Zec[1], Emanuel Vega[1], Victor Reyes[1(✉)], Ignacio Araya[1], and Eduardo Olguín[5]

[1] Pontificia Universidad Católica de Valparaíso, Valparaíso, Chile
{ricardo.soto,broderick.crawford,ignacio.araya}@ucv.cl,
fernando.paredes@udp.cl, emanuel.vega@usm.cl, victor.reyes.r@mail.pucv.cl
[2] Universidad Autónoma de Chile, Santiago, Chile
[3] Universidad Cientifica del Sur, Lima, Peru
[4] Universidad Central de Chile, Santiago, Chile
[5] Facultad de Ingeniería y Tecnología, Universidad San Sebastián, Bellavista 7,
8420524 Santiago, Chile
eduardo.olguin@uss.cl

Abstract. Manufacturing Cell Design is a problem that consist in distributing machines in cells, in such a way productivity is improved. The idea is that a product, build up by using different parts, has the least amount of travel on its manufacturing process. To solve the MCDP we use the Bat Algorithm, a metaheuristic inspired by a feature of the microbats, the echolocation. This feature allows an automatic exploration and exploitation balance, by controlling the rate of volume and emission pulses during the search. Our approach has been tested by using a well-known set of benchmark instances, reaching optimal values for most of them.

Keywords: Bio-inspired systems · Bat algorithm · Metaheuristic · Manufacturing Cell Design Problems

1 Introduction

The Manufacturing Cell Design Problem (MCDP) consist in grouping components under the following statement: "Similar things should be manufactured in the same way". Then, the design of an optimal production plant is achieved through the organization of the different machines that process parts of a given product in production cells. The goal of the MCDP consist in minimize movements and exchange of material between these cells.

Different metaheuristics have been used for cell formation. Aljaber et al. [1] made use of Tabu Search. Wu et al. [7] presented a Simulated Annealing (SA) approach. Durán et al. [4] combined Particle Swarm Optimization (PSO), which consists of particles that move through a space of solutions and that are accelerated in time, with a data mining technique. Venugopal and Narendran [10] proposed using the Genetic Algorithms (GA). Gupta et al. [5] also used

© Springer International Publishing Switzerland 2016
Y. Tan et al. (Eds.): ICSI 2016, Part I, LNCS 9712, pp. 184–191, 2016.
DOI: 10.1007/978-3-319-41000-5_18

GA, but focusing on a different multi-objective optimization, consisting in the simultaneous minimization of the total number of movements between cells and load variation between them. It is also possible to find hybrid techniques in the problem resolution. Such is the case of Wu et al. [11], who combined SA with GA. James et al. [6] introduced a hybrid solution that combines local search and GA. Nsakanda et al. [8] proposed a solution methodology based on a combination of GA and large-scale optimization techniques. Soto et al. [9], utilized Constraint Programming (CP) and Boolean Satisfiability Technology (SAT) for the resolution of the problem, developing the problem by applying five different solvers, two of which are CP solvers, two SAT solvers and a CP and SAT hybrid. Also, some research has been done by combining approximate and global strategies. Such is the case of Boulif and Atif [3], who combined branch-and-bound techniques with GA.

In this paper, we propose a Bat Algorithm (BA) to solve the MCDP. BA is an optimization algorithm based on the natural behavior of microbats and their echolocation system, which it is used to determine the distance to their prey or food. The bats move through the search space attempting to make the least amount of parts movements between the different cells, employing frequency, speed and location the bats attempt to find the optimal solution.

The rest of this paper is organized as follows. In Sect. 2 the mathematical model of the MCDP is described. The Bat Algorithm is explained in Sect. 3. Finally, Sect. 4 illustrates the experimental results, followed by conclusions and future work.

2   Manufacturing Cell Design Problem

The MCDP consists in organizing a manufacturing plant or facility into a set of cells, each of them made up of different machines meant to process different parts of a product, that share similar characteristics. The goal is to minimize movements and exchange of material between cells, in order to reduce production costs and increase productivity. We represent the processing requirements of machine parts through an incidence zero-one matrix A known as the *machine-part matrix*. The optimization model is stated as follows. Let:

- M: the number of machines.
- P: the number of parts.
- C: the number of cells.
- i: the index of machines ($i = 1, ..., M$).
- j: the index of parts ($j = 1, ..., P$).
- k: the index of cells ($k = 1, ..., C$),
- M_{max}: the maximum number of machines per cell.
- $A = [a_{ij}]$: the binary machine-part incidence matrix, where:

$$a_{ij} = \begin{cases} 1 & \text{if machine } i \text{ process the part } j \\ 0 & \text{otherwise} \end{cases}$$

- $B = [y_{ik}]$: the binary machine-cell incidence matrix, where:

$$y_{ik} = \begin{cases} 1 & \text{if machine } i \text{ belongs to cell } k \\ 0 & \text{otherwise} \end{cases}$$

- $C = [z_{jk}]$: the binary part-cell incidence matrix, where:

$$z_{jk} = \begin{cases} 1 & \text{if part } j \text{ belongs to cell } k \\ 0 & \text{otherwise} \end{cases}$$

Then, the MCDP is represented by the following mathematical model [2]:

$$\min \sum_{k=1}^{C} \sum_{i=1}^{M} \sum_{j=1}^{P} a_{i_j} z_{j_k} (1 - y_{i_k}) \tag{1}$$

Subject to the following constraints:

$$\sum_{k=1}^{C} y_{ik} = 1 \quad \forall i \tag{2}$$

$$\sum_{k=1}^{C} z_{jk} = 1 \quad \forall j \tag{3}$$

$$\sum_{i=1}^{M} y_{ik} \leq M_{max} \quad \forall k \tag{4}$$

3   Bat Algorithm

The BA is based on a feature of the microbats, the echolocation. Echolocation is the capability of some animals to map their environments by emitting sounds and interpreting the echo produced by the surrounding objects. This allows them to detect their prey, avoid obstacles and locate crevices in the dark where they can sleep. Their properties may vary according to the hunting strategy of each species. Microbats can emit between 10 and 20 sound bursts per second, while their pulse rate can accelerate to 200 pulses per second as it approaches to its prey [12,13].

Two important features used on this algorithm are the *frequency adjustment* and the *automatic zoom* techniques. While the first is related to guarantee diversity in solutions, the second one allows a balance between exploration and exploitation during the search process.

3.1   Main Features

Based on the description of echolocation and the bat characteristics, the BA is developed according to the three following rules [14]:

- All bats use echolocation to detect distance, as well as somehow knowing the difference between food or prey, and background barriers.
- Bats fly randomly at a speed v_i, a position x_i, and with a frequency f_{min}, varying their wavelength and volume A_0 to search for their prey. They can automatically adjust the wavelength (or frequency) of the emitted pulses and adjust the pulse emission rate $r \in [0,1]$ as they approach their objective.
- Despite there being several ways the volume can vary, it supposedly varies from whole (positive) A_0 to a constant minimum value A_{min}.

Movement of the Bat: Each bat is related to a speed v_i^t, a location x_i^t and an iteration t in a search space. Among all the bats, there is a better current solution x_*. Therefore, the three rules mentioned before can be modeled as the following equations [14]:

$$f_i = f_{min} + (f_{max} - f_{min})\beta \qquad (5)$$

$$v_i^{t+1} = v_i^t + (x_i^t - x_*)f_i \qquad (6)$$

$$x_i^{t+1} = x_i^t + v_i^t \qquad (7)$$

Where $\beta \in [0,1]$ correspond to a random vector generated by using an uniform distribution. For our implementation, a f_{min} and f_{max} will be used, which will initially have a random value selected in a uniform manner from $[f_{min}, f_{max}]$. Because of this, the bat algorithm can be considered a frequency adjustment algorithm, as it provides a balanced combination of exploration and exploitation. Meanwhile, the volume and pulse rate provide an automatic zoom mechanism in the region with promising solutions.

3.2 Volume and Pulse Rate Variations

In order to provide an effective mechanism to control exploration and exploitation, the volume A_i and the pulse emission rate r_i must change during the execution process. Because the volume is generally reduced once the bat has located its prey and the pulse emission rate increases, the value for the volume can be selected as any convenient value between A_{min} and A_{max}. Then, if $A_{min} = 0$, it means that a bat has just found a prey and will temporarily stop emitting any sound [14]. With these preconditions, the following applies:

$$A_i^{t+1} = \alpha A_i^t, \quad r_i^{t+1} = r_i^0[1 - exp(-\gamma t)] \qquad (8)$$

Where α and γ are constants. For any $0 < \alpha < 1$ and $\gamma > 0$, the following applies:

$$A_i^t \to 0, \quad r_i^t \to r_i^0, \quad \text{as } t \to \infty \qquad (9)$$

In the most simple case, $\alpha = \gamma$ can be used.

4   Experimental Results

Our approach has been implemented in Java, on an 2.2 GHz CPU Intel Core i7-3632QM with 8 gb RAM computer using Windows 8 64 bits and tested out by using a well-known set of 10 incidence matrices from [2]. Tests were carried out based on these 10 test problems, employing 2 cells and a maximum of machines (mMax) between 8 and 12 for each of them, remaining constant through the whole execution of the algorithm. The employed parameters, that were established based on other problems already solved through the use of the BA which had yielded satisfying results, are defined as follows:

- Minimum frequency (f_Min): 0.25
- Maximum frequency (f_Max): 0.50
- Alpha constant (*alpha*): 0.99
- Gamma constant (*gamma*): 0.99
- Epsilon constant (*epsilon*): 0.99
- No. of machines (M): 16
- No. of parts (P): 30
- No. of iterations (T): 300
- No. of microbats (n): 50

Tables 1 and 2 show detailed information about the results obtained by using our approach with the previous configuration. We compare our results with the optimal values reported in [2] (Opt), Simulated Annealing (SA) [2] and with Particle Swarm Optimization (PSO) [4] as well. A convergence plot can be see in Fig. 1.

Table 1. Results of BA using 2 cells

Pblm	Mmax = 8				Mmax = 9				Mmax = 10			
	Opt	SA	PSO	BA	Opt	SA	PSO	BA	Opt	SA	PSO	BA
1	11	11	11	11	11	11	11	11	11	11	11	11
2	7	7	7	7	6	6	6	6	4	10	5	5
3	4	5	5	4	4	4	4	4	4	4	5	4
4	14	14	15	14	13	13	13	13	13	13	13	13
5	9	9	10	9	6	6	8	6	6	6	6	6
6	5	5	5	5	3	3	3	3	3	5	3	3
7	7	7	7	7	4	4	5	4	4	4	5	4
8	13	13	14	13	10	20	11	10	8	15	10	8
9	8	13	9	8	8	8	8	8	8	8	8	8
10	8	8	9	8	5	5	8	5	5	5	7	5

Table 2. Results of BA using 2 cells

Pblm	Mmax = 11				Mmax = 12			
	Opt	SA	PSO	BA	Opt	SA	PSO	BA
1	11	11	11	11	11	11	11	11
2	3	4	4	4	3	3	4	4
3	3	4	4	3	1	4	3	1
4	13	13	13	13	13	13	13	13
5	5	7	5	5	4	4	5	4
6	3	3	4	3	2	3	4	2
7	4	4	5	4	4	4	5	4
8	5	11	6	5	5	7	6	5
9	5	8	5	5	5	8	8	5
10	5	5	7	5	5	5	6	5

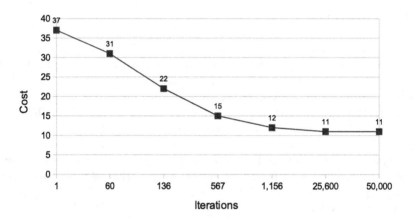

Fig. 1. Convergence chart for Problem 1 solved by SFLA with Mmax = 8.

Table 3 shows the relative percentage derivation (RPD), which is computed as follows:

$$RDP = \frac{(Z - Z_{opt})}{Z_{opt}} \times 100 \qquad (10)$$

Experimental results show that the proposed BA provides high quality solutions and good performance within 2 cells, where 94 % of the expected results were achieved, namely, 47 out of 50 of the analyzed instances.

Table 3. Average and relative percentage derivation

Pblm	Mmax = 8		Mmax = 9		Mmax = 10		Mmax = 11		Mmax = 12	
	Avg	RPD(%)	Avg	RPD(%)	Avg	RPD(%)	Avg	RPD(%)	Avg	RPD(%)
1	11	0	11	0	11	0	11	0	11	0
2	7	0	6	0	5	25	4	33	4	33
3	4	0	4	0	4	0	3	0	1	0
4	14	0	13	0	13	0	13	0	13	0
5	9	0	6	0	6	0	5	0	4	0
6	5	0	3	0	3	0	3	0	2	0
7	7	0	4	0	4	0	4	0	4	0
8	13	0	10	0	8	0	5	0	5	0
9	8	0	8	0	8	0	5	0	5	0
10	8	0	5	0	5	0	5	0	5	0

5   Conclusions

Manufacturing cell design is a well-known problem within manufacturing facto-
ries, as they seek to increase productivity by reducing times and costs. For this
reason we have proposed a BA, a novel method to tackle this type of problem
reporting optimum values for 47 of the 50 tested instances. BA has shown excellent results and high convergence rates in early stages of the search. One of the
most important features that was denoted along the experimental result is the
automatic zoom, which gives two mayor advantages. While the first one allows
to exploit a region intensely finding promising solutions, the second one performs
control over the parameters, giving an advantage over other metaheuristics where
most of the parameters are set before the execution. As a future work, we plan to
implement new modern metaheuristics for solving the MCDP. An Autonomous
Search integration for setting the parameters would be another direction for
research as well.

Acknowledgements. Ricardo Soto is supported by grant CONICYT/FONDECYT/
REGULAR/1160455, Broderick Crawford is supported by grant CONICYT/
FONDECYT/REGULAR/1140897, Victor Reyes is supported by grant INF-
PUCV 2015, and Ignacio Araya is supported by grant CONICYT/FONDECYT/
INICIACION/11121366.

References

1. Aljaber, N., Baek, W., Chen, C.L.: A tabu search approach to the cell formation
   problem. Comput. Ind. Eng. **32**(1), 169–185 (1997)
2. Boctor, F.F.: A linear formulation of the machine-part cell formation problem. Int.
   J. Prod. Res. **29**(2), 343–356 (1991)

3. Boulif, M., Atif, K.: A new branch-&-bound-enhanced genetic algorithm for the manufacturing cell formation problem. Comput. Oper. Res. **33**(8), 2219–2245 (2006)
4. Durán, O., Rodriguez, N., Consalter, L.A.: Collaborative particle swarm optimization with a data mining technique for manufacturing cell design. Expert Syst. Appl. **37**(2), 1563–1567 (2010)
5. Gupta, Y., Gupta, M., Kumar, A., Sundaram, C.: A genetic algorithm-based approach to cell composition and layout design problems. Int. J. Prod. Res. **34**(2), 447–482 (1996)
6. James, T.L., Brown, E.C., Keeling, K.B.: A hybrid grouping genetic algorithm for the cell formation problem. Comput. Oper. Res. **34**(7), 2059–2079 (2007)
7. Lozano, S., Adenso-Diaz, B., Eguia, I., Onieva, L., et al.: A one-step tabu search algorithm for manufacturing cell design. J. Oper. Res. Soc. **50**(5), 509–516 (1999)
8. Nsakanda, A.L., Diaby, M., Price, W.L.: Hybrid genetic approach for solving large-scale capacitated cell formation problems with multiple routings. Eur. J. Oper. Res. **171**(3), 1051–1070 (2006)
9. Soto, R., Kjellerstrand, H., Durán, O., Crawford, B., Monfroy, E., Paredes, F.: Cell formation in group technology using constraint programming and boolean satisfiability. Expert Syst. Appl. **39**(13), 11423–11427 (2012)
10. Venugopal, V., Narendran, T.: A genetic algorithm approach to the machine-component grouping problem with multiple objectives. Comput. Ind. Eng. **22**(4), 469–480 (1992)
11. Wu, T.H., Chang, C.C., Chung, S.H.: A simulated annealing algorithm for manufacturing cell formation problems. Expert Syst. Appl. **34**(3), 1609–1617 (2008)
12. Yang, X.S.: Nature-Inspired Metaheuristic Algorithms. Luniver Press, United Kingdom (2010)
13. Yang, X.S.: Bat algorithm for multi-objective optimisation. Int. J. Bio-Inspired Comput. **3**(5), 267–274 (2011)
14. Yang, X.S., He, X.: Bat algorithm: literature review and applications. Int. J. Bio-Inspired Comput. **5**(3), 141–149 (2013)

Mammographic Mass Classification Using Functional Link Neural Network with Modified Bee Firefly Algorithm

Yana Mazwin Mohmad Hassim[(✉)] and Rozaida Ghazali

Faculty of Computer Science and Information Technology,
Universiti Tun Hussein Onn Malaysia (UTHM),
86400 Batu Pahat, Johor, Malaysia
{yana,rozaida}@uthm.edu.my

Abstract. Functional Link Neural Network (FLNN) is a type of Higher Order Neural Networks (HONNs) known to have the modest architecture as compared to other multilayer feedforward networks. FLNN employs less tunable weights which make the learning method in the network less complicated. The standard learning method used in FLNN network is the Backpropagation (BP) learning algorithm. This method however, is prone to easily get trapped in local minima which affect the performance of the FLNN network. Thus an alternative learning method named modified Bee-Firefly (MBF) algorithm is proposed for FLNN. This paper presents the implementation FLNN trained with MBF on mammographic mass classification task. The result of the classification made by FLNN-MBF is compared with the standard FLNN-BP model to examine whether the MBF learning algorithm is capable of training the FLNN network and improve its performance for the task of classification.

Keywords: Classification · Functional Link Neural Network · Learning scheme · Modified Bee-Firefly algorithm · Mammographic

1 Introduction

Functional Link Neural Network (FLNN) is a type of Higher Order Neural Networks (HONNs) introduced by Giles and Maxwell [1]. FLNN was also known as an alternative approach to standard multilayer feed forward networks in Artificial Neural Networks (ANNs) by Pao [2]. It is a type of flat network (a network without hidden layers) where it reduced the neural architectural complexity and at the same time reduced the learning complexity on the network training. In neural classifications, network training is essential for building a classification model (classifier). The network is trained on the basis of data (training data) facilitated by the error correcting learning method where the connection weights of each neuron are adjusted until the network error reaches an acceptable minimum value [3] which can be viewed as an optimization task [4].

The most widely used error correcting learning method for network training is the Backpropagation (BP) learning algorithm [5–7] which is a type of gradient descent optimization method. However one of the crucial problems with the standard BP-learning

© Springer International Publishing Switzerland 2016
Y. Tan et al. (Eds.): ICSI 2016, Part I, LNCS 9712, pp. 192–199, 2016.
DOI: 10.1007/978-3-319-41000-5_19

algorithm is that it can easily get trapped in local minima which affect the performance of FLNN network particularly when dealing with highly non-linear problems [8]. Thus, an alternative learning scheme named Modified Bee-Firefly (MBF) which combined the benefit of Artificial Bee Colony algorithm, known to have good exploration capabilities for their global search strategy [9] and Firefly algorithm, renowned to have good ability in local search exploitation strategy [10, 11] is implemented for training the FLNN network to overcome the standard gradient based BP-learning algorithm handicaps. In this work, FLNN trained with MBF is used as a classification model for mammographic mass classification task. The result of the classification made by FLNN-MBF is compared with the standard FLNN-BP model to examine whether the MBF learning algorithm is capable of training the FLNN network and improve its performance for the task of classification.

2   Related Works

In this section, the properties and learning scheme of FLNN as well as the proposed swarm optimizations method are briefly discussed.

2.1   Functional Link Neural Network Training

The mostly used learning algorithm for FLNN training is the BP-learning algorithm [8, 12, 13]. The BP-learning algorithm is s gradient descent optimization method used for tuning the weight parameters in FLNN. The FLNN with tensor model architecture is used in this work. In tensor model, each component of the input features multiplies the entire input features vector and generates an entire vector from each of the individual components. This transformation enhanced the input features representation by adding all interaction terms between input values [14]. For example, if a classification data with 3 input features is fed into FLNN, the original input features of this network are $\{x_1, x_2, x_3\}$ known as the first order terms. With tensor model, the original input features can be enhanced into $\{x_1x_2, x_1x_3, x_2x_3\}$ known the second order terms and up to $\{x_1x_2x_3\}$ which is the third order terms input enhancement. Both second order terms and third order terms act as supplementary inputs to the FLNN network.

Let, y be the actual output corresponds to the input pattern x. In the case of tensor model, the tensor of x can be represented as:

$$x_t = <x_1, x_2, \ldots, x_n, x_1x_2, x_1x_3, \ldots, x_{n-1}x_n> \tag{1}$$

where n is the number of features. The summation of weights vector between input layer and output layer is represented as:

$$s = wx_t + \theta \tag{2}$$

where, wx_t is the aggregate value of the weight vector, w and x_t, while θ is the bias.
In this work the logistic sigmoid is used as the activation function f, therefore the output node is define as:

$$s = wx_t + \theta \qquad (3)$$

where $f(s)$ is the logistic sigmoid activation function;

$$f(s) = \frac{1}{1 + \exp(-s)} \qquad (4)$$

The FLNN network has the ability to learn through training where the network is repeatedly presented with training data and the weights are adjusted by the learning algorithm from time to time until the desired input-output mapping is attained. The estimated output is then computed as:

$$\hat{y}_i = f(s_i), \quad i = 1, 2, \ldots, N \qquad (5)$$

where N is the number of training pattern. The objective of network training is to minimize the network error, E as:

$$minimize, \ E(w(t)) = \frac{1}{N} \sum_{i=1}^{N} (y_i - \hat{y}_i)^2 \qquad (6)$$

where $E(w(t))$ is the error of the t_{th} iteration, $w(t)$ is the weights value at the t_{th} iteration, y_i is the desired output, $\hat{y}_i$ is the network output and N is the number of training pattern.

2.2   Artificial Bee Colony (ABC)

Artificial Bees Colony (ABC) Algorithm is an optimization tool, which provides a population-based search procedure. The algorithm simulates the intelligent foraging behavior of a honey bee swarm and was proposed to solve a multidimensional and multimodal optimization problem by Karaboga [15]. The ABC algorithm mimics the foraging behavior of honey bee swarm (employed, onlooker and scout bees). The foraging behavior (search strategy) of the employed and onlooker from original ABC is presented as:

$$v_{ij} = x_{ij} + \emptyset_{ij}(x_{ij} - x_{kj}) \qquad (7)$$

where $k \in \{1, 2, \ldots, SN\}$ and $j \in \{1, 2, \ldots, D\}$ are randomly chosen indexes, and k has to be different from i. The $\emptyset_{ij}$ is a random generated number in the range $[-1, 1]$. The $\emptyset_{ij}$ controls the production of neighbor food sources around x_{ij} and represents the comparison of two food positions visualized by the bees. The solutions, SN denotes the size of employed bees or onlooker bees. Each solution $x_i(i = 1, 2, \ldots, SN)$ is a D-dimensional vector where, D is the number of optimization parameters. For training the FLNN, D is denoted as the weight parameters in the weights vector.

In order to determine whether or not the ABC algorithm can be used as a learning scheme for FLNN, a pilot experiment was conducted on Boolean function classification [16]. The results showed that the FLNN trained with ABC (FLNN-ABC) able to perform the classification task with better accuracy results than the standard FLNN-BP model. However, when implementing on real-life classification data, the FLNN-ABC had some difficulties in producing a better accuracy results. Hence, a modified ABC (MABC) learning algorithm for FLNN training was introduced.

2.3 Modified Artificial Bee Colony

A modified Artificial Bee Colony (MABC) was implemented as an alternative learning scheme on FLNN training for solving classification problems [17, 18]. The modification was driven by the drawbacks of ABC learning algorithm with respect to inefficiency in tuning the FLNN weight parameters when dealing with a large number of optimization parameters. This is due to their random behavior on exploiting only one parameter indexed by j in D-dimensional vector of their food solution (SN) using:

$$j = fix(rand * D) + 1; \tag{8}$$

In MABC, the random behavior of selecting one parameter in the solution vector represented by j is removed from the Eq. (7). The employed and onlooker bee are directed to visit all weight parameters in D for exploitation process before evaluating the vector x_i. The new modified local search strategy is now represented by Eq. (9):

$$v_i = x_i + \emptyset_i(x_i - x_k) \tag{9}$$

where $\emptyset_i$ is a random generated number, $\emptyset \in [-1, 1]$, while k is referring to the chosen neighboring food source, $k \in \{1, 2, \ldots, SN\}$ and $k \neq i$, v_i is a candidate food position and x_i is the current food position.

2.4 Firefly Algorithm

Firefly algorithm (FA) was developed by Yang [19] which is inspired by the flashing behavior of fireflies. The purpose of flashing behavior is to act as signal system to attract other fireflies. In FA, the attractiveness of firefly is determined by its bright-ness or light intensity. The attractiveness, β of a firefly is defined in Eq. (10):

$$\beta(r) = \beta_0 e^{-\gamma r^2} \tag{10}$$

where βo is the attractiveness at distance, $r = 0$. The movement of a firefly i attracted to brighter firefly j at x_i and x_j is determined by Eq. (11).

$$x_i = x_i + \beta_0 e^{-\gamma r^2}(x_j - x_i) + \alpha \varepsilon_i \tag{11}$$

In Eq. (11), the second term of the equation is due to the attraction while the third term is randomization with α being the randomization parameter and ε_i is random number being drawn from a Gaussian distribution. The parameter γ describes the variation of the attractiveness and it typically varies from 0.01 to 100. The parameter γ value is important in determining the speed of convergence and how the FA algorithm behaves.

3   Modified Bee-Firefly Algorithm

The MBF learning scheme algorithm is based on the incorporation of modified ABC optimization algorithm and FA algorithm inspired by the robustness and flexibility of ABC algorithm [9, 20] and the efficiency of FA algorithm in handling local search [11, 21]. The proposed learning scheme was expected to overcome the disadvantages caused by gradient descent BP-learning in the FLNN training procedure and improve the classification accuracy.

   In MABC algorithm Eq. (9) is used by both modified employed bee and modified onlooker bee, with x_k is a random selected neighboring solution in the population, the solution search dominated by Eq. (9) may become random enough which lead to a tendency to overpass the optimal value in the search space. For that reason, the MABC algorithm needs to be improved to balance the exploration and exploitation capability. Inspired by the firefly (FA) algorithm by Yang [22] which renowned to have good ability in local search [10, 11], this search strategy is adopted into the modified onlooker bees phase to balance the exploration and exploitation capability of MABC algorithm. The learning scheme is named as Modified Bee-Firefly (MBF) algorithm. The flow chart of MBF learning scheme for FLNN is given in Fig. 1.

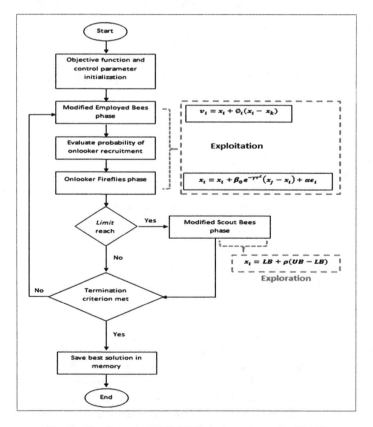

Fig. 1. The flow chart of MBF learning scheme for FLNN

4   FLNN-MBF for Mammographic Mass Classification

4.1   Mammographic Mass Data Set

The mammographic mass dataset was obtained from the Institute of Radiology, Gynecological Radiology, and University Erlangen-Nuremberg available from the UCI machine learning repository [23]. It contains a BI-RADS assessment, the patient's age, and three BI-RADS attributes together with the ground truth (the severity field) for 516 benign and 445 malignant masses that have been identified on a full field digital mammograms collected between the year of 2003 and 2006. The task is to predict the severity (benign or malignant) of mammographic mass lesion from BI-RADS attributes and the patient's age. Each instance has associated BI-RADS assessment ranging from one (definitely benign) to five (highly suggestive of malignancy) assigned in a double-review process by physicians.

4.2   Experiment Setting

The Simulation experiments were performed on the training of FLNN with standard BP-learning (FLNN-BP), FLNN with original standard ABC (FLNN-ABC), FLNN with modified ABC (FLNN-mABC) and FLNN with our proposed modified bee-firefly algorithm (FLNN-MBF). The result of the FLNN-MBF also was compared with standard model of Multilayer Perceptron (MLP-BP) to assess the classification performance in term of network complexity. Table 1 summarized parameters setting for the experiment.

Table 1. The parameter setting for the experiment

Parameters	MLP-BP	FLNN-BP	FLNN-ABC	FLNN-MABC	FLNN-MBF
Learning rate	0.1–0.5	0.1–0.5	–	–	–
Momentum	0.1–0.9	0.1–0.9	–	–	–
Epoch	1000	1000	1000	1000	1000
Minimum error	0.001	0.001	0.001	0.001	0.001

The best accuracy is recorded from these simulations. The activation function used for the both MLP and FLNN network output is Logistic sigmoid function. For the sake of convenience we set our FLNN input enhancement up to second order.

4.3   Simulation Results

Experimental results are presented as in Tables 2 and 3 with each learning scheme is observed base on accuracy, precision, sensitivity and F-measure. The classification accuracy results presented in Table 2 shows the FLNN-MBF gives better accuracy than MLP-BP with less tunable weights. Also, the FLNN-MBF steadily exhibited lower MSE and generated better accuracy result than the FLNN-BP.

Table 2. Result of FLNN-MBF with FLNN-BP and MLP-BP

Learning scheme	Best Network structure	Number of tunable weights	MSE	Accuracy (%)
MLP-BP	5-5-1	36	0.24798	69.82
FLNN-BP	15-1	16	0.17850	58.50
FLNN-MBF	**15-1**	**16**	**0.06679**	**83.45**

Table 3. Performance of network models on unseen data

Learning scheme	Precision	Sensitivity	F-Measure
MLP-BP	0.18415	0.35424	0.24233
FLNN-BP	0.49719	0.72850	0.59102
FLNN-ABC	0.82018	0.79913	0.80952
FLNN-MABC	0.83414	0.81026	0.82202
FLNN-MBF	**0.83939**	**0.82055**	**0.82987**

From Table 3, FLNN-MBF also shows higher sensitivity and precision value and obtained the highest F-measure thus suggested that in term of balance between precision and sensitivity the FLNN-MBF is a better model for solving the classification task on Mammographic mass classification task as compared to the MLP-BP, FLNN-BP, FLNN-ABC, and FLNN-MABC.

5   Conclusion

In this work, we implemented the FLNN-MBF model for the task of classification on mammographic mass data. The experiment demonstrated that FLNN-MBF can successfully perform the classification task with better accuracy result on unseen data with less neural architectural complexity. On the basis of these results it can be concluded that FLNN-MBF can be a competitive alternative classifier for mammographic mass classification task. In future, we will conduct an experiment on other medical dataset in order to explore and to evaluate the feasibility of the model.

Acknowledgement.   The authors would like to thank University Tun Hussein Onn Malaysia and Ministry of High Education (MOHE) for supporting this research under the Fundamental Research Grant Scheme (FRGS), vot. 1235.

References

1. Giles, C.L., Maxwell, T.: Learning, invariance, and generalization in high-order neural networks. Appl. Opt. **26**, 4972–4978 (1987)
2. Pao, Y.H., Takefuji, Y.: Functional-link net computing: theory, system architecture, and functionalities. Computer **25**, 76–79 (1992)

3. Samarasinghe, S.: Neural Networks for Applied Sciences and Engineering. Auerbach Publications, Boca Raton (2006)
4. Karaboga, D., Basturk, B.: On the performance of artificial bee colony (ABC) algorithm. Elsevier Appl. Soft Comput. **8**, 687–697 (2007)
5. Rojas, R.: Neural networks: a systematic introduction. Springer Science & Business Media, New York (2013)
6. Eberhart, R.C.: Neural network PC tools: a practical guide. Academic Press, San Diego (2014)
7. Rubio, J.D., Angelov, P., Pacheco, J.: Uniformly stable backpropagation algorithm to train a feedforward neural network. Neural Netw. IEEE Trans. **22**, 356–366 (2011)
8. Misra, B.B., Dehuri, S.: Functional link artificial neural network for classification task in data mining. J. Comput. Sci. **3**, 948–955 (2007)
9. Zhu, G., Kwong, S.: Gbest-guided artificial bee colony algorithm for numerical function optimization. Appl. Math. Comput. **217**, 3166–3173 (2010)
10. Fister, I., Fister Jr, I., Yang, X.-S., Brest, J.: A comprehensive review of firefly algorithms. Swarm Evol. Comput. **13**, 34–46 (2013)
11. Bacanin, N., Tuba, M.: Firefly algorithm for cardinality constrained mean-variance portfolio optimization problem with entropy diversity constraint. Sci. World J. **2014**, 721521 (2014)
12. Dehuri, S., Cho, S.-B.: A comprehensive survey on functional link neural networks and an adaptive PSO–BP learning for CFLNN. Neural Comput. Appl. **19**, 187–205 (2010)
13. Sierra, A., Macias, J.A., Corbacho, F.: Evolution of functional link networks. Evol. Comput. IEEE Trans. **5**, 54–65 (2001)
14. Sumathi, S., Paneerselvam, S.: Computational Intelligence Paradigms: Theory and Applications using MATLAB. CRC Press Inc, Boca Raton (2010)
15. Karaboga, D.: An Idea Based on Honey Bee Swarm for Numerical Optimization. Erciyes University, Engineering Faculty, Computer Science Department, Kayseri/Turkiye (2005)
16. Hassim, Y.M.M., Ghazali, R.: Using artificial bee colony to improve functional link neural network training. Appl. Mech. Mater. **263**, 2102–2108 (2013)
17. Hassim, Y.M.M., Ghazali, R.: A modified artificial bee colony optimization for functional link neural network training. In: Herawan, T., Deris, M.M., Abawajy, J. (eds.) Proceedings of the First International Conference on Advanced Data and Information Engineering. LNEE, vol. 285, pp. 69–78. Springer, Heidelberg (2014)
18. Hassim, Y.M.M., Ghazali, R.: Optimizing functional link neural network learning using modified bee colony on multi-class classifications. In: S. Obaidat, M. (ed.) Advanced in Computer Science and Its Applications. LNEE, vol. 279, pp. 153–159. Springer, Heidelberg (2014)
19. Yang, X.-S.: Firefly algorithms for multimodal optimization. In: Watanabe, O., Zeugmann, T. (eds.) SAGA 2009. LNCS, vol. 5792, pp. 169–178. Springer, Heidelberg (2009)
20. Karaboga, D., Ozturk, C.: A novel clustering approach: Artificial Bee Colony (ABC) algorithm. Appl. Soft Comput. **11**, 652–657 (2011)
21. Guo, L., Wang, G.-G., Wang, H., Wang, D.: An effective hybrid firefly algorithm with harmony search for global numerical optimization. Sci. World J. **2013**, 9 (2013)
22. Yang, X.S.: Firefly algorithm. Eng. Optim. **29**, 221–230 (2010)
23. Lichman, M.: UCI Machine Learning Repository [http://archive.ics.uci.edu/ml]. Irvine, CA. University of California, School of Information and Computer Science (2013)

Detecting Firefly Algorithm for Numerical Optimization

Yuchen Zhang[1], Xiujuan Lei[1(✉)], and Ying Tan[2]

[1] School of Computer Science, Shaanxi Normal University,
Xi'an 710119, China
xjlei68@163.com
[2] School of Electronics Engineering and Computer Science,
Peking University, Beijing 100871, China

Abstract. Firefly Algorithm (FA) is a stochastic optimization algorithm inspired by the swarm intelligence. It has the advantages of simple implementation, high efficiency and so on. However, the algorithm is easy to come into premature convergence and fall into local optimum. To address this problem, we proposed a novel firefly algorithm, Detecting Firefly Algorithm (DFA), in which we use a detecting firefly that flies round certain target points to improve the search path of standards FA. Moreover, the influence of the brightest firefly and the second brightest firefly is taken into consideration to optimize the movement strategy of the single firefly. The example illustrates that the higher precision and better convergence features of the proposed algorithm in numerical optimization.

Keywords: Firefly algorithm · Detecting firefly · Global optima

1   Introduction

Firefly algorithm (FA) is a stochastic optimization algorithm inspired by the swarm intelligence introduced by XinShe Yang [1] at Cambridge University. The algorithm is of higher convergence rate especially in function optimization. A lot of numerical examples have illustrated the superiority of this method. FA has been applied to many different fields. Nggam J. Cheung *et al.* analyzed of the relevant parameters of the algorithm [2]. Xinshe Yang *et al.* also made a further research on the algorithm that some unique test functions were introduced [3]. And the advanced nature of the algorithm was described in the optimization problem [4]. Xinshe Yang believed that it was more effective in solving NP-hard problems [5]. A new method using FA for hybrid data clustering was explained by Maheshwar *et al.* [6]. Shaik Farook used hybrid genetic-firefly algorithm for regulating LFC regulations in a deregulated power system [7].

However, standard FA is easy to fall into the local optimal solution in the global search. In order to overcome this defect of the algorithm, Arora *et al.* proposed mutated firefly algorithm to improve the convergence rate and accurateness [8]. Shuhao Yu *et al.* used a variable strategy for step size α to modify firefly algorithm (VSSFA) [9]. A new disturbance model of the brightest firefly was improved in the paper of S.L. Tilahun *et al.* [10]. And a peculiar movement strategy was described by adding inertia

© Springer International Publishing Switzerland 2016
Y. Tan et al. (Eds.): ICSI 2016, Part I, LNCS 9712, pp. 200–210, 2016.
DOI: 10.1007/978-3-319-41000-5_20

weight in literature [11]. At the same time, Mahdi Bidar *et al.* described a new method called JFA which promoted the low brightness firefly [12]. The intelligent firefly algorithm (IFA) was introduced by Seif-Eddeen K. Fateen *et al.* in which a firefly acted intelligently by basing its move on top ranking fireflies [13]. Milan Tuba *et al.* also upgraded FA for portfolio optimization problem [14]. In addition, Le'vy Flights was combined with search strategy of FA in literature of XinShe Yang *et al.* [15]. Some scholars also introduced the chaos mechanism into the algorithm [16, 17]. As well, a kind of adaptive firefly optimization algorithm based on stochastic inertia weight was discussed by Changnian Liu [18].

In the above improved methods, the search path of the firefly was not mentioned. In standard FA, a firefly in the search process is not clear about the direction of movement, so that it is not conducive to find the global optimal solution. Therefore, we presented a kind of detecting firefly which surrounds a target point to do a spiral motion, and gradually converges to the target point. When the scope of the definition domain is larger, the more dispersed fireflies will lead to lower convergence rate. In order to solve this problem, we add the item of brightness to the definition of attractiveness. And the influence of the global optimum and the suboptimum also was considered to join in the process of the movement of the firefly.

The rest of the paper is structured as follows. The second section is a review of the standard FA. DFA is described in third section. In fourth section, the proposed algorithm is tested on 28 benchmark functions. Moreover, the simulation result is compared with the standard FA and several intelligent optimization algorithms and then the performance of DFA is verified. The last section summarizes the paper.

2 Standard Firefly Algorithm

For simplicity in describing FA which was developed by XinShe Yang, There are three idealized assumptions as following [1]:

- Each firefly will be attracted to other fireflies only depend on their brightness regardless of their sex or other.
- Any two flashing fireflies, the less bright one will be attracted and move towards the brighter one. The attractiveness of fireflies both decrease as their distance increases. If there is no brighter one than a particular firefly, it will move randomly.
- The brightness of a firefly is affected or determined by the value of the objective function.

In standard FA, the brightness of a firefly is proportional to the value of objective function. The absolute brightness of a firefly at a particular location x is chosen as:

$$I(x) \propto f(x) \tag{1}$$

where $I(x)$ is absolute brightness when the attenuation of brightness in the medium is not considered. As attractiveness of a firefly is inverse proportional to the distance seen by adjacent fireflies, we define the attractiveness β of a firefly by:

$$\beta = \beta_0 e^{-\gamma r^2} \tag{2}$$

where β_0 is the attractiveness at $r = 0$, γ is a light absorption coefficient. The movement strategy of a firefly i is attracted to another more attractive (or brighter) firefly j is determined by:

$$x_i = x_i + \beta(x_j - x_i) + \alpha\varepsilon_i \tag{3}$$

where the second term is due to the attraction. The third term is randomized with α being the randomization parameter, and ε_i is a vector of random numbers drawn form a Gaussian distribution or uniform distribution.

3   Detecting Firefly Algorithm

3.1   Detecting Firefly

Inspired by detecting particle swarm optimization that was proposed by Y Zhang [19], we introduced the detecting firefly to FA. In standard FA, the search path of single firefly is disorganized. It is easy to make the algorithm fall into local optimal solution. The aim of this research is to help the firefly escape from local optimal trap. We selected a small portion of the n fireflies whose number was marked as n_s. These fireflies which are known as detecting firefly try to find a better potential optimal value along the spiral convergence line path (a spiral center is required in the domain of definition). Using the detecting firefly can expand the search scope of the algorithm. In the meantime, the rest of fireflies search the optimal value in accordance with the standard FA. That natural firefly and detecting firefly complement each other effectively to maintain population diversity and avoid premature phenomenon in the evolution.

In order to make the detecting firefly move around the target point (such as current global optimal value) along the path of the spiral line, we used an offset ΔX which is equal to the speed of particle in PSO. When detecting firefly i is going to move, its location update formula is defined as follows:

$$\Delta X_i^{k+1} = C_{s1}\Delta X_i^k + C_{s2}(X_{gbest} - X_i^k) \tag{4}$$

$$X_i^{k+1} = X_i^k + \Delta X_i^{k+1} \tag{5}$$

where X_{gbest} is the position of the brightest firefly, coefficient k represents the k_{th} iteration, C_{s1} and C_{s2} respectively represent the proportion of ΔX_i^k and $(X_{gbest} - X_i^k)$ in movement direction of the detecting firefly and the initial position X_i^0 and initial offset ΔX_i^0 are random. We selected the position of current optimal firefly as the target. $n_s(n_s = 5)$ detecting fireflies travel in the spiral track which were given in Fig. 1 around the position of current optimal firefly. This process will be iterated for T ($T = 10$) times. For example, we selected firefly i as a detecting firefly. The original brightness I_i of firefly i and the brightness I_i^* of detecting firefly i^* which is the brightest in T iterations

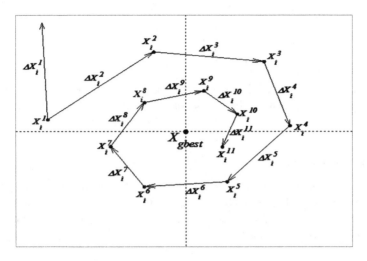

Fig. 1. The search path of the detecting firefly

are compared. If the brightness I_i^* is greater than the original brightness I_i, the algorithm will use the position of detecting firefly i^* to update the original position of firefly i.

3.2 Randomization Parameter (Step Length) and Attractiveness

In fact, the randomization parameter α of the standard FA is fixed in the formula of movement of a firefly. It cannot be perfectly suited to the search process of firefly. When α is large, the algorithm has a satisfied convergence rate. But it is easy to ignore the potential of the optimal value, so that it reduces the accuracy in late search. On the other hand, when α is smaller, the algorithm has the high precision in local. However, reduced search space can result in that the algorithm cannot find the global optimal value. In order to balance the search range and precision, we used an adaptive strategy to generate α [1] as follow:

$$\alpha = \alpha_e + (\alpha_s - \alpha_e)e^{-t} \tag{6}$$

where α_s is initial value, α_e is the final value specified, coefficient t represents the t_{th} iteration. The formula shows that α decreases exponentially with the increase of the iteration.

In standard FA, the movement direction of the firefly is determined by the brightness of each firefly. But the degree of the attractiveness between the fireflies is only related to the distance. It does not reflect the impact of the brightness in the movement of the firefly [9]. If the function optimization problem has a larger scope of definition, a brighter firefly which is far from the others will not have enough attractiveness to the others. So in DFA, a new method of calculating attractiveness was designed depending on both brightness and direction. If firefly j is brighter than firefly i, the attractiveness β_{ij} of firefly j to firefly i is given by:

$$\beta_{ij} = \begin{cases} \dfrac{I_j}{I_i} e^{-\gamma r^2} & I_i > 0 \\ \beta_o & I_i = 0 \end{cases} \tag{7}$$

where I_i and I_j are absolute brightness of firefly i and firefly j respectively. In this paper, it is important to note that the brightness of all fireflies is greater than zero which depends on the objective function. Therefore, the minimum brightness is zero.

3.3   Firefly Movement

In the position update process of original firefly, the current brightest firefly has not been concerned in the algorithm. XinShe Yang introduced to add an extra item about the brightest firefly into Eq. 3 [1]. But this strategy is likely to cause that the movement direction of the firefly becomes single.

Table 1. The pseudo code of DFA

Detecting Firefly Algorithm:
Objective function $F(X)$, $X = (x_1, \dots, x_d)^T$
Generate initial positions of all fireflies $X_i (i = 1, \dots, n)$
Brightness I_i at X_i is determined by $F(X_i)$
Define light absorption coefficient ⌐ number of detecting firefly n_s
while($t <$ MaxGeneration)
Randomly select n_s fireflies as detecting fireflies
for $i = 1:n$ all n fireflies
if (firefly i is detecting firefly)
Randomly generate initial position of detecting firefly i
for $k = 1:T$
Update position detecting firefly i *via* Eq.4-5
Evaluate new solutions
end for k
Determine the brightest firefly i^* in T iterations
if ($I_i^* > I_i$)
Firefly i is replaced by detecting firefly i^*
end if
else
for $j=1:n$ all n fireflies
if ($I_j > I_i$)
Attractiveness varies with brightness and distance *via* Eq.7
Move firefly i towards j in d-dimension according to Eq.8
end if
Evaluate new solutions and update brightness
end for j
end if
end for i
Rank the fireflies and find the current optimum and the suboptimum
end while
Post process result and visualization

DFA was continued to add another item about second brightest in our improvement. When the firefly is moving, the space between the optimum and the suboptimum
will be searched. It avoids the premature convergence that is caused by adding the
optimal location only. Suppose that the brightness of firefly i is not as good as firefly j,
firefly i is attracted by firefly j. The updated location of firefly i is defined by:

$$x_i = x_i + \beta_{ij}(x_j - x_i) + \alpha\varepsilon_{i1} + \lambda_1\varepsilon_{i2}(x_{gbest} - x_i) + \lambda_2\varepsilon_{i3}(x'_{gbest} - x_i) \qquad (8)$$

where x_{gbest} is the location of the brightest firefly and x'_{gbest} is the location of the second
brightest firefly. ε_{i1}, ε_{i2} are vectors of random numbers in [-0.5,-0.5]. λ_1, λ_2 are inertia
weight. If a firefly is the brightest firefly, it will move randomly. We used the best
firefly after disturbing to replace the original firefly. According to Eqs. 4–8, the pseudo
code of DFA is given by Table 1.

4   Simulation and Experiments

The algorithm was compared with other intelligent optimization algorithms, such as
standard particle swarm algorithm, genetic algorithm, SFA, and VSSFA [9]. The
parameters of these algorithm were listed by Table 2. The proposed DFA was tested by
28 benchmark functions in CEC 2013 which be summarized in paper [20]. There are
five unimodal functions (from f_1 to f_5), 20 basic multimodal functions (from f_6 to f_{20})
and 8 composition functions (from f_{21} to f_{28}). All problems are the minimum optimization. And the dimensions of all test functions were 30. The scope of each
dimension was in [−100, 100]. We adopted 50 independent runs for each of test
function and then calculated the average of 50 results.

Table 2. The parameters' setting

Parameter	Variable	GA	SPSO	FA	VSSFA	DFA
Cross coefficient	q_c	0.9	–	–	–	–
Mutation coefficient	q_m	0.1	–	–	–	–
Learning factor	c_1	–	2	–	–	–
Learning factor	c_2	–	2	–	–	–
Maximum velocity	v_{max}	–	0.5	–	–	–
weight	w	–	1	–	–	–
Step length	α_0	–	–	0.2	–	–
Initial step length	α_s	–	–	–	–	1
Final step length	α_e	–	–	–	–	0.1
Maximal attractiveness	β_0	–	–	1	1	1
Light absorption coefficient	γ	–	–	1	1	1
Offset weight	C_{s1}	–	–	–	–	0.8
Target weight	C_{s2}	–	–	–	–	0.8
Optimal factor	λ_1	–	–	–	–	0.6
Sub optimal factor	λ_2	–	–	–	–	0.6

The average values on the optimization were presented in Table 3, where the top of the ranked value was highlighted in boldface. From the results, the average ranking of DFA is 1.46 that is far higher than other algorithms in ranking.

Table 3. Simulation result of the 28 benchmark functions

f	Function Average Value and Rank									
	SPSO	Rank	GA	Rank	SFA	Rank	VSSFA	Rank	DFA	Rank
f_1	**-1.400E + 03**	**1**	1.234E + 04	5	-1.392E + 03	3	-1.389E + 03	4	**-1.400E + 03**	**1**
f_2	**1.249E + 06**	**1**	2.683E + 08	5	7.971E + 06	3	7.105E + 06	2	8.433E + 06	4
f_3	6.361E + 08	2	7.164E + 10	5	1.835E + 09	4	9.300E + 08	3	**5.088E + 08**	**1**
f_4	**1.615E + 04**	**1**	6.880E + 04	5	2.697E + 04	2	3.534E + 04	3	3.944E + 04	4
f_5	-9.999E + 02	2	1.018E + 03	5	-9.972E + 02	4	-9.964E + 02	4	**-1.000E + 03**	**1**
f_6	**-8.581E + 02**	**1**	3.138E + 02	5	-8.421E + 02	3	-8.368E + 02	4	-8.454E + 02	2
f_7	3.070E + 05	5	-5.373E + 02	4	-6.615E + 02	3	-7.074E + 02	2	**-7.215E + 02**	**1**
f_8	-6.790E + 02	3	**-6.791E + 02**	**1**	-6.786E + 02	5	-6.786E + 02	4	-6.790E + 02	2
f_9	-5.602E + 02	5	-5.609E + 02	4	-5.718E + 02	3	-5.741E + 02	2	**-5.754E + 02**	**1**
f_{10}	**-4.989E + 02**	**1**	1.725E + 03	5	-4.943E + 02	3	-4.953E + 02	4	-4.978E + 02	2
f_{11}	6.374E + 01	5	-2.301E + 01	4	-3.487E + 02	3	-3.490E + 02	2	**-3.609E + 02**	**1**
f_{12}	1.598E + 02	5	1.463E + 02	4	-1.128E + 02	2	-9.948E + 01	3	**-2.109E + 02**	**1**
f_{13}	3.683E + 02	5	2.632E + 02	4	8.368E + 01	3	5.537E + 01	2	**-4.610E + 01**	**1**
f_{14}	5.604E + 03	4	6.241E + 03	5	1.477E + 03	2	1.563E + 03	3	**1.186E + 03**	**1**
f_{15}	6.363E + 03	4	7.475E + 03	5	4.554E + 03	2	**4.512E + 03**	**1**	5.455E + 03	3
f_{16}	2.026E + 02	2	2.025E + 02	2	2.058E + 02	4	2.060E + 02	5	**2.016E + 02**	**1**
f_{17}	7.763E + 02	4	1.162E + 03	5	5.610E + 02	3	5.475E + 02	2	**4.447E + 02**	**1**
f_{18}	8.940E + 02	4	1.264E + 03	5	6.900E + 02	2	6.927E + 02	3	**5.748E + 02**	**1**
f_{19}	5.526E + 02	4	1.038E + 04	5	5.170E + 02	3	5.163E + 02	2	**5.103E + 02**	**1**
f_{20}	6.149E + 02	4	6.149E + 02	5	6.148E + 02	3	6.145E + 02	2	**6.141E + 02**	**1**
f_{21}	1.043E + 03	3	3.114E + 03	5	1.038E + 03	2	1.059E + 03	4	**9.906E + 02**	**1**
f_{22}	6.439E + 03	4	7.542E + 03	5	2.311E + 03	2	2.427E + 03	3	**1.949E + 03**	**1**
f_{23}	7.584E + 03	4	8.883E + 03	5	6.269E + 03	2	**6.114E + 03**	**1**	6.607E + 03	3
f_{24}	1.398E + 03	5	1.322E + 03	4	1.278E + 03	3	1.275E + 03	2	**1.266E + 03**	**1**
f_{25}	1.509E + 03	5	1.440E + 03	4	1.400E + 03	3	1.397E + 03	2	**1.381E + 03**	**1**
f_{26}	1.550E + 03	5	1.426E + 03	4	1.508E + 03	2	1.531E + 03	3	**1.407E + 03**	**1**
f_{27}	2.808E + 03	5	2.685E + 03	4	2.336E + 03	3	2.263E + 03	2	**2.242E + 03**	**1**
f_{28}	5.453E + 03	5	4.508E + 03	4	2.199E + 03	3	2.139E + 03	2	**1.806E + 03**	**1**
Mean	SPSO	3.57	GA	4.39	FA	2.82	VSSFA	2.71	DFA	**1.46**

Though the performance of DFA is not as good as SPSO on function f_2, f_4, f_6, f_{10}, and is worse than GA and VSSFA respectively on function f_8 and f_{15}, f_{23}, the algorithm has an absolute advantage on the rest of test functions. The simulation and experiments show that the performance of the algorithm is significant which beyond other in finding the global optimal value.

To amplify the convergence property of DFA, the convergence curves were displayed at Figs. 2, 3, 4 and 5. Due to space limitations, our paper only described four convergence curves on functions $f_1, f_{11}, f_{13}, f_{28}$. Figures 3 and 4 show that convergence rate of DFA was greater than other algorithms on Sphere function and Rastrigin's function. On Sphere function, DFA began to converge at the 400_{th} iteration. But SFA and VSSFA still fluctuated at the 800_{th} iteration. DFA began to converge at the 300_{th}

iteration on Rastrigin's function. However, SFA and VSSFA were not convergent in 1000_{th} iterations. On non-continuous rotated Rastrigin's function, DFA is better than FA, not as good as VSSFA. And on the composition function $8(n = 5$, rotated), the convergence rate of DFA is as high as that of SFA, which is higher than VSSFA.

The experiments displayed that DFA has well accuracy and convergence rate on different function optimization problems. Adding the detecting firefly made the algorithm escape from the local optimal trap and premature convergence. The decline of convergence rate that detecting firefly caused was blocked by adding optimum and suboptimum.

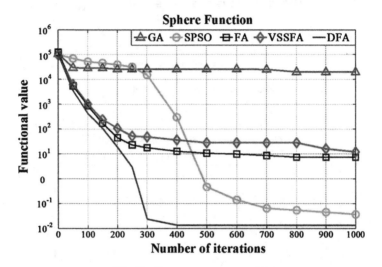

Fig. 2. Convergence curve of f_1

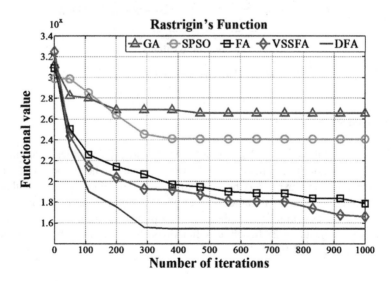

Fig. 3. Convergence curve of f_{11}

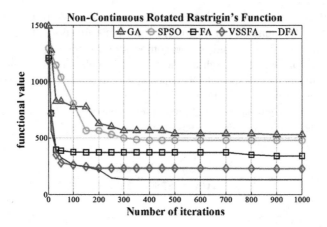

Fig. 4. Convergence curve of f_{13}

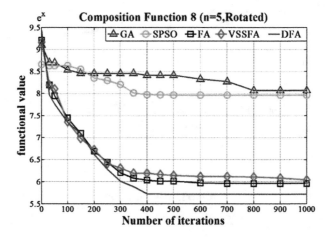

Fig. 5. Convergence curve of f_{28}

5 Conclusion

In the paper, to address the problem that SFA easy to fall into local optimum, an improved firefly algorithm, named as DFA, was proposed. The proposed algorithm based on detecting firefly expanded the searching range of the firefly swarm, so that it is easier to find the global optimal value. Further, that the optimal location was added to the movement strategy highlighted the role of optimal location in the algorithm. Finally, the experiments on 28 benchmark functions show that DFA has a significant improvement on standard FA.

Acknowledgments. This paper is supported by the National Natural Science Foundation of China (61502290, 61401263), Industrial Research Project of Science and Technology in Shaanxi Province(2015GY016), the Fundamental Research Funds for the Central Universities, Shaanxi Normal University (GK201501008) and China Postdoctoral Science Foundation (2015M582606).

References

1. Yang, X.S.: Firefly algorithm. Nature-Inspired Metaheuristic Algorithms Second Edition. Luniver Press, Bristol (2010)
2. Cheung, N.J., Ding, X.M., Shen, H.B.: Adaptive firefly algorithm: parameter analysis and its application. PLoS ONE **9**(11), e112634 (2014)
3. Yang, X.S.: Firefly algorithm, stochastic test functions and design optimization. Int. J. Bio-Inspired Comput. **2**(2), 78–84 (2010)
4. Johari, N.F., Zain, A.M., Mustaffa, N.H.: Firefly algorithm for optimization problem. Appl. Mech. Mater. **421**, 512–517 (2013)
5. Yang, X.S.: Firefly algorithms for multimodal optimization. Stochast. Algorithms Found. Appl. **5792**, 169–178 (2010)
6. Maheshwar Kaushik, K., Arora, V.: A hybrid data clustering using firefly algorithm based improved genetic algorithm. Procedia Comput. Sci. **58**, 249–256 (2015)
7. Farook, S.: Regulating LFC regulations in a deregulated power system using hybrid genetic-firefly algorithm. In: IEEE International Conference on Electrical, Computer and Communication Technologies (ICECCT), IEEE International Conference, Coimbatore, India, pp. 1–7 (2015)
8. Arora, S., Singh, S., Singh, S., Sharma, B.: Mutated firefly algorithm. In: International Conference on Parallel, Distributed and Grid Computing, pp. 33–38 (2014)
9. Yu, S., Zhu, S., Ma, Y., Mao, D.: A variable step size firefly algorithm for numerical optimization. Appl. Math. Comput. **263**, 214–220 (2015)
10. Tilahun, S.L., Ong, H.C.: Modified firefly algorithm. J. Appl. Math. **467631**(12), 2428–2439 (2012)
11. Tian, Y., Gao, W., Yan, S.: An improved inertia weight firefly optimization algorithm and application. Int. Conf. Control Eng. Commun. Technol. Liaoning China **4**, 64–68 (2012)
12. Bidar, M., Kanan, H.R.: Jumper firefly algorithm. In: 3rd International Conference on Computer and Knowledge Engineering (ICCKE), Mashhad, Iran, pp. 267–271 (2013)
13. Fateen, S.E.K.: Intelligent firefly algorithm for global optimization. In: Yang, X.-S. (ed.) Cuckoo Search and Firefly Algorithm. SCI, vol. 516, pp. 315–330. Springer, Heidelberg (2014)
14. Tuba, M., Bacanin, N.: Upgraded firefly algorithm for portfolio optimization problem. In: UK Sim-AMSS 16th International Conference on Computer Modelling and Simulation, Cambridge, USA, pp. 113–118 (2014)
15. Yang, X.S.: Firefly algorithm, L'evy flights and global optimization. In: Bramer, M., Ellis, R., Petridis, M. (eds.) Research and Development in Intelligent Systems XXVI, Springer London, pp. 209–218 (2010)
16. Gandomi, A.H., Yang, X.S., Talatahari, S., Alaiv, A.-H.: Firefly algorithm with chaos. Commun. Nonlinear Sci. Numer. Simul. **18**(1), 89–98 (2013)
17. Baykasoglu, A., Ozsoydan, F.B.: Adaptive firefly algorithm with chaos for mechanical design optimization problems. J Appl. Soft Comput. **36**(c), 152–164 (2015)

18. Liu, C.N., Tian, Y.F., Zhang Q., Yuan J., Xue, B.B.: Adaptive firefly optimization algorithm based on stochastic inertia weight. In: Sixth International Symposium on Computational Intelligence and Design, Hangzhou, China, pp. 334–337 (2013)
19. Zhang, Y.N., Teng, H.F.: Detecting particle swarm optimization. Concurrency Comput. Pract. Experience **21**(4), 449–473 (2009)
20. Liang, J.J., Qu, B.Y., Suganthan, P.N., Hernández-Díaz, A.G.: Problem Definitions and Evaluation Criteria for the CEC 2013 Special Session on Real-Parameter Optimization

Dragonfly Algorithm Based Global Maximum Power Point Tracker for Photovoltaic Systems

Gururaghav Raman$^{(\boxtimes)}$, Gurupraanesh Raman,
Chakkarapani Manickam, and Saravana Ilango Ganesan

Department of Electrical and Electronics Engineering,
National Institute of Technology, Tiruchirappalli 620015, India
gururaghav.raman@gmail.com, gurupraanesh@gmail.com,
chakra_nit@yahoo.com, gsilango@nitt.edu

Abstract. This paper presents the application of the Dragonfly Algorithm (DA) for tracking the Global Maximum Power Point (GMPP) of a photovoltaic (PV) system. Optimization techniques employed for GMPP tracking (GMPPT) are required to be fast and efficient in order to reduce the tracking time and energy loss respectively. The DA, being a meta-heuristic algorithm with good exploration and exploitation characteristics, is a suitable candidate for this application. Due to its simplicity, the DA is implemented on a low cost microcontroller, and is proven to track the GMPP effectively under various irradiation conditions. The performance of the proposed DA based GMPPT scheme is compared with that of the conventional PSO based GMPPT scheme, and proves to be superior in terms of tracking time and energy loss during tracking.

Keywords: Dragonfly algorithm · Global maximum power point tracking · Photovoltaic systems · Swarm intelligence

1 Introduction

Photovoltaic (PV) modules exhibit a non-linear Voltage-Current (V-I) characteristic that changes with solar irradiance and the module temperature [1]. Due to the high installation cost, it becomes imperative to utilize these modules to the fullest extent. Therefore, Maximum Power Point Tracking (MPPT) is essential in order to extract the maximum possible power under all operating conditions [2]. A single PV module has a very small voltage and current rating that is not sufficient to drive most loads. Hence, many modules are connected in series and parallel, forming a PV Array, to increase the voltage and current capability of the PV system respectively. If all the modules in the array experience the same irradiation level, then the Power-Voltage (P-V) characteristic of the overall array has a single maximum. In other words, it is a unimodal function. This curve can easily be tracked to obtain the Maximum Power Point (MPP) using conventional MPPT algorithms such as the Perturb and Observe (P&O) [3], and Incremental Conductance (INC) [4] techniques. However, if non-uniform irradiation falls on the system due to passing clouds, nearby buildings, poles, trees or other structures, the P-V curve is multi-modal [5], and has one Global Maximum Power

© Springer International Publishing Switzerland 2016
Y. Tan et al. (Eds.): ICSI 2016, Part I, LNCS 9712, pp. 211–219, 2016.
DOI: 10.1007/978-3-319-41000-5_21

Point (GMPP) and several Local Maximum Power Points (LMPPs). In such a case, it is very likely that conventional tracking schemes such as the P&O and INC algorithms get stuck in a LMPP, thereby resulting in sustained power losses. Thus, there is a requirement for Global MPPT (GMPPT) techniques.

In literature, several swarm intelligence-based GMPPT algorithms have been proposed. The Particle Swarm Optimization (PSO) is one of the first and most popular algorithms due to its simplicity, and ease of implementation [6]. Subsequently, other algorithms have been proposed, such as the Artificial Bee Colony [7], and Grey-Wolf Optimization [8]. Recently, Mirjalili [9] has proposed a new meta-heuristic swarm intelligence-based optimization technique called the Dragonfly Algorithm (DA). This algorithm was inspired by the swarming behaviour of dragonflies in nature, as they navigate, search for food, and evade enemies. In this paper, the dragonfly algorithm has been applied to determine the GMPP of a PV string under uniform and partial shading conditions. The performance of the DA is compared with that of PSO, and is demonstrated to be superior in terms of tracking time and the energy loss incurred during the tracking process.

2   PV System Characteristics and the Objective Function

To simulate this behaviour, the PV module is assumed to be represented by the single diode model shown in Fig. 1, and its specifications are given in Table 1. The corresponding equation for the current of the module is

$$I_{PV} = I_{ph} - I_o \left[\exp \left(\frac{V_{PV} + R_s I_{PV}}{A} \right) - 1 \right] - \frac{V_{PV} + R_s I_{PV}}{R_{sh}} \tag{1}$$

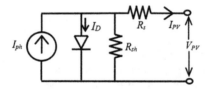

Fig. 1.   Equivalent circuit of a PV module.

Table 1.   PV Module specifications at 1000 W/m^2, 25°C.

PV Power	40 W
Open Circuit Voltage (V_{oc})	11 V
Short Circuit Current (I_{sc})	4.7A
MPP Voltage (V_m)	9 V
MPP Current (I_m)	4.44A

Here, I_{ph} is the light generated current, I_o is the reverse saturation current, and R_s and R_{sh} are the series and shunt resistances respectively. The parameter $A = \eta mkT/q$, where T is the module temperature in Kelvin, k is the Boltzmann's constant, q is the elementary charge, m is the number of cells in series, and η is the ideality factor of the diode. Consider a PV string consisting of four series-connected modules. Depending on the irradiation conditions over the modules, the GMPP location can either be the unique MPP, or, in case of partial shading, it could be one of the local MPPs. Since these modules are connected in series, the voltage of the string is given by the summation of voltages of all the modules, while the current of the string is equal to that of each module. The current-voltage (I-V) and the power-voltage (P-V) characteristics of this string, for different irradiation conditions as detailed in Table 2, are shown in Fig. 2. This P-V curve becomes the objective function for this one-dimensional maximization problem.

Table 2. Irradiation profiles for the PV string.

Curve	Irradiation (W/m^2)			
	Module 1	Module 2	Module 3	Module 4
A	900	900	900	900
B	1000	1000	1000	800
C	1000	1000	900	500

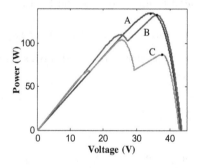 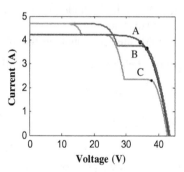

Fig. 2. P-V and I-V characteristics for the PV string.

When all the modules receive the same irradiation of 900 W/m^2, the P-V curve exhibits a unique MPP, as shown in curve A of Fig. 2. However, if partial shading occurs, as shown by curves B and C of Fig. 2, multiple MPPs are formed, of which one is the GMPP. It is clear that the conventional P&O technique cannot determine the GMPP. Hence, global MPPT algorithms are employed.

3 The Dragonfly Algorithm

The Dragonfly Algorithm (DA) is a global optimization technique based on swarm intelligence. It was developed based on the static and dynamic behaviour of dragonflies in nature. It models the social interaction amongst the dragonflies while they navigate, look for food, or evade enemies. This provides it with the exploration and exploitation characteristics that are so essential for optimization.

A static swarm of dragonflies refers to a small group of dragonflies that hunt for prey in a small locality. Their motion in this locality is characterized by small movements and abrupt changes in their flight direction. To the contrary, a dynamic swarm comprises of a large number of dragonflies that maintain a constant direction of motion with an objective of migrating to another place over a long distance. Considering an optimization problem to be analogous to the objectives of the dragonflies in a swarm, it can be seen that the static swarming behaviour represents the exploitation phase, while the dynamic swarming behaviour demonstrates the exploration phase of the swarm. This forms the basis of the DA.

To derive a mathematical formulation to represent the motion of the dragonflies in a swarm, the following five behaviours of dragonflies are established:

Separation: Any dragonfly does not collide with any other fly in a static swarm within a given neighbourhood.

Alignment: The velocity of an individual matches with that of the other dragonflies in the same neighbourhood.

Cohesion: All individuals tend to move towards the centre of mass of the neighbourhood.

Food: All the individuals in the swarm tend to move towards the food as this is essential for survival.

Enemy: All the individuals in the swarm tend to move away from an enemy.

These five behaviours of the dragonflies are represented mathematically as follows. The separation S_i of the ith individual is given by

$$S_i = -\sum_{k=1}^{N} X - X_k \tag{2}$$

Here, X represents the position of the current dragonfly, while X_k is the position of the kth neighbouring dragonfly, with N being the total number of neighbouring individuals. The alignment term is calculated as

$$A_i = \frac{\sum_{k=1}^{N} V_k}{N} \tag{3}$$

V_k represents the velocity of the kth neighbouring individual. The cohesion is calculated as

$$C_i = \frac{\sum\limits_{k=1}^{N} X_k}{N} - X \tag{4}$$

Attraction towards food located at a position X_F is modelled as

$$F_i = X_F - X \tag{5}$$

while the distraction from the enemy located at a position X_E is modelled as

$$E_i = X_E + X \tag{6}$$

All these five motions, together, influence the behaviour of the dragonflies in the swarm. The updation of the position is done using the step (ΔX), which is calculated as follows.

$$\Delta X_i = (sS_i + aA_i + cC_i + fF_i + eE_i) + w \, \Delta X_i \tag{7}$$

Here, s, a, and c are respectively the separation, alignment, and cohesion weights, while f and e are the food, and enemy factors respectively. S_i, A_i, C_i, F_i, and E_i indicate the separation, alignment, cohesion, position of the food, and position of the enemy respectively for the ith individual, and are treated to be scalar quantities. w represents the inertia weight. Having calculated the step ΔX, the position X_i of the individual is updated according to the following equation.

$$X_i = X_i + \Delta X_i \tag{8}$$

With different values of the weights, varied explorative or exploitative behaviours of the swarm can be realised. Static swarming demands high cohesion, and low alignment, and hence for exploitation, the value of c is kept large, while a is kept low. On the other hand, in a dynamic swarm, for good exploration characteristics, c is kept small, and a is increased.

3.1 Implementation of DA for GMPPT of PV Systems

In the context of GMPPT of PV systems, the position of the dragonfly is the duty ratio at which the DC-DC converter operates. This problem is one-dimensional in nature. The objective function to be optimized is the power output of the PV array, and the DA is used for maximizing this quantity. The following steps are carried out for implementing the DA for GMPPT:

Step 1: The dragonflies are distributed in the one dimensional search space between d_{max} and d_{min}, which are the limits of the search space in terms of the duty cycle of operation. In this paper, the value of d_{max} and d_{min} are respectively chosen as 98 % and 2 % respectively. The value of the step (ΔX_i) for the population is initialized.

Step 2: The PV power at each location (duty ratio) of the dragonflies is determined. Based on this, the food source and the enemy locations are updated.

Step 3: The values of s, a, c, f, and e are updated.

Step 4: S, A, C, F, and E for all the individuals are calculated according to the Eqs. (2) to (6).

Step 5: The neighbourhood radius is updated.

Step 6: The steps are calculated according to (7). The positions of the dragonflies are updated using (8). If (8) yields a position that lies outside the boundaries d_{max} and d_{min}, the flies are positioned at the opposite boundary.

Step 7: Check the termination condition. Exit the program if satisfied. Otherwise, go to step 2.

Step 8: Restart the search when the sensed power level changes.

For this application, the radius of the neighbourhood around any individual is increased, and the weights s, a, c, f, and e, decreased in proportion to the iteration count. This aids the swarm to achieve good exploration in the initial iterations. In the later iterations, as the algorithm proceeds towards convergence, the dragonflies are more influenced by each other. This would ensure accurate convergence, and prevent the dragonflies from getting stuck in local maxima. The food sources and the enemy are respectively chosen as the best and the worst solutions as found by the swarm until that moment. This allows the swarm to move towards the best position found so far, while it moves away from non-promising areas in the search space. For this problem, the termination criterion is chosen to be that condition when the mean deviation of all the dragonflies about the best position is less than 5 %. This value can be selected by the user in accordance to the required accuracy of convergence.

4   Results and Discussions

For testing the proposed GMPPT strategy, the hardware setup shown in Fig. 3 is considered. The PV string is emulated using the Agilent E4360A solar simulator. The output of the string is connected to the terminals of a DC-DC boost converter with a 100Ω load resistor. The switching frequency is set at 20 kHz, and the converter specifications are $L = 475\mu H$ and $C_{in} = C_{out} = 120\mu F$. The control strategy is implemented using the MSP430G2553 microcontroller. The DA parameters used are: number of dragonflies = 4, $s = 0.1$, $a = 0.1$, $c = 0.7$, $f = 1$, and $e = 1$.

A comprehensive evaluation of the DA in comparison with existing swarm intelligence techniques against various benchmark functions is provided in [9]. For this particular application, however, it is difficult to select a fair benchmark for comparison with other GMPPT techniques [10] owing to the use of varied power converter topologies, ratings and irradiance patterns. Therefore, the authors in this work consider various types of irradiance patterns to prove the effectiveness of the DA for GMPPT. Three different curves A, B and C are used for testing its performance, the specifications of which are given in Table 2. These curves represent a variety of irradiation patterns, namely uniform irradiation, and non-uniform irradiation with first and second peak as the global peak. For each of these curves, the performance of the DA-based GMPPT is compared with that of the PSO-based GMPPT with the performance indices

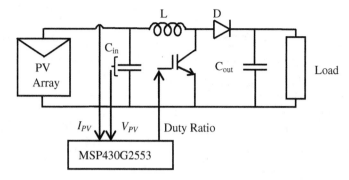

Fig. 3. Hardware implementation of the proposed GMPPT control strategy.

being tracking time and energy lost. The PSO parameters w, c_1 and c_2 are respectively set as 0.7, 1.5 and 1.5. To have a fair comparison with the DA, the number of particles in the PSO is taken as 4. The tracking waveforms of the PV voltage, current and power for the curves A, B, and C, are presented in Figs. 4, 5 and 6 respectively. The performance parameters obtained from the tracking results are tabulated in Table 3. In Fig. 4, the irradiance is 900 W/m^2 uniformly over all the modules. For this P-V curve, the DA takes 307.5 ms and 4 iterations to track the GMPP at 134.8 W, 34.2 V. For the same irradiance pattern, the PSO algorithm takes 910 ms and 6 iterations to track the GMPP. Further, it is seen that the energy loss due to the tracking of the DA is 13.9 J, while it is 46.55 J for PSO. The difference is 32.65 J, and this value increases as the power rating of the string increases. From Table 3, it is clear that irrespective of the type of irradiation, the DA-based tracker exhibits a smaller tracking time, with lower energy losses when compared to the PSO-based tracker.

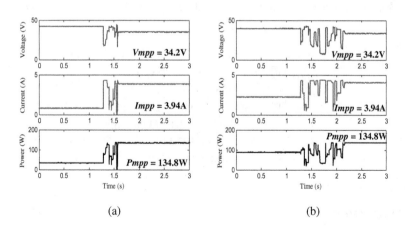

(a)   (b)

Fig. 4. Experimental waveforms for tracking Curve A using (a) DA (b) PSO.

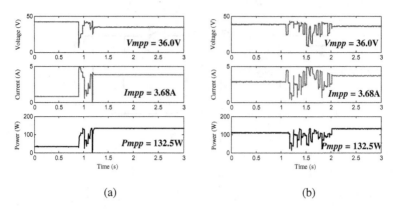

(a)   (b)

Fig. 5. Experimental waveforms for tracking Curve B using (a) DA (b) PSO.

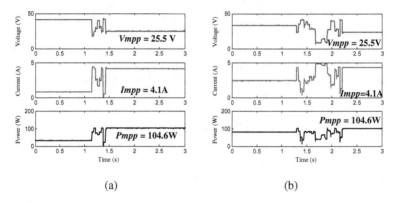

(a)   (b)

Fig. 6. Experimental waveforms for tracking Curve C using (a) DA (b) PSO.

Table 3. Performance of the DA and PSO-based GMPPT controllers.

Curve	Iteration count		Tracking time (ms)		Tracking loss (J)	
	DA	PSO	DA	PSO	DA	PSO
A	4	6	307.5	910.0	13.90	46.55
B	5	6	360.0	927.5	14.76	38.90
C	4	7	302.5	965.0	9.15	29.18

5   Conclusions

This paper proposes the use of the Dragonfly Algorithm for GMPPT of PV systems because of its good exploration and exploitation characteristics. This technique is fast and system independent, and is therefore very suitable for GMPPT under fast-changing environmental conditions. The proposed control is implemented using a low-cost microcontroller to control the DC-DC boost converter. The DA is compared with the PSO and is proven to be superior in terms of tracking speed and reduced energy losses.

References

1. Patel, H., Agarwal, V.: Maximum power point tracking scheme for PV systems operating under partially shaded conditions. IEEE Trans. Ind. Electron. **55**, 1689–1698 (2008)
2. Farivar, G., Asaei, B., Mehrnami, S.: An analytical solution for tracking photovoltaic module MPP. IEEE J. Photovoltaics **3**, 1053–1061 (2013)
3. Fernia, N., Petrone, G., Spagnuolo, G., et al.: Optimization of perturb and observe maximum power point tracking method. IEEE Trans. Power Electron. **20**, 963–973 (2005)
4. Hussein, K.H., Muta, I.: Maximum photovoltaic power tracking: An algorithm for rapidly changing atmospheric conditions. Proc. Inst. Electr. Eng. Gener. Transm. Distrib. **142**, 59–64 (1995)
5. Balasubramanian, I.R., Ganesan, S.I., Chilakapati, N.: Impact of partial shading on the output power of PV systems under partial shading conditions. Power Electron. IET **7**, 657–666 (2014)
6. Ishaque, K., Salam, Z., Amjad, M., et al.: An improved particle swarm optimization (PSO)-based MPPT for PV with reduced steady state oscillation. IEEE Trans. Power Electron. **27**, 3627–3638 (2012)
7. Bilal, B.: Implementation of artificial bee colony algorithm on maximum power point tracking for PV modules. In: Proceedings of the 8th International Symposium on Advanced Topics Electrical Engineering (ATEE), Bucharest, Romania, pp. 1–4 (2013)
8. Mohanty, S., Subudhi, B., Ray, P.K.: A new MPPT design using grey wolf optimization technique for photovoltaic system under partial shading conditions. IEEE Trans. Sust. Energy **7**, 181–188 (2016)
9. Mirjalili, S.: Dragonfly algorithm: a new meta-heuristic optimization technique for solving single-objective, discrete, and multi-objective problems. Neural Computing and Applications. Springer, Heidelberg (2015)
10. Liu, Y.H., Chen, J.H., Huang, J.W.: A review of maximum power point tracking techniques for use in partially shaded conditions. Renew. Sustain. Energy Rev. **41**, 436–453 (2015)

Traffic Aware Based Tail Optimization of Browsing Applications for Energy Saving

Chao Wang$^{(\boxtimes)}$ and Wenneng Ma

Department of Automation, Shanghai Jiao Tong University, Shanghai, China
{sydt,wennengma}@sjtu.edu.cn

Abstract. It is challenging to save energy in smartphones by shortening simply tail time for user real-time interactive applications. Because users' behavior is stochastic and random so that it is difficult to learn network data traffic characteristic of transmission. In this paper, we propose a novel scheme TATO, which effectively forecast long idle time between active network activities in advance to reduce unnecessary extra tail time. The core idea is to learn recent interval pattern of traffic transfer by SVM model to predict next interval time whether exceed predefined threshold thus to adjust radio interface from high power state to idle state by leveraging Fast Dormancy mechanism. Traffic pattern may be different from user to user but we design temporal correlation features to train SVM model to achieve predict accuracy up to 90 % that it can save to 70 % optimal saving energy on average without much influence on user experience.

Keywords: Cellular networks · Radio tail · Traffic pattern · Smartphones · SVM · User experience

1 Introduction

Smartphones with cellular network provide quantities of facilitation for people life. But it is still a big challenging to improve the energy consumption efficiency because of tail [1,3,9] state in radio resource control protocol. The key challenge is determining when to release resources. Clearly, the best time to do so is when the smartphones is about to experience a long idle time period [2], otherwise the incurred resource allocation overhead (i.e., signaling load) might be unacceptably high. In this paper, we collect real trace of network traffic of real-time browsing applications such as browser and news. By analyzing the empirical data, we confirm that the tail duration accounts for nearly 50 % of channel holding time and contributes more than 60 % total power consumption of radio interface. Furthermore, we examine the traffic interval characteristic of burst cluster to learning the traffic pattern. Inspired by observation, we propose innovative, practical scheme-TATO (Traffic Aware Tail Optimal). Once predicting the long idle time, TATO would transmit RRC (Radio Resource Control) state by Fast Dormancy after data transfer completed to cut trashy tail time.

© Springer International Publishing Switzerland 2016
Y. Tan et al. (Eds.): ICSI 2016, Part I, LNCS 9712, pp. 220–227, 2016.
DOI: 10.1007/978-3-319-41000-5_22

Firstly it save significant energy without little affect user experience and performance; Secondly, TATO is lightweight and simple to implement without any other applications changes or influence. We implement a prototype application on Android-based smartphones and conduct some experience. Moreover, we conduct trace-driven simulations based on traffic information collected and result show that TATO can save average 70 % optimal saving energy. we highlight our main contributions in this paper as follows:

1. We have proposed a novel scheme, TATO, which effectively utilizes traffic pattern to search long idle time to reduce energy caused by extra tail time with little user perspective influence.
2. We have developed application on Android based smartphones. The prototype has verified TATO is application-level and without other changes or influence to upper applications.
3. We have conducted trace-driven simulations with our collected traffic information, the result demonstrate TATO can achieve 85 % prediction accuracy and 5 % error ratio.

The rest of paper is organized as follows. After providing sufficient related work in Sect. 2, we discuss energy model of RRC state in Sect. 3 and analyse the tail affect in Sect. 4. In Sect. 5, we detail our system design and evaluate the our proposal in Sect. 6. We summarize our work in Sect. 7.

2   Related Work

The tail effect in RRC protocol plays an important role for balance between energy consumption and performance. However, the tail energy cost contribute 50 % [7] wasted energy of total radio energy consumption. How to decrease the tail effect has significant sense for smartphones.

RRC state machine and energy utilization. The radio interface in mobile device has different power state according to current radio resource allocation. 3GPP gives the definite RRC specification for each cellular system while each carrier has their own implementation. Pathak et al. [1] use system call traces to perform fine-grained modeling of energy consumption. Qian et al. [4] design an active measurement to infer parameters of RRC state machine. These measurements showed tail time results in large part energy drain. Huang et al. [10] extended its investigation to 4G LTE network and discovered tail time was nearly 5 s. Authors [11] compute the energy cost of using different inactivity timers through modeling traffic and 3G radio characteristic.

Tail optimization for interactive applications. Tail Optimization [9] for these applications is more difficult than former because it has interaction with users. Furthermore, to predict user behavior is difficult. What we can do is to study traffic pattern to find available information. Xue et al. [7] find the temporal correlation of packets and apply ARMA model to prediction interval time of data burst. If the predict value exceed threshold, it adopt fast dormancy to actively transmit RRC state. This scheme adopts time sequence method to train predictor about interval time of burst. It is restricted to packets action in short time scale which increase the uncertainty.

Applying Fast Dormancy. Fast dormancy had been discussed by 3GPP editions and implemented by some handset manufactures [5,6,8]. RadioJockey [12] uses program execution traces to predict the end of communication spurts and invoke fast dormancy when necessary. It however has several limitations. (1) It needs heavy instrumentation i.e., requiring complete system call traces in addition to packet traces. (2) RadioJockey only works for background app without user interaction.

3   Analyzing Energy Waste

In this section, we discuss the energy consumption of RRC state including DCH (Dedicated Channel), FACH (Forward access channel). It controls the establishment and release of radio resource between equipment and base station and energy model is show in Fig. 1. We first arrange packets of real-time applications according to the arrival time of packets. Then we consider the inter-packet time which refers to the time interval between two consecutive packets along time. In practice, the interval time of immediate packets is very short and they are transferred in one burst. We begin to consider the burst rather than individual packets. The interval of bursts decides the how much tail time. Figure 4 shows the interval time for browser. We estimate the tail energy consumption by multiplying the duration of tail and corresponding time with burst cluster sequence. We plot the CDF of ratio extra energy cost to the total energy consumption in Fig. 2. It is can be seen that for real-time applications web browser and instance message application, almost 80 % user have more 70 % energy spent on tails. To minimize the energy consumption, radio should switch to IDLE immediately after transmission if next transmission not arrive in the following $t_{threshold}$ of inactivity timer value.

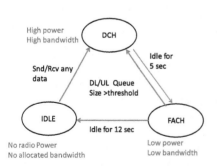

Fig. 1. The RRC state machine for the 3G UMTS network of carrier

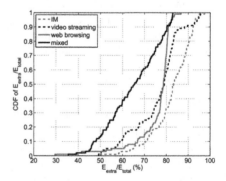

Fig. 2. Ratio of extra energy against total energy

4   System Design

The architecture of TATO is shown in Fig. 3. The TATO runs in applications level. Traffic capture is responsible for collecting traffic information from other applications. Model Training takes charge of learning traffic to train SVM model. Predictor is responsible for burst merge and predicting whether next idle time exceed threshold.

4.1   Analysis of Traffic Pattern

As Fig. 4 show, for a period of time, users may have much interaction with browse application such as slide screen, trigger another web link to load new content and so on and these processes may last longer time about 20 s–50 s or more. When this process is over, the next time will be idle time that no data transmissions happen. Between the network active periods, interval time is long enough to distinguish separate network activity active period time. Although internal transmission is difficult to learning however the comprehensive idle time pattern is able to be traced.

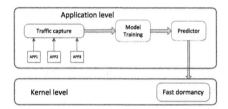

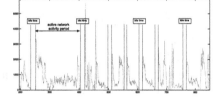

Fig. 3. The architecture of TATO

Fig. 4. Real time traffic for browse application

Burst Cluster. As above, for a period time, the data transmission is high dense and last certain time. During this time, each burst is closed with others or the data transmission is continuous for tens seconds. Here we define these continuous transmissions as burst cluster that interval time of each adjacent data transmission is less 2 s. Otherwise, data transmissions belong to separate cluster. Each cluster will last for a long period time. Figure 5 shows cluster duration distribution for users.

Characteristic of Interval Time. As Fig. 6 shows, for a period time the interval time between clusters maintain a certain range. Although fluctuated but it is still related with others. We utilize relationship of interval time to predict next value. TATO would adjust RRC state actively and allocated radio resource will be released without tail state. Here we set the threshold $t_{threshold}$ 4.5 s including cluster disguising time. This value adjust the energy saving and error ratio.

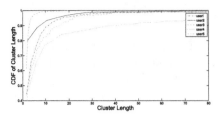
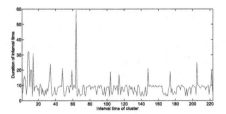

Fig. 5. The CDF of cluster duration (Color figure online)

Fig. 6. The duration of interval time between clusters

4.2 SVM Model

For the traffic pattern learning, we adopt SVM classification model to accomplish our task. As for linear classification problem, feature input vector $x = \{x_i, x_2, ..x_n\}$ and classification label $y = \{y_1, y_2, ...y_n\}$ SVM use super plane

$$w^T x + b = 0 \tag{1}$$

to classify sample data. Where w is support vector, b is parameter. This question can be solved by search the saddle point of Lagrange function

$$\Phi(w, b, \alpha) = \frac{1}{2}\|w\|^2 - \sum_{i=1}^{n} \alpha_i(y_i(w^T x_i + b) - 1) \tag{2}$$

For linear non separable problem, it adopts kernel $\kappa < x_1, x_2 >$ function to avoid the dimension explosion problem. The last optimal judge function is

$$f(x) = \sum_{i=1}^{n} \alpha_i y_i \kappa < x_i, x > + b \tag{3}$$

For feature choose, we adopt following features: (1) previous 4 interval time and the mean value, standard deviation value of previous interval time (2) applications name. The features representation is $x = (t_1, t_2, t_3, t_4, app_{number})$, t_i denotes the adjacent interval time value from now and app_number is digit which represent different applications. To train our SVM model, we put previous a series of interval value and application label as train data to compute current support vector. It is different from our expectation that the more history data and more accurate the result is. Figure 7 show the effectiveness with previous M interval data. With the M increase, the average prediction accuracy increase. However when M exceeds 6, the accuracy begin to decline. It means more older data damage the pattern learning. Experiment demonstrates past 4 or 5 interval data is efficient to represent recent traffic pattern.

5   Evaluation

5.1   Trace Collection

We implement the prototype of TATO in Android OS smartphones. We develop the application that runs in background with android services monitoring

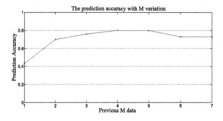
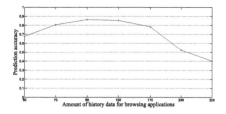

Fig. 7. The previous M data variation

Fig. 8. Average prediction accuracy with history train data

the traffic. It collects traffic information from different applications including browser, new reader and social applications per 1.5 s. We let 40 students to install our applications to play smartphones as usual in WiFi and Cellular network. The result shows the basic tend for cumulative distribution of cluster duration and interval time of burst is similar as Fig. 5 shows.

5.2 Simulation

Prediction Result. Figures 8 and 9 shows the predict accuracy defined as ratio of accurate classification for next idle to corresponding real classification. We gradually increase the number of past interval time to find the optimal training history data size. It isn't more past data for training and more better predict result is. When the training size increases to 90, the predict accuracy increase to 85 %. However, as the training size continues to increase, the predict accuracy begins to decline. The error case is that TATO predict next interval exceed $t_{threshold}$ but in actually next transmission arrive within $t_{threshold}$. Figure 10 shows the error classification ratio variation. It is can be seen that with classification depth of next 70 interval time, the error is limited in about 5 %. However when prediction depth exceeds the 70, the error ratio begin to fluctuate and change violently in our experiment which means prediction classification is not reliable. The Fig. 8 also shows that when classification depth exceed 70, the accuracy decline dramatically. Figure 10 shows another scheme prediction error for next interval sequence-SmartCut [7] which use ARMA model to calculate next interval value. Its prediction error is about 10 % while our prediction error is about 5 %.

Energy Saving. To quantize the energy saving, we introduce the optimal saved energy conception. This conception is quantization how much energy can be saved if we grasp future information. Now we have collect traffic information and can calculate ideal saved energy offline. If interval time is exceed 4 s we think energy is able to save energy from this transmission. According to this standard, to calculate all time when if RRC has demotion to IDLE and extra

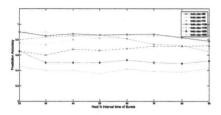
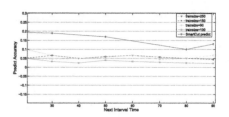

Fig. 9. Prediction accuracy of different history train data

Fig. 10. Prediction error of different history train data size

tail energy consumption is avoided. The calculate equation:

$$E_{opt} = \sum_{i=1} if(inter(burstcluster_{i+1}, burstcluster_i) >= 4)$$

To evaluate our predict result, we calculate the ratio between predict result and ideal saving energy:

$$\frac{E_{pre}}{E_{opt}} = \frac{\sum_{i=1} if(predict(interval(i)) == true)}{E_{opt}}$$

This ratio is about 50 %–70 % in our experiment and the average value is about 60 %. Obvious, the ideal saved energy is dependent on traffic pattern that how much idle time are useful. And this is influenced by user behavior, application category.

The simulation result shows that TATO can accurately predict next interval time to achieve signification amount energy with little impact on user experience.

6 Conclusion

In this paper, through analyzing the network traffic trace, we have confirmed that tail effect wastes signification amount energy. But based our observation, we design a lightweight scheme TATO, which learns relationship between interval time of active network transfer to predict whether next idle time is long enough to transmit RRC state. This scheme modifies current inactivity timer waiting scheme and does not increase other signaling overhead and side-effect to user experience. Our trace-driven experiment demonstrated TATO is very efficient to distinguish long idle time which means energy-efficient. However, we just consider the foreground activity without thinking of other background network activity but RRC state transmits is the result of all data transfer. In future, we are planning to solve this situation.

References

1. Pathak, A., Hu, Y., Zhang, M., Bahl, P., Wang, Y.: Fine-grained power modeling for smartphones using system call tracing. In: Proceedings of Sixth Conference on Computer Systems, pp. 153–168. ACM (2011)

2. Huang, J., Qian, F., Mao, Z.M., Sen, S., Spatscheck, O.: RadioProphet: intelligent radio resource deallocation for cellular networks. In: Faloutsos, M., Kuzmanovic, A. (eds.) PAM 2014. LNCS, vol. 8362, pp. 1–11. Springer, Heidelberg (2014)

3. Qian, F., Wang, Z., Gerber, A., Mao, Z.M., Sen, S., Spatscheck, O.: Top tail optimization protocol for cellular radio resource allocation. In: ICNP, pp. 285–294 (2010)

4. Qian, F., Wang, Z., Gerber, A., Mao, Z., Sen, S., Spatscheck, O.: Characterizing radio resource allocation for 3G networks. In: Proceedings of the 10th Annual Conference on Internet Measurement, pp. 137–150. ACM (2010)

5. Qian, F., Wang, Z., Gerber, A., Mao, Z., Sen, S., Spatscheck, O.: Profiling resource usage for mobile applications: a cross-layer approach. In: Proceedings of the International Conference on MobiSys, pp. 321–334 (2011)

6. Qian, F., et al.: Periodic transfers in mobile applications: network-wide origin, impact, and optimization. In: Proceedings of the 21st International Conference on World Wide Web, pp. 51–60. ACM (2012)

7. Xue, G.T., et al.: SmartCut: mitigating 3G radio tail effect on smartphones. IEEE Trans. Mob. Comput. **14**(1), 169–179 (2015)

8. Liu, H., Zhang, Y., Zhou, Y.: Tailthef: leveraging the wasted time for saving energy in cellular communications. In: Proceedings of the Sixth International Workshops on MobiArch, pp. 31–36. ACM (2011)

9. Deng, S., Balakrishnan, H.: Traffic-aware techniques to reduce 3G/LTE wireless energy consumption. In: Proceedings of the 8th International Conference on Emerging Networking Experiments and Technologies, pp. 181–192. ACM (2012)

10. Huang, J., Qian, F., Gerber, A., Mao, Z.M., Sen, S., Spatscheck, O.: A close examination of performance and power characteristics of 4G LTE. In: Proceedings of the International Conference on Cummunication, pp. 225–238. ACM (2012)

11. Balasubramanian, N., Balasubramanian, A., Venkataramani, A.: Energy consmption in mobile phones: a measurement study and implications for network application. In: Proceedings of IMC, pp. 280–293. ACM (2009)

12. Athivarapu, P.K., et al.: RadioJockey: mining program execution to optimize cellular radio usage. In: Proceedings of International Conference on MobiCom, pp. 101–112. ACM (2012)

Linear ODE Coefficients and Initial Condition Estimation with Co-operation of Biology Related Algorithms

Ivan Ryzhikov[✉], Eugene Semenkin, and Shakhnaz Akhmedova

Siberian State Aerospace University, Krasnoyarsky rabochy avenue 31,
660037 Krasnoyarsk, Russia
{ryzhikov-88, eugenesemenkin}@yandex.ru,
shahnaz@inbox.ru

Abstract. The inverse mathematical modelling problem for a linear dynamic system is considered. The parameter and initial condition identification were reduced to an optimization problem. The proposed approach is based on the simultaneous estimation of linear differential equation coefficients and initial condition vector coordinates. The mathematical model is determined by the vector of equation parameters and the state coordinate of the model. The initial value problem solution is required to fit the sample data. The complexity and multimodality of criterion for the reduced problem leads to the implementation of an efficient optimization technique. The meta-heuristic optimization algorithm called Co-Operation of Biology Related Algorithms (COBRA) was used for this purpose. Its high efficiency had been proven in previous studies. Investigation results show that COBRA is a high-performance and reliable technique for current extremum problem class solving. The usefulness of the proposed approach is confirmed with the investigation results based on experiments made for different sample characteristics and different dynamic systems.

Keywords: Bionic algorithms · Biogeography · Optimization · Initial condition estimation · Parameters identification

1 Introduction

The solution of the dynamic system identification problem is a mathematical model that approximates not only the real object output but the law of the system behaviour. The output is the reaction of a dynamic process on some control action. From a practical point of view the problem is significant and has plenty of applications.

The current study focuses on the dynamical processes of mathematical modelling in the form of linear differential equations. The mathematical model of such a form is useful in further work for solving control and stability problems and predicting the reaction of the system on different control inputs. Moreover, ordinary differential equation models are useful in different scientific fields: engineering, physics, chemistry, econometrics and biology.

The mathematical model in the form of a differential equation requires an initial condition. Generally, the initial condition is not known and, in the case of small or

© Springer International Publishing Switzerland 2016
Y. Tan et al. (Eds.): ICSI 2016, Part I, LNCS 9712, pp. 228–235, 2016.
DOI: 10.1007/978-3-319-41000-5_23

rarefied samples is hard to estimate. Also the identification problem is related to some real process, and observations are made of system that, probably, was not in a stable state. Thus derivatives are not equal to 0 and are to be estimated too.

In many studies different approaches to dynamic system identification are described. One class of approaches is based on the pre-processing of data to make the estimation of system states and its derivatives [1]. After that the dynamic system parameters are to be identified. A similar idea can be found in studies [2, 3]. Another class of approaches is based on shooting or multiple shooting ideas, i.e. [4]. But in a case when system state observations are represented with a sample of small size or highly distorted data, there is a need for different approaches.

The proposed approach is based on the reduction of the initial problem to the extremum problem on the real value vector field: parameters and initial point coordinates are estimated simultaneously. Some studies of linear dynamic system identification concluded that the tuned or modified optimization technique is required for the successful solving of the generated problems [5]. Some modifications were made to sufficiently improve the algorithm efficiency for the order and coefficients estimation. In the current study the order of differential equation is based on an assumption, but the initial condition is unknown. That is the reason why another powerful algorithm was used in the current study.

The application of evolutionary optimization algorithms in the parameters estimation problem is widespread but commonly focuses on specific problems, i.e. [6–8]. The COBRA meta-heuristic that is used in this study is based on the cooperation of five nature-inspired algorithms (Particle Swarm Optimization (PSO) [9], Wolf Pack Search (WPS) [10], the Firefly Algorithm (FFA) [11], the Cuckoo Search Algorithm (CSA) [12] and the Bat Algorithm [13]). The workability and reliability of COBRA for optimization problems with real-valued variables was shown in [14] on a set of benchmark functions with up to 50 variables and later confirmed in [15] on ANN weight coefficients adjustment with up to 110 real-valued variables and structure selection with 100 binary variables.

The aim of this paper is to investigate the performance of the algorithm in its application in the reduced optimization problem and its different sample characteristic and different characteristics of an object. It is necessary for further studies and in order to develop the dynamic system identification problem.

2 Problem Statement

Let a set $Y = \{y_i\}$, $i = \overline{1, s}$ be a sample of s measurements of some dynamical process output, made, respectively, at times $T = \{t_i\}$, $i = \overline{1, s}$. Let the control action be known and determined by a function $u(t) : R^+ \rightarrow R$, which is a piecewise continuous function. It is also assumed that the process can be determined with a linear differential equation.

The system state coordinate measurements are assumed to be distorted: the observed output is the real data with additive noise. Additive noise is a random value $\xi : E(\xi) = 0$, $D(\xi) < \infty$ and

$$y_i = x(t_i) + \xi_i, \ i = \overline{1,s} \tag{1}$$

where the function $x(t) : R^+ \to R$ is the real system state, the solution of the initial value problem

$$a_k \cdot x^{(k)} + a_{k-1} \cdot x^{(k-1)} + \ldots + a_0 \cdot x = b \cdot u(t),$$
$$x(0) = x_0. \tag{2}$$

In the current study, it is proposed to investigate the estimation of the coefficients and initial condition problem in known order case. Thus one can determine the dynamic process with the $\tilde{k}$-th linear differential equation

$$\tilde{a}_{\tilde{k}} \cdot \tilde{x}^{(\tilde{k})} + \tilde{a}_{\tilde{k}-1} \cdot \tilde{x}^{(\tilde{k}-1)} + \ldots + \tilde{a}_1 \cdot \tilde{x}' + \tilde{a}_0 \cdot \tilde{x} = \tilde{b} \cdot u(t), \tag{3}$$

which can be simplified, since the assumption was made and $\tilde{a}_{\tilde{k}} \neq 0$,

$$\tilde{x}^{(\tilde{k})} + \hat{a}_{\tilde{k}} \cdot \tilde{x}^{(\tilde{k}-1)} + \ldots + \hat{a}_2 \cdot \tilde{x}' + \hat{a}_1 \cdot \tilde{x} = \hat{a}_0 \cdot u(t). \tag{4}$$

The estimation of coefficients of differential Eq. (4) requires an initial point. Because of the sample characteristics, i.e. the sample size can be small, data can be strongly distorted; the only way to solve the problem without making one more assumption is by solving both estimation problems simultaneously.

Let $\tilde{x}(t)|_{\hat{a}=c,x(0)=x_0^e}$ be a solution of the initial value problem

$$\tilde{x}^{(\tilde{k})} + c_{\tilde{k}} \cdot \tilde{x}^{(\tilde{k}-1)} + \ldots + c_2 \cdot \tilde{x}' + c_1 \cdot \tilde{x} = c_0 \cdot u(t)$$
$$\tilde{x}(0) = x_0^e \tag{5}$$

Now it is possible to estimate the model (4) adequacy by the following functional

$$F(a, x_0) = \frac{\sum_{i=1}^{s} \left| y_i - \tilde{x}(t_i)|_{\hat{a}=a,x(0)=x_0} \right|}{s} \to \min_{a \in R^{\hat{k}+1}, x_0 \in R^{\hat{k}}}. \tag{6}$$

Due to proposition was made, the estimation of the equation order leads to solve problem (6) for different order assumptions. The best solution of the identification problem is the one that brings the minima for the criteria (6) among all of the solutions found with different equation orders.

3 Optimization Technique

According to criteria (6) and problem definition, every solution is determined by a point on the real value vector field $R^{\hat{k}+1} \times R^{\hat{k}}$. Every candidate solution is a vector

$v \in R^{2 \cdot \hat{k}+1}$ that determines $a|v = \begin{pmatrix} v_1 & \cdots & v_{\hat{k}+1} \end{pmatrix}^T$ and $x_0|v = \begin{pmatrix} v_{\hat{k}+2} \cdots v_{2 \cdot \hat{k}+1} \end{pmatrix}^T$, and has an estimation for the fitting value $f_a(v) = F(a, x_0|v)$. In terms of evolutionary optimization, every individual is a point $v \in R^{2 \cdot \hat{k}+1}$ with fitness $fit(v) = \frac{1}{1+f_a(v)}$.

The proposed optimization algorithm is a self-tuning meta-heuristic. One can find more basic information about this technique in studies [15]. In this paper we briefly discuss some features of the algorithm. First of all, meta-heuristics is a way that some implementation controls algorithm parameters: in the current case we do not have to choose the population size for each algorithm. The number of individuals in the population of each algorithm can be increased or decreased depending on whether the fitness value improves or not.

Besides, each population can "grow" by accepting individuals removed from other populations. The population size grows only if its average fitness value is better than the average fitness values of all other populations. As a result, the biggest computational resource (population size) fits to the most appropriate (in the current generation) algorithm. This property can be very useful in the case of hard optimization problems when, as is known, there is no single best algorithm for all stages of the optimization process execution.

The most important implementation of the suggested meta-heuristic is the migration operator that creates a cooperation environment for component algorithms. All populations communicate with each other: they exchange individuals in such a way that a part of the worst individuals of each population is replaced by the best individuals of other populations. It brings up to date information on the best achievements to all component algorithms and prevents their preliminary convergence to its own local optimum. That improves the group performance of all algorithms.

The proposed algorithm was validated and compared with its component algorithms. Simulations and comparison showed that COBRA is superior to these existing algorithms when the complexity of the problems to be solved increases.

4 Experimental Results

For different assumptions of the linear differential equation order for every distinct identification problem and in the case of there being no error in the measurements, $\xi = 0$, the efficiency of the optimization algorithm was investigated. We need to mention that dynamic systems were simulated by stable linear differential equations. For every distinct order experiments were made for 10 different equations. Every experiment was run 25 times. Samples are generated in a following way: we discretize the solution of the Cauchy problem (2) with discretization step 0.05, randomly choose distinct points and add noise if it is necessary. In this study the number of function evaluation was set to 50000.

First, it is important to explore how a wrong assumption of the system order reacts on the quality of the identification problem solution. The experiments included data of identification of 2nd order differential equations, as well as 3rd, 4th and 5th order equations. Figure 1 demonstrates the average best fitness function values, those were

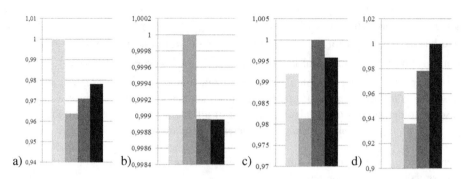

Fig. 1. Identification problem solution for different real system orders: (a) second, (b) third, (c) forth, (d) fifth order, and different model orders.

normed for better representation. On these pictures the lightest bar matches the estimation of the 2nd order model, and the darkest bar matches the estimation of the 5th order model.

As one can see, in the case of there being no noise in measurements the real order assumption gives models those fit the sample data more. It is also expected that with the same computational efforts, the efficiency of the algorithm will decrease as the system order increases. The maxima of the average functional (6) values for the problems mentioned above are 0.995, 0.966, 0.964 and 0.957, respectively.

For better understanding of the identification problem features it was suggested to add two more characteristics. First is a metric for calculating the difference between the model and the system state.

$$Error_1(x, a, x_0) = \int_{t_0}^{t_s} \left| x(t) - \hat{x}(t)|_{\hat{a}=a,\, x(t_0)=x_0} \right| dt. \tag{7}$$

We used its numerical approximation. The second characteristic is a metric on a vector field to calculate the distance between real and estimated parameters and coordinates:

$$Error_2(a, x_0, v) = \frac{\|a - a|v\| + \|x_0 - x_0|v\|}{\dim(v)}. \tag{8}$$

The aim of the next experiment is to estimate the influence of the sample size on algorithm efficiency. The sample size was varied from 5 to 80 for systems of 2nd, 3rd and 4th order. It needs to be mentioned that statistics depend more on the system state features than on the sample size value. Here and later statistics made for the 2nd order systems are represented by a dashed line, for 3rd order systems by a solid line, and for 4th order system by a dash-dotted line. Criteria values for these experiments are presented in Fig. 2 with its linear approximation.

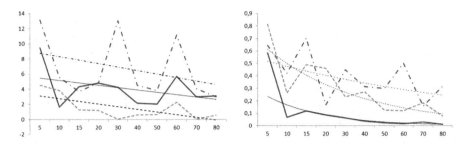

Fig. 2. Average criteria $Error_1$ value (left figure) and $Error_2$ value (right figure) for different sample sizes.

Of course, small samples may not be an adequate representation of the system's dynamics, since system behaviour cannot be observed. This may cause significant errors in estimating real parameters. However experimental results allow the hypothesis to be put forward that the proposed approach helps to achieve satisfactory results for a different sample size in the case of there being no distortion of system state measurements and a known system order.

The next experiment was made to estimate the algorithm performance in the case of different noise amplitudes. We generated an error in measurements as a uniformly distributed value. In this case the noise is $\xi \sim U(-L, L)$, where $L \geq 0$ is the level of noise and $U(a, b)$ is uniform distribution on $[a, b]$. The size of samples was set to 80. The criteria values for this experiment are presented in Fig. 3.

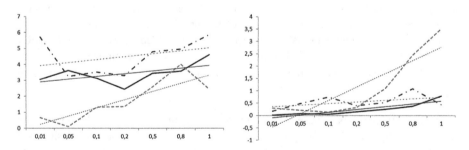

Fig. 3. Average criteria $Error_1$ value (left figure) and $Error_2$ value (right figure) for different noise levels.

The correlation coefficient of noise level and average value of criterion (6) is close enough to 1: 0.9945, 0.9963, and 0.987 for systems of 2nd, 3rd and 4th order, respectively. This indicates the adequacy of approximation, because there is linear dependence between the noise level and "state-observations error" sum.

The other statistic that is related to fitting distorted data value and the benefit in criterion (6) that model gives in compare with criterion value for real system state. The probability estimation of receiving better solution increases with noise level increasing. The same happens to beneficial value, it increases similarly. Its behavior is represented

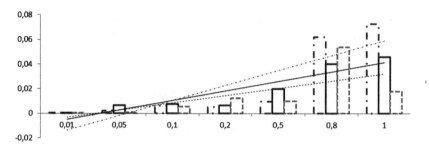

Fig. 4. Average model benefit value of criterion (4) for different noise levels.

in Fig. 4. This fact also confirms the efficiency and usefulness of the proposed approach.

As one can see, the proposed approach allows an inverse modelling problem to be successfully solved using optimization of the reduced problem with the COBRA algorithm.

5   Conclusion

The proposed approach of reducing the dynamic identification problem to an optimization problem requires a powerful extremum seeking technique. In this study it was suggested to use COBRA as this technique and different experiments were performed to inspect the problem features and investigate the algorithm performance. Different experiments were run to estimate the algorithm efficiency and investigate the features of the problem. Obviously, the complicity of the problem requires a lot of computational resources to achieve a satisfactory result. We can still conclude that the COBRA algorithm is a workable and reliable tool for solving this class of optimization problems and its meta-heuristics solves the problem of tuning or estimating the most efficient settings. In further work influence of both the noise level and sample size will be investigated.

Further study will be focused on the extension of this approach to the multi-input and multi-output identification problem. For this case it is necessary to design and implement a scheme of estimating the order, so the problem statement will become more general. Investigation of the algorithm behaviour and problem features, and its analysis will be used to extract some knowledge about how the algorithm or approach can be improved, so that difficulties connected to the method of solution transformation could be avoided.

The proposed approach can be easily transformed to solve identification problems for when the initial condition coordinates are known, which means that the identification problem is reduced to a parameter estimation one.

Acknowledgements. Research is performed with the financial support of the Russian Foundation of Basic Research, the Russian Federation, contract №20 16-01-00767, dated 03.02.2016.

References

1. Fang, Y., Wu, H., Zhu, L.-X.: A two-stage estimation method for random coefficient differential equation models with application to longitudinal HIV dynamic data. Stat. Sinica **21**(3), 1145–1170 (2011)
2. Wu, H., Xue, H., Kumar, A.: Numerical discretization-based estimation methods for ordinary differential equation models via penalized spline smoothing with applications in biomedical research. Biometrics **68**, 344–352 (2012)
3. Brunel, N.J.-B.: Parameter estimation of ODE's via nonparametric estimators. Electronic J. Stat. **2**, 1242–1267 (2008)
4. Peifer, M., Timmer, J.: Parameter estimation in ordinary differential equations for biochemical processes using the method of multiple shooting. IET Syst. Biol. **1**, 78–88 (2007)
5. Ryzhikov, I., Semenkin, E.: Evolutionary strategies algorithm based approaches for the linear dynamic system identification. In: Tomassini, M., Antonioni, A., Daolio, F., Buesser, P. (eds.) ICANNGA 2013. LNCS, vol. 7824, pp. 477–484. Springer, Heidelberg (2013)
6. Wang, J., Zhou, B., Zhou, S.: An improved Cuckoo search optimization algorithm for the problem of chaotic systems parameter estimation. Comput. Intell. Neurosci. **2016**, Article ID 2959370, 8 (2016)
7. Sun, J., Palade, V., Cai, Y., Fang, W., Wu, X.: Biochemical systems identification by a random drift particle swarm optimization approach. BMC Bioinform. **15**(Suppl. 6), S1 (2014)
8. Parmar, G., Prasad, R., Mukherjee, S.: Order reduction of linear dynamic systems using stability equation method and GA. Int. J. Comput. Inf. Eng. **1**(1), 26–32 (2007)
9. Kennedy, J., Eberhart, R.: Particle swarm optimization. In: IEEE International Conference on Neural Networks, pp. 1942–1948 (1995)
10. Yang, C., Tu, X., Chen, J.: Algorithm of marriage in honey bees optimization based on the wolf pack search. In: International Conference on Intelligent Pervasive Computing, pp. 462–467 (2007)
11. Yang, X.S.: Firefly algorithms for multimodal optimization. In: 5th Symposium on Stochastic Algorithms, Foundations and Applications, pp. 169–178 (2009)
12. Yang, X.S., Deb, S.: cuckoo search via levy flights. In: World Congress on Nature & Biologically Inspired Computing, IEEE Publications, USA, pp. 210–214 (2009)
13. Yang, X.-S.: A new metaheuristic bat-inspired algorithm. In: González, J.R., Pelta, D.A., Cruz, C., Terrazas, G., Krasnogor, N. (eds.) NICSO 2010. SCI, vol. 284, pp. 65–74. Springer, Heidelberg (2010)
14. Akhmedova, S., Semenkin, E.: Co-operation of biology related algorithms. In: IEEE Congress on Evolutionary Computation, pp. 2207–2214 (2013)
15. Akhmedova, S., Semenkin, E.: Co-operation of biology related algorithms meta-heuristic in ANN-Based classifiers design. In: IEEE World Congress on Computational Intelligence, pp. 867—873 (2014)

On the Constraint Normalization:
An Empirical Study

Chengyong Si[1(✉)], Jianqiang Shen[1], Xuan Zou[1], Lei Wang[2],
and Qidi Wu[2]

[1] Shanghai-Hamburg College, University of Shanghai for Science
and Technology, 200093 Shanghai, China
sichengyong_sh@163.com
[2] College of Electronics and Information Engineering,
Tongji University, 201804 Shanghai, China
wanglei@tongji.edu.cn

Abstract. The form of constraint violation plays an important role when using
the constraint handling techniques for solving Constrained Optimization Problems. As the measures for different constraint violation may have different
numerical ranges, some normalization forms have already been proposed to
balance the contribution of each constraint. This paper first systematically
analyzes the effect of normalization, and then tries to study it from the aspect of
problem characteristics. The experimental results verify the relationship between
the two forms of constraint violation and problem characteristics, which can
give some guide for future research.

Keywords: Constrained optimization · Constraint handling techniques ·
Differential evolution · Constraint normalization · Ranking methods

1 Introduction

Constrained Optimization Problems (COPs) are very important and common in
real-world applications. The general COPs can be formulated as follows:

$$\begin{aligned}
\text{Minimize} \quad & f(\vec{x}) \\
\text{Subject to}: \quad & g_j(\vec{x}) \le 0, \qquad j = 1, \cdots, l \\
& h_j(\vec{x}) = 0, \quad j = l+1, \cdots, m.
\end{aligned} \tag{1}$$

where $\vec{x} = (x_1, \cdots, x_n)$ is the decision variable which is bounded by the decision
space S. S is defined by the constraints:

$$L_i \le x_i \le U_i, \quad 1 \le i \le n. \tag{2}$$

Here, l and $m-l$ is the number of inequality and equality constraints respectively.

© Springer International Publishing Switzerland 2016
Y. Tan et al. (Eds.): ICSI 2016, Part I, LNCS 9712, pp. 236–243, 2016.
DOI: 10.1007/978-3-319-41000-5_24

As for COPs, equality constraints are usually transformed into inequality constraints as

$$|h_j(\vec{x})| - \delta \leq 0. \tag{3}$$

where $j = l+1,\ldots,m$ and δ is a positive tolerance value.

The Evolutionary Algorithms (EAs), as the unconstrained search techniques and solution generating strategies, are not suitable enough to solve COPs without additional mechanisms to deal with the constraints. Consequently, many constrained optimization evolutionary algorithms (COEAs) have been proposed [1–3]. The three most popular constraint handling techniques in COEAs are based on penalty functions, biasing feasible over infeasible solutions and multi-objective optimization [4–7].

As a rule without parameters, Deb [6] proposed a feasibility-based rule to pair-wise compare individuals:

(1) Any feasible solution is preferred to any infeasible solution.
(2) Among two feasible solutions, the one having better objective function value is preferred.
(3) Among two infeasible solutions, the one having smaller constraint violation is preferred.

Among these COEAs, the degree of constraint violation plays an important role when determining the feasibility of an individual. Usually, the degree of constraint violation of individual $\vec{x}$ on the jth constraint is calculated as

$$G_j(\vec{x}) = \begin{cases} \max\{0, g_j(\vec{x})\} & 1 \leq j \leq l \\ \max\{0, |h_j(\vec{x})| - \delta\} & l+1 \leq j \leq m. \end{cases} \tag{4}$$

A common form of the overall constraint violation suggested in [8, 9] has been widely adopted as an evaluation criterion, as

$$G(\vec{x}) = \sum_{i=1}^{l} G_i(\vec{x}) + \sum_{j=l+1}^{m} H_j(\vec{x}). \tag{5}$$

where $G_i(\vec{x}) = \begin{cases} g_i(\vec{x}) & \text{if } g_i(\vec{x}) > 0 \\ 0 & \text{if } g_i(\vec{x}) \leq 0 \end{cases}$, and $H_j(\vec{x}) = \begin{cases} |h_j(\vec{x})| & \text{if } |h_j(\vec{x})| - \varepsilon > 0 \\ 0 & \text{if } |h_j(\vec{x})| - \varepsilon \leq 0 \end{cases}$.

Some researchers noticed that the measures for different constraint violation may have different numerical ranges, and to balance the contribution of each constraint in the problem, some normalized forms were proposed.

Venkatraman *et al.* [10] proposed a scalar constraint violation method. In this method, each constraint violation is normalized through dividing it by the largest violation of that constraint in the population, so as to treat each constraint equally. This process can be expressed as:

$$G_j^{\max} = \max_{\vec{x} \in S} G_j(\vec{x}). \tag{6}$$

$$G(\vec{x}) = \frac{\sum\limits_{j=1}^{m} \frac{G_j(\vec{x})}{G_j^{\max}}}{m}.$$ (7)

Wang et al. [11, 12] employed these two criteria to compute the degree of constraint violation of an individual in $(\mu + \lambda)$-CDE and ICDE based on the problems' characteristics. The information of $G_j^{\max}$ in the initial population, which can reflect the problem characteristics, was used to measure the constraint difference. If the constraint violation difference is significant, i.e., $\max\limits_{j=1,\ldots,p} G_j^{\max} - \min\limits_{j=1,\ldots,p} G_j^{\max} > \varepsilon$, then the constraints should be normalized; otherwise, the normalization is not needed.

This is a very good idea, but the parameter ε defined by the users is a little difficult to decide.

Mallipeddi and Suganthan [13] adopted a similar normalization way as

$$V(\vec{x}) = \frac{\sum_{j=1}^{m} w_j (G_j(\vec{x}))}{\sum_{j=1}^{m} w_j}.$$ (8)

Here, $w_j (= 1/G_j^{\max})$ is a weight parameter. It can be seen that w_i varies during the evolution and the role of each constraint changes accordingly.

For a certain generation, as the values of $G_j^{\max}$ are fixed, there is no difference of these two normalization methods.

Besides the above form of normalization, Deb et al. [14] proposed another normalization process. They found the constraints can usually be transformed as a left side term $(g_j'(\vec{x}))$ restricted with a least value b_j. Accordingly, the following normalization process was proposed:

$$\hat{g}_j(\vec{x}) = g_j'(\vec{x})/b_j - 1 \geq 0.$$ (9)

These authors think that by this way, each constraint violation can be approximately on the same scale, and consequently a single penalty parameter R can be adopted. This is a good effort, however, the purpose to make the constraint violation have similar values in their magnitudes may not always be satisfied. For example, suppose $g_j'(\vec{x}) = 10000$, $b_j = 0.01$, then the original and normalized constraint violation will be 9999.99 and 999999 respectively, with a larger difference.

Following [14], an adaptive normalization where constraints are normalized adaptively during the optimization process was proposed [15].

Unlike the aforementioned methods, in this work, we try to study the features of two different forms of constraint violation, and get the corresponding relationship between problem characteristics and the forms.

The rest of this paper is organized as follows. Section 2 presents the systematic analysis of the normalization. The experimental results and analysis are presented in

Sect. 3. Finally, Sect. 4 concludes this paper and provides some possible paths for future research.

2 Systematical Analysis of the Constraint Normalization

In this section, we try to find out the influence of the constraint normalization based on Deb's feasibility-based rule (i.e., the ranking difference using Deb's feasibility-based rule with or without constraint normalization, taking two members for example). Here, the forms of the constraint violation with and without normalization are expressed as (5) and (7).

Given two population members, $\vec{x}_s$ and $\vec{x}_t$, where s and t are randomly selected from $[1, NP]$ and satisfying: $s \neq t$. Suppose there are p items in $\Delta G(\vec{x}_s, \vec{x}_t)$ with the value larger than 0, i.e., $\Delta G_1, \Delta G_2, \ldots, \Delta G_p > 0$, $\Delta G_{p+1}, \Delta G_{p+2}, \ldots, \Delta G_m < 0$.

$$G_P^{\max} = \max(G_1^{\max}, G_2^{\max}, \ldots, G_p^{\max}), \quad G_P^{\min} = \min(G_1^{\max}, G_2^{\max}, \ldots G_p^{\max}),$$
$$G_M^{\max} = \max(G_{p+1}^{\max}, G_{p+2}^{\max}, \ldots, G_m^{\max}), \quad G_M^{\min} = \min(G_{p+1}^{\max}, G_{p+2}^{\max}, \ldots, G_m^{\max}).$$

Then under the situation without constraint normalization, the constraint violation difference between $\vec{x}_s$ and $\vec{x}_t$ is:

$$\begin{aligned}
\Delta G(\vec{x}_s, \vec{x}_t) &= G(\vec{x}_s) - G(\vec{x}_t) \\
&= \sum_{j=1}^{m} G_j(\vec{x}_s) - \sum_{j=1}^{m} G_j(\vec{x}_t) \\
&= \sum_{j=1}^{m} \Delta G_j(\vec{x}_s, \vec{x}_t).
\end{aligned} \tag{10}$$

Similarly, under the situation of constraint normalization, the constraint violation difference is:

$$\begin{aligned}
\Delta G_{nor}(\vec{x}_s, \vec{x}_t) &= G_{nor}(\vec{x}_s) - G_{nor}(\vec{x}_t) \\
&= \frac{\sum_{j=1}^{m} \frac{G_j(\vec{x}_s)}{G_j^{\max}}}{m} - \frac{\sum_{j=1}^{m} \frac{G_j(\vec{x}_t)}{G_j^{\max}}}{m} \\
&= \sum_{j=1}^{m} \frac{\Delta G_j(\vec{x}_s, \vec{x}_t)}{m \cdot G_j^{\max}}.
\end{aligned} \tag{11}$$

According to (10) and (11), if $\Delta G(\vec{x}_s, \vec{x}_t)$ and $\Delta G_{nor}(\vec{x}_s, \vec{x}_t)$ has the same symbol (i.e., positive or negative), then we can conclude that there is no ranking difference with or without constraint normalization. Two cases are listed as follows:

(1) $\Delta G(\vec{x}_s, \vec{x}_t) > 0$: if $G_P^{\max} \leq G_M^{\min}$, then

$$\Delta G_{nor}(\vec{x}_s, \vec{x}_t) = \sum_{j=1}^{m} \frac{\Delta G_j(\vec{x}_s, \vec{x}_t)}{m \cdot G_j^{\max}}$$

$$= \sum_{j=1}^{p} \frac{\Delta G_j(\vec{x}_s, \vec{x}_t)}{m \cdot G_j^{\max}} + \sum_{j=p+1}^{m} \frac{\Delta G_j(\vec{x}_s, \vec{x}_t)}{m \cdot G_j^{\max}}$$

$$\geq \frac{1}{m \cdot G_P^{\max}} \sum_{j=1}^{p} \Delta G_j(\vec{x}_s, \vec{x}_t) + \frac{1}{m \cdot G_M^{\min}} \sum_{j=p+1}^{m} \Delta G_j(\vec{x}_s, \vec{x}_t) \qquad (12)$$

$$\geq \frac{1}{m \cdot G_P^{\max}} \sum_{j=1}^{m} \Delta G_j(\vec{x}_s, \vec{x}_t)$$

$$> 0.$$

(2) $\Delta G(\vec{x}_s, \vec{x}_t) < 0$: $G_P^{\min} \geq G_M^{\max}$, then

$$\Delta G_{nor}(\vec{x}_s, \vec{x}_t) = \sum_{j=1}^{m} \frac{\Delta G_j(\vec{x}_s, \vec{x}_t)}{m \cdot G_j^{\max}}$$

$$= \sum_{j=1}^{p} \frac{\Delta G_j(\vec{x}_s, \vec{x}_t)}{m \cdot G_j^{\max}} + \sum_{j=p+1}^{m} \frac{\Delta G_j(\vec{x}_s, \vec{x}_t)}{m \cdot G_j^{\max}}$$

$$\leq \frac{1}{m \cdot G_P^{\min}} \sum_{j=1}^{p} \Delta G_j(\vec{x}_s, \vec{x}_t) + \frac{1}{m \cdot G_M^{\max}} \sum_{j=p+1}^{m} \Delta G_j(\vec{x}_s, \vec{x}_t) \qquad (13)$$

$$\leq \frac{1}{m \cdot G_P^{\min}} \sum_{j=1}^{m} \Delta G_j(\vec{x}_s, \vec{x}_t)$$

$$< 0.$$

As can be seen from the above analysis, if some conditions are satisfied (i.e., $\Delta G(\vec{x}_s, \vec{x}_t) > 0$ with $G_P^{\max} \leq G_M^{\min}$, or $\Delta G(\vec{x}_s, \vec{x}_t) < 0$ with $G_P^{\min} \geq G_M^{\max}$), there is no difference on the ranking using Deb's feasibility-based rules between normalization and without normalization. This condition can also be reflected by the problem characteristics, and that is what we will discuss in the following experimental study.

3 Experimental Study

3.1 Experimental Settings

18 benchmark functions with 10-D developed in IEEE CEC2010 were employed in our experiment and the detailed characteristics can be found in [9].

The evolutionary algorithm used in this paper is *DE/rand/1/bin*, and the boundary constraint is reset as [16]. The parameters in DE are set as follows: the population size (*NP*) is set to 100; the scaling factor (*F*) is randomly chosen between 0.5 and 0.6, and the crossover control parameter (*Cr*) is randomly chosen between 0.9 and 0.95.

Table 1. Feasibility study for the problems C01–C18 at 10D-Deb's feasibility-based rule

Func.	Form	Best	Median	Worst	Mean	Std.	FR
C01	N = 1	1	1	1	1	0	1
	N = 0	1	1	1	1	0	1
C02	N = 1	144	680	1745	8.1313E+02	5.2444E+02	0.64
	N = 0	15	993	1808	8.7742E+02	6.1965E+02	0.76
C03	N = 1	512	552	1395	5.9916E+02	1.7221E+02	1
	N = 0	335	373	**405**	3.7504E+02	1.4146E+01	1
C04	N = 1	1579	1770	1995	1.7540E+03	1.5037E+02	0.36
	N = 0	242	267	283	2.6512E+02	1.0795E+01	**1**
C05	N = 1	1210	1210	1210	1210	0	0.04
	N = 0	1012	1874	1912	1.5993E+03	5.0900E+02	**0.12**
C06	N = 1	1740	1740	1740	1740	0	0.04
	N = 0	1884	1884	1884	1884	0	0.04
C07	N = 1	1	1	1	1	0	1
	N = 0	1	1	1	1	0	1
C08	N = 1	1	1	1	11	0	1
	N = 0	1	1	1	11	0	1
C09	N = 1	849	849	849	849	0	0.04
	N = 0	1350	1350	1350	1350	0	0.04
C10	N = 1	924	924	924	924	0	0.04
	N = 0	88	88	88	88	0	0.04
C11	N = 1	–	–	–	–	–	–
	N = 0	708	875	1548	1069	3.2964E+02	**0.36**
C12	N = 1	–	–	–	–	–	–
	N = 0	264	296	430	3.1552E+02	4.9688E+01	**1**
C13	N = 1	19	27	37	2.6920E+01	4.2810E+00	1
	N = 0	11	18	26	1.7800E+01	3.7193E+00	1
C14	N = 1	1	3	14	3.9200E+00	3.5346E+00	1
	N = 0	1	2	13	3.1200E+00	2.8769E+00	1
C15	N = 1	5	90	533	1.3864E+02	1.4167E+02	1
	N = 0	1	33	242	6.4320E+01	6.6242E+01	1
C16	N = 1	382	1166	1449	1.0523E+03	4.3612E-01	0.24
	N = 0	912	1301	1649	1.3280E+03	2.8080E+02	0.32
C17	N = 1	304	912	1925	1.0308E+03	5.7299E+02	**0.56**
	N = 0	258	658	1592	8.3580E+02	4.4671E+02	0.40
C18	N = 1	3	646	1685	7.1435E+02	4.9922E+02	**0.92**
	N = 0	27	628	1451	6.4668E+02	4.6814E+02	0.76

3.2 Experimental Results

25 independent runs were performed for each test function using 2×10^5 function evaluations (i.e., FES) at maximum, as suggested by Mallipeddi *et al.* [9].

Table 1 presents the feasibility study for different forms of the constraint violation. The feasible rate (i.e., FR) is the percentage of runs where at least one feasible solution is found during the 2×10^5 FES. "-" means there is no feasible solution in 25 runs. "N = 1" and "N = 0" stands for with or without normalization respectively.

It should be noticed that the main aim of this paper is not to propose an improved algorithm or constraint handling technique, but to study the features of normalization from the aspect of problem characteristics, and the following discussion will focus more on this.

From Table 1, it can be observed that some results by different constraint violations vary greatly. A much better FR in C04, C05, C11, and C12, also a better overall performance in C03 were obtained with N = 0; while a much better FR in C17 and C18 were obtained with N = 1. Both forms show a similar performance in other test functions.

It should also be pointed out that N = 1 (i.e., normalization) can not obtain any feasible solution in C11 and C12, indicating that normalization is not always useful.

For page limited, considering both the feasibility study and the objective function values, N = 0 (i.e., without normalization) shows a better performance in C03, C04, C05, C06, C09, C10, C11, and C12 while N = 1 shows a better performance in C17 and C18.

With the consideration of problem characteristics, N = 0 is good at solving the problems with only equality constraints, and N = 1 is good at solving problems with the non-separable objectives with both the equality and inequality constraints.

From the analysis, we can conclude that the element of problem characteristics plays an important role in choosing the right constraint violation form, which can give some guide for the normalization selection.

4   Conclusion

This paper tries to get the corresponding relationship between constrained optimization problems and the two different forms of constraint violation (i.e., normalization or not). 18 benchmark functions with 10-D developed in IEEE CEC2010 were utilized to verify the corresponding relationship. Before the experimental study, this paper also gives some systematic analysis of the normalization.

As different setting of algorithms will also influence the performance, in the future research we will try to study this influence in a more detailed way. Besides, more different kinds of problems are needed for testing the relationship, and this paper just gives some examples in this direction.

As the form of constraint violation can be seen as the first step to use a constraint handling technique, how to make the original information of the solutions better utilized, together with the algorithms and the problem characteristics (e.g., more real world problems) considered, needs to be further studied, and will be our future work.

Acknowledgments. This work was supported in part by the National Natural Science Foundation of China under Grants 71371142, Shanghai Young Teachers' Training Program under Grants ZZslg15087.

References

1. Michalewicz, Z., Schoenauer, M.: Evolutionary algorithm for constrained parameter optimization problems. Evol. Comput. **4**(1), 1–32 (1996)
2. Coello Coello, C.A.: Theoretical and numerical constraint-handling techniques used with evolutionary algorithms: a survey of the state of the art. Comput. Methods Appl. Mech. Eng. **191**(11/12), 1245–1287 (2002)
3. Mezura-Montes, E., Coello Coello, C.A.: Constraint-handling in nature-inspired numerical optimization: past, present and future. Swarm. Evol. Comput. **1**(4), 173–194 (2011)
4. Cai, Z., Wang, Y.: A multiobjective optimization-based evolutionary algorithm for constrained optimization. IEEE Trans. Evol. Comput. **10**(6), 658–675 (2006)
5. Runarsson, T.P., Yao, X.: Search biases in constrained evolutionary optimization. IEEE Trans. Syst. Man Cybern. Part C (Appl. Rev.) **35**(2), 233–243 (2005)
6. Deb, K.: An efficient constraint handling method for genetic algorithms. Comput. Methods Appl. Mech. Eng. **186**(2–4), 311–338 (2000)
7. Runarsson, T.P., Yao, X.: Stochastic ranking for constrained evolutionary optimization. IEEE Trans. Evol. Comput. **4**(3), 284–294 (2000)
8. Liang, J.J., Runarsson, T.P., Mezura-Montes, E., Clerc, M., Suganthan, P.N., Coello Coello, C.A., Deb, K.: Problem definitions and evaluation criteria for the CEC 2006. Technical report, Special Session on Constrained Real-Parameter Optimization (2006)
9. Mallipeddi, R., Suganthan, P.N.: Problem definitions and evaluation criteria for the CEC 2010 competition on constrained real-parameter optimization. Technical report, Special Session on Constrained Real-Parameter Optimization (2010)
10. Venkatraman, S., Yen, G.G.: A generic framework for constrained optimization using genetic algorithms. IEEE Trans. Evol. Comput. **9**(4), 424–435 (2005)
11. Wang, Y., Cai, Z.: Constrained evolutionary optimization by means of $(\mu + \lambda)$-differential evolution and improved adaptive trade-off model. Evol. Comput. **19**(2), 249–285 (2011)
12. Jia, G., Wang, Y., Cai, Z., Jin, Y.: An improved $(\mu + \lambda)$-constrained differential evolution for constrained optimization. Inf. Sci. **222**, 302–322 (2013)
13. Mallipeddi, R., Suganthan, P.N.: Ensemble of constraint handling techniques. IEEE Trans. Evol. Comput. **14**(4), 561–579 (2010)
14. Deb, K., Datta, R.: A fast and accurate solution of constrained optimization problems using a hybrid bi-objective and penalty function approach. In: Proceedings of CEC, pp. 1–8 (2010)
15. Deb, K., Datta, R.: An adaptive normalization based constrained handling methodology with hybrid bi-objective and penalty function approach. In: Proceedings of CEC, pp. 1–8 (2012)
16. Menchaca-Mendez, A., Coello Coello, C.A.: Solving multiobjective optimization problems using differential evolution and a maximin selection criterion. In: Proceedings of CEC, pp. 3143–3150 (2012)

Logic Gates Designed with Domain Label Based on DNA Strand Displacement

Qianhao Yang, Changjun Zhou$^{(\boxtimes)}$, and Qiang Zhang$^{(\boxtimes)}$

Key Laboratory of Advanced Design and Intelligent Computing,
Dalian University, Ministry of Education, Dalian 116622, China
`zhou-chang231@163.com, zhangq26@126.com`

Abstract. The construction of DNA logic gates plays a very significant role in solving NP-complete problems, because DNA computer applied to solving NP-complete problems consists of DNA logic gates. Although AND Gate module and OR Gate module with dual-rail logic constructed by Winfree avoided the instability caused by NOT Gate, the scale of dual-rail logic circuit is two times that of single-rail logic circuit. In this paper, domain t and domain f are applied to representing signal 1 and signal 0 respectively instead of high concentration and low concentration, and AND Gate, OR Gate, NOT Gate with domain label (domain t and f) are constructed. AND Gate, OR Gate, NOT Gate with domain label have good stability and encapsulation, which can be applied to DNA computing in the future.

Keywords: DNA strand displacement · Domain label · DNA logic gates · Binary 4 × 4 array multiplier · Visual DSD

1 Introduction

Many different methods for computing began to attract researcher's attention because NP-complete problems can't be solved by electronic computer in polynomial time. Compared with electronic computing, DNA computing is a new biological method, which can solve many NP-complete problems due to its large storage capacity, high parallelism and other advantages. DNA logic gates are the basic arithmetic structure of DNA computer, the construction of which is the prerequisite for solving NP-complete problems. Molecular logic gates are constructed based on complementary base pairing. Molecular logic gates can be divided into two categories, one involving enzymes and the other not involving enzymes. The latter mainly involves DNA strand displacement reaction because DNA strand displacement involves not enzymes due to its autonomy [1].

In 1994, Adleman [2] solved directed Hamilton Path Problem with 7 vertices via DNA computing, which presented a new approach for solving NP-complete problem. Baron [3] chose many different enzymes and then applied them to designing eight different logic gates, including XOR, INHIBIT A, INHIBIT B, AND, OR, NOR, Identity and inverter gates. Half adder and half subtractor can be designed by operating on AND and XOR or on XOR and INHIBIT A. In 2012, the first realization of a biomolecular OR gate function with double-sigmoid response (sigmoid in both inputs) was reported [4]. There are also many logic gates involving enzymes, including

© Springer International Publishing Switzerland 2016
Y. Tan et al. (Eds.): ICSI 2016, Part I, LNCS 9712, pp. 244–255, 2016.
DOI: 10.1007/978-3-319-41000-5_25

AND GATE [5–8], OR GATE [8], NAND GATE [9], NOR GATE [9], CNOT GATE [10], XOR GATE [8, 11, 12], INHIBIT [8]. Georg Seeling [13] designed AND Gate, OR Gate, and NOT Gate based on DNA strand displacement, but they can't be applied to large-scale stable logic circuit due to the instability caused by NOT Gate. In 2011, Winfree [14] solved the instability caused by NOT Gate via dual-rail logic, and used dual-rail logic to construct AND Gate module and OR Gate module. The two modules were equivalent to AND Gate, OR Gate and NOT Gate of single-rail logic circuit, and four-bit square root circuit was designed via the two modules. Qian [15] designed four interconnected artificial neurons which can implement associative memory based on the same dual-rail logic as [14].

Although Winfree [14] solved the instability caused by NOT Gate and then designed many stable biological logic circuits, dual-rail logic brought many new problems. The scale of dual-rail logic circuit is two times that of single-rail logic circuit, which increases the material cost, complexity and difficulty in designing logic circuit. The most fundamental problem of the instability caused by NOT Gate and the scale of dual-rail logic circuit is that concentration is applied to presenting logic 1 and logic 0. The concentration of reactant has an important effect on reaction rate, because the reaction rate of high concentration is faster than that of low concentration and reaction works along from high concentration side to low concentration side under the same circumstances [16]. It appears that changing the conditions of NOT Gate of single-rail logic circuit goes against designing large-scale single-rail logic circuit. In this paper, AND Gate, OR Gate and NOT Gate with domain label are constructed by domain t and f representing logic 1 and logic 0 respectively. The three modules can be applied to implementing large-scale DNA computing due to their stability and encapsulation. This paper is arranged as follows so that it can be presented clearly: the development status of DNA strand displacement is presented briefly in the first part; the background of DNA logic gates is presented in the second part; AND Gate, OR Gate and NOT Gate with domain label are presented in the third part; the simulation of binary 4 × 4 array multiplier via Visual DSD is presented in the fourth part; conclusion is presented in the fifth part.

2 Background

2.1 DNA Strand Displacement

DNA strand displacement is the reaction that one DNA single strand hybridizes with DNA double strand and completely displaces another DNA single strand from the double strand. The reaction process is showed in Fig. 1 [17].

Fig. 1. DNA strand displacement

In Fig. 1, every letter represents a domain, which is a sequence consisting of A, G, C, T. The red domains, T and T*, are complementary short domains, called toehold. The length of the short domain is assumed to be sufficiently small so that it can react reversibly. In contrast to short domain, other letters represent long domains, indicated by superscript ^, the length of which is assumed to be sufficiently large so that they can react irreversibly. The DNA strand displacement reaction in Fig. 1 is reversible, which is represented by ↔. The short domain T on the left DNA single strand, <jL^ t jR^ T^ iL^ f iR^> (code of Visual DSD), binds with T* on DNA double strand, {T^*}[iL^ f iR^ T^]<kL^ t kR^>. Then <jL^ t jR^ T^ iL^ f iR^> hybridizes with other domains on {T^*}[iL^ f iR^ T^]<kL^ t kR^> and displaces another DNA single strand, < iL^ f iR^ T^ kL^ f kR^>, from the DNA double strand. The reaction can work along from right side to left side due to reversible reaction.

3   Methods

In this paper, four kinds of DNA signal strand are constructed, which are showed in Fig. 2. Signal 0 is represented by < iL^ t iR^ T^ jL^ f jR^> and < iL^ f iR^ T^ jL^ f jR^> due to long domain f on the right side of DNA signal strand. In contrast, signal 1 is represented by < iL^ t iR^ T^ jL^ t jR^> and < iL^ f iR^ T^ jL^ t jR^> due to long domain t on the right side of DNA signal strand. The long domains, t and f, on the left part of DNA signal strand represent the logical value of upstream DNA signal strand.

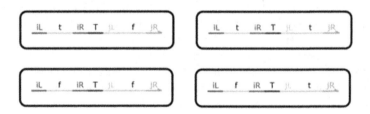

Fig. 2. Signal 0 and signal 1

3.1   NOT Gate with Domain Label

NOT Gate with domain label is constructed in this paper, which consists of two DNA double strands. The two DNA double strands are the same, except for the location of t and f, the structure of which are showed in Fig. 3.

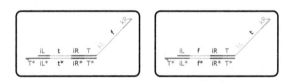

Fig. 3. NOT Gate module with domain label

The NOT Gate with domain label can react with four kinds of DNA signal strand we just mentioned above. If a DNA signal strand with domains, iL^, t, iR^, on the right hybridizes with NOT Gate with domain label, they will release DNA signal strand, < iL^ t iR^ T^ kL^ f kR^>, the right part of which is different from that of DNA input strand. The process is showed in Fig. 4. If a DNA signal strand with domains, iL^, f, iR^, on the right hybridizes with NOT Gate with domain label, they will release DNA signal strand, < iL^ f iR^ T^ kL^ t kR^>, the right part of which is different from that of DNA input strand. Hence, NOT Gate with domain label can transform signal into its opposite. The process is showed in Fig. 4.

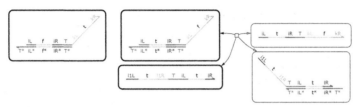

(a). transforming DNA signal from logic 1 into logic 0

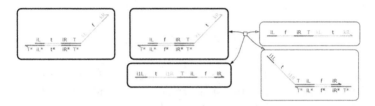

(b). transforming DNA signal from logic 0 into logic 1

Fig. 4. Reaction with NOT Gate module with domain label

3.2 AND Gate with Domain Label

AND Gate with domain label is constructed in this paper, which consists three DNA double strands denoted by $G_{ik,f}$, $G_{jk,f}$ and $G_{ijk,t}$ respectively ($G_{ik,f}$ denotes {T^*}[iL^ f iR^ T^]<kL^ f kR^>; $G_{jk,f}$ denotes {T^*}[jL^ f jR^ T^]<kL^ f kR^>; $G_{ijk,t}$ denotes {T^*}[iL^ t iR^ T^]:[jL^ t jR^ T^]<kL^ t kR^>). The structure of the three DNA double strands is showed in Fig. 5.

The AND Gate with domain label can react with four kinds of DNA signal strand showed in Fig. 2. Given two DNA input strands, it is assumed that one DNA signal strand have domains, iL^ and iR^, on the right part, and the other DNA signal strand have domains, jL^ and jR^, on the right part. If the two DNA input strands have domain t on their right part, they will hybridize with $G_{ijk,t}$ in sequence. First, the DNA input strand with iL^ and iR^ on the right hybridizes with $G_{ijk,t}$ via exposed toehold T^* and then displaces <iL^ t iR^ T^> from $G_{ijk,t}$. After the first displacement, T^* in the middle of $G_{ijk,t}$ unbinds so that the DNA input strand with jL^ and jR^ can hybridize with the product via the exposed toehold T^* in the middle of $G_{ijk,t}$ and

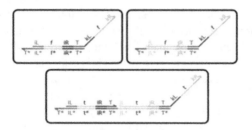

Fig. 5. AND Gate module with domain label

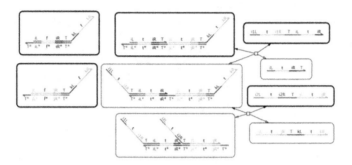

Fig. 6. Reaction releasing signal 1

release <jL^ t jR^ T^ kL^ t kR^>. The reaction process indicates that the AND Gate with domain label can transform two signal strands, representing logic 1, into a signal strand representing logic 1, which is showed in Fig. 6.

If a DNA input strand have iL^, f, iR^ on the right part, it will hybridize with $G_{ik,f}$ via the exposed toehold T^*. <iL^ f iR^ T^ kL^ f kR^> will be displaced from $G_{ik,f}$. In contrast, a DNA input strand with jL^, f, jR^ on the right part will hybridize with $G_{jk,f}$ via the exposed toehold T^*. <jL^ f jR^ T^ kL^ f kR^> will be displaced from $G_{jk,f}$. The reaction process indicates that the AND Gate with domain label can transform two signal strands, one or two of them representing logic 0, into a signal strand representing logic 0, which is showed in Fig. 7.

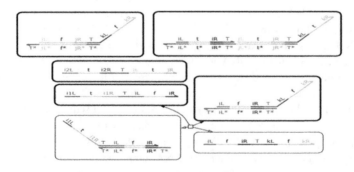

Fig. 7. Reaction releasing signal 0

3.3 OR Gate with Domain Label

OR Gate with domain label is constructed in this paper, which consists of three DNA double strands denoted by $G_{ik,t}$, $G_{jk,t}$ and $G_{ijk,f}$ respectively ($G_{ik,t}$ denotes {T^*}[iL^ t iR^ T^]<kL^ t kR^>; $G_{jk,t}$ denotes {T^*}[jL^ t jR^ T^]<kL^ t kR^>; $G_{ijk,f}$ denotes {T^*}[iL^ f iR^ T^]:[jL^ f jR^ T^]<kL^ f kR^>). The structure of the three DNA double strands is showed in Fig. 8.

The OR Gate with domain label can react with four kinds of DNA signal strand showed in Fig. 2. Given two DNA input strands, it is assumed that one DNA signal strand have domains, iL^ and iR^, on the right part, and the other DNA signal strand have domains, jL^ and jR^, on the right part. If the two DNA input strands have domain f on their right part, they will hybridize with $G_{ijk,f}$ in sequence. First, the DNA input strand with iL^ and iR^ on the right hybridizes with $G_{ijk,f}$ via exposed toehold T^* and then displaces <iL^ f iR^ T^> from $G_{ijk,f}$. After the first displacement, T^* in the middle of $G_{ijk,f}$ unbinds so that the DNA input strand with jL^ and jR^ can hybridize with the product via the exposed toehold T^* in the middle of $G_{ijk,f}$ and release <jL^ f jR^ T^ kL^ f kR^>. The reaction process indicates that the OR Gate with domain label can transform two signal strands, representing logic 0, into a signal strand representing logic 0, which is showed in Fig. 9.

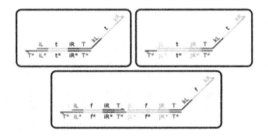

Fig. 8. OR Gate module with domain label

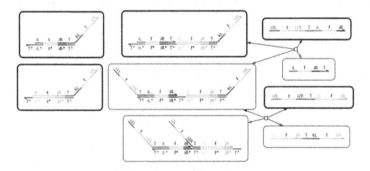

Fig. 9. Reaction releasing signal 0

If a DNA input strand have iL^, t, iR^ on the right part, it will hybridize with $G_{ik,t}$ via the exposed toehold T^*. <iL^ t iR^ T^ kL^ t kR^> will be displaced from $G_{ik,t}$. In contrast, a DNA input strand with jL^, t, jR^ on the right part will hybridize with $G_{jk,t}$ via the exposed toehold T^*. <jL^ t jR^ T^ kL^ t kR^> will be displaced from $G_{jk,t}$. The reaction process indicates that the OR Gate with domain label can transform two signal strands, one or two of them representing logic 1, into a signal strand representing logic 1, which is showed in Fig. 10.

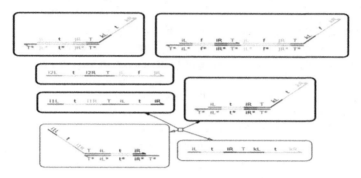

Fig. 10. Reaction releasing signal 1

4 Simulation

4.1 One-Bit Half Adder

One-bit half adder is constructed based on AND Gate, OR Gate and NOT Gate with domain label in this paper, acting as a module of binary 4 × 4 array multiplier. One-bit half adder can implement the addition between addend and augend, no taking adjacent low-order carry into account. It has two DNA input strands and two DNA output strands, the biological logic circuit of which is showed in Fig. 11. Two DNA input strands are denoted by x1 and x2 respectively, and two DNA output strands, sum and carry, are denoted by y and c respectively. Simulation of x1x2=11 and cy=10 is taken for example below.

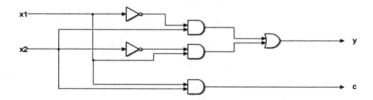

Fig. 11. One-bit half adder

The simulation result of x1x2=11 and cy=10 is showed in Fig. 12. On the right frame of Fig. 12, DNA input strands <iL^ t iR^ T^ S1L^ t S1R^>, <iL^ t iR^ T^ S2L^ t S2R^> correspond to x1, x2 respectively and represent logic 1 due to domain t on their right part. DNA output strands <SUML^ f SUMR^ fluor>, <CARRYL^ t CARRYR^ fluor> correspond to y, c respectively and represent logic 0 and logic 1 due to domain f and t. The concentration of DNA input strands decrease swiftly to less than 2000 nM, and the concentration of DNA output strands increase to almost 10000 nM in 600 s. The simulation result in Fig. 12 corresponds to x1x2=11 and cy=10 correctly, which sufficiently indicates that biological one-bit half adder designed in this paper possesses encapsulation and stability. The simulation result also indicates that the biological one-bit half adder can be applied to simulating electronic one-bit half adder.

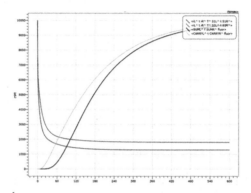

Fig. 12. Simulation result of one-bit half adder (Color figure online)

4.2 One-Bit Full Adder

Compared to one-bit half adder, one-bit full adder takes adjacent low-order carry into account. One-bit full adder has DNA input strands denoted by c1 (low-order carry), x1, x2 respectively, and has two DNA output strands denoted by y (sum), c2 (carry). It can implement the addition of three binary logic values, and simulate electronic one-bit full adder. The logic circuit is showed in Fig. 13. Simulation of c1x1x2=111 and c2y=11 is taken for example below.

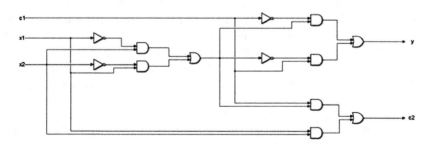

Fig. 13. One-bit full adder

The simulation result of c1x1x2=111 and c2y=11 is showed in Fig. 14. One the right frame of Fig. 14, DNA input strands <iL^ t iR^ T^ S1L^ t S1R^>, <iL^ t iR^ T^ S2L^ t S2R^>, <iL^ t iR^ T^ S3L^ t S3R^> correspond to c1, x1, x2 respectively, and represent logic 1 due to domain t on their right part. DNA output strands <SUML^ t SUMR^ fluor>, <CARRYL^ t CARRYR^ fluor> correspond to y and c2 respectively, and represent logic 1 due to domain t on their right part. The simulation result in Fig. 14 corresponds to c1x1x2=111 and c2y=11 correctly, which sufficiently indicates that the biological one-bit full adder designed in this paper possesses encapsulation and stability. The simulation result also indicates that the biological one-bit full adder can be applied to simulating electronic one-bit full adder.

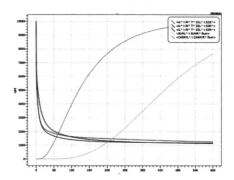

Fig. 14. Simulation result of one-bit full adder (Color figure online)

4.3   Binary 4 × 4 Array Multiplier

Binary 4 × 4 array multiplier is constituted by four biological one-bit half adders and eight biological one-bit full adders, which can implement multiplication between 4-bit binary multiplier and 4-bit binary multiplicand and output an 8-bit binary number [18]. The biological Binary 4 × 4 array multiplier designed in this paper is showed in Fig. 15. Simulation of 1111 × 1111 = 11100001 [19] is taken for example below.

The simulation result of 1111 × 1111 = 11100001 is showed in Fig. 16. On the right frame of Fig. 16, DNA input strands <A0L^ t A0R^ T^ A0L^ t A0R^>, <A1L^ t A1R^ T^ A1L^ t A1R^>, <A2L^ t A2R^ T^ A2L^ t A2R^>, <A3L^ t A3R^ T^ A3L^ t A3R^>, <B0L^ t B0R^ T^ B0L^ t B0R^>, <B1L^ t B1R^ T^ B1L^ t B1R^>, <B2L^ t B2R^ T^ B2L^ t B2R^>, <B3L^ t B3R^ T^ B3L^ t B3R^> correspond to A0, A1, A2, A3, B0, B1, B2, B3 respectively, represent logic 1 due to domain t on their right part. DNA output strands <J0L^ t J0R^ fluor>, <J3L^ f J3R^ fluor>, <J10L^ f J10R^ fluor>, <J20L^ f J20R^ fluor>, <J29L^ f J29R^ fluor>, <J35L^ t J35R^ fluor>, <J38L^ t J38R^ fluor>, <J39L^ t J39R^ fluor> correspond to S0, S1, S2, S3, S4, S5, S6, S7 respectively, represent logic 1, 0, 0, 0, 0, 1, 1, 1 depending on whether domain t or f is on their right part. The concentration of DNA input strands decrease very swiftly to almost zero and that of DNA output strands increase to almost saturation level, which sufficiently indicates that the biological binary 4 × 4 array multiplier possesses good

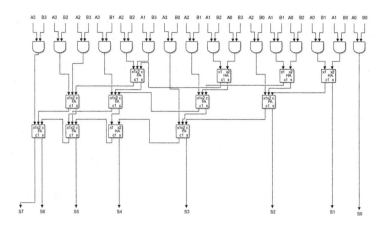

Fig. 15. Binary 4 × 4 array multiplier

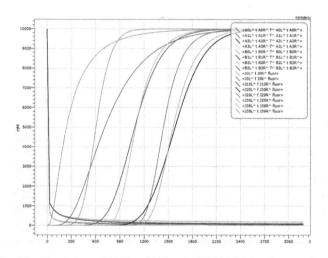

Fig. 16. Simulation of $1111 \times 1111 = 11100001$ (Color figure online)

encapsulation and stability. The simulation result in Fig. 16 corresponds to $1111 \times 1111 = 11100001$ correctly, which indicates that the biological binary 4 × 4 array multiplier can be applied to simulating electronic binary 4 × 4 array multiplier.

5   Conclusion

In this paper, AND Gate, OR Gate, NOT Gate with domain label are constructed. One-bit half adder and one-bit full adder, acting as a module of binary 4 × 4 array multiplier, are formed by combining the three elementary modules. Biological binary 4 × 4 array multiplier is formed by four biological one-bit half adders and eight biological one-bit full adders. The good encapsulation and stability of the three elementary modules is further verified by simulating biological one-bit half adder, biological one-bit full adder and biological binary 4×4 array multiplier. Except for biological logic circuit designed in this paper, AND Gate with domain label, OR Gate with domain label, NOT Gate with domain label can also be applied to large-scale biological logic circuit and DNA computing due to the great potential of the three elementary modules.

AND Gate, OR Gate, NOT Gate with domain label can solve the instability caused by NOT Gate and the problem about the scale of dual-rail logic circuit [14]. Hence, many functional large-scale biological logic circuits are expected to be designed by combining the three elementary modules in the near future.

Acknowledgment. This work is supported by the National Natural Science Foundation of China (Nos. 31370778, 61370005, 61425002, 61402066, 61402067), the Basic Research Program of the Key Lab in Liaoning Province Educational Department (Nos. LZ2014049, LZ2015004), the Project Supported by Natural Science Foundation of Liaoning Province (No. 2014020132), the Project Supported by Scientific Research Fund of Liaoning Provincial Education (No. L2014499), and by the Program for Liaoning Key Lab of Intelligent Information Processing and Network Technology in University.

References

1. Green, S.J., Lubrich, D., Turberfield, A.J.: DNA hairpins: fuel for autonomous DNA devices. Biophys. J. **91**(8), 2966–2975 (2006)
2. Adleman, L.M.: Molecular computation of solutions to combinatorial problems. Science **266** (5187), 1021–1024 (1994)
3. Baron, R., Lioubashevski, O., Katz, E., Niazov, T., Willner, I.: Logic gates and elementary computing by enzymes. J. Phys. Chem. A **110**(27), 8548–8553 (2006)
4. Zavalov, O., Bocharova, V., Privman, V., Katz, E.: Enzyme-based logic: OR gate with double-sigmoid filter response. J. Phys. Chem. B **116**(32), 9683–9689 (2012)
5. Bakshi, S., Zavalov, O., Halámek, J., Privman, V., Katz, E.: Modularity of biochemical filtering for inducing sigmoid response in both inputs in an enzymatic AND gate. J. Phys. Chem. B **117**(34), 9857–9865 (2013)
6. Privman, V., Fratto, B.E., Zavalov, O., Halámek, J., Katz, E.: Enzymatic AND logic gate with sigmoid response induced by photochemically controlled oxidation of the output. J. Phys. Chem. B **117**(25), 7559–7568 (2013)
7. Halámek, J., Zavalov, O., Halámková, L., Korkmaz, S., Privman, V., Katz, E.: enzyme-based logic analysis of biomarkers at physiological concentrations: AND gate with double-sigmoid "filter" response. J. Phys. Chem. B **116**(15), 4457–4464 (2012)
8. Strack, G., Pita, M., Ornatska, M., Katz, E.: Boolean logic gates that use enzymes as input signals. ChemBioChem **9**(8), 1260–1266 (2008)
9. Zhou, J., Arugula, M.A., Halamek, J., Pita, M., Katz, E.: Enzyme-based NAND and NOR logic gates with modular design. J. Phys. Chem. B **113**(49), 16065–16070 (2009)
10. Moseley, F., Halámek, J., Kramer, F., Poghossian, A., Schöning, M.J., Katz, E.: An enzyme-based reversible CNOT logic gate realized in a flow system. Analyst **139**(8), 1839–1842 (2014)

11. Halamek, J., Bocharova, V., Arugula, M.A., Strack, G., Privman, V., Katz, E.: Realization and properties of biochemical-computing biocatalytic XOR gate based on enzyme inhibition by a substrate. J. Phys. Chem. B **115**(32), 9838–9845 (2011)
12. Privman, V., Zhou, J., Halámek, J., Katz, E.: Realization and properties of biochemical-computing biocatalytic XOR gate based on signal change. J. Phys. Chem. B **114**(42), 13601–13608 (2010)
13. Seelig, G., Soloveichik, D., Zhang, D.Y., Winfree, E.: Enzyme-free nucleic acid logic circuits. Science **314**(5805), 1585–1588 (2006)
14. Qian, L., Winfree, E.: Scaling up digital circuit computation with DNA strand displacement cascades. Science **332**(6034), 1196–1201 (2011)
15. Qian, L., Winfree, E., Bruck, J.: Neural network computation with DNA strand displacement cascades. Nature **475**(7356), 368–372 (2011)
16. Levine, R.D.: Molecular Reaction Dynamics. Cambridge University Press, Cambridge (2009)
17. Lakin, M.R., Youssef, S., Cardelli, L., Phillips, A.: Abstractions for DNA circuit design. J. R. Soc. Interface **9**(68), 470–486 (2012)
18. Jayaprakasan, V., Vijayakumar, S., Bhaaskaran, V.K.: Evaluation of the conventional vs. ancient computation methodology for energy efficient arithmetic architecture. In: 2011 International Conference on Process Automation, Control and Computing (PACC), pp. 1–4. IEEE Press (2011)
19. Baugh, C.R., Wooley, B.A.: A two's complement parallel array multiplication algorithm. IEEE Trans. Comput. **12**, 1045–1047 (1973)

Hybrid Search Optimization

Missing Data Estimation in High-Dimensional Datasets: A Swarm Intelligence-Deep Neural Network Approach

Collins Leke$^{(\boxtimes)}$ and Tshilidzi Marwala

University of Johannesburg, Johannesburg, South Africa
{collinsl,tmarwala}@uj.ac.za

Abstract. In this paper, we examine the problem of missing data in high-dimensional datasets by taking into consideration the Missing Completely at Random and Missing at Random mechanisms, as well as the Arbitrary missing pattern. Additionally, this paper employs a methodology based on Deep Learning and Swarm Intelligence algorithms in order to provide reliable estimates for missing data. The deep learning technique is used to extract features from the input data via an unsupervised learning approach by modeling the data distribution based on the input. This deep learning technique is then used as part of the objective function for the swarm intelligence technique in order to estimate the missing data after a supervised fine-tuning phase by minimizing an error function based on the interrelationship and correlation between features in the dataset. The investigated methodology in this paper therefore has longer running times, however, the promising potential outcomes justify the trade-off. Also, basic knowledge of statistics is presumed.

Keywords: Missing data · Deep learning · Swarm intelligence · High-dimensional data · Supervised learning · Unsupervised learning

1 Introduction

Previous research across a wide range of academic fields suggests that decision-making and data analysis tasks are made nontrivial by the presence of missing data. As such, it can be assumed that decisions are likely to be more accurate and reliable when complete/representative datasets are used instead of incomplete datasets. This assumption has led to a lot of research in the data mining domain, with novel techniques being developed to perform this task accurately [1–9]. Research suggests that applications in various professional fields such as in medicine, manufacturing or energy that use sensors in instruments to report vital information and enable decision-making processes may fail and lead to incorrect outcomes due to the presence of missing data. In such cases, it is very important to have a system capable of imputing the missing data from the failed sensors with high accuracy. The imputation procedure will require the approximation of missing values taking into account the interrelationships that exist between the

© Springer International Publishing Switzerland 2016
Y. Tan et al. (Eds.): ICSI 2016, Part I, LNCS 9712, pp. 259–270, 2016.
DOI: 10.1007/978-3-319-41000-5_26

data from sensors in the system. Another instance where the presence of missing data poses a threat in decision-making is in image recognition systems, whereby the absence of pixel values renders the image prediction or classification task difficult and as such, systems capable of imputing the missing values with high accuracy are needed to make the task more feasible.

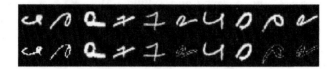

Fig. 1. Sample of MNIST dataset. Top row - real data: bottom row - data with missing pixel values

Consider a high dimensional dataset such as the Mixed National Institute of Standards and Technology (MNIST) dataset with 784 feature variables being the pixel values as shown in Fig. 1 above. Assuming that pixel values are missing at random as observed in the bottom row and a statistic analysis is required to classify the above dataset, the questions of interest would be: (i) Can we impute with some degree of certainty the missing data in high dimensional datasets with high accuracy? (ii) Can new techniques be introduced for approximation of the missing data when correlation and interrelationships between the variables are considered? This paper therefore aims to use a Deep Learning (DL) technique built with Restricted Boltzmann machines stacked together to form an autoencoder in tandem with a swarm intelligence (SI) algorithm to estimate the missing data with the model created which would cater to the mechanisms of interest and the arbitrary pattern. The dataset used is the MNIST database of handwritten digits by LeCun [10]. It has a training set of 60,000 sample images and a test set of 10,000 sample images with 784 features. These images show handwritten digits from 0 to 9. Due to the fact that the research discussed in this paper was conducted at a time when there was little or no interest in the DL-SI missing data predictors on high dimensional data, this paper seeks to exploit the use of this technique on the MNIST dataset.

The remainder of this paper is structured as follows, Sect. 2 introduces missing data, the deep learning techniques used as well as the swarm intelligence algorithm implemented. This section also presents related work in the domain. Section 3 presents the experimental design and procedures used, while Sect. 4 focuses on the results and key findings from the experiments conducted in this article. Discussions, concluding remarks and suggestions for future research are further presented in Sect. 5.

2   Background

This article implements a Deep Learning technique referred to as a Stacked Autoencoder built using Restricted Boltzmann machines, all of which have been individually trained using the Contrastive Divergence algorithm and stacked together in a bottom-up manner. The estimation of missing values is performed by using the Firefly Algorithm, which is the swarm intelligence method. However, this article will first briefly discuss the methods used and the problem it aims to solve.

2.1 Missing Data and Deep Learning

Missing data is a situation whereby some features within a dataset are lacking components [11]. With this ensues problems in application domains that rely on the access to complete and quality data which can affect every academic/professional fields and sectors. Techniques aimed at rectifying the problem have been an area of research in several disciplines [11–13]. The manner in which data points go missing in a dataset determines the approach to be used in estimating these values. As per [13], there exist three missing data mechanisms. This article focuses on investigating the Missing Completely at Random (MCAR) and Missing at Random (MAR) mechanisms. Previous research suggests that MCAR scenario arises when the chances of there being a missing data entry for a feature is not dependent on the feature itself or on any of the other features in the dataset [4]. This implies a lack of correlation or cross-correlation between features including the feature of interest [11]. MAR on the other hand arises when missingness in a specific feature is reliant upon the other features within the dataset, but not the feature of interest itself [4]. According to [13], there are two main missing data patterns. These are the arbitrary and monotone missing data patterns. In the arbitrary pattern, missing observations may occur anywhere and the ordering of the variables is of no importance. In monotone missing patterns, the ordering of the variables is of importance and occurrence is not random. Based upon this realization, this article will go on to focus on the arbitrary missing pattern.

Deep Learning comprises of several algorithms in machine learning that make use of a cataract of nonlinear processing units organized into a number of layers that extract and transform features from the input data [14,15]. Each of the layers use the output from the previous layer as input and a supervised or unsupervised algorithm could be used in the training phase. With these come applications in supervised and unsupervised problems like classification and pattern analysis, respectively. It is also based on the unsupervised learning of multiple levels of features or representations of the input data whereby higher-level features are obtained from lower level features to yield a hierarchical representation of the data [15]. By learning multiple levels of representations that depict different levels of abstraction of the data, we obtain a hierarchy of concepts. In this article, the Deep Learning technique used is the Stacked AutoEncoder.

2.2 Restricted Boltzmann Machine (RBM)

Firstly, a Boltzmann machine (BM) is an undirected network with nodes possessing stochastic traits that can be described as a neural network. It is used amongst other things to extract vital information from an unknown probability distribution using samples from the distribution, which is generally a difficult process [16]. This learning process is made simple by implementing restrictions on the network structure leading to Restricted Boltzmann machines (RBMs). An RBM can be described as an undirected, probabilistic, parameterized graphical model also known as a Markov random field (MRF). RBMs became techniques of interest after being suggested as components of multi-layer topologies termed deep networks [16]. The idea is that hidden nodes extract vital information from the observations, which subsequently represent inputs to the next RBM. Stacking these RBMs together has as objective, obtaining high level representations of data by learning features from features. An RBM which is also an MRF associated with a bipartite undirected graph consists of m visible nodes, $V = (V_1, \ldots, V_m)$ representing input data, and n hidden nodes, $H = (H_1, \ldots, H_n)$ capturing interdependencies between features in the input layer [16]. In this article, the features V have as values, $v \in [0, 1]^{m+n}$, while H have as values, $h \in \{0, 1\}^{m+n}$. The distribution given by the Gibbs distribution has as energy function [16]:

$$E\left(v, h\right) = -h^T W v - b^T v - c^T h . \tag{1}$$

In scalar form, (1) is expressed as [16]:

$$E\left(v, h\right) = -\sum_{i=1}^{n} \sum_{j=1}^{m} w_{ij} h_i v_j - \sum_{j=1}^{m} b_j v_j - \sum_{i=1}^{n} c_i h_i . \tag{2}$$

In (2), w_{ij}, which is the most important part of an RBM model is a real valued weight between units V_j and H_i, while b and c are the bias terms for the visible and hidden variables, respectively. If w_{ij} is negative, and v_j and h_i are equal to one, the probability decreases leading to a high energy. On the contrary, if w_{ij} is positive, and v_j and h_i are equal to zero, the probability increases leading to a lower energy. If b_j is negative and $v_j = 1$, E increases leading to a low probability. Therefore, there is a preference for $v_j = 0$ instead of $v_j = 1$. However, if b_j is positive and $v_j = 0$, E decreases leading to a high probability, and a preference for $v_j = 1$ instead of $v_j = 0$. A negative b_j value decreases the second term in (2), while a positive value for b_j increases this second term. The same applies for c_i and the third term in (2). The Gibbs distributions or probabilities from (1) or (2) are then obtained by [16]:

$$p\left(v, h\right) = \frac{e^{-E(v,h)}}{Z} = \frac{e^{(h^T W v + b^T v + c^T h)}}{Z} = \frac{e^{(h^T W v)} e^{(b^T v)} e^{(c^T h)}}{Z} . \tag{3}$$

Here, the exponential terms are factors of a markov network with vector nodes, while Z is the intractable partition function. It is intractable courtesy of the exponential number of values it can take. For an RBM, $Z = \sum_{v,h} e^{-E(v,h)}$. Another key aspect of RBMs is that h is conditionally independent of v and vice versa, due to the fact that there are no connections between nodes in the same layer. This property is expressed mathematically as [16]:

$$p(h|v) = \prod_{i=1}^{n} p(h_i|v) \; and \; p(v|h) = \prod_{i=1}^{m} p(v_i|h) \; . \qquad (4)$$

2.3   Contrastive Divergence (CD)

The objective in training an RBM is to minimize the average negative log-likelihood (loss) without regularization using a stochastic gradient descent algorithm as it scales well with high-dimensional datasets. Achieving this objective requires the partial derivative of any parameter, θ, of the loss function as per the following equation:

$$\frac{\partial \left(-logp \left(v^{(t)}\right)\right)}{\partial \theta} = E_h \left[\frac{\partial E \left(v^{(t)}, h\right)}{\partial \theta}|v^{(t)}\right] - E_{v,h} \left[\frac{\partial E \left(v, h\right)}{\partial \theta}\right] \; . \qquad (5)$$

The first term in (5) is the expectation over the data distribution and is referred to as the positive phase, while v and h represent the same variables as in (1)–(4). The second term, which is the expectation over the model distribution is termed the negative phase. This phase is hard to compute and also intractable because an exponential sum is required over both h and v. Furthermore, many sampling steps are needed to obtain unbiased estimates of the log-likelihood gradient. However, it has been shown recently that running a markov chain for just a few steps leads to estimates that are sufficient for training a model [16]. This approach has led to the contrastive divergence (CD) algorithm. CD is a training method for undirected probabilistic graphical models with the idea being to do away with the double expectations in the negative phase in (5) and instead focus on estimation. It basically implements a Monte-Carlo estimate of the expectation over a single input data point. The idea of k-step CD (CD-k) is that rather than the second term being approximated in (5) by a sample from the model distribution, k steps of a Gibbs chain is run, with k frequently set to 1 [16]. The Gibbs chain starts with a training sample $v^{(0)}$ of the training data and returns $v^{(k)}$ after k steps [16]. Each step, t, entails sampling $h^{(t)}$ from $p(h|v^{(t)})$, then obtaining samples $v^{(t+1)}$ from $p(v|h^{(t)})$ [16]. For one training pattern, $v^{(0)}$, the log-likelihood gradient w.r.t. θ is approximated by [16]:

$$CD_k(\theta, v^{(0)}) = -\sum_h p(h|v^{(0)})\frac{\partial E(v^{(0)}, h)}{\partial \theta} + \sum_h p(h|v^{(k)})\frac{\partial E(v^{(k)}, h)}{\partial \theta} \; . \qquad (6)$$

Due to the fact that $v^{(k)}$ is not a obtained from the stationary model distribution, the approximation (6) is biased. The bias in effect fades away as $k \longrightarrow \infty$ [16]. Another aspect that points to CD being biased is that it maximizes the difference between two Kullback-Liebler (KL) divergences [16]:

$$KL(q|p) - KL(p_k|p) \; . \qquad (7)$$

Here, the experimental distribution is q and the distribution of the visible variables after k steps of the Markov chain is p_k [16]. If stationarity in the execution of the chain is already attained, $p_k = p$ holds, and therefore $KL(p_k|p) = 0$, and the error of the approximation by CD fades away [16].

2.4   Autoencoder (AE)

An Autoencoder is an artificial neural network that attempts to reproduce its input at the output layer. The basic idea behind autoencoders is that the mapping from the input to the output, $x^{(i)} \mapsto y^{(i)}$ reveals vital information and the essential structure in the input vector $x^{(i)}$ that is otherwise abstract. An autoencoder takes an input vector x and maps it to a hidden representation y via a deterministic mapping function f_θ of the form $f_\theta(x) = s(Wx + b)$ [17]. The θ parameter comprises of the matrix of weights W and the vector of offsets/biases b. s is the sigmoid activation function expressed as:

$$s = \frac{1}{1 + e^{-x}} \; . \tag{8}$$

The hidden representation y is then mapped to a reconstructed vector z which is obtained by the functions [16]:

$$z = g_{\theta'}(y) = s(W'y + b') \; or \; z = g_{\theta'}(y) = W'y + b' \; . \tag{9}$$

Here, the parameter set θ' comprises of the transpose of the matrix of weights and vector of biases from the encoder prior to the fine-tuning phase [17]. When the aforementioned transposition of weights is done, the autoencoder is said to have tied weights. z is not explained as a rigorous regeneration of x but instead as the parameters of $p(X|Z = z)$ in probabilistic terms, which may yield x with high probability [17]. This thus leads to:

$$p(X|Y = y) = p(X|Z = g_{\theta'}(y)) \; . \tag{10}$$

From this, we obtain a reconstruction error which is to be optimized by the optimization technique and is of the form $L(x, z) \propto -logp(x|z)$. This equation as per [18] could also be expressed as:

$$\delta_{AE}(\theta) = \sum_t L\left(x^{(t)}, g_\theta\left(f_\theta\left(x^{(t)}\right)\right)\right) \; . \tag{11}$$

2.5   Firefly Algorithm (FA)

FA is a nature-inspired metaheuristic algorithm based on the flashing patterns and behavior of fireflies [19]. It is based on three main rules being: (i) Fireflies are unisex so all fireflies are attracted to all other fireflies, (ii) Attractiveness is proportional to the brightness and they both decrease as the distance increases. The idea is the less brighter firefly will move towards the brighter one. If there is no obvious brighter firefly, they move randomly, and, (iii) Brightness of a firefly is determined by the landscape of the objective function [19]. Considering that attractiveness is proportional to light intensity, the variation of attractiveness can be defined with respect to the distance as [19]:

$$\beta = \beta_0 e^{-\gamma r^2} \; . \tag{12}$$

In (12), β is the attractiveness of a firefly, β_0 is the initial attractiveness of a firefly, and r is the distance between two fireflies. The movement of a firefly towards a brighter one is determined by [19]:

$$x_i^{t+1} = x_i^t + \beta_0 e^{-\gamma r_{ij}^2} \left(x_j^t - x_i^t \right) + \alpha_t \epsilon_i^t . \tag{13}$$

Here, x_i and x_j are the positions of two fireflies, and the second term is due to the attraction between the fireflies. t and $t+1$ represent different time steps, α is the randomization parameter controlling the step size in the third term, while ϵ is a vector with random numbers obtained from a Gaussian distribution. If $\beta_0 = 0$, the movement is then a simple random walk [19]. If $\gamma = 0$, the movement reduces to a variant of the particle swarm optimization algorithm [19]. The parameters used in this research are: (i) n = number of missing cases per sample, (ii) 1000 iterations, (iii) $\alpha = 0.25$, (iv) $\beta = 0.2$ and (v) $\gamma = 1$. The parameters were selected as they yielded the more optimal results after experimentation with different permutations and combinations of values. The FA algorithm is used because although it has been successfully applied in a number of domains such as digital image compression, eigenvalue optimization, feature selection and fault detection, scheduling and TSP, etc., its efficiency has not been investigated in missing data estimation tasks on high-dimensional datasets.

2.6   Related Work

We present some of the work that has been done by researchers to address the problem of missing data. The research done in [1] implements a hybrid genetic algorithm-neural network system to perform missing data imputation tasks with varying number of missing values within a single instance while [2] creates a hybrid k-Nearest Neighbor-Neural Network system for the same purpose. In [4], a hybrid Auto-Associative neural network or autoencoder with genetic algorithm, simulated annealing and particle swarm optimization model is used to impute missing data with high levels of accuracy in cases where just one feature variable has missing input entries. In some cases, neural networks were used with Principal Component Analysis (PCA) and genetic algorithm as in [5,6]. In [7], they use robust regression imputation for missing data in the presence of outliers and investigate its effectiveness. In [8], it is suggested that information within incomplete cases, that is, instances with missing values be used when estimating missing values. A nonparametric iterative imputation algorithm (NIIA) is proposed that leads to a root mean squared error value of at least 0.5 on the imputation of continuous values and a classification accuracy of at most 87.3 % on the imputation of discrete values with varying ratios of missingness. In [9], the shell-neighbor method is applied in missing data imputation by means of the Shell-Neighbor Imputation (SNI) algorithm which is observed to perform better than the k-Nearest Neighbor imputation method in terms of imputation and classification accuracy as it takes into account the left and right nearest neighbors of the missing data as well as varying number of nearest neighbors contrary to k-NN that considers just fixed k nearest neighbors. In [20], a multi-objective genetic algorithm approach is presented for missing data imputation.

It is observed that the results obtained outperform some of the well known missing data methods with accuracies in the 90 percentile. Novel algorithms for missing data imputation and comparisons between existing techniques can be found in papers such as [20–27].

3   Experimental Design and Procedure

In the design of the experiments, MATLAB R2014a software was used on a Dell Desktop computer with Intel(R) Core(TM) i3-2120 CPU @ 3.30 GHz processor, 4.00 GB RAM, 32 GB virtual RAM, 64-bit Operating System running Windows 8.1 Pro. Additionally, the MNIST database was used and it contains 60,000 training images and 10,000 test images. Each of these images is of size $28 \times 28 = 784$ pixels. This results in a training set of size 60000×784 and a test of size 10000×784. Data preprocessing was performed normalizing all pixel values in the range [0, 1]. The individual network layers of the Deep AE were pretrained using RBMs and CD to initialize the weights and biases in a good solution space. The individual layers pretrained were of size $784 - 1000$, $1000 - 500$, $500 - 250$, and $250 - 30$. These are stacked and subsequently transposed to obtain the encoder and decoder parts of the autoncoder network, respectively. The resulting network architecture is of size, $784 - 1000 - 500 - 250 - 30 - 250 - 500 - 1000 - 784$, with an input and output layer with the same number of nodes, and seven hidden layers with varying number of nodes. The network is then fine-tuned using backpropagation, minimizing the mean squared network error. The error value obtained after training is 0.0025. The training is done using the entire training set of data that are divided into 600 balanced mini-batches. The weight and bias updates are done after every mini-batch. Training higher layers of weights is achieved by having the real-valued activations of the visible nodes in preceding RBMs being transcribed as the activation probabilities of the hidden nodes in lower level RBMs. The Multilayer Perceptron (MLP) AE has an input and output layer, both consisting of 784 nodes, and one hidden layer consisting of 400 nodes obtained by experimenting with different numbers of nodes in the hidden layer, and observing which architecture leads to the lowest mean squared network error. A $784 - 400 - 784$ network architecture led to the lowest mean squared network error value of 0.0032. The hidden and output layer activation function used is the sigmoid function. The training is done using the scaled conjugate gradient descent algorithm for 1000 epochs. Missingness in the test set of data is then created at random according to the MAR and MCAR mechanisms, as well as the arbitrary pattern, and these missing values are approximated using the swarm intelligence algorithm which has as objective function minimizing the loss function of the fine-tuned network. The tolerance error is initially set to 0.05 (5 %) in one of the networks, and is considered reasonable for a first time investigation of the proposed method. The overall approach consist of four consecutive steps being:

1. Train the individual RBMs on a training set of data with complete records using the greedy layer-by-layer pre-training algorithm described in [28]

starting from the bottom layer. Each layer is trained for 50 epochs with the learning rate for the weights, visible unit biases and hidden unit biases set to 0.1. The initial and final momentum are set to 0.5 and 0.9, respectively. The final parameter is the weight cost which is set to 0.0002.

2. Stack the RBMs to form the Encoder and Decoder phases of a Deep Autoencoder with tied weights.

3. Fine-tune the Deep Autoencoder using back-propagation for 1000 epochs through the entire set of training data.

4. Estimate the missing data with the fine-tuned deep network as part of the objective function in the Firefly Algorithm parsing the known variable values to the objective function, while first estimating the unknown values before parsing these estimates to the objective function. The estimation procedure is terminated when a stopping criterion is achieved, which is either an error tolerance of 5 % (0.05), or the maximum number of function evaluations being attained.

4   Experimental Results

In the investigation of the imputation technique, we used the test set of data which contained missing data entries accounting for approximately 10 % of the data. We present in Tables 1 and 2, Actual, Estimate and Squared Error values from the proposed Deep Autoencoder system without tolerance (Table 1), and from MLP Autoencoder system (Table 2). The distance, ϵ, from the estimate to the actual value, added to the squared error are parameters that determine the performance of the method. In all cases presented in both tables, the Deep Autoencoder system shows $\epsilon_d = 0, 0.0608, 0, 0.0275, 0, 0.0922, 0.0009, 0.0283$, while for the same entries (actual values), the MLP Autoencoder shows that $\epsilon_m = 0.0246, 0.2646, 0.0149, 0.1643, 0, 0.1982, 0.0509, 0.0473$, respectively. They show better performance of the proposed technique without a set error tolerance when compared to the existing MLP Autoencoder. This knowledge is validated by the squared error which is always smaller for the proposed technique, for all cases presented in Tables 1 and 2. We could consider this enough to conclude of on the performance of both compared techniques, but we need to analyse the processing time, which seems to be better for the existing method when compared to the proposed Deep Autoencoder system. This is demonstrated by Fig. 3, where we compare processing times for both techniques. It is evident that setting an error tolerance value makes the estimation process faster as observed in Fig. 3. However, this is at the expense of accuracy which is the main aspect in such a task as seen in Fig. 2. The bigger the error tolerance value, the faster the estimation of the missing data.

Table 1. Actual, estimated and squared error values from deep autoencoder system without set tolerance.

Actual	Estimate	Squared error
0	0	0
0.3216	0.3824	0.0037
0	0	0
0.9725	1	0.0008
0	0	0
0.9961	0.9039	0.0085
0.0509	0.0500	8.38e-07
0.5765	0.6048	0.0008

Table 2. Actual, estimated and squared error values from MLP autoencoder system without set tolerance.

Actual	Estimate	Squared error
0	0.0246	0.0006
0.3216	0.5862	0.0700
0	0.0149	0.0002
0.9725	0.8082	0.0270
0	0	0
0.9961	0.7979	0.0393
0.0509	0	0.0026
0.5765	0.5292	0.0022

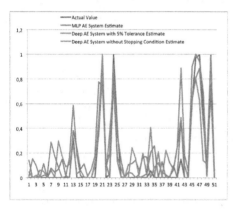

Fig. 2. Actual vs estimated values (Color figure online).

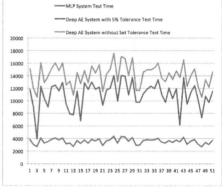

Fig. 3. Test times per sample (Color figure online).

5   Discussion and Conclusion

This paper investigates the use of a deep neural network with a swarm intelligence algorithm to impute missing data in a high-dimensional dataset. According to the arbitrary missing data pattern, MAR and MCAR mechanisms, missing data could occur anywhere in the dataset. The experiment in this paper considers a scenario in which 10 % of the test set of data is missing. These values are to be estimated with a set error tolerance of 5 %, as well as no set error tolerance. Also, the proposed method is compared to an MLP Autoencoder estimation system. The results obtained reveal that the proposed system yields the more accurate estimates, especially when there is no set error tolerance value. This is made evident when the distance and squared error values are considered. The AE systems both yield better estimates than the MLP system. However, with these accurate estimates come longer running times which are observed to become smaller when error tolerance values are set. The bigger the tolerance value, the smaller the running time. Based on the findings in this article, we intend to perform an in-depth parameter analysis in any future research in order to observe which parameters are optimal for the task and we will generalize this aspect using several datasets. Another obstacle faced in this research was the computation time to estimate the missing values and to address this, we will parallelize the process on a multi-core system to observe whether parallelizing the task does indeed speed up the process and maintain efficiency.

References

1. Abdella, M., Marwala, T.: The use of genetic algorithms and neural networks to approximate missing data in database. In: 3rd International Conference on Computational Cybernetics, ICCC 2005, pp. 207–212. IEEE (2005)
2. Aydilek, I.B., Arslan, A.: A novel hybrid approach to estimating missing values in databases using k-nearest neighbors and neural networks. Int. J. Innovative Comput. Inf. Control 7(8), 4705–4717 (2012)
3. Koko, E.E.M., Mohamed, A.I.A.: Missing data treatment method on cluster analysis. Int. J. Adv. Stat. Probab. 3(2), 191–209 (2015)
4. Leke, C., Twala, B., Marwala, T.: Modeling of missing data prediction: computational intelligence and optimization algorithms. In: International Conference on Systems, Man and Cybernetics (SMC), pp. 1400–1404. IEEE (2014)
5. Mistry, F.J., Nelwamondo, F.V., Marwala, T.: Missing data estimation using principle component analysis and autoassociative neural networks. J. Systemics Cybernatics Inform. 7(3), 72–79 (2009)
6. Nelwamondo, F.V., Mohamed, S., Marwala, T.: Missing data: A comparison of neural network and expectation maximisation techniques. arXiv preprint arXiv:0704.3474 (2007)
7. Rana, S., John, A.H., Midi, H., Imon, A.: Robust regression imputation for missing data in the presence of outliers. Far East J. Math. Sci. 97(2), 183 (2015). Pushpa Publishing House
8. Zhang, S., Jin, Z., Zhu, X.: Missing data imputation by utilizing information within incomplete instances. J. Syst. Soft. 84(3), 452–459 (2011). Elsevier
9. Zhang, S.: Shell-neighbor method and its application in missing data imputation. Appl. Intell. 35(1), 123–133 (2011). Springer
10. LeCun, Y.: The MNIST database of handwritten digits. http://yann.lecun.com/exdb/mnist/. Accessed 1 Jan 2016
11. Rubin, D.B.: Multiple imputations in sample surveys-a phenomenological Bayesian approach to nonresponse. In: Proceedings of the Survey Research Methods Section of the American Statistical Association, vol. 1, pp. 20–34. American Statistical Association (1978)
12. Allison, P.D.: Multiple imputation for missing data: a cautionary tale. Sociol. Methods Res. 28, 301–309 (1999). Philadelphia
13. Little, R.J., Rubin, D.B.: Statistical Analysis with Missing Data. Wiley, New York (2014)
14. Deng, L., Li, J., Huang, J.-T., Yao, K., Yu, D., Seide, F., Seltzer, M., Zweig, G., He, X., Williams, J.: Recent advances in deep learning for speech research at Microsoft. In: International Conference on Acoustics, Speech and Signal Processing (ICASSP), pp. 8604–8608. IEEE (2013)

15. Deng, L., Yu, D.: Deep learning: methods and applications. Found. Trends Sig. Process. **7**(3–4), 197–387 (2014). Now Publishers Inc
16. Fischer, A., Igel, C.: An introduction to restricted Boltzmann machines. In: Alvarez, L., Mejail, M., Gomez, L., Jacobo, J. (eds.) CIARP 2012. LNCS, vol. 7441, pp. 14–36. Springer, Heidelberg (2012). ISBN 978-3-642-33275-3
17. Isaacs, J.C.: Representational learning for sonar ATR. In: Proceedings SPIE 9072, Detection and Sensing of Mines, Explosive Objects, and Obscured Targets XIX, p. 907203, 9 June 2014. http://dx.doi.org/10.1117/12.2053057
18. Bengio, Y., Courville, A., Vincent, P.: Representation learning: a review and new perspectives. Trans. Pattern Anal. Mach. Intell. **35**(8), 1798–1828 (2013). IEEE
19. Yang, X.-S.: Firefly algorithm, Levy flights and global optimization. In: Bramer, M., Ellis, R., Petridis, M. (eds.) Research and Development in Intelligent Systems XXVI, pp. 209–218. Springer, London (2010)
20. Lobato, F., Sales, C., Araujo, I., Tadaiesky, V., Dias, L., Ramos, L., Santana, A.: Multi-objective genetic algorithm for missing data imputation. Pattern Recogn. Lett. **68**(1), 126–131 (2015). Elsevier
21. Jerez, J.M., Molina, I., García-Laencina, P.J., Alba, E., Ribelles, N., Martín, M., Franco, L.: Missing data imputation using statistical and machine learning methods in a real breast cancer problem. Artif. Intell. Med. **50**(2), 105–115 (2010). Elsevier
22. Kalaycioglu, O., Copas, A., King, M., Omar, R.Z.: A comparison of multiple-imputation methods for handling missing data in repeated measurements observational studies. J. R. Stat. Soc. Ser. A (Stat. Soc.) (2015). Wiley Online Library
23. Lee, K.J., Carlin, J.B.: Multiple imputation for missing data: fully conditional specification versus multivariate normal imputation. Am. J. Epidemiol. **171**(5), 624–632 (2010). Oxford Univ Press
24. Liew, A.W.-C., Law, N.-F., Yan, H.: Missing value imputation for gene expression data: computational techniques to recover missing data from available information. Brief. Bioinform. **12**(5), 498–513 (2011). Oxford Univ Press
25. Myers, T.A.G.: Listwise deletion: presenting hot deck imputation as an easy and effective tool for handling missing data. Commun. Methods Measures **5**(4), 297–310 (2011). Taylor & Francis
26. Schafer, J.L., Graham, J.W.: Missing data: our view of the state of the art. Psychol. Methods **7**(2), 147 (2002). American Psychological Association
27. Van Buuren, S.: Flexible Imputation of Missing Data. CRC Press, Boca Raton (2012)
28. Hinton, G.E., Salakhutdinov, R.R.: Reducing the dimensionality of data with neural networks. Science **313**(5786), 504–507 (2006). American Association for the Advancement of Science

A Hybrid Search Optimization Technique Based on Evolutionary Learning in Plants

Deblina Bhattacharjee and Anand Paul$^{(\boxtimes)}$

Department of Computer Science and Engineering, Kyungpook National
University, 80 Daehak-ro, Bukgu, Daegu 702701, South Korea
{deblina0210,anand}@knu.ac.kr

Abstract. In this article, we have proposed a search optimization algorithm based on the natural intelligence of biological plants, which has been modelled using a three tier architecture comprising Plant Growth Simulation Algorithm (PGSA), Evolutionary Learning and Reinforcement Learning in each tier respectively. The method combines the heuristic based PGSA along with Evolutionary Learning with an underlying Reinforcement Learning technique where natural selection is used as a feedback. This enables us to achieve a highly optimized algorithm for search that simulates the evolutionary techniques in nature. The proposed method reduces the feasible sets of growth points in each iteration, thereby reducing the required run times of load flow, objective function evaluation, thus reaching the goal state in minimum time and within the desired constraints.

Keywords: Plant growth simulation algorithm · Evolutionary learning · Reinforcement learning · Plant intelligence · Search optimization

1 Introduction

In this paper, we have discussed the decision making process in plants guided by an underlying learning mechanism which is both evolutionary and adaptive to solve an important search optimization problem, unravelling a form of natural intelligence. Plants gather and continually update diverse information about their surroundings, combining this with internal information about their internal state and making decisions that reconcile their well-being with their environment. The life time goal of a plant is to maximize this fitness which it does by adapting to non-stationary environment by competing vigorously with the surrounding plants for resources, and as the individuals grow along with the competitors the resources get exhausted rapidly. Thus, a search for new resources must be immediately undertaken. This requires the plants to perceive an

This study was supported by the Brain Korea 21 Plus project (SW Human Resource Development Program for Supporting Smart Life) funded by Ministry of Education, School of Computer Science and Engineering, Kyungpook National University, Korea (21A20131600005) and the Institute for Information & communications Technology Promotion (IITP) grant funded by the Korea government (MSIP). [No. 10041145, Self-Organized Software platform (SoSp) for Welfare Devices].

© Springer International Publishing Switzerland 2016
Y. Tan et al. (Eds.): ICSI 2016, Part I, LNCS 9712, pp. 271–279, 2016.
DOI: 10.1007/978-3-319-41000-5_27

information spectrum in a continuous flux and to also maintain a network that can manipulate their own information flow. Different models depicting plant learning has been earlier displayed by [1, 3]. Interestingly, to model the plant learning and cognition many have used the neural network model of Hopfield [2] giving enough proof that plants too learn like animals although they lack a neural structure. This neural like mechanism is brought about by the complex cellular and biochemical reactions leading to wave like signal transductions [7]. In all the models mentioned so far, the plant behavior has been described as either evolutionary (concerned with the very gradual change of the behavior of the entire plant population over a very long time) or adaptively learnt (concerned with a behavioral change in an individual plant in a short span of time). However, the basis to model the type of learning in such systems should be both evolutionary and adaptive learning, as the decisions that are learnt via experiences in individual plants over a short span is passed on across generations to make plants adapt naturally to such rewarding decisions. These decisions are then constantly modified and evaluated based on fitness functions in unfavorable environments. In plants, learning is mainly used for morphological functions like growth and survival. Thus, growth should occur to give rise to an optimal configuration for effective resource use. The selection of these growth points is done via a search technique in plants. In Fig. 1a we have outlined the existing model of plant intelligence and the centre of the plants decision making, i.e. the adaptive representational framework. In our proposed model in Fig. 1b we have shown a three tier architecture that simulates the inner mechanism of this adaptive representational framework by forming a hybrid learning model, applied and solved in context of shoot growth (a search optimization problem seen in plants).

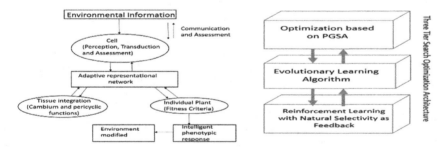

Fig. 1. a. An overview of the existing plant learning and decision making model. **b.** The proposed three tier architecture

PGSA has made huge progress of late to provide methods to find the global optimum of a function search space by overcoming the problem of falling into a local minimum [4, 10]. Initial works of Lindenmayer and Prusinkiewicz in the field of graphics, fractals and the generation of L-System Grammar [5], threw little light on the natural optimization models that simulate plant growth.e.g.: phototropism, gravitropism. Therefore, very little is known about the search optimization algorithms which simulate the natural selectivity based on evolution that is demonstrated by nature while giving globally optimal solutions, except for Genetic Algorithms. Genetic Algorithms have been long attached to evolutionary techniques due to its inherent genetic parameters which form an integral part in evolution. Although, Genetic Algorithms can be applied to unstructured and discrete optimization problems, it searches from a population of points where the objective function coefficient, mutation rate and crossover rate need to be mentioned [10]. Thus, there have been many refined hybrid versions of Genetic Algorithm till date to simulate the evolutionary learning mechanism in nature, in order to modify the weights associated in any learning network. PGSA on the other hand doesn't need any external parameters. Moreover, the objective function and constraint condition is treated separately [4, 10]. It has a search mechanism with ideal direction and randomness balancing properties which are determined by the morphactin concentration in plants and hence, it finds global optimal solution quickly. However, PGSA leads to a large number of growth points and its efficiency needs to be enhanced. To solve this problem, this article proposes a refined PGSA with Evolutionary Learning technique (where the weights are evolved using a hybrid version of Genetic Algorithm and a feedforward reinforcement learning method using natural selectivity as the reward), as a possible approach.

2   Methodology

In this section, we present a detailed overview of our proposed scheme. The proposed scheme as mentioned in Fig. 1(b) has a three tier architecture and is divided into three phases namely (1) PGSA (2) The Evolutionary Learning Algorithm (ELA) (3) The Reinforcement Learning Algorithm. The above Fig. 1a shows the overall learning and decision making model of the plant with an adaptive representational network where all the information processing is done [8]. This representational framework which is refined using our three tier architecture that comprises a hybrid learning network that we have used in context of shoot branching, is shown in detail in Fig. 2.

2.1   Tier 1 - Plant Growth Simulation Algorithm

Based on plant phototropism, the PGSA regards the feasible region of Integer programming as plant growth environment and evaluates the probability on different growth points according to the changes in the objective function [10]. It then grows towards the global optimal solution – light source. Biological experiments prove the following plant growth laws:

First, in the growth process of a plant, when it has more than one node starting at the root; the node with the higher morphactin concentration will have a higher growth probability to grow into a new branch. Second, the morphactin concentrations of the nodes in a plant are not predetermined or fixed but varies based on environmental information of the nodes that depend on their respective positions. While the new nodes appear, the morphactin concentrations of all plant nodes will be freshly allotted as per the new environment.

Mathematical Model for Plant Growth. According to [4, 9], in a plant system, if the length of a trunk is M and the number of growth points is I, then SM = (SM1, SM2,..., SMI), corresponding respectively to the morphactin concentration PM = (PM1,PM2, ...PMI). All the growth points which are inferior to the root, that is, those which have a poorer growth function than the initial feasible solution, if there are G growth points such that G lesser than I, these growth points and their corresponding morphological concentration is SM = (SM1,SM2,..SMG) and PM = (PM1,PM2,..PMG). If the length of the branches are m (m < M) and the available growth point numbers are g, these growth points and their concentrations were SM = (Sm1, Sm2,...,Smg) and Pm = (Pm1,Pm2,...,Pmg). We then calculate the morphactin concentration of every point using (1) and (2);

$$P_{Mi} = \frac{f(a) - f(S_{Mi})}{\sum_{i=1}^{G}(f(a) - f(S_{Mi})) + \sum_{i=1}^{g}(f(a) - f(S_{Mi}))} \tag{1}$$

$$P_{mk} = \frac{f(a) - f(M_{mk})}{\sum_{i=1}^{G}(f(a) - f(S_{Mi})) + \sum_{i=1}^{g}(f(a) - f(S_{Mi}))} \tag{2}$$

Where, a is the root (initial feasible solution), f (...) is the environmental information of a growth point (objective function).

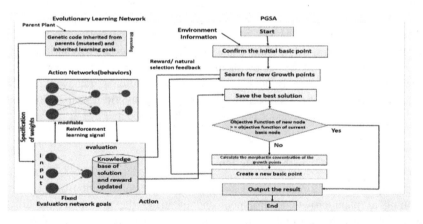

Fig. 2. The detailed model of the proposed three tier architecture based on PGSA and Evolutionary Learning to solve the search optimization problem of optimal shoot growth.

Add all the morphactin concentrations of growing points, we get the result as one. Now, a random number in state space [0, 1] is randomly located which grows out a new generation branch as shown in Fig. 3. This process is repeated until no new branches are generated in the state space. Our approach is to use the learning algorithm with natural selectivity as a feedback to select this random point in the state space thus finding the most optimum growth point in each iteration.

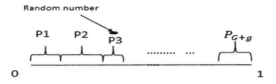

Fig. 3. Morphactin Concentration Space

2.2   Tier 2 - Evolutionary Learning Algorithm (ELA)

The PGSA takes in the sensor inputs and confirms the initial basic point (root). Thereafter, it iterates to search for new growth points as per the mathematical model discussed above. The control then goes to the learning network which takes the state from the recent environment as the input and instructs the PGSA to perform the new branch growth action from the naturally selected growth point based on the reward from the current environment. The problem of training the weights are solved by specifying not only the inherited behaviors but also the inherited goals that are used to facilitate learning. This is done by constructing a genetic code specifying two major components (1) a set of initial values for the weights of the action network that maps from sensory input to what action needs to be taken and (2) the evaluation network that maps from the inputs to a fitness value of the current decision. The weights in the learning network are trained using a modified genetic algorithm model with reinforcement. The initial weights of the action network are specified genetically. However, they are adjusted over time by a reinforcement learning mechanism that uses natural selectivity as its reward.

Input: Plant p , Seed s, EnvironmentalVariables I_t, GeneticCode gco
Output: Evaluation E_t, ReinforcementSignal R_t , NewAction X_t
   if $age_s = 0$ **then**
     $gco_s = gco_p$
     $cgco_s = Crossover(gco_s , 1)$
     $mgco_s = Mutate(cgco_s, mutation_{probabilty})$
     $[W]_{mgco} = Translate(mgco, [E], [A])$
   end if
   if $age_s \neq 0$ **then**
     $E_t = Evaluate(I_t)$
     if $age_p = 0$ **then**
       $X_t = Learn\ (Reinforce(I_t))$
       Output X_t
     else
       $R_t = E_t - E_{t-1}$
       $Reinforce(Update([A], X_{t-1}, I_{t-1})$
     end if
   end if
  end

In the above ELA, a parent plant P's genetic code is copied to the genetic code of seed S at birth (t = 0), followed by crossover and mutation of the bit encoded genetic bits. The genetic code is now translated into weights for S's evaluation network and the initial weights for S's action network. At time t, when there is a living plant P and a new current input vector from the environment I_t, the plant starts at the evaluation network, propagating I_t through the evaluation network to produce a scalar evaluation E_t. If this is P's day of birth, the algorithm invokes the reinforcement algorithm to generate a new action (branch) based on I_t otherwise, a reinforcement signal R_t is produced by comparison with the previous evaluation. The reinforcement algorithm is invoked to update the action network with respect to the previous action and input.

Here a specific reinforcement function suggested by Sutton has been used for learning to proceed [6]. However, the algorithm has been modified as described in Sect. 2.3. Learning is mainly brought about by the inherited evaluation function which converts long time scale feedback (natural selection over lifetimes) into short time scale feedback (reinforcement signals over moments). Thus we see that randomization is regressed here.

2.3   Tier 3 - Reinforcement Learning Algorithm

In this Reinforcement algorithm, the supervised training like back propagation network is refined which provides a memory model for the plant. This makes the plant rehearse the same positively rewarding decision into its knowledge base as shown in Fig. 2.

Input: ReinforcementNetwork $[R]_n$ with input dimensionality n and output
dimensionality m, and a ReinforcementFunction $f(R^n, R^n) \rightarrow r$.
Output: BinaryOutputVector o,   ReinforcementNetwork $[R]_m$
Initialize $t = 0$, Receive vector $i_t \in R^n$.
repeat
    $r = ComputeReinforcement f(i_t, i_{t-1})$.
    GenerateOutputErrors( )
        begin
            if $r > 0$ **then**
                $e_j = (o_j - s_j) \, s_j \, (1 - s_j)$
            else
                $e_j = (1 - o_j - s_j) \, s_j \, (1 - s_j)$
            end if
        end *Procedure{GenerateOutputErrors( )}*
        BackpropogateErrors(e_j)
    UpdateWeights( )
        begin
            if $r \geq 0$ **then**
                $\eta = \eta_+$
            else
                $\eta = \eta_-$
            end if
            $\Delta w_{jk} = \eta e_k s_j$,
        end *Procedure{UpdateWeights ( )}*
        $s_j = ForwardPropogate([R]_n)$
        GenerateTemporaryOutput(o^o)
        Perform action associated with o.
            t=t+1
end

---

3 Implementation and Analysis

For the artificial simulation of this learnt behavior of plants in context of search optimization, we considered a heterogeneously unfavorable environment, i.e. almost all input parameters are non-stationary and unknown to the plant. The simulations were performed in a rectangular cell of length $l = d \, x \, n$, n is the number of seeds that are subjected to branch out into shoots growing into plants according to the varying environmental cues and d is the separation between two adjacent seeds. Through the simulations, we calculated the shoot growth rate, the delay in finding the global solution, the direction of root growth for a resource optimal configuration and the energy cost or resource use that the plants foraged during the growth simulation. We have analyzed the results across a sample of 10 generations. Initially, the axiom growth point searches are fed to PGSA alone and then to our proposed scheme. In the Evolutionary Learning Algorithm (ELA), the population of growth points show regressive behavior for the initial few iterations. Thereafter, the algorithm shows a 3 point convergence of the population space, showing the fastest approach. For just the PGSA mechanism, the population seemed to behave in a much similar way as ELA except it was comparatively slower and resource costly. The comparison of the two models has been shown in Fig. 4(a) and (b). While Fig. 4(a), shows the net energy cost or total number of growth points in unfavorable environments for both the models, Fig. 4(b) shows the rate of shoot growth across 10 generations as studied. Clearly, the number of growth points or Energy Cost of applying the proposed model by the plant turned out to be optimal. Thus, the major problem of too many growth points in PGSA was solved by the proposed scheme. The comparative successes in growth point search in both the models has been analyzed and seen to be almost similar as seen in Fig. 5(a). Therefore, while the two models are successful in locating the optimal morphactin concentration in the shoot and in finding the optimal growth point, yet the proposed model is clearly more cost efficient and fast as compared to the existing model (Fig. 5a and b). Thus, this leads to optimal solutions of search problems within desired constraints.

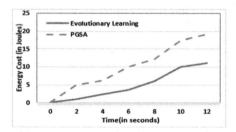 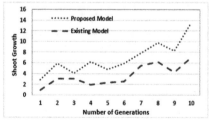

Fig. 4. a. Analysis of the 2 models 1: The proposed model labeled as Evolutionary Learning 2: Plant Growth Simulation Algorithm with respect to Energy Cost (No. of growth points) vs Time. **b.** Analysis of shoot growth over 10 generations using both the models. (Color figure online)

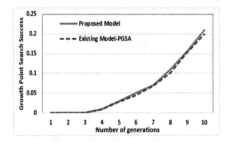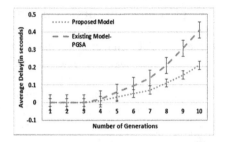

Fig. 5. a. The comparative analysis from the simulation of the two models with respect to Growth Point Search Success vs Number of generations. It shows that both the models of PGSA and Evolutionary Learning give optimal search solutions but PGSA is slow and energy costly. **b.** Analysis of Average Delay vs. Number of Generations in both the proposed and existing model. (Color figure online)

4 Conclusion

In this paper, a new bionic random search algorithm has been proposed that makes use of the objective function's value as an input to the learning model while simulating a plant's phototropism. The learner evaluates the various growth points and then directs the new branch to grow from the most optimal growth point, by giving a simulated natural selectivity feedback. Therefore, the proposed algorithm can exactly simulate the natural evolutionary process by combining it with learning algorithms to give fast globally optimal solutions of multimode functions. Therefore, the hybrid approach of learning (that evolves through generations) and PGSA, proves to be efficient and optimal to solve complex search optimization problems under desired constraints and time. Thus, this algorithm when applied to search optimization problems, can effectively reduce load factor, search space as well as help the system to evolve on its own via information processing, maintaining a memory base and ultimately learning through its experiences. The applications range from optimal network architecture, modelling of hydro-plants motivated from water absorption in trees to architectural designs of robust structures etc. The future scope of this study will incorporate the ELA scheme with respect to Baldwin Effect, Shielding Effect and Genetic Drift which were not covered in the scope of this paper.

References

1. Bose, I., Karmakar, R.: Simple models of plant learning and memory. Phys. Scr. **2003** (T106), 1–10 (2003)
2. Hopfield, J.J.: Neural Networks and Physical Systems with emergent collective computational abilities. Proc. Natl. Acad. Sci. U.S.A. **79**, 2554–2558 (1982)
3. Inoue, J., Chakrabarti, B.K.: Competition between ferro-retrieval and antiferro orders in Hopfield- like network model plant intelligence. Physica A **346**, 58–67 (2004)

4. Tong, L., Zhong-tuo, W.: Application of plant growth simulation algorithm on solving facility location problem. Syst. Eng. Theory Pract. **28**(12), 107–115 (2008)
5. Prusinkiewicz, P., Lindenmayer, A.: The Algorithmic Beauty of Plants. Springer, New York (1990)
6. Sutton, R.S.: Temporal credit assignment in reinforcement learning. Ph.D. Thesis, Department of Computer and Information Science, University of Massachusetts (1984)
7. Trewavas, A.J.: Calcium makes waves. Plant Physiol. **120**, 1–6 (1999)
8. Trewavas, A.J.: Green plants as intelligent organisms. Trends Plant Sci. **10**(9), 413–419 (2005)
9. Xu, L., Tao, M., Ming, H.: A hybrid Algorithm based on genetic algorithm and plant growth simulation. In: International Conference on Measurement, Information and Control (MIC), vol. 1, pp. 445–448 (2012)
10. Jing, Y., Wang Zong F.: A refined plant growth simulation algorithm for distribution network configuration. In: IEEE International Conference on Intelligent Computing and Intelligent Systems, vol. 1, pp. 357–361 (2009)

Development of Hybrid Memetic Algorithm and General Regression Neural Network for Generating Iterated Function System Fractals in Jewelry Design Applications

Somlak Wannarumon Kielarova[✉]

iD3 - Industrial Design, Decision and Development Research Unit,
Department of Industrial Engineering, Faculty of Engineering,
Naresuan University, Phitsanulok 65000, Thailand
somlakw@nu.ac.th

Abstract. This paper proposes a hybrid memetic algorithm and a general regression neural network for generating decorative elements. This tool is aimed to be used in jewelry design applications. Local search used is the greedy hill-climbing algorithm. Decorative elements are represented using iterated function systems (IFS) fractals. The aesthetic evaluation used in the design system is modeled using a general regression neural network with multiple perception feed-forward back propagation network to evaluate aesthetics of generated decorative elements. Although this paper demonstrates the application in jewelry design, the proposed algorithm is applicable to other product designs. The results of this study were compared to the results obtained with a genetic algorithm. This comparison implies that the proposed memetic algorithm can obtain better fitness and more variety, but requires larger amount of computational time than the genetic algorithm. The results prove that the proposed algorithm can be applied in design applications.

Keywords: Memetic algorithm · Decorative element · Jewelry design · Iterated function system · Fractal · General regression neural networks · Local search

1 Introduction

In conceptual design stage, designers commonly handle the activities such as generating and recording ideas, deciding to continue to generate more ideas, or exploring more possibilities of some existing favorite ideas [1]. Unfortunately, the available CAD packages are not suitable for use at this stage.

In jewelry design, decorative elements are used as a part of creating jewelry items. Jewelry designers, in conceptual design, usually cope with creating design elements, generating decorative elements, and spatially arranging them to form the design patterns.

Generative Design (GD) system is a computer-based tool, which has characteristics and capabilities to be used for supporting designers in design exploration and design

© Springer International Publishing Switzerland 2016
Y. Tan et al. (Eds.): ICSI 2016, Part I, LNCS 9712, pp. 280–289, 2016.
DOI: 10.1007/978-3-319-41000-5_28

generation. It can support designers in their attempts to explore a large design space, which provides a larger range of design possibilities than accessible by manual production [2]. Among generative techniques [3], genetic algorithm (GA) is a popular technique more suited to specific design purposes. It is based on modifications, combinations, and other operations. In GA, design exploration and selection occur at each cycle. Genetic operations are able to generate random designs; as a result, it is easier to model and modify. GD systems could be further studied from the following references [4–9].

Iterated function system (IFS) geometry has been used in various design applications. It is interesting to use IFS as decorative element representation, because of its variety, compactness, and productiveness.

This paper aims to develop a generative design system. It is extended from our previous research [6, 10] in which IFS fractals were used to represent art forms and the desirable IFS fractals were stored in a library. It was found that, the resulting generated individuals depend upon the stored IFS fractals, as a result, the generated results are similar from batch to batch, and do not improve. Therefore, in this paper, new IFS fractals work as individuals and are always randomly generated in the initialization stage with the proposed memetic algorithm (MA) to guarantee the similarity and compactness of the IFS fractals.

This paper also aims to improve the objective function of the previous generative design system by using the general regression neural network (GRNN) as a replacement of the nonlinear regression techniques used before. In this paper, IFS fractals are used as decorative elements in jewelry design without the consideration of gem setting.

2 Related Works

2.1 Memetic Algorithm

Memetic algorithm (MA) is an approach to evolutionary computation that attempts to imitate cultural evolution in order to solve optimization problems. It combines the concepts from local search techniques and population-based search methods. MA consists of two main parts [11]. One is an evolutionary structure and the other one is a set of local improvement methods, which work during the evolutionary cycle. In other words, MA is an extension of GA with the use of local search methods to reduce premature convergence [12]. MAs typically have small population sizes from 10 to 40 individuals, because local search is expensive [13]. Neri and Cotta [11] provide the basic MA as follows. MA starts from the Initialize procedure, which attempts to set up a good starting point by using superior solutions with the use of some mechanisms to supply good solutions in the initial population, or use of local search methods to improve random solutions. Next, the Cooperate and Improve procedures establish the core of MA. The Cooperate procedure consists of selection and recombination of the selected solutions with various operators. Next, the Improve procedure provides the local search techniques to solutions in the population. After that, the Compete procedure uses the previous population and the freshly generated population to reconstruct the current population. The two ways for the reconstruction are comma strategy and plus strategy. The last procedure is the Restart, which will be invoked when the population comes to a degenerate state, to generate new solutions to complete the population by using several methods such as the random-immigrant strategy, and strong mutation operator.

2.2   Iterated Function System

In this research, fractal geometry first coined by Mandelbrot [14] was used to create decorative elements used in jewelry design. Iterated Function Systems (IFS) introduced by Barnsley [15] is one of the interesting methods for creating fractals. An IFS fractal is defined by a finite number of affine transformations. The affine transformations used in IFS are translation, rotation, scaling, and shearing. Such transformations have significant geometric and algebraic properties including image compression with a compact set of numbers.

IFS of affine transformations in two-dimensional space $\mathbb{R}^2$ can be represented as $\{\mathbb{R}^2; w_1, w_2, \ldots w_N\}$, where w_i are affine transformations. IFS of affine maps can be written down as:

$$w_i(x, y) = w_i \begin{bmatrix} x \\ y \end{bmatrix} = \begin{bmatrix} a_i & b_i \\ c_i & d_i \end{bmatrix} \begin{bmatrix} x \\ y \end{bmatrix} + \begin{bmatrix} e_i \\ f_i \end{bmatrix} \tag{1}$$

With probabilities p_i where $\{p_i : i = 1, 2, \ldots, N\}$ associate with w_i for $i = 1, 2, 3, \ldots, N$ that obey

$$p_1 + p_2 + p_3 + \ldots + p_n = 1 \tag{2}$$

and $p_i > 0$ for $i = 1, 2, 3, \ldots, N$ Find p_i by

$$p_i \approx \frac{|\det A_i|}{\sum_{i=1}^{N} |A_i|} = \frac{|a_i d_i - b_i c_i|}{\sum_{i=1}^{N} |a_i d_i - b_i c_i|} \tag{3}$$

Random Iteration Algorithm (RIA) is used for rendering the images of IFS attractors, and in this paper it is used for mapping IFS genotype to fractal phenotype: Let $\{X; w_1, w_2, \ldots w_N\}$ be an IFS, where probability $p_i > 0$ assigned to w_i for $i = 1, 2, \ldots, N$, and $\sum_{i=1}^{n} p_i = 1$.

2.3   General Regression Neural Network

General Regression Neural Network (GRNN) is a type of Artificial Neural Network (ANN) invented by D. Specht in 1991 [16]. It is a probabilistic neural network, which uses data training as a back propagation (BP) neural network. In case data from observations or experiments are not sufficient for back propagation neural networks, using probabilistic neural networks can be useful, because they need small amount of training data for convergence of function [17]. As a result, GRNN is useful for forecasting and comparing systems' efficiencies. GRNN has regression pattern as Probability Density Function; pdf, which is defined from observational or experimental data by using nonparametric estimator without hypothesis testing. Multilayered perception (MLP) [18] is the most popular ANN model adopted in various industries. Several works [19–21] used BP for supervised learning methods to train the MLP model. The number of hidden neurons is recommended in [21, 22], and is calculated by Eq. (4) as follows:

$$H = \frac{2}{3}(I + O) \tag{4}$$

where H is number of hidden neurons, I is number of input nodes, and O is number of output nodes used in the MLP model.

The MLP algorithm for GRNN will be further used as fitness function or aesthetic evaluator in the proposed MA.

3 Development of Generative Design System Using Memetic Algorithm and General Regression Neural Network

3.1 The Generative Design System: System Architecture

The proposed generative design system is developed based on MA, while the fitness function is developed using GRNN. The proposed system is improved from the previous work [6, 10], which developed GA as a core of the system and the fitness function was derived by using nonlinear regression. In this paper, the core of the system was improved by developing the MA to generate decorative elements, which are represented using IFS Fractals. GRNN is used to search for the regression equation to be used as the fitness function in the proposed MA.

The proposed MA and GRNN were developed in MATLAB and linked to Rhinoceros 3D modeler to demonstrate the designed jewelry items from the resulting generated 3D decorative elements. The system architecture is shown in Fig. 1.

The proposed MA works as a shape optimization tool and a shape generator, whereas the fitness function evaluates the aesthetics of the generated shapes. The resulting generated shapes are rendered in 3D using Rhinoceros-Grasshopper platform to display the results to users. Users then make decisions on whether they are satisfied with the results or prefer to continue the process.

3.2 Decorative Element Representation

In this evolutionary design system, the compact sets of numbers (IFS codes) are encoded in one-dimensional chromosomes to represent artistic forms as shown in Fig. 2. During the evolutionary process only 2D point clouds are generated to reduce computational time, the 3D model of the point cloud will be generated later when the process reaches to the final stage.

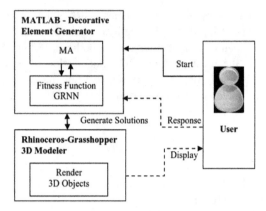

Fig. 1. System architecture of the proposed generative design system.

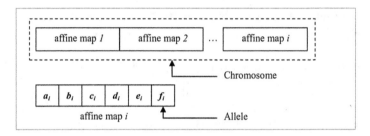

Fig. 2. IFS genotype.

The RIA explained in Sect. 2.2 is used to map a genotype to its phenotype as shown in Fig. 3. The phenotype in this case is the IFS fractal, which is rendered with the selection probability of each affine map by Eq. 3.

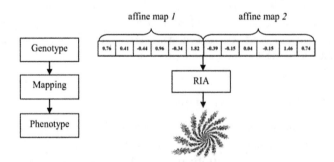

Fig. 3. Mapping IFS genotype to fractal phenotype.

3.3 The Proposed Memetic Algorithm

The proposed MA was developed to improve local optimization in generating IFS fractals. The proposed MA is shown in Fig. 4. The MA process begins with the creation of an initial population set with the population size $Popsize = 10$. This is followed with the execution of the greedy hill-climbing algorithm [23] to supply good individuals for the population (Pop). Next, the parent selection is based on their fitness. The fitter candidates are selected and randomly paired for recombination using the modified arithmetic crossover [6], and then the multi-Gaussian mutation [6] is applied to introduce new candidates into the population configuration. These operators are applied to generate the new population ($newPop$). Next, the greedy hill-climbing algorithm is executed for local search of solutions in the new population to improve the population ($newPop$). The current population is then reconstructed using $Pop \cup newPop$ with the common strategy of the population size of $newPop = 40$, where $|newPop| > Popsize$, in order to set the selective pressure on the evolutionary process. The occurrence of premature convergence is judged from whether the population is at a degenerate stage, by measuring of information diversity using Shannon's entropy approach [24, 25]. If the population is in a degenerate state, the restart procedure is invoked and implemented with a heavy mutation operator to drive the population away from its current position. Next, the members of the population would be evaluated for their fitness by the fitness function, further explained in the next section. The process is run iteratively until one of the termination criteria is met.

The termination criteria used in this work include the predefined maximum number of iterations = 20; a maximum number of iterations without improvement of the fitness; the process performs a predefined number of population restarts; the maximum fitness of 1 is reached; and the user terminates the process.

3.4 Fitness Function: Multilayered Perception Algorithm for General Regression Neural Network

This paper has studied ANN and GRNN to improve aesthetic evaluation module developed in previous works [6, 10]. Aesthetic evaluation of jewelry art forms developed by Wannarumon et al. [6] and Wannarumon [10] uses eight aesthetic variables: compactness; connectivity; Lyapunov exponent; complexity; mirror symmetry; rotational symmetry; golden ratio; and logarithmic spiral symmetry. Statistical analysis techniques such as factor analysis and multiple regressions [26] were mainly used to analyze the relationships among these variables.

This paper proposes a new approach to develop the aesthetic evaluation module based on ANN and GRNN for integrating it into the proposed MA. Eight aesthetic variables mentioned above [6, 10] are used as inputs and aesthetic scores are considered as outputs (Y) for training the GRNN. The GRNN then calculates weights in a hidden layer. The MLP with three-layer BP network used in this study consists of input layer, hidden layer, and output layer. Two hundred data sets of eight aesthetic variables and the related survey aesthetic scores were used to train the MLP-GRNN model.

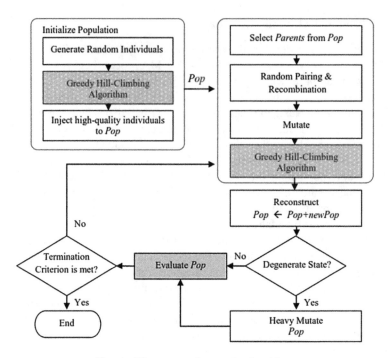

Fig. 4. The proposed memetic algorithm.

The proposed GRNN has the number of input nodes equal to the number of aesthetic variables. A single output node represents aesthetic value of the corresponding input parameters.

The proposed GRNN model is a feed-forward back propagation network, which consists of three layers using the Dot product weight function. The transfer function of inputs is log-sigmoid, while linear function is used for transfer function of outputs. The ANN model was developed with 8 input nodes, 1 output node, and 6 hidden neurons. The number of hidden neurons is calculated by Eq. (4). Transfer function of inputs used is Log-sigmoid. Transfer function of outputs used is Pure linear, while training function used is Levenberg-Marquardt backpropagation.

4   Results and Discussions

The prototype system was developed using MATLAB and linked with Rhinoceros 5.0 platform for rendering jewelry items on a computer workstation with Intel Xeon CPU Processor 1.8 GHz Dual and 4.0 GB of RAM, working on 64 bit.

From the comparison experiments between the previous GA and the proposed MA with the same $pop\_size = 10$, the crossover probability $p_c = 0.75$, and the mutation rate $p_m = 0.25$ can be seen. The average computational time of the proposed MA and the GA is 5,306 s and 1,935 s, respectively. The MA used five times more computational time than the GA, because the MA performed two sets of local searches during the evolutionary process and maximum number of restarts used is 5. In each generation, MA uses more time than GA. The greedy hill-climbing local search can help to improve the individuals but it is time-consuming. Nevertheless, the MA is able to obtain more variety and higher fitness than the GA. In the experiments, the maximum values of fitness obtained by the MA are higher than those obtained by the GA, because the GA is trapped in a local maximum. The results of some experiments are shown in Fig. 5.

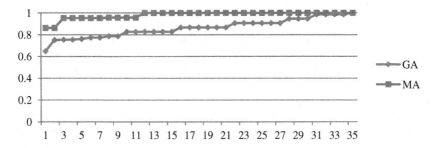

Fig. 5. Comparison of the previous GA and the proposed MA

There are some reasons for keeping a small population size. In generative design system, user needs to interact with the system therefore short response time is required. The individuals can improve their fitness faster with the smaller population size. The evolutionary art and design system typically works with encoding and decoding with computer graphics and 3d modeling, which are time-consuming. With the small population size the system is able to run from one generation to the next one to improve the fitness in shorter time.

The developed GRNN model is used as fitness function in the MA to evaluate aesthetic values of decorative elements or fractals. Comparing to the non linear regression model developed in [6] with $R^2 = 0.655$, the GRNN model obtains $R^2 = 0.957$, which improves the evaluation algorithm by about 46 %.

Some examples of the generated decorative elements for jewel items are shown in Fig. 6.

Fig. 6. Examples of decorative elements generated by the system and used in jewelry designs.

5 Conclusions and Future Works

An MA for generating decorative elements to be used in jewelry design applications was developed and integrated with GRNN aesthetic evaluation. Greedy hill-climbing algorithm was used for local search at the initialization and at the improvement steps. This can result in a longer computational time of the MA. Two genetic operators used are modified arithmetic crossover and multi-Gaussian mutation to generate new populations. Shannon entropy was used as a measure of information diversity, and to determine whether the population is at a degenerate stage. The heavy mutation operator was used to drive the population away from the convergence location.

The aesthetic evaluation is modeled using GRNN with MLP feed-forward back propagation network. ANN is used for regression issues. The results support the integration of MA and GRNN into a generative design system for generating decorative elements. Furthermore, the results of the proposed algorithm support the MA as a tool in product design applications.

In future works, it is our aim to reduce the computational time of the MA as in generative design systems, a short response time between the user and the system is necessary.

Acknowledgments. The research has been carried out as part of the research projects funded by National Research Council of Thailand and Naresaun University with Contract No. R2559B094. The author would like to gratefully thank all participants for their collaboration in this research. Finally, the author would like to thank Dr. Filip Kielar for correcting the manuscript.

References

1. Kolli, R., Pasman, G.J., Hennessey, J.M.: Some considerations for designing a user environment for creative ideation. In: Interface 1993, pp. 72–77. Human Factors and Ergonomics Society, Santa Monica (1993)
2. Krish, S.: A practical generative design method. Comput. Aided Des. **43**, 88–100 (2011)
3. Singh, V., Gu, N.: Towards an integrated generative design framework. Des. Stud. **33**, 185–207 (2012)
4. Brintrup, A.M., Ramsden, J., Takagi, H., Tiwari, A.: Ergonomic chair design by fusing qualitative and quantitative criteria using interactive genetic algorithms. IEEE Trans. Evol. Comput. **12**, 343–354 (2008)
5. Hu, Z.-H., Ding, Y.-S., Zhang, W.-B., Yan, Q.: An interactive co-evolutionary CAD system for garment pattern design. Comput. Aided Des. **40**, 1094–1104 (2008)
6. Wannarumon, S., Bohez, E.L.J., Annanon, K.: Aesthetic evolutionary algorithm for fractal-based user-centered jewelry design. Artif. Intell. Eng. Des. Anal. Manuf. **22**, 19–39 (2008)
7. Gong, D.-W., Yuan, J., Sun, X.-Y.: Interactive genetic algorithms with individual's fuzzy fitness. Comput. Hum. Behav. **27**, 1482–1492 (2011)
8. Sun, X., Gong, D., Zhang, W.: Interactive genetic algorithms with large population and semi-supervised learning. Appl. Soft Comput. **12**, 3004–3013 (2012)

9. Kielarova, S.W., Pradujphongphet, P., Bohez, E.L.J.: New interactive-generative design system: hybrid of shape grammar and evolutionary design - an application of jewelry design. In: Tan, Y., Shi, Y., Buarque, F., Gelbukh, A., Das, S., Engelbrecht, A. (eds.) ICSI-CCI 2015. LNCS, vol. 9140, pp. 302–313. Springer, Heidelberg (2015)

10. Wannarumon, S.: An aesthetic approach to jewelry design. Comput. Aided Des. Appl., Special Issue: CAD in the Arts **7**, 489–503 (2010)

11. Neri, F., Cotta, C.: A primer on memetic algorithm. In: Neri, F., Cotta, C., Moscato, P. (eds.) Handbook of Memetic Algorithms. Studies in Computational Intelligence, vol. 379, pp. 43–52. Springer-Verlag, Berlin Heidelberg (2012)

12. Garg, P.: A comparison between memetic algorithm and genetic algorithm for the cryptanalysis of simplified data encryption standard algorithm. Int. J. Netw. Secur. Appl. **1**, 34–42 (2009)

13. Merz, P.: Memetic algorithms and fitness landscapes in combinatorial optimization. In: Neri, F. (ed.) Handbook of Memetic Algorithms. SCI, vol. 379, pp. 111–142. Springer, Heidelberg (2011)

14. Mandelbrot, B.: The Fractal Geometry of Nature. W.H. Freeman, New York (1982)

15. Barnsley, M.: Fractal Everywhere. Academic Press, San Francisco (1993)

16. Specht, D.F.: A general regression neural network. IEEE Trans. Neural Netw. **2**, 568–576 (1991)

17. Fausett, L.: Fundamentals of Neural Networks Architectures, Algorithms and Applications. Prentice Hall, New Jersey (1994)

18. Rosenblatt, F.: Principles of Neurodynamics. Spartan Books, New York (1962)

19. Lai, H.-H., Lin, Y.-C., Yeh, C.-H.: Form design of product image using grey relational analysis and neural network models. Comput. Oper. Res. **32**, 2689–2711 (2005)

20. Lai, H.-H., Lin, Y.-C., Yeh, C.-H., Wei, C.-H.: User-oriented design for the optimal combination on product design. Int. J. Prod. Econ. **100**, 253–267 (2006)

21. Tang, C.Y., Fung, K.Y., Lee, E.W.M., Ho, G.T.S., Siu, K.W.M., Mou, W.L.: Product form design using customer perception evaluation by a combined superellipse fitting and ANN approach. Adv. Eng. Inform. **27**, 386–394 (2013)

22. Heaton, J.: Introduction to Neural Networks with Java. Heaton Research, Inc. (2005)

23. Montes de Oca, M.A., Cotta, C., Neri, F.: Local search. In: Neri, F. (ed.) Handbook of Memetic Algorithms. SCI, vol. 379, pp. 31–49. Springer, Heidelberg (2011)

24. Shannon, C.E., Weaver, W.: The mathematical theory of communication. Bell Syst. Tech. J. **27**, 379–423 (1948)

25. Stone, J.V.: Information Theory: A Tutorial Introduction. Sebtel Press, Sheffield (2015)

26. Hill, T., Lewicki, P.: Statistics: Methods and Applications. StatSoft, Tulsa (2006)

Particle Swarm Optimization

Heterogeneous Vector-Evaluated Particle Swarm Optimisation in Static Environments

Dieter Doman, Mardé Helbig$^{(\boxtimes)}$, and Andries Engelbrecht

Department of Computer Science, University of Pretoria, Hatfield 0128, South Africa
dieter@labs.epiuse.com, {mhelbig,engel}@cs.up.ac.za

Abstract. Particle swarm optimisation (PSO) is a population-based stochastic swarm intelligence (SI) optimization algorithm that converges very fast and thus lacks diversity. Heterogeneous vector evaluated particle swarm optimisation (HVEPSO) tries to introduce the ability to balance exploration and exploitation by increasing diversity of the particles' behaviour. This study evaluates the performance of different HVEPSO configurations in static multi-objective environments. The particles of each sub-swarm of HVEPSO use different position and velocity update approaches selected from a behaviour pool. Strategies to determine when to change the particles' behaviour are investigated for various knowledge transfer strategies (KTSs). Results indicate that the parent-centric crossover (PCX) KTS using the dynamic heterogeneous PSO (dHPSO) behaviour selection strategy with periodic window management performed the best. However, HVEPSO experienced problems converging to the optimal solutions and finding a diverse set of solutions for certain benchmarks, such as WFG1, which is a separable unimodal function with a convex Pareto-optimal front.

Keywords: Multi-objective optimisation · Heterogeneous VEPSO

1 Introduction

Optimisation is the process of finding the best solution of a certain function, subject to given constraints [1]. Swarm intelligence (SI) algorithms are optimisation algorithms based on the study of individual behaviour in swarms or colonies. Particle swarm optimisation (PSO) is a population-based stochastic SI algorithm that is modelled on the social behaviour of bird flocks [4]. PSO was first introduced by Eberhart and Kennedy [3], where individuals are referred to as particles and the population is referred to as a swarm. Each particle's behaviour is influenced by its position and velocity. When the particle's behaviour (position and velocity) is updated, it causes the particle to move through the search space. Knowledge is kept on where the best position for the neighbourhood was (referred to as the neighbourhood best or nbest) and each particle's position that leads to its best solution (referred to as the personal best or pbest) [4]. When all particles in a swarm use the same position and velocity update equations, the

© Springer International Publishing Switzerland 2016
Y. Tan et al. (Eds.): ICSI 2016, Part I, LNCS 9712, pp. 293–304, 2016.
DOI: 10.1007/978-3-319-41000-5_29

PSO is referred to as a homogeneous PSO. However, if particles use different approaches to update their position and velocity, the PSO is referred to as a heterogeneous PSO (HPSO).

Most real world problems are defined by multiple conflicting objectives and therefore require the simultaneous optimisation of a number of objectives. Parsopoulos and Vrahatis [18] developed the vector-evaluated particle swarm optimisation (VEPSO) algorithm based on the vector-evaluated genetic algorithm (VEGA) [19]. VEPSO uses sub-swarms, where each sub-swarm only solves one objective. Knowledge is then shared amongst the sub-swarms by using this knowledge in the velocity update of the particles. The approach used to share knowledge between the sub-swarms are referred to as the knowledge transfer strategy (KTS).

According to Engelbrecht [4], the basic PSO algorithm converges very fast and thus lacks diversity maintenance. Heterogeneous VEPSO (HVEPSO) tries to introduce the ability of balancing exploration (search over the whole search space) and exploitation (search more locally where the best solution might be) [5] by increasing diversity of the particles behaviour when solving multi-objective optimisation problems (MOOPs). Each sub-swarm of HVEPSO is a HPSO. Each particle of the HPSO selects a behaviour (approach to update its position and velocity) from a behaviour pool. Different strategies exist to determine when a particle's behaviour should change, i.e. when a new behaviour should be selected from the behaviour pool.

A dynamic version of HVEPSO has been evaluated on dynamic environments [14] and the results indicate that HVEPSO can successfully solve both dynamic single-objective optimisation problems (DSOOPs) and dynamic MOOPs (DMOOPs). However, no comprehensive study has been conducted to evaluate HVEPSO on static MOOPs. Therefore, this paper focuses on static multi-objective optimisation (MOO) environments.

The goal of this paper is to investigate the effect of the various behaviour selection mechanisms on the performance of heterogeneous vector evaluated particle swarm optimisation (HVEPSO) when using a specific KTS. Therefore, for each KTS, the study determines which selection mechanism is the best to use with the specific KTS.

The rest of the paper's layout is as follows: Sect. 2 provides background information on HPSO and MOOPs. The experiments conducted for this study are discussed in Sect. 3. Section 4 discusses the results optained from the experiments. Conclusions are discussed in Sect. 5.

2   Background

This section provides definitions with regards to MOO and background information about HPSOs and the KTSs implemented for this study.

2.1   Multi-objective Optimisation

The conflicting objectives of a MOOP results in no single optimum existing for the problem. However, an optimal set of trade-off solutions exists. Therefore, the goal of a multi-objective algorithm (MOA) is to find the optimal set of trade-off solutions. In the objective space, these solutions are referred to as the Pareto-optimal front (POF) and in the decision variable space they are referred to as the Pareto-optimal set (POS). To compare two solutions the principle of Pareto-dominance is used, i.e. a decision vector x_1 dominates another decision vector x_2 if x_1 is not worse than x_2 in all objectives and x_1 is strictly better than x_2 in at least one objective. If none of the decision vectors dominate the other, then both are non-dominated with regards to each other.

An archive contains the set of non-dominated solutions found by a MOA throughout the run. If a new solution dominates the current solutions in the archive, then the new solution is added to the archive and all solutions dominated by the new solution are removed from the archive. However, a new solution that is dominated by any solution in the archive is not added to the archive. If a new solution is non-dominated with regards to all the solutions in the archive, the new solution is added to the archive if there is space in the archive [4]. Due to the computational cost of an unrestricted archive, the size of the archive is normally restricted [4]. When the archive reaches its capacity, the new solution replaces a solution in the archive that is situated in the highest populated grid cell, i.e. a highly populated area.

2.2   Heterogeneous Particle Swarm Optimisation

Engelbrecht [9] proposed a HPSO, where each particle has a different behaviour (approach used to update the particle's position and velocity) that is selected from a behaviour pool. This section discusses the following three elements of HPSO: which behaviour to select for the behaviour pool, when to change the behaviour of a particle, and how to select the next behaviour.

Selecting Behaviours for the Behaviour Pool: Behaviours are selected for the behaviour pool in such a way that they differ in their exploration and exploitation capabilities in order to balance the exploration and exploitation capabilities of the MOA. The following behaviours are included in the behaviour pool:

- **Traditional gbest PSO** [3]
- **The cognitive-only PSO** [16]: The social component of the traditional PSO is removed and the same position update is used as in traditional PSO. In the cognitive-only model each particle becomes a hill-climber, resulting in more exploration [9].
- **The social-only PSO** [16]: The cognitive component of the traditional PSO is removed and the same position update is used as in traditional PSO. In the social-only model the entire swarm becomes a stochastic hill-climber, resulting in faster exploitation [9].

- **Barebones PSO** [17]: Formal proofs show that each particle converges to a point that is the weighted average between the pbest and gbest positions [11]. This convergence supports Kennedy's proposal to replace the entire velocity by Gaussian random numbers [17]. The position update follows that of the standard PSO. Due to large deviations at the start of the search the barebones PSO explores more at the beginning of the search and as the number of iterations increases, the deviation becomes very small and approaches zero. The barebones PSO then exploits more around the average of the pbest and gbest positions [9].
- **Modified Barebones** [17]: Kennedy improved the barebones velocity equation's exploration by modifying the velocity update such that 50 % of the time the search focusses on pbest positions. The pbest positions initially differ significantly due to uniform random initialization of particles, resulting in more exploration. As the number of iterations increases, the focus moves to exploitation, since all the pbest positions converge around the gbest position [9].

Behaviour Changing Schedules: Behaviour changing schedules determine when a particle should change its behaviour. In this study the following schedules are used [7]:

- **Personal Best (pbest) Stagnation:** The next behaviour is selected when the fitness of a particle's pbest does not change for a specified number of iterations.
- **Periodic:** A new behaviour is selected every m iterations.
- **Random:** The next behaviour is selected according to a certain probability. This approach is similar to periodic, but change orrurs at an irregular interval.

Behaviour Selection Strategies: The following strategies are investigated to select a new behaviour from the pool:

- **Dynamic HPSO (dHPSO)** [9]: A behaviour is randomly selected from the behaviour pool.
- **The frequency-based HPSO (f_k-HPSO)** [8]: f_k-HPSO is a self-adaptive HPSO, which gives each behaviour a success counter. The success counter considers only the previous k iterations to determine which current behaviour has brought improvement in the particle's fitness. Behaviours are selected using tournament selection based on the success counter.
- **Pheromone-based HPSO constant strategy (pHPSO-const)** [6]: PHPSO-const was inspired by the foraging behaviour of ants. Each behaviour is initialised with a pheromone concentration and at each iteration the pheromone is updated (incremented if the particle's new position improves the objective function value using the behaviour and decremented if the particle's new position weakens the objective function value using the behaviour). Evaporation is used to maintain behavioural diversity by subtracting a constant value from the behaviour's pheromone value. A particle selects a behaviour using roulette wheel selection.

2.3   Knowledge Transfer Strategies Pool

VEPSO uses a PSO sub-swarm for each objective function and each sub-swarm optimises only one objective function. The sub-swarms transfer knowledge they have gained to one another through the global guide(s) in the velocity update. The same KTSs that have been developed for dynamic environments was applied in this study [14], namely:

- **Random KTS** [12]: A sub-swarm is selected randomly (different sub-swarm or the same sub-swarm) and a global guide is chosen from the selected swarm using tournament selection for knowledge transfer.
- **Ring KTS** [18]: The sub-swarm whose gbest is used as the global guide is selected as the sub-swarm that is next in a ring topology structure. Therefore, the same sub-swarm's gbest is used for every iteration.
- **Parent-centric crossover (PCX) KTS** [13]: The global guide is computed as the offspring of three randomly selected solutions (from the sub-swarm) on which PCX is performed.

For each sub-swarm a HPSO was used as discussed in Sect. 2.2. The gbest and pbest is updated similar to those of the default PSO (using objective function evaluation value).

3   Experimental Setup

This section discusses the experimental setup used for this study.

3.1   Parameter Tuning

Parameters were tuned by factorial design (fix all other parameters, test different values of one, find the best value, and do it for all the parameters) over the whole WFG [15] and ZDT [22] benchmark suites and over all KTSs. This process was followed multiple times with different order of parameters to remove sequential dependencies. The parameter values that produced the best wins value (refer to Sect. 3.4) over all the benchmarks suites and knowledge transfer strategies were used in this study.

HPSO Parameters: For this study, 30 particles were used for each HPSO and a maximum of 1000 iterations were used for each run. HPSO has two types of parameters:

- **Behaviour parameters:** Behaviour parameters are used by the behaviours in the behaviour pool. These parameters are chosen according to those used in the work published on the respective algorithms, since the behaviours are chosen for specific search capabilities.

– **Algorithm parameters:** Algorithm parameters are used by the algorithm that controls the behaviour pool. These parameters need to be optimized. The algorithm parameters were tuned and it was found that the optimal values were very close to the values proposed by Engelbrecht and Nepomuceno [7] for HPSO in static single-objective environments. Table 1 summarises the algorithm parameters for dHPSO, f_k-HPSO and pHPSO-const. In Table 1 Random, Periodic and Pbest refers to the behaviour changing probability for the random schedule, the period for the periodic schedule and the pbest stagnation threshold, respectively. Table 2 summarises additional algorithm parameters for f_k-HPSO and pHPSO-const.

Table 1. Algorithm parameters

Algorithm	Random	Periodic	Pbest
dHPSO	0.44	29	19
f_k-HPSO	0.08	7	32
pHPSO-const	1.0	7	29

Table 2. Additional algorithm parameters

Algorithm	Parameter	Random	Periodic	PBest
f_k-HPSO	Tournament Size	2	2	2
f_k-HPSO	k iterations	29	19	248
pHPSO-const	Better Score	1.08	2.09	1.49
pHPSO-const	Same Score	−0.39	−0.36	−0.45
pHPSO-const	Worse Score	−0.08	−0.02	−0.17
pHPSO-const	Min pheromone	0.04	0.07	0.04

VEPSO Parameters: Only two parameters (σ_n and σ_e) for VEPSO with parent-centric crossover (PCX) KTS had to be optimized. Through factorial design it was determined that $\sigma_n = \sigma_e = 0.1$ should be used. In addition, an archive size of 200 was used for this study.

3.2 Benchmark Functions

Seven benchmark functions with specific properties where selected from the WFG [15] and ZDT [2] benchmark suites to evaluate HVEPSO on:

1. **WFG1:** A separable unimodal function that has a convex POF.
2. **WFG2:** A non-separable unimodal function where POF is convex and disconnected.

3. **WFG5:** A separable deceptive function with a concave POF.
4. **WFG7:** A separable unimodal function that is concave.
5. **ZDT1:** A convex Pareto-optimal front and the solutions are uniformly distributed in the search space.
6. **ZDT3:** A disconnected Pareto-optimal front that consists of several non-contiguous convex parts and the search space is unbiased.
7. **ZDT6:** Includes two difficulties caused by the non-uniformity of the search space: firstly, the Pareto-optimal solutions are non-uniformly distributed along the global Pareto front and secondly, the density of the solutions is the least near the Pareto-optimal front and highest away from the front.

The jMetal4.5 framework was used to implement the benchmark functions, available at http://jmetal.sourceforge.net/.

3.3 Performance Measures

The following three performance measures were used to compare the HVEPSO configurations' performance against one another:

1. **Inverse generational distance (IGD):** [20] This measure uses the true POF as a reference and then compares each of its elements with each of the elements in the front produced by a MOA. Lower values of IGD indicate better performance, since they show your solutions are close to the true POF, i.e. good convergence. True POF values are available for all benchmark problems at: http://jmetal.sourceforge.net/problems.html.
2. **Maximum spread (MS):** MS [21] quantifies the spread of the POF solutions in each objective. Higher values of MS means a higher spread of solutions, i.e. higher diversity.
3. **Hypervolume ratio (HVR):** Zitzler [21] proposed the hypervolume (HV) measure. The HV calculates the volume that is dominated by the POF found by a MOA. A higher value of HV indicates better performance. The code of Fonseca *et al.* [10] was used to calculate the hypervolume. The HVR calculates the HV of the found POF in relation to the HV of the true POF.

3.4 Statistical Analysis

For each benchmark function, environment and performance measure, the following process was followed for the statistical analysis of the results: The average performance measure value over 30 independent runs was calculated. A Kruskal-Wallis test was performed on these values obtained by the various HVEPSO configurations to determine whether there is a statistical significant difference in their performance. A pair-wise Mann-Whitney U test was performed if the Kruskal-Wallis test indicated that there was a statistical significant difference between the performance of any two of these configurations. If the Mann-Whitney U test indicated a statistical significant difference, the average performance measure value was used to award a win to the algorithm with the best average measure value. A confidence level of 95 % was used for all statistical tests.

4   Results

The procedure discussed in Sect. 3.4 was used for all the performance measures and the benchmarks discussed in Sects. 3.2 and 3.3. Results obtained for the various KTSs are discussed in Sects. 4.1, 4.2 and 4.3. Section 4.4 discusses the overall performance of the KTSs. Table 3 presents the various configurations that are referred to in the rest of the paper. In all tables BNR refers to the benchmark number (refer to Sect. 3.2). Due to space limitations, only performance measure values that indicated a statistical significant difference in performance of the HVEPSO configurations are included in this section.

Table 3. Configuration lookup table and overal wins per configuration

Configuration Number	Behavior strategy	Behavior shedule	Overall #Wins
cfg1	dHPSO	pbest	18
cfg2	dHPSO	Periodic	**22**
cfg3	dHPSO	Random	18
cfg4	f_k-HPSO	pbest	7
cfg5	f_k-HPSO	Periodic	12
cfg6	f_k-HPSO	Random	9
cfg7	pHPSO-const	pbest	21
cfg8	pHPSO-const	Periodic	15
cfg9	pHPSO-const	Random	16

4.1   Random Transfer Strategy

IGD: The Kruskal-Wallis test found that there was no statistical significant difference between the HVEPSO configurations' performance for IGD. All configurations performed well on the WFG benchmarks, and performed the worst on ZDT6 with a non-uniform search space.

MS: According to the Kruskal-Wallis test, there was no statistical significant difference between the performance of the HVEPSO configurations. All configurations performed well on WFG5 and WFG7, which are separable functions that have concave POFs. However, all configurations performed poorly on WFG1 (refer to Fig. 1(a)), a separable unimodal function with a convex POF.

HVR: The Kruskal-Wallis test indicated that there was a statistical significant difference in performance between the HVEPSO configurations for HVR. The mean HVR values and the wins and losses (refer to Sect. 3.4) obtained by the configurations for HVR are presented in Table 4. Configuration 1 (dHPSO & pbest) performed the best for HVR, with the highest number of wins. The random selection of behaviour used by this configuration has a high diversity, since all behaviours have equal probability to be selected.

Table 4. Mean HVR values and number of wins for Random KTS

BNR	1	2	3	4	5	6	7	#Wins
cfg1	0.00158	**0.2052**	0.66486	0.78311	1.13246	0.97071	0.27375	**13**
cfg2	0.00164	0.15538	0.75954	0.826	0.95917	0.71738	0.06776	3
cfg3	0.00094	0.15528	**0.7668**	0.78442	**1.58653**	0.91463	0.22958	6
cfg4	0.00196	0.16773	0.68036	0.86803	1.18212	0.74911	0.14635	2
cfg5	0.0017	0.13745	0.73032	**0.93939**	1.06011	**1.00113**	0.05662	4
cfg6	**0.002018**	0.16163	0.67804	0.89185	1.13847	0.60155	0.21316	2
cfg7	0.00158	0.14501	0.70884	0.85557	1.15314	0.84197	**0.38873**	10
cfg8	0.00133	0.10695	0.75917	0.77762	1.31747	0.6461	0.089006	2
cfg9	0.00173	0.10671	0.74029	0.76053	1.39006	0.67657	0.31814	6

4.2   Ring Transfer Strategy

IGD and MS: Results similar to those observed for the random KTS for IGD and MS were also observed for the ring KTS.

HVR: The Kruskal-Wallis test indicated that there was no statistical significant difference in performance between the HVEPSO configurations for HVR. All configurations performed well on ZDT1, which has a convex POF with solutions that are uniformly distributed in the search space. All configurations struggled with WFG1 (refer to Fig. 1(a)). The configurations' low HVR values indicate poor performance with regards to convergence and spread. These results differ from the results obtained for the random KTS.

4.3   PCX Transfer Strategy

IGD and MS: Table 5 presents the average IGD and MS values, as well as the number of wins obtained by the PCX KTS. The Kruskal-Wallis test indicated that there was a statistical significant difference between the configurations at each benchmark for both IGD and MS. Configuration 2 (dHPSO & Periodic) performed the best for IGD and MS, with the highest number of wins. The worst performing configuration for both IGD and MS was configuration 1 (dHPSO & pbest), obtaining no wins for either of the performance measures.

HVR: The Kruskal-Wallis test found that there was no statistical significant difference between the configurations' performance for HVR. All configurations performed well on WFG7 (refer to Fig. 1(b)), which is concave and a unimodal function. Similar to the ring KTS, all configurations performed the worst on WFG1 (refer to Fig. 1(a)).

Table 5. Mean IDG and MS values, and number of wins for PCX KTS

Measure	BNR	1	2	3	4	5	6	7	#Wins
IGD	cfg1	0.08449	0.10965	0.02167	0.011711	0.12084	0.12287	0.32	0
IGD	cfg2	0.08405	**0.08859**	**0.02158**	0.0117	0.1186	0.11408	0.32949	**7**
IGD	cfg3	0.08372	0.09037	0.02264	0.01167	0.11698	0.1148	0.31985	4
IGD	cfg4	0.08403	0.09443	0.02175	0.01129	0.118	**0.11315**	0.32174	1
IGD	cfg5	0.08338	0.10052	0.02278	**0.01107**	**0.11642**	0.11961	0.32028	1
IGD	cfg6	0.08306	0.09428	0.02167	0.01182	0.11878	0.11879	0.32032	1
IGD	cfg7	0.08341	0.091645	0.02304	0.01138	0.1167	0.11905	0.31975	4
IGD	cfg8	**0.08266**	0.09245	0.02286	0.01119	0.11863	0.11649	**0.31903**	3
IGD	cfg9	0.08321	0.09326	0.02232	0.01158	0.11756	0.11685	0.32263	3
MS	cfg1	0.09627	1.15066	2.2577	2.40933	1.56964	1.38516	0.40042	0
MS	cfg2	0.10149	**1.6595**	**2.32932**	2.50982	1.4543	**1.61986**	0.33087	**7**
MS	cfg3	0.11627	1.4788	2.06945	2.33593	1.39789	1.57052	0.5152	3
MS	cfg4	0.11072	1.38583	2.10053	2.40641	1.488992	1.46394	0.58658	1
MS	cfg5	0.12338	1.1592	2.17588	2.50952	1.61364	1.49078	0.54739	1
MS	cfg6	**0.13464**	1.4833	2.26166	2.35933	1.44948	1.58424	**0.62058**	4
MS	cfg7	0.12038	1.43789	2.08498	2.51935	1.5939	1.45246	0.57318	2
MS	cfg8	0.14924	1.50889	2.09975	**2.54498**	**1.65845**	1.48448	0.4552	4
MS	cfg9	0.13408	1.2392	2.19457	2.38694	1.58361	1.39775	0.52673	2

4.4 Overall

To get an overview of which configuration performed the best over all KTSs and over all performance measures, the number of wins were calculated over all benchmarks and measures (Table 3). When there was no statistical significant difference, the configuration with the best average performance measure value for the specific benchmark was awarded a win.

Configuration 2 (dHPSO & Periodic) performed the best overall (over all benchmarks, all KTSs and all performance measures), being awarded 22 wins. Configuration 7 (pHPSO-const & pbest) also performed well, obtaining only one less win than configuration 2. Therefore, configuration 2 is used to compare the 3 different KTSs. The worst performing configuration with only 7 wins was configuration 4 (f_k-HPSO & pbest). In order to analyse the various KTSs for configuration 2, the wins obtained by various KTSs for configuration 2 were calculated.

The PCX KTS outperformed the other KTSs with 15 wins. The Random and Ring KTSs obtained 4 and 3 wins respectively. Thus, the best performing HVEPSO configuration for these benchmarks over all 3 performance measures was the PCX KTS with the dHPSO behaviour selection strategy and periodic window management. The POF found by all KTSs for WFG1 is presented in Fig. 1(a). The line on the left is the true POF. The KTSs found only a few solutions a few solutions on the right that are quite far from the true POF. Therefore, it can be seen that the KTSs struggled to converge to the true POF of WFG1. Figure 1(b) presents the approximated POF of the PCX KTS using configuration 2 on WFG7 benchmark. From the figure it can be seen that the approximated POF is a good estimation of the true POF.

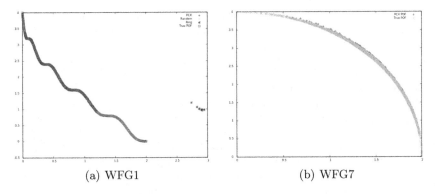

(a) WFG1   (b) WFG7

Fig. 1. POFs of WFG1 found by all KTSs and POF of WFG7 found by HVEPSO configuration 2 with PCX KTS

5   Conclusion

This paper investigated the performance of various HVEPSO configurations for static MOO environments. Three knowledge transfer strategys (KTSs) were used and the results showed that the PCX KTS performed the best over all benchmarks and all performance measures. Nine different configurations for HVEPSO were used and the results indicated that dHPSO behaviour selection strategy with periodic window management performed the best over all benchmarks and performance measures. The results also indicated that HVEPSO had problems with convergence and spread of solutions on benchmarks such as WFG1, a separable unimodal function with a convex Pareto-optimal front (POF).

References

1. Clapham, C., Nicholson, J.: The Concise Oxford Dictionary of Mathematics, vol. 4. Oxford University Press, USA (2009)
2. Deb, K.: Multi-objective genetic algorithms: Problem difficulties and construction of test functions. Technical report CI-49/98, Department of Computer Science/XI, University of Dortmund (1998)
3. Eberhart, R., Kennedy, J.: A new optimizer using particle swarm theory. In: Proceedings of the International Symposium on Micro Machine and Human Science, Nagoya, Japan, pp. 39–43 (1995)
4. Engelbrecht, A.: Computational Intelligence, 2nd edn. John Wiley & Sons Ltd., Hoboken (2007)
5. Engelbrecht, A.: CIlib: A component-based framework for plug-and-simulate. In: Proceedings of International Conference on Hybrid Computational Intelligence Systems, Barcelona, Spain (2008)
6. Nepomuceno, F.V., Engelbrecht, A.P.: A self-adaptive heterogeneous PSO inspired by ants. In: Dorigo, M., Birattari, M., Blum, C., Christensen, A.L., Engelbrecht, A.P., Groß, R., Stützle, T. (eds.) ANTS 2012. LNCS, vol. 7461, pp. 188–195. Springer, Heidelberg (2012)

7. Engelbrecht, A., Nepomuceno, F.: Behavior changing schedules for heterogeneous particle swarms. In: Proceedings of 1st Computational BRICS Countries Intelligence Congress, pp. 112–118 (2013)
8. Engelbrecht, A., Nepomuceno, F.: A self-adaptive heterogeneous PSO for real-parameter optimization. In: Proceedings of the IEEE International Congress on Evolutionary Computation (2013)
9. Engelbrecht, A.P.: Heterogeneous particle swarm optimization. In: Dorigo, M., et al. (eds.) ANTS 2010. LNCS, vol. 6234, pp. 191–202. Springer, Heidelberg (2010)
10. Fonseca, C., Paquete, L., López-Ibáñez, M.: An improved dimension - sweep algorithm for the hypervolume indicator. In: Proceedings of the IEEE Congress on Evolutionary Computation, Vancouver, Canada, pp. 1157–1163, 16–21 July 2006
11. Van Den Bergh, F.: An analysis of particle swarm optimizers. Ph.D. thesis, Department of Computer Science University of Pretoria (2002)
12. Greeff, M., Engelbrecht, A.: Dynamic multi-objective optimisation using PSO. In: Nedjah, N., dos Santos Coelho, L., de Macedo Mourelle, L. (eds.) Multi-Objective Swarm Intelligent Systems. SCI, vol. 261, pp. 105–123. Springer, Heidelberg (2010)
13. Harrison, K.R., Ombuki-Berman, B., Engelbrecht, A.P.: Knowledge transfer strategies for vector evaluated particle swarm optimization. In: Purshouse, R.C., Fleming, P.J., Fonseca, C.M., Greco, S., Shaw, J. (eds.) EMO 2013. LNCS, vol. 7811, pp. 171–184. Springer, Heidelberg (2013)
14. Helbig, M., Engelbrecht, A.: Using heterogeneous knowledge sharing strategies with dynamic vector-evaluated particle swarm optimisation. In: Proceedings of the IEEE Symposium on Swarm Intelligence, Orlando, USA, pp. 1–8, 9–12 December 2014
15. Huband, S., Hingston, P., Barone, L., While, L.: A review of multiobjective test problems and a scalable test problem toolkit. IEEE Trans. Evol. Comput. 10(5), 477–506 (2006)
16. Kennedy, J.: The particle swarm: social adaptation of knowledge. In: IEEE International Conference on Computation, Indianapolis, USA, pp. 303–308, 13–16 April 1997
17. Kennedy, J.: Bare bones particle swarms. In: Proceedings of the IEEE Swarm Intelligence Symposium. pp. 80–87, April 2003
18. Parsopoulos, K., Vrahatis, M.: Recent approaches to global optimization problems through particle swarm optimization. Nat. Comput. 1(2), 235–306 (2002)
19. Schaffer, J.: Multiple objective optimization with vector evaluated genetic algorithms. In: Proceedings of the International Conference on Genetic Algorithms, pp. 93–100 (1985)
20. Sierra, M.R., Coello, C.A.C.: Improving PSO-based multi-objective optimization using crowding, mutation and ϵ-dominance. In: Coello Coello, C.A., Hernández Aguirre, A., Zitzler, E. (eds.) EMO 2005. LNCS, vol. 3410, pp. 505–519. Springer, Heidelberg (2005)
21. Zitzler, E.: Evolutionary algorithms for multiobjective optimization: methods and applications. Ph.D. thesis, Swiss Federal Institute of Technology (ETH) Zurich Switzerland (1999)
22. Zitzler, E., Deb, K., Thiele, L.: Comparison of multiobjective evolutionary algorithms: empirical results. Evol. Comput. 8(2), 173–195 (2000)

Heterogeneous Bare-Bones Particle Swarm Optimization for Dynamic Environments

Yuanxia Shen$^{(\boxtimes)}$, Jian Chen, Chuanhua Zeng, and Linna Wei

School of Computer Science and Technology,
Anhui University of Technology, Maanshan 243002, China
yuanxiashen@163.com

Abstract. Particle swarm optimization is an effective technique to track and find optimum in dynamic environments. In order to improve convergence accuracy of solutions, a heterogeneous bare-bones particle swarm optimization (HBPSO) is proposed in which several master swarms and a slaver swarm are employed to exploration search and exploitation search, respectively. When detecting environments change, a new strategy is used to update the position of particles for keeping swarm diversity. If the search areas of two swarms are overlapped, the worse swarm will be initialized. Experimental results on moving peaks benchmark (MPB) functions show that the proposed algorithm is effective and easy to implement.

Keywords: Particle swarm optimization · Dynamic environments · Learning parameters

1 Introduction

Many real-world optimization problems have dynamic characteristics due to environmental changes. For static optimization problems, the goal is to find a global optimum. Dynamic optimization problems (DOPs) require algorithms to not only find the global optimal solution under a specific environment but also to continuously track the changing optima over different dynamic environments [1].

Particle Swarm Optimization (PSO), firstly proposed by Kennedy and Eberhart in 1995, was inspired by the simulation of simplified social behaviours including fish schooling, bird flocking, etc. Although PSO has been shown to perform well for many static problems, it faces the difficulty when addressing DOPs. The difficulty includes diversity loss due to swarm convergence and outdated memory because of the changing environment [2].

Recently, several PSO variants have been proposed to address DOPs. Blackwell [2,3] developed a multi-swarm PSO in which the population of particles is split into a set of interacting swarms and the exclusion and anti-convergence mechanisms are used to keep the population diversity.

Du and Li [4] presented a multi-strategy ensemble PSO algorithm to achieve the balance between exploration and exploitation by the cooperative searching

© Springer International Publishing Switzerland 2016
Y. Tan et al. (Eds.): ICSI 2016, Part I, LNCS 9712, pp. 305–313, 2016.
DOI: 10.1007/978-3-319-41000-5_30

of two swarms. In [4], the Gaussian local search and differential mutation are introduced into the exploration swarm and exploitation swarm.

Esquivel and Coello [5] proposed a hybrid PSO in which a dynamic macro-mutation operator is used to maintain the swarm diversity. Experimental results indicate that hybrid PSO is efficient for searching the optimum in a dynamic environment.

Liu et al. [6] presented a composite particle swarm optimizer in which a new velocity-anisotropic reflection (VAR) scheme and a "fitness-and-distance" based pioneer particle identification (PPI) method are introduced to search the changed optima.

Parvin et al. [7] proposed a multi-swarm algorithm for dynamic environments to address the diversity loss problem. In [7], a parent swarm is used to explore the search space to find promising area containing local optima and several child swarms are employed to exploit a promising area found by the parent swarm.

Yazdani et al. [8] proposed an improved multi-swarm approach in which one finder swarm and several similar tracker swarms are used to find peaks in the problem space and track them after an environment change, respectively. In addition, modified updating equations of the velocity and the position of particles were proposed to increase the diversity of swarms. The multi-population method has been widely used to solve DOPs with the aim of finding different peaks and tracking multiple changing peaks simultaneously [9].

To obtain desirable solutions, the learning parameters in the PSO need to set, such as the inertia weight, acceleration coefficients and the velocity clamping. Empirical results have shown that values of those learning parameters have effect on the performance of PSO [10]. In 2003, Kennedy [11] proposed a bare-bones PSO (BPSO) in which the traditional velocity term of PSO is eliminated and the position of each particle is randomly updated from the Gaussian distribution based on two leaders, i.e., the personal best position (*pbest*) and the global best position of particles (*gbest*). Compared to the original PSO, BPSO is may be the simplest version PSO since it does not involve the inertia weight, the acceleration coefficient and the velocity. BPSO has been applied successfully in some real problems.

In this paper, a heterogeneous bare-bones particle swarm optimization (HBPSO) based on the multi-populations method is developed where several master swarms are employed to the global search and a slaver swarm is used to the local search. In the master swarm, each particle can learn from several exemplars according to learning probability, which can enhance the exploration capability of the particle. In a slave swarm, each particle adopts the learning strategy of basic BPSO, which is beneficial for the fast convergence. If the environment is changed, a new strategy is used to update the position of particles for keeping swarm diversity. If the search areas of two swarms are overlapped, the worse swarm will be initialized.

The remainder of this paper is organized as follows. PSO and BPSO are introduced in Sect. 2. HBPSO for DOPs is presented in Sect. 3. Experimental results on benchmark optimization problems are discussed in Sect. 4. Conclusions are drawn in Sect. 5.

2   BPSO

Assuming that the search space is D-dimensional, each particle i represents a potential solution of the problem and is represented by a position vector $\boldsymbol{x_i} = (x_{i1}, x_{i2}, \cdots, x_{iD})$. A Gaussian sampling based on *pbest* and *gbest* is used to update the position of the particles. For particle i, its position $\boldsymbol{x_i}(t+1)$ and at iteration $t+1$ are updated as follows:

$$x_{id}(t+1) = N\left(\frac{P_{id}(t) + G_d(t)}{2}, |P_{id}(t) - G_d(t)|\right) \tag{1}$$

where $N(.)$ denotes the Gaussian distribution, $d = 1, 2, \cdots, D$; $P_i = (p_{i1}, p_{i2}, \cdots, p_{iD})$, called the personal best position (*pbest*), denotes the best position found by the i-th particle itself so far. $G = (g_1, g_2, \cdots, g_D)$, called the global best position (*gbest*), denotes the global best position found by neighbors of this particle so far. In BPSO, the position of each particle is selected from the Gaussian distribution with the mean $\mu = 0.5(P_{id}(t) + G_d(t))$ and the standard deviation $\delta = |P_{id}(t) - G_d(t)|$. From Eq. (1), BPSO is parameter-free.

3   HBPSO for DOPs

In this section, we describe HBPSO for DOPS in detail. In HBPSO, several master swarms and a slaver swarm are employed to perform the exploration search and the exploitation search, respectively. After detecting environment change, a tracking mechanism is introduced to address outdate memory.

3.1   Learning Strategies

Before changing of the environment, the environment can be had as a static one. In the basic BPSO, each particle in the swarm learns from the arithmetic average of *pbest* and *gbest*. The learning mechanism can lead to the fast convergence, but it can result in the premature convergence when solving multimodal problems. In order to avoid the premature convergence, each particle in the master swarm can learn from the superior particles, its own *pbest* and the *gbest* of swarm according to different probabilities.

The individuals are sorted by the fitness value, and the superior particles are the top $100pr\%$ individuals in the current swarm with $pr \in (0,1]$. The position updating formula for each particle in the master swarm is given by:

$$x_{ij}(t+1) = \begin{cases} P_{ij}^*(t) + |P_{ij}^*(t) - P_{ij}(t)| \cdot N(0,1) & 0 < \xi \leq 0.5 \\ P_{ij}(t) + |P_{r3j}(t) - P_{ij}(t)| \cdot N(0,1) & 0.5 < \xi \leq 0.5 + 0.25pc \\ 0.5(P_{ij}(t) + G_j(t)) + |G_j(t) - P_{ij}(t)| \cdot N(0,1) & 0.5 + 0.25pc < \xi \leq 1 \end{cases} \tag{2}$$

where P_i^* is randomly chosen from the set of superior particles. ξ is a random number with a uniform distribution in the range of $[0,1]$. P_{r1} is randomly chosen from the set of personal best particles. $pc = 1 - t/T_{\max}$ is used to adjust the learning rate of the particle learning toward *pbest* and *gbest*. With the iteration increasing, the value of pc desecrates form 1 to 0. Form Eq. (2), the particle learns from the superior particle by the probability 0.5, which is conducive to the balance between the global and the local search. The particle learns from its own *pbest* or the *gbest* by the adaptive probability, respectively. This learning method can maintain the explore search in the early evolutionary phase and enhance the exploitation of the solution.

The slave swarm performs the exploitation search, and hence each particle adopts the learning strategy of basic BPSO. The master swarm and the slaver swarm play different roles in the evolutionary process. The master swarm is used for the global search, while the slaver swarm is employed for the local search. The master swarm doesnt receive information from the slaver swarm in order to keep the ability to perform the global search. When the best particle (G^s) in the slaver swarm is not improved from L successive FEs or the fitness value of G^s is less than the one of the best particle (G^m) in the slaver swarms (the higher of the fitness value, the better performance of the particle), particles in the slaver swarm can learn from the best particle in the master swarm (G^m).

3.2 Tracking Mechanism

After detecting environment change, the master swarm and the slaver swarm might be converged. In order to avoid the diversity loss, the position and velocity of the whole swarm need initialize. For each master swarm and the slaver swarm, the new position of particle i in the k-th subgroup, can be computed by the following equations.

$$x_i^k = G^k + \eta(\overline{x} - \underline{x}) \tag{3}$$

where G^k is the *gbest* of k-th subgroup, $\overline{x}$ and $\underline{x}$ are the upper and lower bounders of the variable x_i^k, η is random vectors with uniform distribution in [-0.5,0.5].

In proposed algorithm, the overlapping check between two subgroups is carried out using the distance of their best particles. The distance $d(G^{k1}, G^{k2})$ between two particles G^{k1} and G^{k2} in the D-dimensional space is defined as the Euclidean distance between them as follows:

$$d_{k1k2} = \sqrt{\frac{1}{D} \sum_{j=1}^{D} (G_j^{k1} - G_j^{k2})^2} \tag{4}$$

If the distance $d(G^{k1}, G^{k2})$ is smaller a constant value defined by users, the swarm with the worse fitness of the best particle will be initialized. Incorporating learning strategies and the tracking mechanism, the pseudo-code for HBPSO for DOPs is summarized in Algorithm 1.

Algorithm 1. HBPSO

1: Randomly initialize each swarm;
2: **for** each swarm k **do**
3:    initialize x^k_{ij}, $i \in 1, \cdots, Ne$ and $j \in 1, \cdots, d$
4:    Evaluate each of the particles in the k-th swarm. $gbest^k \leftarrow \max F(pbest_i)$
5: **end for**
6: **for** $t = 1$ to $t = IterMax$ **do**
7:    **for** each master swarm k **do**
8:       Update the position of particles using Eq. (2);
9:    **end for**
10:   **for** each particle i in the slaver swarm **do**
11:      Update the position of particles using Eq. (1);
12:   **end for**
13:   **for** each swarm k **do**
14:      Update $pbest, gbest$;
15:   **end for**
16:   **for** each swarm k **do**
17:      **for** each particle i **do**
18:         Update the position using Eq. (3);
19:      **end for**
20:   **end for**
21:   Compute the distance $d(G^{k1}, G^{k2})$; Check the overlapping of two swarms
22:   **if** $d(G^{k1}, G^{k2}) < dmin$ **then**
23:      **if** $f(G^{k1}) < f(G^{k2})$ **then**
24:         Initialize G^{k2}
25:      **else**
26:         Initialize G^{k1}
27:      **end if**
28:   **end if**
29: **end for**
30: **return** $gbest$ in the finder subgroup

---

4 Experimental Setup and Simulation Results

4.1 Experimental Setup

Moving Peaks Benchmark (MPB) functions [12] are used to investigate the performance of the proposed algorithm. MPB functions are widely adopted in dynamic environments [3–10]. The performance measure used is the offline error, which is defined in [14]. HBPSO is compared to the following algorithms: (a) mQSO [2], (b) FMSO [13], (c) cellular PSO (CPSO) [14], (d) APSO [7]; In HBPSO, the maximum fitness evaluation (FE) is set at 30000. The number of the master swarms is set at 9 and the number of particles in each swarm is set to 10 particles. The learning rate pr is set at 0.25. Each algorithm was run 30 times independently to reduce random discrepancy.

Table 1. Comparison of offline error of five algorithms on MPB problem with different number of peaks and change frequency=500, 1000, 2500, 5000

Frequency	M	APSO	mQSO	FMSO	CPSO	HBPSO
500	1	4.81±0.14	33.67±3.42	27.58±0.94	13.46±0.7	**0.8607±0.2678**
	5	4.95±0.11	11.91±0.76	19.45±0.45	9.63±0.49	**1.1399±0.2325**
	10	5.16±0.11	9.62±0.34	18.26±0.32	9.42±0.21	**2.5519±0.5656**
	20	5.81±0.08	9.07±0.25	17.34±0.30	8.84±0.28	**3.0933±0.9145**
	30	6.03±0.07	8.80±0.21	16.39±0.48	8.81±0.24	**3.3453±0.5064**
	50	5.95±0.06	8.72±0.20	15.54±0.26	8.62±0.23	**3.5887±1.0042**
	100	6.08±0.06	8.54±0.16	12.87±0.60	8.54±0.21	**4.2979±1.0474**
	200	6.20±0.04	8.19±0.17	11.52±0.61	8.28±0.18	**4.1016±0.9601**
1000	1	2.72±0.04	18.60±1.63	14.42±0.48	6.77±0.38	**0.3856±0.0780**
	5	2.99±0.09	6.56±0.38	10.59±0.24	5.30±0.32	**0.5410±0.0854**
	10	3.87±0.08	5.71±0.22	10.40±0.17	5.15±0.13	**1.0826±0.5171**
	20	4.13±0.06	5.85±0.15	10.33±0.13	5.23±0.18	**3.5702±1.3474**
	30	4.12±0.04	5.81±0.15	10.06±0.14	5.33±0.16	**2.2276±0.4772**
	50	4.11±0.03	5.87±0.13	9.54±0.11	5.55±0.14	**2.9963±1.0225**
	100	4.26±0.04	5.83±0.13	8.77±0.09	5.57±0.12	**3.4233±0.7893**
	200	**4.21±0.02**	5.54±0.11	8.06±0.07	5.50±0.12	5.9274±1.7718
2500	1	1.06±0.03	7.64±0.64	6.29±0.20	4.15±0.25	**0.1879±0.0555**
	5	1.55±0.05	3.26±0.21	5.03±0.12	2.85±0.24	**0.2154±0.0258**
	10	2.17±0.07	3.12±0.14	5.09±0.09	2.80±0.10	**0.7413±0.6618**
	20	2.51±0.05	3.58±0.13	5.32±0.08	3.41±0.14	**1.4910±0.9879**
	30	2.61±0.02	3.63±0.10	5.22±0.08	3.62±0.12	**1.9394±1.0817**
	50	**2.66±0.02**	3.63±0.10	4.99±0.06	3.86±0.10	3.4216±1.9881
	100	**2.62±0.02**	3.58±0.08	4.60±0.05	4.10±0.11	2.9328±1.4756
	200	**2.64±0.01**	3.30±0.06	4.34±0.04	3.97±0.10	2.7004±1.4447
5000	1	0.53±0.01	3.82±0.35	3.44±0.11	2.55±0.12	**0.1118±0.0218**
	5	1.05±0.06	1.90±0.08	2.94±0.07	1.68±0.11	**0.1436±0.0453**
	10	1.31±0.03	1.91±0.08	3.11±0.06	1.78±0.05	**0.1554±0.0408**
	20	1.69±0.05	2.56±0.10	3.36±0.06	2.61±0.07	**0.8944±0.7924**
	30	1.78±0.02	2.68±0.10	3.28±0.05	2.93±0.08	**1.1171±1.0554**
	50	1.95±0.02	2.63±0.08	3.22±0.05	3.26±0.08	**1.8441±1.0933**
	100	**1.95±0.01**	2.52±0.06	3.06±0.04	3.41±0.07	3.3282±1.3631
	200	**1.90±0.01**	2.36±0.05	2.84±0.03	3.40±0.06	2.0326±0.9247

4.2   Experimental Results and Discussions

Table 1 presents the results of the offline error and standard deviation of the proposed algorithms and five PSO variants with various change frequencies and different peak numbers. The experimental results of the peer algorithms are taken from the corresponding papers with the configuration that enables them to achieve their best performance. The best results among the six algorithms are shown in bold.

Form Table 1, we obverse that HBPSO obtains the best searching accuracy on MPB problem with different number of peaks with the low change frequency. The interaction between the master swarm and the slaver swarm can help the slaver swam reach the optimums before environment changes. The performance of all algorithms degrades when the number of peaks increases. Figure 1 shows the chart of the current error of HBPSO in MPB with 10 peaks and the environment change frequencies of 500 and 1000. From Fig. 1, we can observe HBPSO can reach the optimums and track them during environment changes. As can be seen from Table 1, all algorithms performance improve by increasing the change frequency, and the performance of HBPSO outperforms that of five PSO variants in most of cases. Table 2 presents the average ranking of algorithms on MPB problems with different change frequency. From Table 2, the rank of proposed algorithm is first.

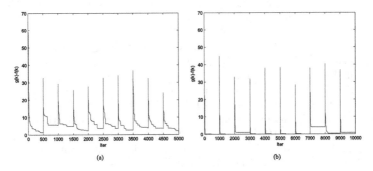

(a)   (b)

Fig. 1. The current error of HBPSO on MPB with different change frequencies. (a) The change frequency of 500. (b) The change frequency of 1000.

Table 2. The rank of algorithms on MPB problem based on different change frequency.

Algorithm	Change frequency				
	500	1000	2500	5000	Overall ranking
APSO	2	2	2	2	2
mQSO	4	4	3	3	3
FMSO	5	5	5	5	5
CPSO	3	3	4	4	3
HBPSO	1	1	1	1	1

5 Conclusion

In this paper, a HBPSO is developed for addressing DOPs. In HBPSO, master swarms are used to find the promising area, and a salver swarm is employed to track the master swarm with the best gbest and to exploit better solutions. To keep the swarm diversity, the master swarm adopts a dynamic exemplars learning strategy, and the whole swarm could be updated according to the current status when environments changes are detected. HBPSO has been evaluated on moving peaks benchmark, and experimental results show that the proposed algorithm is effective. In our future works, the development of the HBPSO is used to solve real dynamic optimization problems.

Acknowledgments. This work was supported by National Natural Science Foundation of China under Grant Nos. 61300059 and 61502010.

References

1. Li, C., Yang, S.: A general framework of multipopulation methods with clustering in undetectable dynamic environments. IEEE Trans. Evol. Comput. **16**(4), 556–577 (2012)
2. Blackwell, T., Branke, J.: Multiswarms, exclusion, and anti-convergence in dynamic environments. IEEE Trans. Evol. Comput. **10**(4), 459–472 (2006)
3. Blackwell, T.: Particle swarm optimization in dynamic environments. In: Yang, S., Ong, Y.S., Jin, Y. (eds.) Evolutionary Computation in Dynamic and Uncertain Environments. SCI, vol. 51, pp. 29–49. Springer, Heidelberg (2007)
4. Du, W., Li, B.: Multi-strategy ensemble particle swarm optimization for dynamic optimization. Inf. Sci. **178**(15), 3096–3109 (2008)
5. Esquivel, S.C., Coello Coello, C.A.: Hybrid particle swarm optimizer for a class of dynamic fitness landscape. Eng. Optim. **38**(8), 873–888 (2006)
6. Liu, L., Yang, S., Wang, D.: Particle swarm optimization with composite particles in dynamic environments. IEEE Trans. Syst. Man Cybern. Part B Cybern. **40**(6), 1634–1648 (2010)
7. Parvin, H., Minaei, B., Ghatei, S.: A new particle swarm optimization for dynamic environments. In: Herrero, Á., Corchado, E. (eds.) CISIS 2011. LNCS, vol. 6694, pp. 293–300. Springer, Heidelberg (2011)
8. Yazdani, D., Nasiri, B., Sepas-Moghaddam, A., Meybodi, M.R.: A novel multi-swarm algorithm for optimization in dynamic environments based on particle swarm optimization. Appl. Soft Comput. **13**(4), 2144–2158 (2013)
9. Li, C., Nguyen, T.T., Yang, M., Yang, S., Zeng, S.: Multi-population methods in unconstrained continuous dynamic environments: the challenges. Inf. Sci. **296**, 95–118 (2015)
10. Van den Bergh, F., Engelbrecht, A.P.: A convergence proof for the particle swarm optimiser. Fundam. Inform. **105**(4), 341–374 (2010)
11. Kennedy, J.: Bare bones particle swarms. In: Proceedings of the 2003 IEEE Swarm Intelligence Symposium, SIS 2003, pp. 80–87. IEEE (2003)
12. Branke, J.: Memory enhanced evolutionary algorithms for changing optimization problems. In: In Congress on Evolutionary Computation CEC 1999. Citeseer (1999)

13. Li, C., Yang, S.: Fast multi-swarm optimization for dynamic optimization problems. In: Fourth International Conference on Natural Computation, ICNC 2008, vol. 7, pp. 624–628. IEEE (2008)
14. Hashemi, A.B., Meybodi, M.R.: Cellular PSO: a PSO for dynamic environments. In: Cai, Z., Li, Z., Kang, Z., Liu, Y. (eds.) ISICA 2009. LNCS, vol. 5821, pp. 422–433. Springer, Heidelberg (2009)

A New Particle Acceleration-Based Particle Swarm Optimization Algorithm

Shailesh Tiwari[1]([⊠]), K.K. Mishra[2], Nitin Singh[2], and N.R. Rawal[2]

[1] CSED, ABES Engineering College, Ghaziabad, India
shail.tiwari@yahoo.com
[2] Motilal Nehru National Institute of Technology Allahabad, Allahabad, India
{kkm,nitins,nrrawal}@mnnit.ac.in

Abstract. Optimization of one or more objective function is a requirement for many real life problems. Due to their wide applicability in business, engineering and other areas, a number of algorithms have been proposed in literature to solve these problems to get optimal solutions in minimum possible time. Particle Swarm Optimization (PSO) is a very popular optimization algorithm, and was developed by Dr. James Kennedy and Dr. Russell Eberhart in 1995 which was inspired by social behavior of bird flocking or fish schooling. In order to improve the performance of PSO algorithm, number of its variants has been proposed in literature. Few variants such as PSO Bound have been designed differently, whereas others use various methods to tune the random parameters. PSO - Time Varying Inertia Weight (PSO-TVIW), PSO Random Inertia Weight (PSO-RANDIW), and PSO-Time Varying Acceleration Coefficients (PSO-TVAC), APSO-VI, LGSCPSOA and many more are based on parameter tuning. On similar principle, the proposed approach improves the performance of PSO algorithm by adding new parameter henceforth called as "acceleration to particle" in its velocity equation. Efficiency of the proposed algorithm is checked against other existing PSO, and results obtained are very encouraging.

Keywords: PSO · PSO-TVAC · Parameter tuning

1 Introduction

An optimization problem is to find the best solution from all feasible solutions, which is an optimal value of objective function and may be either maximum or minimum. The objective function, its search space, and the constraints are all parts of the optimization problem. A local optimum of an optimization problem is a solution that is optimal (either maximal or minimal) within a neighboring set of solutions. The global optimum is the optimum of all local optima, i.e., it is the optimum (either maximal or minimal) in whole search space.

PSO is a population based optimization algorithm that optimizes a problem by updating the velocity and position of each particle according to simple mathematical formulae. Each particle's movement is influenced by its personal best (pbest) position and global best (gbest) position. The values of pbest and gbest are updated in each generation. The performance of PSO depends on many factors in which parameter

© Springer International Publishing Switzerland 2016
Y. Tan et al. (Eds.): ICSI 2016, Part I, LNCS 9712, pp. 314–321, 2016.
DOI: 10.1007/978-3-319-41000-5_31

tuning plays a major role. Tuning of random parameters is very important issue and if properly done, it can avoid premature convergence of PSO on local optimal solutions. Many papers [2–4] suggested solutions for this issue by giving various methods to tune random parameters (Like w, c1, c2, r1 and r2). The latest tuned PSO is PSO-TVAC. However the performance of PSO-TVAC is not as good as of PSO Bound which is designed by improving the traditional operators of PSO.

This paper implements a modified particle swarm optimization algorithm named as "Particle Acceleration based Particle Swarm Optimization (PA-PSO)" in which additional parameter "acceleration to particle" is included in velocity vector, which causes the PSO to converge toward the best solution very fast. The major aim of this paper is to develop an effective variant of particle swarm optimization algorithm based on the concept of particle acceleration for global optimization problems. The rest of the paper is structured in following sections: Sect. 2, describes general background on Particle Swarm Optimization (PSO) and its variants, with brief overview of the related work of PSO. Section 3, provides the complete description of the proposed algorithm, Sect. 4, describes the experimental setup for the benchmarks functions and provides results for the proposed approach on benchmarks functions and finally in Sect. 5, conclusion are drawn and future work is suggested.

2 Background Details and Related Work

PSO is a robust optimization technique that applies the social intelligence of swarms to solve optimization problems. First version of this algorithm was proposed by Kennedy and Eberhart [1]. Although PSO has very good convergence rate but it may face problem of stagnation. To remove basic limitations of PSO many new variants of PSO have been proposed. Here is a brief overview of some variants of PSO.

Discrete PSO algorithm [2] uses discrete values of particle's positions [4]. The velocity and position equation are developed for real (discrete) value and updated in each iteration. GCPSO [5] has been introduced which searches the region around the current global best position, i.e. its local best position is equal to the current global best position.

Niching concept is first induced to PSO in [6] in order to heighten its ability to handle more complex optimization problems that can search for multiple solutions in parallel. The niche concept is applied in many variants of PSO, such as ASNPSO (adaptive sequential niche particle swarm optimization), PVPSO (parallel vector-based particle swarm optimizer) [7]. There are also some enhancements for niche based PSO such as Enhancing the NichePSO [8], adaptively choosing niching parameters in a PSO [9]. RegPSO [10] was proposed to deal with the stagnation problem by building into the PSO algorithm a mechanism capable of automatically triggering swarm regrouping when premature convergence is detected. A new PSO algorithm called PSO with neighborhoods search strategies (NSPSO), which utilizes one local and two global neighborhood search strategies has been defined by Yao and Han [11].

Riget et al. proposed Attractive Repulsive Particle Swarm Optimization (ARPSO) algorithm to remove the drawback of PSO in premature convergence [12]. Ye and Chen [13] proposed an Alternative KPSO clustering (AKPSO) algorithm using a novel alternative metric. BRPSO algorithm [14] has been designed by Alviar et al. (2007) to optimize multi-modal functions. COPSO algorithm was given by Aguirre et al. to solve constrained single objective problem [15]. Lin et al. [16] has proposed a Complementary Cyber Swarm algorithm, which is a combination of Particle Swarm Optimization (PSO) with strategies of scatter search and path re-linking to create more efficient variant of PSO. Mohammed El-Abd et al. introduced PSO-Bounds that are based on the concept of Population-Based Incremental Learning (PBIL) that allow PSO algorithm to adjust the search space boundary as the search progresses [17, 18]. Xu et al. proposed an adaptive parameter tuning in APSO-VI [20] by introducing velocity information defined as the average absolute value of velocity of all its particles. Lin et al. [16] proposed a local and global search combined PSO (LGSCPSOA), analyzed its convergence, and obtained its convergence qualification. A variant of PSO-TVAC has been proposed and used in software test case generation process in [21].

3   Proposed Approach

Each optimization algorithm is designed to minimize searching time to reach to the global optimal solution. The proposed algorithm PSOM is also designed to fulfill these requirements. To improve the convergence rate of existing PSO, a change has been made in the velocity vector of PSO. In PSO, each solution updates its position only on the basis of current velocity and there is no role of information collected by particle during previous generations. In proposed version, information collected during previous generation is kept and is used to increase or decrease the velocity of particle. This is done with the help of additional parameter which is named as Acceleration (Figs. 1 and 2).

This improved PSO use following formulae to calculate new velocity and the new position of swarm (position vector of particles):

$$v_{id}^{t+1} = \omega * v_{id}^t + c_1 r_1 (P_{id} - x_{id}) + c_2 r_2 (G_{best} - x_{id}^t) + a \qquad (5)$$

$$x_{id}^{t+1} = x_{id}^t + v_{id}^{t+1} \qquad (6)$$

where a is the acceleration

Initially, population of particles is generated randomly and each particle is associated with random velocity and random position in the n-dimensional search space. To calculate the value of acceleration fitness values of pbest and new position of all particles are compared. In each generation, particles are forced to direct to good regions only. To terminate the algorithm, the iteration number is compared with maximum number of iteration. If it is the last generation (termination criteria is met) then position of the global best particle will be the optimal solution. If not, then increase the current iteration by 1 and repeat until the maximum number of iteration is reached (termination criteria) (Fig. 3).

Pseudo code of proposed Algorithm:

Input: Size of population (pop), dimension (dim), fitness function f(x) with its constraints and maximum iteration (I_{max}).

Output: Optimum fitness value of function f(x) and position of particle at which optimum value is found.

For Each particle of ith population

1. Initialize position vector x_{id} randomly of specified dim.
2. Create a pop of velocity vectors v_{id} of dim and initialize it with random numbers between 0 and 1.
3. Calculate fitness vector f(x_{id}).
4. Choose personal best P_{id} and initialize it with fitness vector.
5. Initialize the parameter c_{1min} and c_{2min} with 0.5, c_1max and c_2max with 2.5 while ω_{min} with 0.4 and ω_{max} with 0.9.
6. For t = 1 to I_{max}.
7. Set the parameter of algorithm as follows:

    c_1= c_1min + (c_1max - c_1min) * (I_{max} - $I_{current}$) / I_{max}

    c_2= c_2max + (c_2min - c_2max) * (I_{max} - $I_{current}$) / I_{max}

    ω = ω_{min} + (ω_{max} - ω_{min}) * (I_{max} - $I_{current}$) / I_{max}

8. For id = 1 to pop.
9. Find Personal best P_{id}.
    if f(x_{id}) <P_{id}
    P_{id}= x_{id}.
10. Find Global best Gbest
    if f(x_{id}) <Gbest
    Gbest = x_{id}.
11. Initialize initial velocity of particle:

    $$u_{id}^{t} = v_{id}^{t}$$

12. Update the velocity (v_{id}) of particle by the velocity vector equation:

    $$v_{id}^{t+1} = \omega * v_{id}^{t} + c_1 r_1 (P_{id} - x_{id}^{t}) + c_2 r_2 (G_{best} - x_{id}^{t}) + a$$

13. Calculate Acceleration of particle:

    $$a_{id}^{t+1} = v_{id}^{t+1} - u_{id}^{t}$$

14. Update the position (x_{id}^{t+1}) of particle by the position vector equation:

    $$x_{id}^{t+1} = x_{id}^{t} + v_{id}^{t+1}$$

15. Repeat step 7 to step 13, until the termination criteria is met (id becomes equal to pop).
16. Repeat step 6 to step 15, until the termination criteria is met (t becomes equal to I_{max}).

17. The global best position of the swarm is the optimal solution.

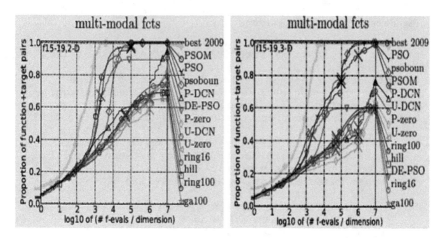

Fig. 1. Improved results of PSOM on 2-D and 3-D

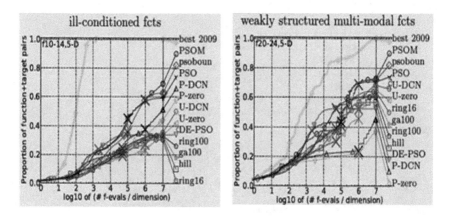

Fig. 2. Improved results of PSOM on 5-D

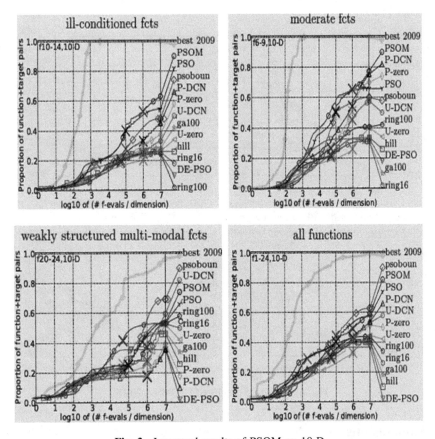

Fig. 3. Improved results of PSOM on 10-D

4 Experimental Setup and Results

To check the performance of proposed algorithm with state of art algorithms, BBOB noiseless test-bed is used. Optimal value of particles position is searched in closed interval -5 to +5. BBOB has 24 benchmark functions and they are checked for six dimensions. The population size is kept 50 and total number of function's evaluation is made dependent on the dimension and it varies as the dimension changes. To calculate the function evaluation formula dimension* 10^6 is used. The total function evaluation is calculated by the formula i.e. *Total function evaluation = Maximum function evaluation * Function evaluation in one iteration.* The data set which is used to compare the performance of the proposed algorithm has been taken from the COCO framework [22] (Fig. 4).

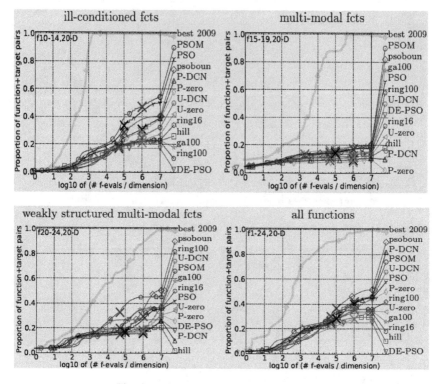

Fig. 4. Improved results of PSOM on 20-D

5   Conclusions

From the results it is clear that on dimension 2 the performance of PSOM has improved over PSO on multimodal functions. However its convergence rate has decreased in separable functions. Similarly the convergence of modified PSO has improved in multimodal functions of 3 dimensions. A slight improvement has seen in its performance on separable functions in 3 dimensions. In 5 dimensions, PSOM has dominated PSO in ill conditioned and weakly structured multimodal functions. On other functions the performance of PSO was good. It has been noticed that in dimension 10 PSOM outperform PSO in All functions. Moreover it has dominated PSO in three functions. Similar performance has been observed in 20 dimensions where it dominated PSO in all except separable and moderate functions.

References

1. Kenndy, J., Eberhart, R.C.: Particle swarm optimization. In: Proceedings of IEEE International Conference on Neural Networks, vol. 4, pp. 1942–1948 (1995)
2. Premalatha, K., Natarajan, A.M.: Discrete PSO with GA operators for document clustering. Int. J. Recent Trends Eng. **1**(1), 20–24 (2009)

3. Parsopoulos, K.E., Vrahatis, M.N.: Recent approaches to global optimization problems through particle swarm optimization. Nat. Comput. **1**(2–3), 235–306 (2002)
4. Laskari, E.C., Parsopoulos, K.E., Vrahatis, M.N.: Particle swarm optimization for integer programming. In: WCCI, pp. 1582–1587. IEEE, May 2002
5. Van Den Bergh, F.: An analysis of particle swarm optimizers (Doctoral dissertation, University of Pretoria) (2006)
6. Brits, R., Engelbrecht, A.P., Van den Bergh, F.: A niching particle swarm optimizer. In: Proceedings of the 4th Asia-Pacific conference on simulated evolution and learning, vol. 2, pp. 692–696. Orchid Country Club, Singapore, November 2002
7. http://en.wikipedia.org/wiki/ Particle swarm optimization
8. Zhang, J., Huang, D.S., Lok, T.M., Lyu, M.R.: A novel adaptive sequential niche technique for multimodal function optimization. Neurocomputing **69**(16), 2396–2401 (2006)
9. Bird, S., Li, X.: Adaptively choosing niching parameters in a PSO. In: Proceedings of the 8th Annual Conference on Genetic and Evolutionary Computation, pp. 3–10. ACM, July 2006
10. Evers, G.I., Ben Ghalia, M.: Regrouping particle swarm optimization: a new global optimization algorithm with improved performance consistency across benchmarks. In: IEEE International Conference on Systems, Man and Cybernetics, SMC 2009, pp. 3901–3908. IEEE, October 2009
11. Yao, J., Han, D.: Improved barebones particle swarm optimization with neighborhood search and its application on ship design. Math. Probl. Eng. **2013**, Article ID 175848, 12 (2013).http://dx.doi.org/10.1155/2013/175848
12. Riget, J., Vesterstrøm, J.S.: A diversity-guided particle swarm optimizer-the ARPSO. Dept. Comput. Sci., Univ. of Aarhus, Aarhus, Denmark, Technical report 2 (2002)
13. Ye, F., Chen, C.Y.: Alternative KPSO-clustering algorithm. Tamkang J. Sci. Eng. **8**(2), 165 (2005)
14. Barrera Alviar, J., Peña, J., Hincapié, R.: Subpopulation best rotation: a modification on PSO. Revista Facultad de Ingeniería Universidad de Antioquia (40), pp. 118–122 (2007)
15. Zavala, A.E., Aguirre, A.H., Diharce, E.R.: Continuous constrained optimization with dynamic tolerance using the COPSO algorithm. In: Mezura-Montes, E. (ed.) Constraint-Handling in Evolutionary Optimization. SCI, vol. 198, pp. 1–23. Springer, Heidelberg (2009)
16. Yin, P.Y., Laguna, M., Zhu, J.X.: A complementary cyber swarm algorithm (2011)
17. El-Abd, M., Kamel, M.S.: Particle swarm optimization with varying bounds. In: Evolutionary Computation, CEC 2007 (2007)
18. El-Abd, M., Kamel, MS.: Particle swarm optimization with adaptive bounds. In: Evolutionary Computation (CEC) (2012)
19. Lin, W., Lian, Z., Gu, X., Jiao, B.: A local and global search combined particle swarm optimization algorithm and its convergence analysis. Math. Probl. Eng. **2014**, 11 (2014)
20. Xu, G.: An adaptive parameter tuning of particle swarm optimization algorithm. Appl. Math. Comput. **219**(9), 4560–4569 (2013)
21. Tiwari, S., Mishra, K.K., Misra, A.K.: Test case generation for modified code using a variant of particle swarm optimization (PSO) algorithm. In: 2013 Tenth International Conference on Information Technology: New Generations (ITNG), pp. 363–368. IEEE (2013)
22. http://coco.gforge.inria.fr/

Dense Orthogonal Initialization
for Deterministic PSO: ORTHOinit+

Matteo Diez[1], Andrea Serani[1], Cecilia Leotardi[1], Emilio Fortunato Campana[1],
Giovanni Fasano[2(✉)], and Riccardo Gusso[2]

[1] National Research Council–Marine Technology Research Institute (CNR-INSEAN),
Via di Vallerano, 139, 00128 Rome, Italy
{matteo.diez,emiliofortunato.campana}@cnr.it,
{andrea.serani,cecilia.leotardi}@insean.cnr.it
[2] Department of Management, University Ca'Foscari of Venice,
S.Giobbe, Cannaregio 873, 30121 Venice, Italy
{fasano,rgusso}@unive.it

Abstract. This paper describes a novel initialization for Deterministic
Particle Swarm Optimization (DPSO), based on choosing specific dense
initial positions and velocities for particles. This choice tends to induce
orthogonality of particles' trajectories, in the early iterations, in order
to better explore the search space. Our proposal represents an improvement, by the same authors, of the theoretical analysis on a previously
proposed PSO reformulation, namely the initialization ORTHOinit. A preliminary experience on constrained Portfolio Selection problems confirms
our expectations.

Keywords: Global optimization · Deterministic PSO · Particles initial
position and velocity

1 Introduction

Particle Swarm Optimization (PSO) is a stochastic approach for the solution of
global optimization problems [1]. A population of particles is initialized in PSO,
and their trajectories in the search space explore potential solutions, in order to
approach a global minimum. There are several convergence studies on PSO in
the literature, focusing on the role of PSO control parameters, as for instance
swarm size, inertia weight, acceleration coefficients, velocity clamping, as well as
particles' initialization. All these studies reveal that PSO parameters must be
confined to specific subsets of values, in order to avoid diverging trajectories, and
some of them provide necessary conditions for the trajectories convergence (see
for instance [2–6], along with [7–10]). Improper initializations of PSO parameters
may yield divergent or even cyclic behavior. The initial position and velocity of
particles may have dramatic consequences on convergence. Moreover, effective
particles initializations are often dependent on the current problem structure
(see for instance [4]).

© Springer International Publishing Switzerland 2016
Y. Tan et al. (Eds.): ICSI 2016, Part I, LNCS 9712, pp. 322–330, 2016.
DOI: 10.1007/978-3-319-41000-5_32

Finally, we consider the deterministic version of PSO, namely DPSO. Earlier studies (see for instance [7–9]) show that coupling proper control parameters with an efficient particles initialization may give strong synergies and improve the algorithm effectiveness. On this guideline, here we want to propose a novel particles initialization, for DPSO, showing the following two features:

(i) we can scatter particles trajectories, at least in the early iterations, in order to better explore the search space
(ii) the resulting initialization is *dense*, i.e. a large portion of the entries of particles position and velocity is nonzero. On some problems (i.e. the portfolio selection problem we report) the latter event partially allows particles trajectory to avoid too sparse approximate solutions.

2   Preliminaries on DPSO Iteration

Let us consider the following PSO iteration

$$\begin{cases} v_j^{k+1} = \chi \left[w^k v_j^k + c_j^k r_j^k (p_j^k - x_j^k) + c_g^k r_g^k (p_g^k - x_j^k) \right], & k \geq 0, \\ x_j^{k+1} = x_j^k + v_j^{k+1}, & k \geq 0. \end{cases} \quad (1)$$

Here $j = 1, ..., P$ represents the j-th particle of the swarm, $v_j^k \in \mathbb{R}^n$ and $x_j^k \in \mathbb{R}^n$ are respectively the vector of *velocity* and the vector of *position* of particle j at step k, whereas $\chi, w^k, c_j^k, r_j^k, c_g^k, r_g^k$ are suitable positive bounded coefficients. Finally, p_j^k is the best position outreached by particle j up to step k, and p_g^k represents the best particle position of the overall swarm, up to step k.

Assumption 1 (DPSO). *Let us be given the iteration (1) and the positive constant values $c, r, \bar{c}, \bar{r}, w$. Then, for any $k \geq 0$ and $j = 1, ..., P$, we assume that $c_j^k = c, r_j^k = r = 1, c_g^k = \bar{c}, r_g^k = \bar{r} = 1$ and $w^k = w$.*

As it was proved in [7], setting $a = \chi w$ and $\omega = \chi(cr + \bar{c}\bar{r})$, with $\omega \neq (1 \pm \sqrt{a})^2$, we can consider the distinct real eigenvalues λ_1 and λ_2 of matrix

$$\Phi(k) = \begin{pmatrix} aI & -\omega I \\ aI & (1-\omega)I \end{pmatrix}.$$

Then, we can set at step k of DPSO iteration the two nonzero parameters

$$\gamma_1(k) = \frac{\lambda_1^k(a - \lambda_2) - \lambda_2^k(a - \lambda_1)}{\lambda_1 - \lambda_2} \qquad \gamma_2(k) = \frac{\omega(\lambda_1^k - \lambda_2^k)}{\lambda_1 - \lambda_2}, \quad (2)$$

along with the $2n$ vectors

$$z_i(k) = \begin{pmatrix} \frac{\gamma_2(k)}{\gamma_1(k)} e_i \\ e_i \end{pmatrix} \in \mathbb{R}^{2n}, \qquad i = 1, \ldots, n, \quad (3)$$

$$z_{n+i}(k) = \begin{pmatrix} -\frac{\gamma_1(k)}{\gamma_2(k)} e_i \\ e_i \end{pmatrix} \in \mathbb{R}^{2n}, \qquad i = 1, \ldots, n, \quad (4)$$

where $e_i \in \mathbb{R}^n$ is the i-th unit vector of $\mathbb{R}^n$. When $k = 0$ the vectors defined in (3) and (4) can be used to fruitfully set the initial particles position and velocity (respectively of the i-th and $(n+i)$-th particle), according with the initializations

$$\begin{pmatrix} v_i^0 \\ x_i^0 \end{pmatrix} = \rho_i^1 z_i(0), \qquad \rho_i^1 \in \mathbb{R} \setminus \{0\}, \ i = 1, \dots, n \tag{5}$$

$$\begin{pmatrix} v_{n+i}^0 \\ x_{n+i}^0 \end{pmatrix} = \rho_i^2 z_{n+i}(0), \qquad \rho_i^2 \in \mathbb{R} \setminus \{0\}, \ i = 1, \dots, n. \tag{6}$$

The latter initializations reveal interesting properties, and proved to be effective on several practical problems. As shown in [7], in case Assumption 1 holds (which also implies that a DPSO iteration is considered), then

1. as long as $P \leq 2n$, setting the sequences $\{v_i^0\}$ and $\{x_i^0\}$ as in (5) and (6), guarantees that the particles trajectories are nearly orthogonal at step 0 (see [7] for details), and the latter property is in practice likely maintained also in the subsequent early iterations;
2. in case $P > 2n$ (i.e. more than $2n$ particles are considered), the initializations (5) and (6) can be adopted for the first $2n$ particles; then, for the remaining $(P - 2n)$ particles an arbitrary initialization can be chosen by the user.

3   A Possible Drawback of the Choice (5) and (6)

Here we detail why, on some real problems, the setting (5) and (6) for the initial DPSO population might be still inadequate. As a preliminary consideration, observe that the proposal (5) and (6) matches item (i) of Sect. 1, but possibly it does not match also (ii). In the following we report a numerical experience on real problems, in order to highlight the latter fact.

We experienced (5) and (6) for the solution of a tough reformulation of a nondifferentiable constrained portfolio selection problem, proposed in [11]. This model uses a coherent risk measure, based on the combination of lower and upper moments of different orders of the portfolio return distribution. Such a measure can manage non-Gaussian distributions of asset returns, to reflect different investors' risk attitudes. The model includes also *cardinality constraints* (minimum and maximum number of assets to trade), along with constraints on the minimum and the maximum capital percentage to invest in each asset. The model uses the following parameters:

- N, number of possible assets;
- r_e, minimum expected return of the portfolio;
- K_d and K_u, minimum and maximum number of assets to trade;
- d and u, minimum and maximum budget (percentage) to invest in each asset;
- r_i, random variable indicating the return of the i-th asset, for $i = 1, \dots, N$;

- p, index of the norm used in the risk measure of the portfolio, with $p \geq 1$;
- a, parameter of the risk measure, with $0 \leq a \leq 1$.

Moreover, the variables in our model are described as follows:

- x_i, percentage of the portfolio invested in the i-th asset, for $i = 1, \ldots, N$;

- $z_i = \begin{cases} 1 \text{ if the } i-th \text{ asset is included in the portfolio, for } i = 1, \ldots, N \\ 0 \text{ otherwise;} \end{cases}$

- r, portfolio return.

In addition, $E[y]$ indicates the expected value of the random argument y, while y^- indicates $\max\{0, -y\}$ and y^+ indicates $(-y)^-$. Finally, we use the symbol $\widehat{r}_i$ for $E[r_i]$. Given the above notation, the expected portfolio return $E[r]$ is equal to $E[r] = \sum_{i=1}^{N} \widehat{r}_i x_i$ so that the overall portfolio selection problem is as follows:

$$\min_{x,z} \quad \rho_{a,p}(r) = a\|(r - E[r])^+\|_1 + (1-a)\|(r - E[r])^-\|_p - E[r] \tag{7}$$

$$\text{s.t.} \quad E[r] \geq r_e \tag{8}$$

$$\sum_{i=1}^{N} x_i = 1 \tag{9}$$

$$K_d \leq \sum_{i=1}^{N} z_i \leq K_u \tag{10}$$

$$z_i d \leq x_i \leq z_i u, \qquad i = 1, \ldots, N \tag{11}$$

$$z_i \in \{0, 1\}, \qquad i = 1, \ldots, N. \tag{12}$$

The function $\rho_{a,p}(r)$ in (7) is a *coherent* risk measure (i.e. it satisfies some formal properties which are appealing for investors).

In order to solve the constrained nonsmooth mixed integer problem (7) to (12) by DPSO, we considered the unconstrained penalty function reformulation

$$\min_{x \in \mathbb{R}^N, z \in \mathbb{R}^N} \quad P(x, z; \varepsilon), \tag{13}$$

$$P(x, z; \varepsilon) = \rho_{a,p}(r) + \frac{1}{\varepsilon_0}\left[\varepsilon_1 \max\left\{0, r_e - \sum_{i=1}^{N} \widehat{r}_i x_i\right\} + \varepsilon_2 \left|\sum_{i=1}^{N} x_i - 1\right|\right.$$

$$+ \varepsilon_3 \max\left\{0, K_d - \sum_{i=1}^{N} z_i\right\} + \varepsilon_4 \max\left\{0, \sum_{i=1}^{N} z_i - K_u\right\} + \varepsilon_5 \sum_{i=1}^{N} \max\{0, z_i d - x_i\}$$

$$+ \varepsilon_6 \sum_{i=1}^{N} \max\{0, x_i - z_i u\} + \varepsilon_7 \sum_{i=1}^{N} |z_i(1 - z_i)|\right] \tag{14}$$

and $\varepsilon = (\varepsilon_0, \varepsilon_1, \ldots, \varepsilon_7)^T > 0$ is a suitable set of penalty parameters, which are adaptively chosen according with the literature.

Note that the portfolio selection problem (13) is NP-hard and its formulation is nonconvex, nondifferentiable, and mixed-integer, so that DPSO was *specifically adopted to provide fast approximate solutions* on several scenarios.

This is indeed a typical application where tradesmen often claim for a quick solution of different scenarios, rather than an accurate solution to propose to their customers. In particular, for a fair comparison the initialization (5) and (6) was tested vs. a standard random initialization of DPSO, with really effective results in terms of fast minimization of the fitness function $P(x, z; \varepsilon)$ (as Fig. 1 shows).

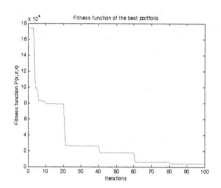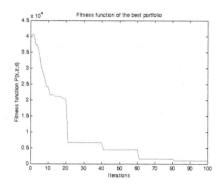

Fig. 1. Fitness function $P(x, z; \varepsilon)$ in (13) vs. the number of iterations: (*left*) when a *random DPSO initialization* is chosen; (*right*) when the *DPSO initialization* (5) and (6) is adopted.

However, we also observed that the initialization (5) and (6) tends to provide sparse solutions, which reduce diversification and might be therefore of scarce interest for several investors. To better appreciate the latter drawback, we monitored the sparsity of the approximate solution provided by the choice (5) and (6) in DPSO, after a relatively small number of iterations (simulating the time required by a tradesman before yielding a possible scenario to investors). After 100 iterations of DPSO, the approximate solution computed by DPSO with (5) and (6) gave a portfolio including just 2 titles (over 32), which is often too restrictive for many investors.

This was a consequence of the corresponding *sparsity* (i.e. a few nonzero entries) of the initialization (5) and (6). On the other hand, though a standard random initialization of particles' position/velocity had a worse performance in terms of minimization for $P(x, z; \varepsilon)$ (see again Fig. 1), nevertheless it yielded a final portfolio with several titles, allowing investors to spread the risk of investment.

4   Dense Modification of the Choice (5) and (6)

Starting from the conclusions of Sect. 3, here we want to propose a modification of PSO initialization (5) and (6), in order to possibly pursue a *dense* final solution.

On this purpose, let be given the $2n$ vectors $z_i(k)$, $i = 1, \ldots, 2n$, in (3) and (4). After some computation, it can be proved that they coincide with the orthogonal eigenvectors of the symmetric matrix $\mathcal{A} \in \mathbb{R}^{2n \times 2n}$, being

$$\mathcal{A} = \begin{pmatrix} \sigma_1 I & \sigma_2 I \\ \sigma_2 I & \sigma_3 I \end{pmatrix}$$

and (see also (2)) $\sigma_1 = [\gamma_1(k)]^2$, $\sigma_2 = -\gamma_1(k)\gamma_2(k)$, $\sigma_3 = [\gamma_2(k)]^2$. In particular, the (singular) matrix $\mathcal{A}$ has just the two eigenvalues $\mu_1 = 0$ and $\mu_2 = [\gamma_1(k)]^2 + [\gamma_2(k)]^2 > 0$, such that respectively

$$\begin{cases} \mathcal{A}z_i(k) = \mu_1 z_i(k), \quad i = 1, \ldots, n, \\ \mathcal{A}z_i(k) = \mu_2 z_i(k), \quad i = n+1, \ldots, 2n. \end{cases} \tag{15}$$

Now, following the motivations suggested in the end of Sect. 3, we can consider the set of (*dense*) vectors $\nu_1(k), \ldots, \nu_{2n}(k)$ as in

$$\nu_i(k) = z_i(k) - \alpha \sum_{j=1,\ j\neq i}^{n} z_j(k) - \gamma \sum_{j=n+1}^{2n} z_j(k), \qquad i = 1, \ldots, n \tag{16}$$

$$\nu_t(k) = z_t(k) - \beta \sum_{j=n+1,\ j\neq t}^{2n} z_j(k) - \delta \sum_{j=1}^{n} z_j(k), \qquad t = n+1, \ldots, 2n, \tag{17}$$

which are obtained by linearly combining the eigenvectors $\{z_i(k)\}$ in (15). We want to compute the real parameters α, β, γ and δ such that

$$\begin{cases} \nu_j(k)^T \mathcal{A}\nu_i(k) = 0, \quad \text{for any } 1 \leq j \neq i \leq n \\ \nu_t^T(k)\mathcal{A}\nu_s(k) = 0, \quad \text{for any } n+1 \leq t \neq s \leq 2n. \end{cases} \tag{18}$$

It can be proved after a tedious computation that conditions (18) allow to satisfy both the properties (i) and (ii) of Sect. 1. Moreover, taking the values

$$\alpha \in \mathbb{R} \setminus \{-1, 1/n\}, \qquad \beta = \frac{2}{n-2}, \qquad \gamma = 0, \qquad \delta \in \mathbb{R} \setminus \{0, 1\}, \tag{19}$$

then the vectors $\nu_1(k), \ldots, \nu_{2n}(k)$ in (16) and (17) both satisfy (18) and are *dense* in comparison with the original vectors $z_i(k)$, $i = 1, \ldots, 2n$ (which satisfy a relation similar to (18) but are not dense).

Now, observe that $z_1(k), \ldots, z_{2n}(k)$ are mutually orthogonal in the extended space $\mathbb{R}^{2n}$, while the latter property in general does not hold also for the vectors $\nu_1(k), \ldots, \nu_{2n}(k)$. Nevertheless, we can prove in the next proposition that $\nu_1(k), \ldots, \nu_{2n}(k)$ are still enough *scattered* in $\mathbb{R}^{2n}$, possibly being uniformly *linearly independent*. The latter fact is of great relevance and is not immediately evident, since the singular matrix $\mathcal{A}$ is only positive semidefinite.

Proposition 1. *Given the vectors* $\nu_1(k), \ldots, \nu_{2n}(k)$ *in (16), (17) and (19), then*

$$\det[\nu_1(k) \vdots \cdots \vdots \nu_{2n}(k)] = \frac{(3n-2)n^n}{(n-2)^{n+1}} \left[(1+\alpha)^n (1-n\alpha) \right] \cdot \det[z_1(k) \vdots \cdots \vdots z_{2n}(k)].$$

As long as $\alpha \notin \{-1, 1/n\}$, then $\det[\nu_1(k) \vdots \cdots \vdots \nu_{2n}(k)] \neq 0$ and the vectors $\nu_1(k), \ldots, \nu_{2n}(k)$ are possibly uniformly linearly independent. Observe that the last limitation on α is definitely unrestrictive, considering that our proposal is expected to be effective when the size n of the problem increases.

As regards the results on the portfolio selection problem analyzed in Sect. 3, we compared the deterministic initialization (5) and (6), with the one obtained replacing $z_1(k), \ldots, z_{2n}(k)$ by $\nu_1(k), \ldots, \nu_{2n}(k)$ (and using the parameters $\alpha = 0.75$, $\beta = 2/(n-2)$, $\gamma = 0$, $\delta = 0.25$). We obtained similar results (see Fig. 2) in terms of decrease of the fitness function $P(x, z; \varepsilon)$ in (13). Moreover, adopting

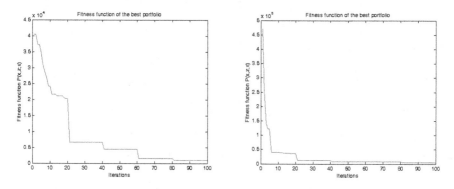

Fig. 2. Fitness function $P(x, z; \varepsilon)$ in (13) vs. the number of iterations: (*left*) when the *DPSO initialization* (5) and (6) is adopted; (*right*) replacing $z_1(k), \ldots, z_{2n}(k)$ by $\nu_1(k), \ldots, \nu_{2n}(k)$ (setting the parameters $\alpha = 0.75$, $\beta = 2/(n-2)$, $\gamma = 0$ and $\delta = 0.25$).

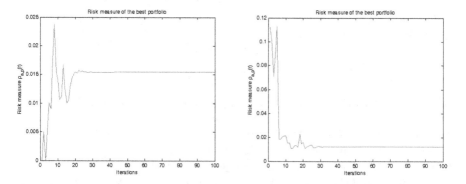

Fig. 3. Risk measure $\rho_{a,p}(r)$ vs. the number of iterations: (*left*) when the initialization (5) and (6) is adopted in DPSO; (*right*) replacing $z_1(k), \ldots, z_{2n}(k)$ by $\nu_1(k), \ldots, \nu_{2n}(k)$ (setting the parameters $\alpha = 0.75$, $\beta = 2/(n-2)$, $\gamma = 0$ and $\delta = 0.25$).

$\nu_1(k), \ldots, \nu_{2n}(k)$, both we improved also the minimization of the risk measure $\rho_{a,p}(r)$ (see Fig. 3), and we drastically improved the density of the final approximate solution, as expected.

5   Conclusions

In this paper we have described a novel *dense* initialization for DPSO, which is based on a reformulation of PSO iteration as a dynamic linear system. Our proposal was tested on tough portfolio selection problems, confirming its effectiveness on large scale problems.

Acknowledgments. The research is partially supported by the Italian Flagship Project RITMARE, coordinated by the Italian National Research Council and funded by the Italian Ministry of Education, University and Research. Matteo Diez is grateful to Dr Woei-Min Lin and Dr Ki-Han Kim of the US Navy Office of Naval Research, for their support through NICOP grant N62909-15-1-2016.

References

1. Kennedy, J., Eberhart, R.C.: Particle swarm optimization. In: Proceedings of the 1995 IEEE International Conference on Neural Networks, Perth, Australia, IV, pp. 1942–1948. IEEE Service Center, Piscataway, NJ (1995)
2. Clerc, M., Kennedy, J.: The particle swarm - explosion, stability, and convergence in a multidimensional complex space. IEEE Trans. Evol. Comput. **6**(1), 58–73 (2002)
3. Ozcan, E., Mohan, C.K.: Particle swarm optimization: surfing the waves. In: Proceedings of the 1999 IEEE Congress on Evolutionary Computation, pp. 1939–1944. IEEE Service Center, Piscataway (1999)
4. Poli, R.: The Sampling Distribution of Particle Swarm Optimisers and their Stability. Technical report CSM-465, University of Essex (2007)
5. Trelea, I.C.: The particle swarm optimization algorithm: convergence analysis and parameter selection. Inf. Process. Lett. **85**, 317–325 (2003)
6. van den Berg, F., Engelbrecht, F.: A study of particle swarm optimization particle trajectories. Inf. Sci. **176**, 937–971 (2006)
7. Campana, E.F., Diez, M., Fasano, G., Peri, D.: Initial particles position for PSO, in bound constrained optimization. In: Tan, Y., Shi, Y., Mo, H. (eds.) Advances in Swarm Intelligence. LNCS, vol. 7928, pp. 112–119. Springer, Heidelberg (2013)
8. Diez, M., Serani, A., Leotardi, C., Campana, E.F., Peri, D., Iemma, U., Fasano, G., Giove, S.: A proposal of PSO particles' initialization for costly unconstrained optimization problems: ORTHOinit. In: Tan, Y., Shi, Y., Coello Coello, C.A. (eds.) Advances in Swarm Intelligence. LNCS, vol. 8794, pp. 126–133. Springer, Switzerland (2014)
9. Campana, E.F., Fasano, G., Pinto, A.: Dynamic analysis for the selection of parameters and initial population, in particle swarm optimization. J. Global Optim. **48**, 347–397 (2010)
10. Zheng, Y., Ma, L., Zhang, L., Qian, J.: On the convergence analysis and parameter selection in particle swarm optimization. In: Proceedings of the International Conference on Machine Learning and Cybernetics, pp. 1802–1807 (2003)

11. Corazza, M., Fasano, G., Gusso, R.: Particle Swarm Optimization with non-smooth penalty reformulation, for a complex portfolio selection problem. Appl. Math. Comput. **224**, 611–624 (2013)

An Improved Particle Swarm Optimization Algorithm Based on Immune System

Xiao Zhang[1], Hong Fan[1(✉)], Huiyu Li[2], and Xiaohu Dang[3]

[1] School of Computer Science, Shaanxi Normal University,
Xi'an 710119, China
zhangxiaosnnu@163.com,
fanhong@snnu.edu.cn
[2] School of Information and Technology,
Northwest University,
Xi'an 710127, China
[3] School of Geology and Environment,
Xi'an University of Science and Technology,
Xi'an 710054, China

Abstract. An improved immune particle swarm optimization is proposed to solve drawbacks of low convergence rate and local optimization. The improved algorithm is to increase the diversity of the population of particles of PSO and to expand the search space of solutions by introducing the immune algorithm's antibody concentration regulation mechanisms and immune selection operation. It is to improve the convergence rate and precision of PSO by using immune memory and immune vaccine. According to the characteristics of specific problem, the improved algorithm defines a new selection operator of antibody concentration and immune vaccines to improve the optimizing ability of the algorithm. The proposed algorithm is applied to TSP problem and results show that the improved algorithm has an effective global optimization and higher convergence rate.

Keywords: Particle swarm optimization · Concentration regulation · Immune memory · Immune vaccine · TSP

1 Introduction

Traveling Salesman Problem (TSP) is also called the shortest path problem. It may be stated as follows: "A salesman is required to visit each of the n given cities once and only once, starting from any city and returning to the original place of departure. What route, or tour, should be chose in order to minimize the total distance traveled?" The TSP is NP-hard and often used as a criterion to measure various optimization methods. Therefore, the method that can optimize TSP will be evaluated and concerned highly. In recent years, many intelligent optimization algorithms have been proposed, such as tabu search algorithm, simulated annealing algorithm, ant colony algorithm, particle swarm optimization and immune algorithm, etc.

© Springer International Publishing Switzerland 2016
Y. Tan et al. (Eds.): ICSI 2016, Part I, LNCS 9712, pp. 331–340, 2016.
DOI: 10.1007/978-3-319-41000-5_33

Particle Swarm Optimization (PSO) is proposed by Dr. Eberhart and Dr. Kennedy. It shares the individual information among groups and guides the migration of groups towards the optimal solution as the feeding behavior of birds [1, 2]. PSO has been applied widely to all areas because of its simple principle, easy to understand, few parameters and easy to implement. The algorithm is also revealed gradually some common drawbacks, such as slow convergence rate, low diversity keeping ability and easy to be trapped in local optimum. Some algorithms have been put forward to improve PSO's abilities [3–5]. Although these algorithms improve the searching ability of PSO to a certain extent, most are based on single group structure strategy or the parameter. So it is difficult to balance the convergence rate and jump out of the local optimum.

Inspired by the processing mechanism about immune information in the biological immune system, some scholars have proposed hybrid immune particle swarm optimization [6–8]. Most of them introduced the cross, mutation and clone and other operations to improve the performance of particle swarm optimization. But these operations will increase the randomness and complexity of the algorithm in some extent. In this paper, Immune Algorithm (IA) and particle swarm algorithm are combined in a new way. Some characteristics and Principle in the immune algorithm are introduced into the PSO which constructed a new algorithm called IPSO. The new algorithm improves the diversity of particle populations and extends the searching space of solutions by introducing concentration regulation mechanism and immune selection to the PSO, and improves the convergence rate of PSO algorithm by adding Immune memory cells and immune vaccine to. Thus, the IPSO can avoid to be trapped in local optimum effectively and improve the convergence performance of the algorithm. Simulation experiment results show that ISPO algorithm has significantly improved in convergence rate, accuracy and global search capability.

2 Particle Swarm Optimization and Immune Algorithm

2.1 Particle Swarm Optimization

Each particle in the population search space is corresponding to a solution of the optimization problem when PSO solves optimization problems. It ignores the influence of the particle's quality and volume on the algorithm. The information of particle i can be expressed by a D-dimensional vector. Position is expressed as $x_i = (x_{i1}, x_{i2} ... x_{iD})$ and speed is expressed as $v_i = (v_{i1}, v_{i2} ... v_{iD})$. The particle updates itself by tracking two extremum in each iterative process. One is the individual optimal solution which finds by itself, called individual extreme points, denoted as P_{best}. Another is the optimal value of the particle population, called global optimal point, denoted as G_{best}. Then the particle's information about location and speed is updated according to the following formula after P_{best} and G_{best} have been found.

$$v_i^{k+1} = w v_i^k + c_1 r_1 \left(P_{best}^k - x_i^k \right) + c_2 r_2 \left(G_{best}^k - x_i^k \right). \tag{1}$$

$$x_i^{k+1} = x_i^k + v_i^{k+1} \,. \tag{2}$$

Where w is the inertia weight factor, r_1 and r_2 are random numbers between [0, 1], c_1 and c_2 are learning factor which the values are non-negative constants. They are used to adjust the function in the optimization process between particle's own experience and social groups' experience. In k-th iteration v_i^k is the speed, x_i^k is the position, P_{best}^k is the individual optimal solution and G_{best}^k is global optimal solution.

2.2 Artificial Immune Algorithm

Artificial immune algorithm (AIA) is a new intelligent optimization algorithm based on immune evolution mechanism and information processing mechanism of biological immune system [9]. It simulates the process of antigen recognition and antibody proliferation of biological immune system, which can promote the high affinity antibody, inhibit the high concentration antibody and can also maintain the diversity of antibody. In the artificial immune algorithm, the antigen can be regarded as the objective function and the constraint condition of the practical problem to be solved. The antibody is a feasible solution to the problem to be solved. The level of affinity between antigen and antibody can be regarded as the degree of match the objective function and the feasible solution. The size of affinity in antibodies indicates the similarity between the feasible solutions. At the same time, the optimal antibody is kept in memory cell unit. The optimal solution can be quickly adapted to the problem when similar problems appear again. AIA is always preferred to select high affinity and low concentration of the antibody to enter a new generation of antibody population in order to maintain the population diversity. This method can effectively improve the algorithm search ability and accelerate the convergence rate [10].

3 Immune Particle Swarm Optimization

3.1 Algorithm Principle

The immune memory, immune regulation and immune vaccine are introduced into the particle swarm algorithm in order to improve the global search ability and convergence rate PSO. The approach to the implementation of the proposed method is: the optimal solution is regarded as antigen, the particle is regarded as antibody, the magnitude of affinity between antibody and antigen means the matching degree between objective function and the optimal solution. Based on concentration regulation mechanism of antibody, those with high affinity and low concentration of particle will be selected and are promoted. On the contrary, those with low affinity and high concentrations of particle will be inhibited. This strategy can maintain the diversity of population and increase the algorithm's global search capability, and improve the convergence rate and convergence precision of the algorithm by using the memory cells and immune vaccine inoculation operations. Where the immune memory cells are the antibody which affinity is bigger, the immune vaccines are some excellent genes in antibody. For TSP problem, the vaccine can be a local distance between adjacent cities of the shortest path.

3.2   Related Definition

Affinity. Affinity is used to measure the matching degree between antigen and antibody. It shows the closeness between this iterative solution and the optimal solution. It can be defined as follows:

$$affinity(i) = \frac{d_{\max} - d_i}{\sum\limits_{i=1}^{n} (d_{\max} - d_i) + \varepsilon} . \tag{3}$$

Where d_i is the path value of antibody i in this iteration process, d_{max} is the maximum value of the path length in this iteration, ε as a non-zero coefficient that to ensure the formula (3) is meaningful. Formula (3) shows that the less d_i is, the higher antibody's affinity is, thus, antibody and antigen are more suitable.

Concentration Regulation Mechanism. The level of concentration is the symbol of the degree of population diversity. In the process of population search, it is hard to guarantee the diversity of antibody when particle concentration is too high. The traditional description method of antibody concentration is based on the distance and it is expressed by the level of difference in the affinity between the various antibodies. The defect of this method is that some antibodies with great potential of development have no chance to be selected. Thus it reduces the diversity of the population. This paper uses a new fragments interval method to describe the concentration of antibody. It can be described as follows: Assuming a total of n antibodies, the affinity of each antibody is $affinity_1$, $affinity_2$, ..., $affinity_n$. After sorting them according to the fitness value, the interval is [min-affinity,max-affinity]. Divide them into k sub ranges, recorded as L_1, L_2, ...,. Then find out the number of each sub region of the antibody, denoted as $sum\ (m)$, m for the interval number. The antibody concentration of each interval is expressed as ($sum\ (m)/k$), denoted as $D\ (L_k)$. Therefore, the concentration of each antibody can be defined as the $D\ (L_k)$ if the antibody is in the interval L_k, recorded as $individual_i$.

$$individual_i = D(L_k) = \frac{sum(m)}{k} . m = 1, 2, \cdots, k; \ i = 1, 2, \cdots, n. \tag{4}$$

It can be seen from the formula (4) that the concentration of the antibody on each subinterval is consistent, which makes that some temporary concentration is very low but that antibodies with great development potential have the chance to be selected and to evolution. Thus, the diversity of the population can be maintained better.

Selection Probability. In order to increase the diversity of population in the process of evolution, we always hope to promote those antibodies which have the high affinity and low concentration and inhibit those antibodies which have the low affinity and high concentration. So this paper defines the selection probability P_d based on antibody concentration and the selection probability P_a based on the affinity of antibody. They are described as follows:

$$P_d(i) = 1 - \frac{individual_i}{\sum\limits_{i}^{n} individual_i} . \quad i = 1, 2, \cdots, n .$$ (5)

$$P_a(i) = \frac{fitnesss(i)}{\sum\limits_{i}^{n} fitnesss(i)} . i = 1, 2, \cdots, n .$$ (6)

Therefore, the antibody's comprehensive selection probability P can be defined as follows:

$$P(i) = \alpha P_a(i) + (1 - \alpha)P_d(i) . \alpha > 0, \ P_a < 1, \ P_d < 1, \ i = 1, 2, \cdots n .$$ (7)

Where α is the coordination coefficient, which is used to coordinate the weights of P_d and P_a. It can be seen from formula (7) that the chance to be selected of antibody which has more concentration and less affinity is less. On the contrary, the antibody which has less concentration and more affinity is more likely to be selected to evolve. This can not only improve the affinity of the antibody, but also ensure the diversity of the population.

Immune Vaccine. Generally, the vaccine is some excellent genes on the individual or a priori knowledge or characteristic information about a particular type of problem. The correct selection of the vaccine will greatly promote the efficiency of the algorithm. For the TSP problem, the optimal solution is bound to have a large probability of including the shortest distance between adjacent cities of path. This is a natural attribute of TSP itself [11], and it can be regarded as a kind of characteristic information of TSP. Therefore, from this perspective, this paper extracts the characteristic information of the TSP problem to construct a vaccine. Specific operation as follows: the two best known antibodies are used to carry out the cross operation and the resulting public subset that is an immune vaccine. Zou Peng pointed out by experiment, in the TSP, the local optimal solution in about 80 % of the edge is an edge in the global optimal solution [12]. Therefore, the efficiency of the algorithm can be greatly accelerated by introducing the immune vaccine.

Immune Vaccination. Immune vaccination operation, that according to the characters of the problem and a certain proportion, using immune vaccine update low affinity antibodies to improve the fitness of antibody population and accelerate the convergence to the global optimal. The antibody A vaccination operation can be described as follows:

```
Vaccination (A)
  { A'= A;
    Randomly select a vaccine V from the vaccine
    library;
    C is the first city in vaccine V, C'is the last
    city in vaccine V;
    i=1,j=i+|V|-1;//(|V| represents the sequence length
    of vaccine V )
    while (i<|A|-|V|+1)
    {if (A(i) ==C && A(j) ==C' && |A(i:j)| == |V|)
        Insert V before A(i), get middle excessive
        antibody A^T; Remove the city number in the
        vaccine A^T which is the same as the number in
        vaccine V , then get the new antibody A^{new};
     i=i+1,j=i+|V|-1;
    }
    if (fitness(A^{new}) > fitness(A'))
      then A= A';
  }
```

3.3   Algorithm Step

The main process of immune particle swarm optimization is as follows:

Step 1.   Initialization related parameters

Step 2.   Generating the initial particle (antibody) population and getting the initial position and speed of the particles

Step 3.   Calculate the affinity of antibodies according to formula (3) and updating P_{best} and G_{best} according to the affinity, output results and stop when the given terminating condition is satisfied. Otherwise continue

Step 4.   Generate immune memory cells: according to the magnitude of affinity, select a part of the affinity of the antibody and put them in the memory library as immune memory cells

Step 5.   Generate vaccines: select two highest affinity antibodies for cross operation and put the common subset in the vaccine library as an immune vaccine

Step 6.   Update the position and speed of particles according to formula (1), (2) and (8) and get n new particles. Then select q antibodies randomly from the memory cell units and form a new group A with $n + q$ antibodies

Step 7.   Promotion and inhibition of antibody: Select n antibodies from the antibody group A according to formula (4) to (7) to form a new antibody group B

Step 8.   According to the characteristics of the problem and a certain proportion, using immune vaccine carry on immunity inoculation to the lower affinity antibody

Step 9.   Immune selection: calculate the fitness value of vaccination particles. Give up the inoculation operation if the fitness value is worse than the values before inoculation; otherwise accept the vaccination operation

Step 10.   Go to Step 3, recalculate the affinity of new antibody and judge the next step

4   The Experimental Simulation

4.1   Experiment Parameter Initialization

For the TSP problem, the optimal solution is bound to have a large probability of including the shortest distance between adjacent cities of path [11]. Therefore, based on these research experiences, the IPSO uses roulette selection operation to initialize the path and thus generated initial solution tends to favor the optimal solution problem. It can save search time and improve the convergence rate.

In the parameter setting of the algorithm, for the inertia weight ω, the researchers have proved that the convergence performance of the algorithm can greatly improve with parameters ω linear decreasing. Therefore, this paper uses the following formula to adjust the weight factor ω:

$$\omega = \omega_{max} - \frac{\omega_{max} - \omega_{min}}{iter_{max}} * iter . \tag{8}$$

Where $iter$ is the current iteration number, $iter_{max}$ is the maximum number of iterations. According to experts' experience, the $\omega_{max}, \omega_{min}$ are generally valued at 0.9 and 0.4. Parameter settings are as follows: $C_1 = C_2 = 2$, $n = 50$, the maximum number of iterations is set to 1000.

4.2   Experimental Results and Analysis

This paper adopts the classic examples of three different types of TSPLIB test library to test the optimal performance of this algorithm. Three different types of classic examples of problems are Burma14, Oliver30 and Eil51. The experimental results compare with reference [13] which is A Self-Adaptive Discrete Particle Swarm Optimization Algorithm (SADPSO). Table 1 shows the results. The data in the table are the results that each algorithm runs 20 times independently in each data set. 'Values of TSPLIB path' in Table 1 show the optimal value of TSPLIB test library. 'Known optimal solution' show the problem has obtained the optimal value at present. And '—'means the paper does not provide this data.

Table 1. Result comparison of two algorithms for solving TSP problem.

TSP	Algorithm	The optimal value	The average value	Known optimal solution (Values of TSPLIB path)
Burma14	IPSO	30.8785	30.8785	30.8785(30.8785)
	SADPSO	30.8785	30.8785	
Oliver30	IPSO	423.7406	423.7406	423.7406(—)
	SADPSO	423.7410	424.8267	
Eil51	IPSO	430.6013	434.0539	426(429.9833)
	SADPSO	436.7730	440.7810	

It can be obtained from Table 1: For solve different types of TSP, the optimization performance of IPSO is significantly better than that of SADPSO. For the Burma14 problem, the IPSO and SADPSO have the same solution performance and every time they can also converge to the optimal solution 30.8785. For the Oliver30 problem, every time IPSO can converge to the optimal solution 423.7406. And the optimal value and the average value is 0.0004 and 1.0861 respectively less than that of the SADPSO. For the Eil51 problem, the performance of IPSO is significantly stronger than SADPSO. Although the two algorithms are not seeking the path optimal value, the optimal value and the average value obtained by the IPSO algorithm are 6.1717 and 6.7271 respectively less than that of the SADPSO. Furthermore, the average value of the IPSO is 2.7191 even less than the optimal values of SADPSO. In addition, it can be seen from the optimal solution and the average values that the IPSO has a better stability. Therefore, by contrast, IPSO has good performance and it is a feasible and effective optimization algorithm.

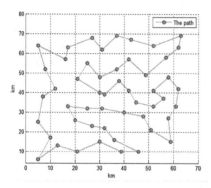

Fig. 1. Simulation result of instance Eil51.

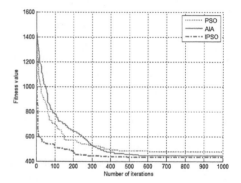

Fig. 2. Convergence process of instance Eil51

Figure 1 shows the optimal path diagram that the IPSO to solve the Eil51 problem. The paper put the IPSO with the particle swarm optimization (PSO) and artificial immune algorithm (AIA) to do a comparison in order to show the advantages of the IPSO more intuitively. Figure 2 shows the convergence curves of the three algorithms for solving the Eil51 problem. It can be seen from Fig. 2 that the optimization performance of the IPSO has been significantly improved. Compared with AIA, the IPSO has faster convergence rate and convergence to the optimal solution at the time of the 200th generation. However, due to the crossover and mutation operation, AIA has a certain degree of randomness, increased uncertainty, resulting in a slower convergence rate until the 500th generations to converge to the optimal solution. For the PSO, the IPSO has a strong ability of diversity and can quickly converge to the global optimal solution. However, PSO in the late stage of convergence, the difference between the individual particles is reduced and tends to be the same. As a result, the diversity of the whole population is reduced and lead to the PSO falls into the local optimum. Therefore, compared to the other two algorithms, IPSO has a significant improvement in convergence rate and accuracy.

5 Conclusions

This paper proposes a novel algorithm to solve the TSP problem. New algorithm combines with the advantages of two kinds of intelligent optimization algorithms. The new algorithm has the following characteristics: (1) It defines a new selection operator of antibody concentration and immune vaccines and improves the rate of convergence effectively. (2) To improve the ability that can maintain the diversity of the algorithm by taking the concentration of self-regulation mechanism of antibody. The high affinity and low concentration of antibodies are promoted and the low affinity and high concentration of antibodies are inhibited. Thus, it avoided successfully to be trapped in local optimum. (3) It improves greatly the global optimization ability and the optimization accuracy by taking immunization vaccination, immune selection operation and the linear decreasing strategy of inertia weight factor. The simulation results show that the improved algorithm has an effective global optimization and higher convergence rate.

Acknowledgments. This paper is supported by the National Natural Science Foundation of China (No. 41271518), and the Natural Science Foundation of Shaanxi Province of China (No. 2014JM2-6115).

References

1. Kennedy, J., Eberhart, R.: Particle swarm optimization. In: IEEE International Conference on Neural Networks, pp. 1942–1948. IEEE Press (1995)
2. Eberhart, R., Kennedy, J.: A new optimizer using particle swarm theory. In: Proceedings of the 6th International Symposium on Micro Machine and Human Science, pp. 39–43. IEEE Press (1995)
3. Li, J., Cheng, Y., Chen, K.: Chaotic particle swarm optimization algorithm based on adaptive inertia weight. In: The 26th Chinese Control and Decision Conference, pp. 1310–1315. IEEE Press (2014)
4. Pan, T.-S., Dao, T.-K., Nguyen, T.-T., Chu, S.-C.: Hybrid particle swarm optimization with bat algorithm. In: Sun, H., Yang, Chin-Yu., Lin, C.-W., Pan, J.-S., Snasel, V., Abraham, A. (eds.) Genetic and Evolutionary Computing. AISC, vol. 329, pp. 37–48. Springer, Heidelberg (2015)
5. Zhan, Z.H., Zhang, J., Li, Y., et al.: Adaptive particle swarm optimization. IEEE Trans. Syst. Man Cybern. B Cybern. **39**(6), 1362–1381 (2009). IEEE
6. Li, X., Xu, H., Cheng, Z.: One immune simplex particle swarm optimization and it's application. In: 2008. The 4th International Conference on Natural Computation, vol. 1, pp. 331–335. IEEE (2008)
7. Li, Z., Tan, G., Li, H.: Hybrid particle swarm optimization algorithm and its application. In: The 6th International Conference on Natural Computation, vol. 8, pp. 4045–4048. IEEE (2010)
8. Xie, M.: An improved hybrid particle swarm optimization algorithm for TSP. J. Taiyuan Univ. Technol. **44**(4), 506–509 (2013)
9. De Castro, L.N., Timmis, J.I.: Artificial immune systems as a novel soft computing paradigm. Soft. Comput. **7**(8), 526–544 (2003)

10. Ulutas, B.H., Kulturel-Konak, S.: An artificial immune system based algorithm to solve unequal area facility layout problem. Expert Syst. Appl. **39**(5), 5384–5395 (2012)
11. Yang, H., Kang, L.S., Chen, Y.P.: A gene-based genetic algorithm for TSP. Chin. J. Comput. **26**(12), 1753–1758 (2003)
12. Zou, P., Zhou, Z., Chen, G.L., Gu, J.: A multilevel reduction algorithm to TSP. J. Softw. **14**(1), 35–42 (2003)
13. Chang-Sheng, Z., Ji-Gui, S., Dantong, O.: A self-adaptive discrete particle swarm optimization algorithm. Acta Electr. Sin. **37**(2), 299–304 (2009)

The Impact of Population Structure on Particle Swarm Optimization: A Network Science Perspective

Wen-Bo Du[1,2], Wen Ying[1,2], and Gang Yan[3(✉)]

[1] School of Electronic and Information Engineering, Beihang University,
Beijing 100191, People's Republic of China
[2] Beijing Key Laboratory for Network-Based Cooperative
Air Traffic Management, Beijing 100191, People's Republic of China
[3] Center for Complex Network Research and Department of Physics,
Northeastern University, Boston, MA 02115, USA
eegyan@gmail.com

Abstract. Particle swarm optimization (PSO) is one of the most important swarm intelligence optimization algorithms due to its ease of implement and outstanding performance. As an information flow system, PSO is influenced by the population structure to a great extent. While previous works considered several classical structure, such as fully-connected and ring structures, here we systematically explore the impact of population structure, including scale-free and small-world networks that have been found in many real-world complex systems. In particular, we examine the influence of average degree, degree distribution and topological randomness of the networks underlying PSO. Our results are not only useful for developing more effective structures to improve the performance of PSO but also helpful in bridging the two fast-growing fields–network science and swarm intelligence.

Keywords: PSO · Swarm intelligence · Network science · Population structure

1 Introduction

PSO is a population-based optimization algorithm by imitating the animals' social behaviors, such as bird flocking and fish schooling [1]. Individuals spread information of their experiences and interact with each other, aiming to find the global best solution cooperatively. Owing to its simple concept and ease of implement, PSO has received much attention in recent years [2–4]. Several variants have been proposed to overcome the weaknesses of the original version of PSO, such as FIPSO, where each particle utilizes all its neighbor's information [5], CLPSO, in which particles are encouraged to learn different exemplars in different dimensions [6], SLPSO, in which each particle adaptively chooses one of four learning strategies on the basis of the different situations [7], and SIPSO, where particles adopt heterogeneous learning strategy based on the heterogeneous structure [8].

Among the studies of PSO, the influence of population structure is one of the most important points of view. In the canonical version of PSO, the fully-connected structure

© Springer International Publishing Switzerland 2016
Y. Tan et al. (Eds.): ICSI 2016, Part I, LNCS 9712, pp. 341–349, 2016.
DOI: 10.1007/978-3-319-41000-5_34

is applied and the algorithm may easily get trapped in local optima due to its full connectedness, where misinformation can easily spread to the whole population. Kennedy and Mendes first indicated the important role of population structure and proposed PSO with ring structure (LPSO) and some other topologies [9]. Ratnaweera *et al.* introduced HPSO-TVAC with self-organizing hierarchical structure and time-varying acceleration coefficients [10]. Liu *et al.* suggested SFPSO where the scale-free structure is adopted [11].

In this paper, we describe the population structure as a network, where the nodes are particles in PSO and the links represent the information flow and interactions between the particles. We consider several classical networks, scale-free and small-world networks that are widely-employed in network science community. Our results show that the average degree, degree distribution, and topological randomness affect the performance of PSO. Specifically, the average degree affects the early stage of the evolution of PSO, while the degree distribution takes effect soon afterwards.

2 Method

2.1 PSO Algorithm

In this paper, we adopt the canonical PSO framework [12]. In PSO, a swarm of N particles fly in the D-dimensional space and interact with each other to search for the global best solution of the D-dimensional objective function $f(x)$. Each particle i is described with a velocity $v_i = [v_i^1, v_i^2, \ldots, v_i^D]$, a position $x_i = [x_i^1, x_i^2, \ldots, x_i^D]$ and a fitness value $f_i = f(x_i)$ to evaluate the position according to the objective function.

In the beginning, each particle is randomly initialized in the search space, and then adjusts its position and velocity in every iteration, following the equations:

$$v_i^d = \chi \times \left[v_i^d + \frac{\varphi}{2} \times r_1 \times \left(p_i^d - x_i^d \right) + \frac{\varphi}{2} \times r_2 \times \left(p_g^d - x_i^d \right) \right]. \tag{1}$$

$$x_i^d = x_i^d + v_i^d. \tag{2}$$

where $p_i = [p_i^1, p_i^2, \ldots, p_i^D]$ and $p_g = [p_g^1, p_g^2, \ldots, p_g^D]$ denote the historical best position of particle i itself and the historical best position in all its neighbors respectively, r_1 and r_2 are two independently stochastic numbers in the interval $[0,1]$. $\varphi > 4$ and χ are the constriction coefficients as follows:

$$\chi = 2 / \left| 2 - \varphi \sqrt{\varphi^2 - 4\varphi} \right|. \tag{3}$$

Here $\varphi = 4.1$ and $\chi = 0.729$, according to common practices [5, 8, 12, 13]. When the process runs a certain number of iterations or finds a solution with an acceptable fitness value, the algorithm terminates.

2.2 Network Structure

In canonical PSO, each particle can learn from all the other particles that are considered as its neighbors. In other words, each particle connects to all the other particles and the population structure is a fully-connected network. The PSO with fully-connected network is denoted as G-PSO in the rest of this paper. Due to the direct connection between any pair of particles, the individual with the best experience will influence the whole population very quickly. If this experience is misleading, such as near a local optimum, the population will move towards this position collectively, yet adverse to explore other regions. In [9], PSO with ring structure (L-PSO) was introduced, where each particle connects only to the nearest n particles. Ring structure weakens the connectivity of population and increases the average distance between the particles, leading to a lower convergence. L-PSO outperforms G-PSO in solving complex multimodal problems with many local optima [9].

To systematically investigate the influence of structure on the performance of PSO, we use the topologies from network science, including small-world [14] and scale-free [15] networks, which can be generated with different topological properties by tuning a few parameters.

To generate a scale-free network, we employ the Barabási-Albert model [15]. As shown in Fig. 1, start from a fully-connected core with a small number m_0 of node, and at each step a new node is added and connected to m existing nodes. The probability that the new node connects to an existing node i is proportional to i's degree (the number of nodes that node i was connecting to). In other words, the new node is more likely to connect to a large-degree node. PSO with scale-free network is denoted as SF-PSO.

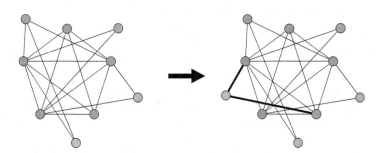

Fig. 1. The generation process of scale free network. The initial core contains 5 nodes (gray). The new node is bright red at the step it was added and turns to gray afterwards gradually. The new edges are marked as bold lines.

We will also apply a homogeneous small-world network (HSWN) in [16]. Starting from a ring structure, at each step two different edges are randomly selected and their end nodes are swapped, as shown in Fig. 2. The edge-swapping procedure keeps running until a certain proportion of edges have been swapped (the proportion is denoted as p). The nodes in HSWN have the same degree and HSWN has small-world

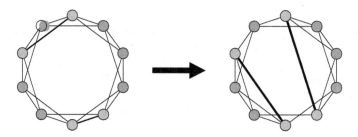

Fig. 2. The generation process of homogeneous small-world network. The initial ring contains 10 nodes. Two edges (bold) are selected and their end nodes (red) are swapped. (Color figure online)

effect (i.e. the average distance between nodes is small while the clustering coefficient is large) [16]. Hence we can adjust p to generate networks with different topological randomness, and sharing the same degree distribution. PSO that adopts HSWN is named HSW-PSO hereafter. Note that when $p = 0$, no edges are swapped and HSW-PSO degenerates to L-PSO.

3   Result

3.1   Functions and Simulation Conditions

Table 1 list the information of 4 widely used test functions we employ. Sphere is a typical unimodal function with a smooth landscape, which is easy for most optimization algorithm, and could investigate the ability of local search of algorithm in a way. Rosenbrock is unimodal, yet it has non-separable dimensions and a much complex landscape, hence it could be also treated as a multimodal function when dimension is high. Rastrigin is a widely used multimodal function with lots of deep local optima. When solving Rastrigin, algorithm may easily get trapped in a local optimum. Griewank is selected due to its asymmetric dimensions and multimodal feature.

In this paper, all experiments are repeated 100 runs, and each run stops at 3000 iterations. The mean and standard deviation (SD), average converge speed and success rate are employed to evaluate the performance of algorithms precisely. The converge speed shows the iteration that algorithm needs to reach the goal fitness. If the fitness could not reach the goal in the end of a run, it fails, otherwise it successes.

3.2   The Influence of Degree Distribution

Firstly, we investigate the influence of degree distribution by comparing G-PSO, L-PSO, SF-PSO. Here $N = 50$, neighbor set $n = 4$ in L-PSO and $m = 2$, $m_0 = 5$ in SF-PSO. Hence all particles share the same degree $k = 49$ in G-PSO and $k = 4$ in L-PSO, the average degree of scale-free network is 4, yet its degree distribution follows power-law distribution [15], differing from the ring structure. The statistic results are shown in Table 2. Figure 3 displays the fitness value as a function of iteration for the three algorithms.

Table 1. Test functions and conditions

Function	Dimension	Range	Goal
Sphere	30	$[-100, 100]^D$	0.01
Rosenbrock	30	$[-30, 30]^D$	100
Rastrigin	30	$[-5.12, 5.12]^D$	100
Griewank	30	$[-600, 600]^D$	0.05

Table 2. Statistic results of three algorithms, including mean, standard deviation (SD), convergence speed (CS), success rate (SR).

Function	Statistics	G-PSO	L-PSO	SF-PSO
Sphere	Mean	1.07E-58	1.04E-33	2.29E-36
	SD	9.84E-58	1.85E-33	7.23E-36
	CS	315	552	495
	SR	1	1	1
Rosenbrock	Mean	2.29E+01	2.93e+-1	4.35E+01
	SD	3.26E+01	2.82E+01	5.85E+01
	CS	528	649	1046
	SR	0.96	1	0.98
Rastrigin	Mean	6.17E+01	6.02E+01	4.83E+01
	SD	1.84E+01	1.22E+01	1.45E+01
	CS	158	295	249
	SR	0.98	0.98	0.99
Griewank	Mean	1.66E-02	1.30E-03	9.33E-03
	SD	2.00E-02	2.78E-03	9.62E-03
	CS	335	534	477
	SR	0.91	1	1

Due to its population structure, G-PSO converges faster than the other two algorithms at the early stage of evolution. However, on multimodal functions, especially (c) and (d), the misleading information of local optima spread to the whole swarm, which causes the deceleration or even stagnation of G-PSO at the end. Hence, the final performance of G-PSO is worse.

The curves of L-PSO and SF-PSO are fairly close in (a). Similar phenomenon occurs at the early stage in (b), (c) and (d), while at the end no algorithm can always beat the others. It demonstrates that the average degree of topology plays an important

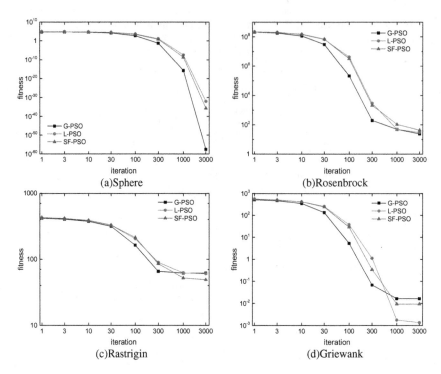

Fig. 3. Comparison of G-PSO, L-PSO, and SF-PSO. All test functions are 30-dimension. The population N is set to 50. Each fitness value is the average of 100 runs with the same parameters.

role on the early stage of optimization, rather than the heterogeneity of structure. The difference emerges as the evolution goes on, caused by the degree distribution of structure, and notably it is function-dependent.

3.3   The Influence of Topological Randomness

Moreover, we examine HSW-PSO with different randomness p. As p does not change the degree distribution, we can focus on the influence of pure topological randomness. The HSWN we employed is based on a ring structure with $k = 4$. Table 3 shows the statistic results of HSW-PSO on four benchmark functions and Fig. 4 displays the performance of HSW-PSO vs p. In (a) and (c), the network with strong randomness performs much better. Especially on Rastrigin function, which has a plenty of local optima, the performance of HSW-PSO with $p = 1$ is outperforms others with great advantage. While in (b) and (d), HSW-PSO with $p = 0$, i.e. L-PSO beats others, indicating that regular structure is better. Yet, note that the curve is not monotonous. In (b), the worst point is $p = 0.9$ while in (d) it is $p = 0.5$. Generally, the effect of topological randomness is also function-dependent.

Table 3. Statistic results of HSW-PSO with varying p, including mean, standard deviation (SD), convergence speed (CS), success rate (SR).

Function	Statistics	p = 0	p = 0.2	p = 0.4	p = 0.6	p = 0.8	p = 1
Sphere	Mean	9.56E-33	3.64E-35	1.70E-35	6.44E-36	6.38E-36	5.49E-36
	SD	1.43E-32	7.35E-35	5.64E-35	1.13E-35	1.21E-35	1.21E-35
	CS	547	513	512	510	505	504
	SR	1	1	1	1	1	1
Rosenbrock	Mean	2.80E+01	3.64E+01	3.41E+01	3.99E+01	3.90E+01	4.14E+01
	SD	2.24E+01	3.20E+01	3.25E+01	4.75E+01	4.45E+01	3.63E+01
	CS	669	763	891	939	902	894
	SR	1	1	1	1	1	1
Rastrigin	Mean	6.17E+01	5.38E+01	5.25E+01	5.12E+01	5.11E+01	4.83E+01
	SD	1.48E+01	1.56E+01	1.30E+01	1.11E+01	1.33E+01	1.42E+01
	CS	357	281	315	268	262	277
	SR	1	1	1	1	1	1
Griewank	Mean	1.11E-03	3.99E-03	5.10E-03	5.24E-03	6.52E-03	8.74E-03
	SD	3.86E-03	7.60E-03	7.01E-03	7.72E-03	8.67E-03	1.04E-02
	CS	531	499	493	486	492	491
	SR	1	1	1	1	1	1

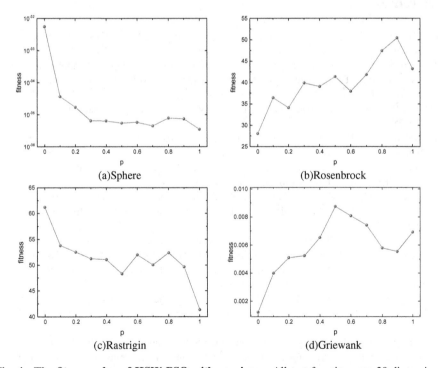

(a)Sphere   (b)Rosenbrock

(c)Rastrigin   (d)Griewank

Fig. 4. The fitness value of HSW-PSO with varying _p_. All test functions are 30-dimension. The population _N_ is set as 50. Each fitness value is the average of 100 runs with the same parameters. Each run terminates at 3000 iterations.

4 Conclusion

In this paper, we systematically explore the influence of networks on the performance of PSO, including population structures with different degree distributions and topological randomness. The average degree of the network influences the optimization process at the early stage of evolution, while the degree distribution is more effective after the beginning evolution. We also showed that the topological randomness is influential to PSO. Our on-going studies have demonstrated that network structure also significantly affects the performance of genetic algorithm (GA) and differential evolution (DE) (not shown here). As there are more and more empirical studies of information networks underlying real-world swarm and flocking recently, we believe that our investigation sheds light on the important role of network topology on swarm intelligence and could lead to a promising field by bridging network science and optimization algorithms.

Acknowledgments. This work is supported by the National Natural Science Foundation of China (Grant no. 61521091), Beijing Higher Education Young Elite Teacher Project (Grant no. YETP1072) and National Key Technology R&D Program of China (Grant no. 2015BAG15B01).

References

1. Kennedy, J., Eberhart, R.: Particle swarm optimization. In: Proceedings IEEE International Conference on Neural Networks, Perth, WA, Australia, vol. 4, pp. 1942–1948 (1995)
2. Eberhart, R.C., Shi, Y.: Particle swarm optimization: development, applications and resources. In: IEEE Proceedings of the 2001 Congress on Evolutionary Computation vol. 1, pp. 81–86 (2001)
3. Poli, R., Kennedy, J., Blackwell, T.: Particle swarm optimization. Swarm Intell. **1**, 33–57 (2007)
4. Jordehi, A.R.: Particle swarm optimisation for dynamic optimisation problems: a review. Neural Comput. Appl. **25**, 1705–1716 (2014)
5. Mendes, R., Kennedy, J., Neves, J.: The fully informed particle swarm: simpler, maybe better. IEEE Trans. Evol. Comput. **8**(3), 204–210 (2004)
6. Liang, J.J., Qin, A.K., Suganthan, P.N., Baskar, S.: Comprehensive learning particle swarm optimizer for global optimization of multimodal functions. IEEE Trans. Evol. Comput. **10**(3), 281–295 (2006)
7. Li, C.H., Yang, S.X., Trung, T.N.: A self-learning particle swarm optimizer for global optimization problems. IEEE Trans. Syst. Man Cybern. Part B **42**, 627–646 (2012)
8. Gao, Y., Du, W.B., Yan, G.: Selectively-informed particle swarm optimization. Sci. Rep. **5**, 9295 (2015)
9. Kennedy, J., Mendes, R.: Population structure and particle swarm performance. In: IEEE Proceedings of the 2002 Congress on Evolutionary Computation vol. 2, pp. 1671–1676 (2002)
10. Ratnaweera, A., Halgamuge, S., Watson, H.C.: Self-organizing hierarchical particle swarm optimizer with time-varying acceleration coefficients. IEEE Trans. Evol. Comput. **8**(3), 240–255 (2004)

11. Liu, C., Du, W.B., Wang, W.X.: Particle swarm optimization with scale-free interactions. PLoS ONE **9**(5), e97822 (2014)
12. Clerc, M., Kennedy, J.: The particle swarm-explosion, stability, and convergence in a multidimensional complex space. IEEE Trans. Evol. Comput. **6**(1), 58–73 (2002)
13. Du, W.B., Gao, Y., Liu, C., Zheng, Z., Wang, Z.: Adequate is better. Appl. Math. Comput. **268**(C), 832–838 (1998)
14. Watts, D.J., Strogatz, S.H.: Collective dynamics of small-world networks. Nature **393**, 440–442 (1998)
15. Barabási, A.L., Albert, R., Jeong, H.: Scale-free characteristics of random networks: the topology of the world-wide web. Physica A Stat. Mech. Its Appl. **281**(1–4), 69–77 (2000)
16. Santos, F.C., Rodrigues, J.F., Pacheco, J.M.: Epidemic spreading and cooperation dynamics on homogeneous small-world networks. Phys. Rev. E **72**(5 Pt 2), 168–191 (2005)

Headless Chicken Particle Swarm Optimization Algorithms

Jacomine Grobler[1]([⊠]) and Andries P. Engelbrecht[2]

[1] Department of Industrial and Systems Engineering,
University of Pretoria, Pretoria, South Africa
jacomine.grobler@gmail.com
[2] Department of Computer Science, University of Pretoria, Pretoria, South Africa
engel@cs.up.ac.za

Abstract. This paper investigates various strategies for implementing the headless chicken macromutation operator in the particle swarm optimization domain. Three different headless chicken particle swarm optimization algorithms are proposed and evaluated against a standard guaranteed convergence PSO algorithm on a diverse set of benchmark problems. Competitive performance is demonstrated by a Von Neumann headless chicken particle swarm optimization algorithm when compared to a classic guaranteed convergence particle swarm optimization algorithm. Statistically significantly superior results are obtained over a number of difficult benchmark problems.

1 Introduction

Effective diversity management is an important requirement for the successful design of an optimization algorithm. Traditionally, management of the diversity of the solution or decision space prevents algorithms from converging too quickly to suboptimal solutions and also ensures effective exploration of a large part of the search space. A large number of strategies have already been developed to improve the diversity of optimization algorithms.

This paper describes a preliminary investigation into the use of the headless chicken macromutation operator to better manage the diversity of a particle swarm optimization algorithm. Three variations of a headless chicken particle swarm optimization algorithm is proposed and performance is evaluated on a set of varied floating-point benchmark problems.

This paper is considered significant because, to the best of the authors' knowledge, it proposes the first investigation into the use of the headless chicken macromutation based PSO (HCPSO) for solving single objective static optimization problems. The paper also describes the first Von Neumann HCPSO and the first guaranteed convergence HCPSO algorithm.

The rest of the paper is organized as follows: Sect. 2 provides an overview of existing literature. Section 3 provides a description of the HCPSO algorithms developed in this paper. Thereafter, Sect. 4 describes the experimental setup and results obtained. Finally, the paper is concluded in Sect. 5.

© Springer International Publishing Switzerland 2016
Y. Tan et al. (Eds.): ICSI 2016, Part I, LNCS 9712, pp. 350–357, 2016.
DOI: 10.1007/978-3-319-41000-5_35

2   The Headless Chicken Macromutation

The headless chicken (HC) macromutation was first introduced by Jones [1] as part of an experiment to test the progress made by a genetic algorithm through crossover, over and above what could be achieved by simple macromutations. The HC macromutation operates as follows: Provided two parents, the HC crossover generates two other random individuals and uses these in crossover with the two parents. The study concluded that for problems which did not contain well defined building blocks, there was no difference in performance between the traditional crossover operators and the headless chicken crossover macromutation.

Since then the HC crossover has been used successfully in a number of optimization algorithms. Some examples include significant application in genetic programming [2,3], playing the game Nim [4], evolving a filter for a braincomputer interface mouse [5] and evolving finite state machines for automatic target detection [6]. Of greater interest to this paper, is the potential of the HC operator to make a positive contribution to managing the exploration-exploitation tradeoff of an optimization algorithm. The HC crossover has the ability to increase the diversity of a population,reducing the chances of prematurely converging to a local minimum.

The only application of headless chicken crossover in the PSO domain that the authors are aware of, is the dynamic vector evaluated particle swarm optimisation algorithm (DVEPSO) with headless chicken crossover [7]. Each particle in a PSO algorithm stores its own previous best solution (*pbest*) as well as the best solution found by the entire swarm (*gbest*). This information is used to guide the particles through the search space. The DVEPSO algorithm made use of a headless chicken based parentcentric crossover operator applied between the particle's position, a randomly generated position in the search space and the sub-swarm's *pbest* or *gbest* positions. Results showed that the operator improved the accuracy of the set of solutions, but fewer solutions were found. The next section provides some important PSO background information and elaborates on further potential applications of the HC crossover in the PSO domain.

3   Particle Swarm Optimization and the Headless Chicken Macromutation Operator

The PSO algorithm [8] represents each potential problem solution by the position of a particle in multi-dimensional hyperspace. Throughout the optimization process velocity and displacement updates are applied to each particle to move it to a different position and thus a different solution in the search space.

The *gbest* model calculates the velocity of particle i in dimension j at time $t + 1$ using

$$v_{ij}(t + 1) = wv_{ij}(t) + c_1 r_{1j}(t)[\hat{x}_{ij}(t) - x_{ij}(t)] + c_2 r_{2j}(t)[x_j^*(t) - x_{ij}(t)] \quad (1)$$

where $v_{ij}(t)$ represents the velocity of particle i in dimension j at time t, c_1 and c_2 are the cognitive and social acceleration constants, $\hat{x}_{ij}(t)$ and $x_{ij}(t)$ respectively denotes the personal best position (*pbest*) and the position of particle i in dimension j at time t. $x_j^*(t)$ denotes the global best position ($gbest$) in dimension j, w refers to the inertia weight, and $r_{1j}(t)$ and $r_{2j}(t)$ are sampled from a uniform random distribution, $U(0,1)$. The displacement of particle i at time t is simply derived from the calculation of $v_{ij}(t+1)$ in Eq. (1) and is given as

$$x_{ij}(t+1) = x_{ij}(t) + v_{ij}(t+1) \tag{2}$$

3.1   The Guaranteed Convergence PSO Algorithm

Unfortunately, it has been shown that the basic PSO swarm can stagnate on a solution which is not necessarily a local optimum [9]. The guaranteed convergence particle swarm optimization (GCPSO) algorithm [9] has been shown to address this problem effectively and has thus been used as basis for the design of the HCPSO algorithms in this paper. The GCPSO algorithm requires that different velocity and displacement updates, defined as

$$v_{\tau j}(t+1) = -x_{\tau j}(t) + x_j^*(t) + wv_{\tau j}(t) + \rho(t)(1 - 2r_j(t)) \tag{3}$$

and

$$x_{\tau j}(t+1) = x_j^*(t) + wv_{\tau j}(t) + \rho(t)(1 - 2r_j(t)), \tag{4}$$

are applied to the global best particle, where $\rho(t)$ is a time-dependent scaling factor, $r_j(t)$ is sampled from a uniform random distribution, $U(0,1)$, and all other particles are updated by means of Eqs. (1) and (2). This algorithm forces the $gbest$ particle into a random search around the global best position. The size of the search space is then adjusted on the basis of the number of consecutive successes or failures of the particle, where success is defined as an improvement in the objective function value.

3.2   The Von Neumann PSO Algorithm

The Von Neumann PSO [10] organizes the particles into a lattice according to the particle indices. Each particle belongs to a neighbourhood consisting of its nearest neighbours in the cubic structure. Instead of being partially attracted to $gbest$, the velocity of a particle is influenced by the best solution found by the other particles in the same neighbourhood. Since these neighbourhoods overlap, information about good solutions is eventually propagated throughout the swarm, but at a much slower rate. In so doing more diversity and subsequent slower convergence is obtained, leading to significantly improved chances of finding a good solution.

3.3 The Headless Chicken PSO Algorithms

The main idea behind the headless chicken PSO algorithm is to introduce a crossover operator into the PSO algorithm aimed at increasing the swarm diversity. Various options exist with regard to how the operator can be implemented. This paper applies arithmetic crossover [11] with a specific probability, p_c, to each particle at each iteration as indicated in Algorithm 1. If p_c is less than a number sampled from a uniform random distribution, $U(0,1)$, $\boldsymbol{x}_i(t)$ is calculated as follows

$$\boldsymbol{x}_i(t) = (1 - \gamma)\boldsymbol{x}_1(t) + \gamma\boldsymbol{x}_2(t) \tag{5}$$

where $\boldsymbol{x}_1(t)$ is a randomly generated position in the search space, $\boldsymbol{x}_2(t)$ is a particle position selected according to the HCGCPSO strategy used and $\gamma \in [0,1]$.

Algorithm 1. The headless chicken PSO algorithm.

---

1 Initialize an n_x-dimensional swarm of n_s particles
2 $t = 1$
3 **while** $t < I_{max}$ **do**
4   **for** *All particles i* **do**
5   **if** $f(\boldsymbol{x}_i(t)) < f(\hat{\boldsymbol{x}}_i)$ **then**
6 | $\hat{\boldsymbol{x}}_i = \boldsymbol{x}_i(t)$
7   **end**
8   **if** $f(\hat{\boldsymbol{x}}_i) < f(\boldsymbol{x}^*)$ **then**
9 | $\boldsymbol{x}^* = \hat{\boldsymbol{x}}_i$
10   **end**
11   **end**
12   **for** *All particles i* **do**
13   **if** $U(0,1) \leq p_c$ **then**
14 | $\boldsymbol{x}_i(t) = (1 - \gamma)\boldsymbol{x}_1(t) + \gamma\boldsymbol{x}_2(t)$
15   **end**
16   **end**
17   **for** *All particles i* **do**
18   Update the particle velocity using Eq. (1)
19   Update the particle position using Eq. (2)
20   **end**
21     $t = t + 1$
22 **end**

Three headless chicken PSO variations were tested in this paper differing with regard to how $\boldsymbol{x}_2(t)$ is selected:

- **HCGCPSO1:** Crossover is performed between a selected particle's position and a randomly generated position in the search space.

- **HCGCPSO2:** Crossover is performed between a selected particle's *pbest* position and a randomly generated position in the search space.
- **HCGCPSO3:** Crossover is performed between a selected particle's *lbest* position and a randomly generated position in the search space. In this variation the particles are arranged in a Von Neumann structure.

4   Empirical Evaluation of the HCPSO Algorithms

The three headless chicken PSO variations were evaluated on the 2015 IEEE Congress of Evolutionary Computation benchmark problem set [12] in 10 and 30 dimensions. Initial algorithm control parameters were selected as specified in [13] and listed in Table 1. These parameters will, however, have an influence on algorithm performance and will thus need to be tuned more thoroughly in future. The notation $m \longrightarrow n$ indicates that the associated parameter is decreased linearly from m to n over 95 % of the total number of iterations, I_{max}.

Table 1. HCPSO algorithm parameters.

Parameter	Value used
Number of particles in swarm (n_s)	27
Maximum number of iterations (I_{max})	$\frac{100000}{n_s}$
Acceleration constant (c_1)	$2.0 \longrightarrow 0.7$
Acceleration constant (c_2)	$0.7 \longrightarrow 2.0$
Inertia weight (w)	$0.9 \longrightarrow 0.4$
Probability of HC crossover (p_c)	$0.9 \longrightarrow 0.01$
Arithmetic crossover constant (γ)	0.5

The results of the first comparison between the various headless chicken particle swarm optimization algorithms are presented in Table 2. For the experiments conducted on the CEC 2015 benchmark problem set, results for each algorithm were recorded over 30 independent simulation runs. The notation, μ and σ, denote the mean and standard deviation associated with the corresponding algorithm. For comparison purposes, results for a standard GCPSO algorithm and the self-adaptive dynamic particle swarm optimizer (sDMS-PSO) [14], the best performing CEC 2015 PSO algorithm, were also included in Table 2.

Statistical tests were also used to evaluate the significance of the results. The results in Table 3 were obtained by comparing each dimension-problem-combination of the strategy under evaluation, to all of the dimension-problem-combinations of the other strategies. For every comparison, a Mann-Whitney U test at 95 % significance was performed (using the two sets of 30 data points of the two strategies under comparison) and if the first strategy statistically significantly outperformed the second strategy, a win was recorded. If no statistical

Table 2. Comparison results of the three HCPSO variations, the GCPSO and the sDMS-PSO algorithm on the CEC 2015 benchmark problem set.

Problem	Dimension	GCPSO		HCGCPSO1		HCGCPSO2		HCGCPSO3		sDMS-PSO	
		μ	σ	μ	σ	μ	σ	μ	σ	μ	σ
1	10	10297	6573.6	$7.36E+05$	$1.01E+06$	$1.03E+06$	$1.07E+06$	$1.62E+05$	$2.15E+05$	34.20724	111.515
1	30	$4.75E+05$	$2.82E+05$	$3.33E+07$	$1.97E+07$	$3.82E+07$	$2.07E+07$	$1.09E+07$	$7.83E+06$	0.014634	0.085067
2	10	9878.7	11549	18633	15793	14992	15095	19018	14451	111.5708	293.4532
2	30	6075	5742.4	$7.49E+08$	$1.06E+09$	$6.11E+08$	$6.74E+08$	8472.5	5801.2	265.8853	941.9696
3	10	20.062	0.097712	20.093	0.075106	20.13	0.10591	20.181	0.091023	19.99886	0.005563
3	30	20.213	0.33324	20.416	0.1292	20.532	0.10992	20.907	0.063308	19.99999	$2.79E-05$
4	10	11.484	4.3028	20.6	8.8109	22.241	9.2812	10.474	5.1137	4.760195	1.932512
4	30	132.46	37.514	122.84	34.26	114.88	34.798	44.12	11.416	41.43707	8.369055
5	10	311.09	183.33	547.92	255.65	541.95	255.83	350.51	178.06	130.9601	86.34699
5	30	3711.3	708.75	4453.1	776.91	4378.2	595.67	3608.6	680.02	2629.81	386.8353
6	10	1986.1	2430.4	8277.2	3479.5	7519	3877.3	6152.8	3938	276.9249	182.9651
6	30	77359	43523	$1.18E+06$	$1.23E+06$	$1.18E+06$	$1.21E+06$	$5.05E+05$	$3.14E+05$	1649.45	662.1082
7	10	1.331	0.64452	3.4253	1.0407	3.4767	0.95164	1.994	0.80836	0.650468	0.378079
7	30	14.997	6.5212	25.168	4.6545	26.667	4.943	14.269	2.4661	8.782088	1.684147
8	10	1628.9	1930.8	4940.4	5768.2	7128.5	7341.1	2580.4	4085.1	61.80629	67.25348
8	30	48491	32250	$3.98E+05$	$2.78E+05$	$3.94E+05$	$2.74E+05$	$2.77E+05$	$1.57E+05$	1676.354	1143.393
9	10	100.24	0.062606	100.35	0.09732	104.58	23.253	100.27	0.053498	100.181	0.037846
9	30	135.3	83.609	119.26	34.057	130.46	52.721	103.31	0.31332	102.9289	0.161441
10	10	1246.2	1444.1	4632.6	4323.1	6142.1	5395	1759.9	2404.3	439.5543	131.626
10	30	36687	20208	$3.04E+05$	$2.99E+05$	$3.01E+05$	$1.87E+05$	$1.34E+05$	$1.10E+05$	5595.516	3833.932
11	10	252.51	134.51	320.1	58.493	329.55	72.879	261.64	102.66	167.4859	147.8587
11	30	838.75	310.21	1026.4	90.655	936.98	245.29	780.31	73.236	316.1506	8.861513
12	10	102.67	0.93814	103.15	1.1602	103.28	1.0935	101.37	0.39856	101.4498	0.389331
12	30	109.03	1.2925	152.13	45.936	143.04	44.408	140.41	46.426	105.3203	0.363941
13	10	35.643	4.3445	0	0	0	0	0	0	28.18082	2.150674
13	30	123.07	5.4094	0	0	0	0	0	0	103.2476	5.121294
14	10	5221.4	2937.2	7018.9	2343.2	6792.8	2382.3	5194.1	1854.2	878.7974	848.9997
14	30	34069	909.01	35519	1929.6	36232	1637.1	34269	829.67	20058.53	3002.365
15	10	100	0	100	0	100	0	100	0	100	$1.32E-13$
15	30	100	0	117.54	4.591	118.9	4.1997	100.55	2.6418	100	$9.14E-14$

difference could be observed a draw was recorded. If the second strategy out-performed the first strategy, a loss was recorded for the first strategy. The total number of wins, draws and losses were then recorded for all combinations of the strategy under evaluation. As an example, (23-4-3) in row 1 column 2, indicates that the normal GCPSO strategy significantly outperformed the HCGCPSO1 algorithm 23 times over the benchmark problem set. Furthermore, 4 draws and 3 losses were recorded.

Table 3. Hypotheses analysis of the performance of the headless chicken PSO variations.

	GCPSO	HCGCPSO1	HCGCPSO2	HCGCPSO3	TOTAL
GCPSO	NA	23-4-3	22-5-3	11-12-7	56-21-13
HCGCPSO1	3-4-23	NA	2-28-0	2-5-23	7-37-46
HCGCPSO2	3-5-22	0-28-2	NA	2-5-23	5-38-47
HCGCPSO3	7-12-11	23-5-2	23-5-2	NA	53-23-15

From the results it can be seen that the standard GCPSO and the sDMS-PSO algorithm performs very well over the benchmark problem set as a whole. However, the performance of the HCGCPSO3 algorithm is also very competitive. A direct comparison between the HCGCPSO3 algorithm and the standard GCPSO algorithm shows that the HCGPSO3 algorithm performed statistically significantly better than the GCPSO algorithm for seven of the problems. Eleven problems were solved better by the GCPSO algorithm and there was no statistically significant difference for twelve of the problems. It is also encouraging to see that it is for the more complicated problems, i.e. Composition function 4 to 7, that the HCGPSO3 algorithm outperforms the basic GCPSO algorithm. The general trend is that the more complicated the search space with regard to multi-modality, asymmetry and non-separability, the better the HCPSO algorithms perform. More complicated search spaces typically require a higher level of diversity in the swarm to solve the problem efficiently, explaining why the HCPSO algorithms perform better for the more complicated types of search spaces. Finally, with regard to overall ranking the HCGCPSO3 algorithm is the best performing HCPSO algorithm outperforming the HCGPSO1 and HCG-PSO2 algorithms over a large number of problems.

5   Conclusion

This paper investigated the application of a headless chicken crossover operator in the PSO domain. Three HCPSO algorithms were implemented, namely GCHCPSO1, which applies the HC operator to particle positions, GCHCPSO2, which applies the HC operator to *pbest* positions, and GCHCPSO3, which applies the HC operator to *lbest* positions in a Von Neumann structure. Of the three algorithms, the GCHCPSO3 was the best performing algorithm over the CEC

2015 benchmark problem set. Competitive performance was shown when compared to a standard GCPSO algorithm with statistically significantly superior results over a number of the most complicated problems in the benchmark set.

Future research opportunities exist in more in depth analysis and benchmarking of the HCPSO algorithm performance, evaluating the impact of different control parameters on algorithm performance and expanding the analysis to dynamic optimization, where the greater diversity management capabilities of the HCPSO algorithm can be more fully utilized.

References

1. Jones, T.: Crossover, macromutation, and population-based search. In: International Conference on Genetic Algorithms, pp. 73–80 (1995)
2. Angeline, P.J.: Subtree crossover: building block engine or macromutation. Genetic Program. **97**, 9–17 (1997)
3. Poli, R., McPhee, N.F.: Exact GP Schema Theory for Headless Chicken Crossover with Subtree Mutation. Cognitive Science Research Papers - University of Birmingham CSRP (2000)
4. Hynek, J.: Evolving strategy for game playing. In: 4th International ICSC Symposium on Engineering Intelligent Systems, pp. 1–6 (2004)
5. Citi, L., Poli, R., Cinel, C., Sepulveda, F.: P300-based BCI mouse with genetically-optimized analogue control. IEEE Trans. Neural Syst. Rehabil. Eng. **16**(1), 51–61 (2008)
6. Benson, K.: Evolving finite state machines with embedded genetic programming for automatic target detection. In: Congress on Evolutionary Computation, pp. 1543–1549 (2000)
7. Helbig, M., Engelbrecht, A.P.: Using headless chicken crossover for local guide selection when solving dynamic multi-objective optimization. In: Pillay, N., Engelbrecht, A.P., Abraham, A., du Plessis, M.C., Snášel, V., Muda, A.K. (eds.) Advances in Nature and Biologically Inspired Computing. Advances in Intelligent Systems and Computing, vol. 419, pp. 381–392. Springer, Switzerland (2016)
8. Kennedy, J., Eberhart, R.: Particle swarm optimization. In: IEEE International Confererence on Neural Networks, pp. 1942–1948 (1995)
9. Van den Bergh, F., Engelbrecht, A.P.: A new locally convergent particle swarm optimiser. In: IEEE International Conference on Systems, Man and Cybernetics, pp. 6–12 (2002)
10. Kennedy, J., Mendes, R.: Population structure and particle performance. In: IEEE Congress on Evolutionary Computation, pp. 1671–1676 (2002)
11. Michalewicz, Z.: Genetic Algorithms + Data Structures = Evolutionary Programs. Springer, Heidelberg (1996)
12. Liang, J.J., Qu, B.Y., Suganthan, P.N., Chen, Q.: Problem definitions and evaluation criteria for the CEC 2015 competition on learning-based real-parameter single objective optimization. Technical report201411A, Computational Intelligence Laboratory, Zhengzhou University, Zhengzhou China and Nanyang Technological University, Singapore (2014)
13. Grobler, J., Engelbrecht, A.P., Kendall, G., Yadavalli, V.S.S.: Heuristic space diversity control for improved meta-hyper-heuristic performance. Inf. Sci. **300**, 49–62 (2015)
14. Liang, J.J., Guo, L., Liu, R., Qu, B.Y.: A self-adaptive dynamic particle swarm optimizer. In: Congress on Evolutionary Computation, pp. 3206–3213 (2015)

On the Hybridization of Particle Swarm Optimization Technique for Continuous Optimization Problems

Akugbe Martins Arasomwan and Aderemi Oluyinka Adewumi[(⊠)]

School of Mathematics, Statistics and Computer Science, University of
Kwazulu-Natal, Private Bag X54001, Durban 4000, South Africa
accuratesteps@yahoo.com, adewumia@ukzn.ac.za

Abstract. A hybrid particle swarm optimization (PSO) algorithm is proposed. In literature, the optimization algorithms that hybridize one PSO variant with another PSO variant are rare. In this paper, linear decreasing inertia PSO (LPSO) and random inertia weight PSO (RPSO) are hybridized to form a new hybrid PSO (NHPSO) algorithm. This algorithm addresses premature convergence associated with PSO technique when handling continuous optimization problems. RPSO periodically makes NHPSO jump out of any local optima and strengthens its searching ability while LPSO enhances the convergence ability of NHPSO. The performance of NHPSO is experimentally tested to verify the practicability and profitability of hybridizing two separate existing PSO variants to effectively handle continuous optimization problems. Results show that NHPSO is very successful, compared to some existing PSO variants. This implies that many more efficient algorithms could be built from hybridizing two or more existing PSO variants.

Keywords: Particle swarm optimization · Hybridization · PSO variants · Global best · Local best · Continuous optimization

1 Introduction

Particle swarm Optimization (PSO) was introduced as optimization technique in 1995 [1]. It is one of the nature-inspired techniques that have been applied to solve simple and complex continuous optimization problems [2, 3]. A number of variations to the original PSO have been proposed to make it faster and more reliable. Such improvements include hybridizing it with local search [2–4] or other evolutionary techniques [5–7]. Hybridization is the combination of principles (elements) from different methods so as to give rise to a new method that displays desirable properties of the original methods. For example, global features of a global method could be combined with local features of a local method. This type of combination often results to methods that are more efficient, accurate and reliable in finding the global minimum to simple and complex optimization problems.

The hybridization of PSO with Genetic Algorithm, Differential Evolution and Ant Colony Optimization are popular choices among researchers [8]. However, among numerous research articles on hybrid PSOs [8, 9], the hybridization of one PSO variant

© Springer International Publishing Switzerland 2016
Y. Tan et al. (Eds.): ICSI 2016, Part I, LNCS 9712, pp. 358–366, 2016.
DOI: 10.1007/978-3-319-41000-5_36

with another PSO variant is lacking. Although, there are some reported cases that may seem to contradict this claim. For example, Unified PSO variant proposed in [10] combines the exploration and exploitation capabilities of the global and local PSO variants by aggregating their velocity updating formulas using a unification factor; it was experimentally validated and confirmed efficient. Also reported in [11] is hybrid PSO (HTPSO)[1] based on multi-neighbourhood topologies. This variant also combined the idea of global and local PSO variants of the original PSO and mixes the use of the traditional velocity and position update rules of star, ring and Von Neumann topologies together. It was also experimentally validated to be efficient. However, these two cases do not really involve distinct PSO variants, because the global variant is actually a generalization of the local variant. They are only distinguished due to their different exploration and exploitation properties. Therefore, in this paper, we propose a PSO algorithm built on two distinct PSO variants namely, linear decreasing inertia weight PSO (LPSO) and random inertia weight PSO (RPSO), to build a new hybrid PSO (NHPSO). This new algorithm addresses the problem of premature convergence associated with PSO technique when handling complex continuous optimization problems. RPSO periodically makes NHPSO jump out of local optima and strengthens its searching ability while LPSO enhances the convergence ability of NHPSO. It is expected that the implementation of NHPSO will show that hybridization of PSO variants is a potential area researchers should explore.

The rest of this paper is organized as follows: Sect. 2 gives a brief description of PSO methodology. The proposed algorithm is described in Sect. 3. The experiment is described in Sect. 4. Results and discussions are reported in Sect. 5. Section 6 concludes the paper, highlighting some areas for future research.

2 PSO Methodology

A number of agents called particles (swarm) are required to implement the PSO technique. The particles are randomly initialized in the search space defined by the upper and lower limits of decision variables before PSO begins execution. Each particle is characterized with position (X) and velocity (V) which are n-dimension vectors. The position of each particle represents a potential solution to the optimization problem being solved and the quality of the solution is determined by the objective function of the problem. Usually, other particles follow the one that has found the best solution. As the particles move around in the search space their respective position and velocity are adjusted according to Eqs. (1) and (2).

$$V_i(t+1) = \omega V_i(t) + c_1 r_1 (P_i - X_i) + c_2 r_2 (P_g - X_i). \tag{1}$$

$$X_i(t+1) = X(t) + V_i(t+1). \tag{2}$$

---

[1] The name used by the authors that proposed this variant is "*Hybrid topology*".

Where P_i and P_g are vectors representing the ith particle personal best position and swarm global best position respectively; g is the index of the global (or local) best; r_1, $r_2 \in [0,1]$ while c_1 and c_2 are acceleration coefficients. The parameter t represents iteration index and ω is the inertia weight (commonly believed in the PSO community to be) responsible for striking a balance between exploration and exploitation activities. Further description of PSO technique can be found in [1, 2].

3 Proposed Hybrid PSO Algorithm

It has been shown in Sect. 1 that PSO variants which hybridize two (or more) distinct PSO variants are lacking in literature. To bridge this gap, we hybridize LPSO and RPSO. The major reason for selecting these variants for hybridization is that, they are among the earliest variants of the original PSO technique and have been claimed to perform poorly in handling complex continuous optimization problems and their level of their reliability and robustness is very low [12, 13].

LPSO implements the inertia weight strategy stated in Eq. (3). It starts with a large initial value (ω_{start}) and then linearly decreases to a smaller final value (ω_{end}), with the belief that large inertia weights facilitate global search while small inertia weights facilitate local search. Further descriptions of LPSO can be found in [13].

$$\omega_t = (\omega_{start} - \omega_{end})\left(\frac{T_{max} - t}{T_{max}}\right) + \omega_{end}. \tag{3}$$

In Eq. (3), t is the current iteration number, T_{max} is the maximum iteration number and $\omega_t \in [0, 1]$ is the inertia weight value in the tth iteration. LPSO does global search at early stage but lacks enough momentum to do local search as it approaches its terminal point of execution.

RPSO was introduced in [14]. It implements a random strategy which produces values that randomly vary between 0 and 1 (see Eq. (4)). The strategy increases convergence in RPSO and could find good results. However, its performance is also affected by premature convergence.

$$\omega_t = random(0, 1). \tag{4}$$

Parallel hybridization is employed to hybridize LPSO and RPSO (i.e., RPSO starts before LPSO ends). This method is adopted because it will facilitate the collaboration between LPSO and RPSO. The proposed algorithm, new hybrid PSO (NHPSO), seeks to *(i)* Reduce the number of function evaluations and increase convergence speed, *(ii)* Increase success rate, *(iii)* Increase robustness and stability, *(iv)* Improve solution accuracy. The operation of NHPSO is described in Fig. 1. In the figure, NHPSO begins with LPSO because it can quickly locate promising region in the search space. The specified maximum number of function evaluations (FEs) is divided into n equal parts and at every FEs/n there is a switch of execution between LPSO and RPSO until the stopping criteria for NHPSO is met. Different values (2, 4, 8, 10, 20, etc.) were tested to determine the proper value for n. NHPSO generally performed better when the value of

20 was used. When $n < 20$, solution quality deteriorates and when $n > 20$, FE increases with insignificant improvement in solution quality. RPSO intermittently enables NHPSO jump out of any local optima to improve and strengthen the searching ability of the algorithm as it approaches terminal point. On the other hand, LPSO is used to quickly locate promising region in the search space and enhance the convergence ability of NHPSO.

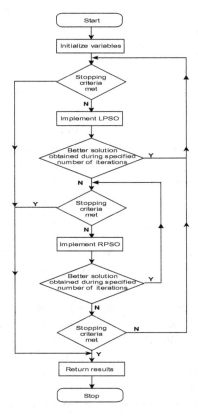

Fig. 1. The flowchart for NHPSO algorithm

4 Experiments

Defined below are 9 scalable minimization benchmark problems obtained from [2, 11, 15] which are used to validate NHPSO. US = unimodal separable, UN = unimodal non-separable, MS = multimodal separable and MN = multimodal non-separable. The global optimal value is f_{min}, "range" is the search space and "accept" is the solution accuracy level. Solution between the acceptable value and the actual global optimum is considered successful otherwise it is considered unsuccessful.

1. Noisy Quadric: $f_1(\vec{x}) = \sum_{i=1}^{d} ix_i^4 + random(0,1)$, type = US, range = ± 1.28, $f_{min} = 0$, accept = 10^{-4}.

2. Rosenbrock: $f_2(\vec{x}) = \sum_{i=1}^{d-1} \left(100(x_{i+1} - x_i^2)^2\right) + (x_i - 1)^2$, type = UN, range = ± 30, $f_{min} = 0$, accept = 30.0.

3. Sphere: $f_3(\vec{x}) = \sum_{i=1}^{d} x_i^2$, type = US, range = ± 100, $f_{min} = 0$, accept = 10^{-6}.

4. Ackley: $f_4(\vec{x}) = -20exp\left(-0.2\sqrt{\frac{1}{n}\sum_{i=1}^{d} x_i^2}\right) - exp\left(\frac{1}{n}\sum_{i=1}^{d} \cos(2\pi x_i)\right) + 20 + e$, type = MN, range = ± 32, $f_{min} = 0$, accept = 10^{-6}.

5. Griewank: $f_5(\vec{x}) = \frac{1}{4000}\left(\sum_{i=1}^{d} x_i^2\right) - \left(\prod_{i=1}^{d} \cos\left(\frac{x_i}{\sqrt{i}}\right)\right) + 1$, type = MN, range = ± 600, $f_{min} = 0$, accept = 10^{-6}.

6. Rastrigin: $f_6(\vec{x}) = \sum_{i=1}^{d} \left(x_i^2 - 10\cos(2\pi x_i) + 10\right)$, type = MS, range = ± 5.12, $f_{min} = 0$, accept = 10^{-6}.

7. Shifted Rosenbrock: $f_7(\vec{x}) = \sum_{i=1}^{d-1} \left(100(z_{i+1} - z_i^2)^2\right) + (z_i - 1)^2 + f\_bias_9$, where $z = x - o$, $o = [o_1, o_2 ..., o_d]$: the shifted global optimum, M is an orthogonal matrix; type = MN, range = ± 100, $f_{min} = 390$, accept = 420.

8. Shifted Rastrigin: $f_8(\vec{x}) = \sum_{i=1}^{d} \left(z_i^2 - 10\cos(2\pi z_i) + 10\right) + f\_bias_9$, where $z = x - o$, $o = [o_1, o_2 ..., o_d]$: the shifted global optimum; type = MS, range = ± 5, $f_{min} = -330$, accept = -330.

9. Shifted Rotated Rastrigin: $f_9(\vec{x}) = \sum_{i=1}^{d} \left(z_i^2 - 10\cos(2\pi z_i) + 10\right) + f\_bias_9$, where $z = (x - o) * M$, $o = [o_1, o_2 ..., o_d]$: the shifted global optimum, M is an orthogonal matrix; type = MN, range = ± 5, $f_{min} = -330$, accept = -330.

NHPSO is compared with the individual LPSO and RPSO. The same swarm size of 40 and random initializations are used. The maximum and minimum velocity limits are dynamically set to the minimum and maximum search space limits of each problem. The values for c_1 and c_2 are individually set to 1.494 in RPSO [14] and 2.0 in LPSO [13]; ω is made to linearly vary from $\omega_{max} = 0.9$ to $\omega_{min} = 0.4$ for LPSO but $\omega \in [0,1]$ for RPSO. The problems are scaled to 10 and 30 dimension sizes and the maximum number of function evaluations is set to 80,000 and 120,000 respectively. Each test problem is independently simulated 30 times. Fully connected topology is used for LPSO and RPSO, which is inherited by NHPSO.

The performance of each algorithm is measured in terms of *Mean Fitness* (MF), *Standard Deviation* (SD), *Function Evaluations* (FE) and *Success Ratio* (SR). MF measures solution accuracy; SD measures algorithm's stability and robustness; FE measures algorithm's efficiency and SR measures the reliability of algorithm. Wilcoxon signed rank statistical tool is another measurement used for comparisons.

5 Results and Discussions

Results obtained from all the experiments are presented and discussed in this section. Table 1 shows the MF, SD, FE and SR across the problems dimensions for RPSO, LPSO and NHPSO over 30 independent runs. The best results are marked in bold face. Table 2 shows the results of the statistical analyses as R^+ (win), R^- (loss), p-value (probability value or significant level) and z values obtained in the analysis. In all comparisons, a 5 % level of significance was used. In the tables, "PB" and "PM" represent benchmark problem and performance measurement respectively.

Table 1. Results obtained by the various PSO variants tested in the experiments

PB	Variant	Dimension = 10				Dimension = 30			
		MF	SD	FE	SR	MF	SD	FE	SR
f_1	RPSO	2.41e-03	1.54e-03	80000	0	6.57e–02	4.13e-02	120000	0
	LPSO	2.86e-01	2.97e-01	80000	0	1.73e+01	8.71e+00	120000	0
	NHPSO	**6.23e-05**	**3.63e-05**	**40241**	**0.83**	**5.86e-05**	**3.37e-05**	**43060**	**0.97**
f_2	RPSO	2.27e+01	5.39e+00	3027	1	3.03e+03	1.62e+04	40260	0.83
	LPSO	3.09e+05	4.43e+05	80000	0	2.14e+07	1.16e+07	120000	0
	NHPSO	**1.81e+01**	**5.31e+00**	4615	1	**2.95e+01**	**2.39e-01**	7013	1
f_3	RPSO	7.73e-07	**1.45e-07**	3091	1	7.66e-07	2.69e-07	9848	1
	LPSO	7.54e+02	5.18e+02	80000	0	1.14e+04	2.95e+03	120000	0
	NHPSO	**5.49e-07**	2.49e-07	5635	1	**5.88e-07**	**2.37e-07**	7819	1
f_4	RPSO	4.34e-01	7.57e-01	24944	0.73	3.72e+00	1.58e+00	120000	0
	LPSO	7.69e+00	2.81e+00	80000	0	1.49e+01	1.20e+00	120000	0
	NHPSO	**6.84e-07**	**2.21e-07**	6533	1	**7.28e-07**	**1.47e-07**	8713	1
f_5	RPSO	9.73e-02	4.76e-02	80000	0	4.90e-02	5.86e-02	90592	0.27
	LPSO	6.32e+00	4.31e+00	80000	0	9.41e+01	2.56e+01	120000	0
	NHPSO	**5.25e-07**	**2.75e-07**	5776	1	**5.21e-07**	**2.91e-07**	7905	1
f_6	RPSO	8.22e+00	3.16e+00	80000	0	6.62e+01	1.42e+01	120000	0
	LPSO	2.52e+01	1.27e+01	80000	0	1.57e+02	2.35e+01	120000	0
	NHPSO	**4.77e-07**	**2.42e-07**	5655	1	**5.61e-07**	**2.64e-07**	7833	1
f_7	RPSO	1.09e+03	2.51e+03	15071	0.87	3.54e+04	1.79e+05	64155	0.57
	LPSO	3.87e+07	7.09e+07	77337	0.03	4.99e+09	3.31e+09	120000	0
	NHPSO	**4.09e+02**	**6.68e+00**	4796	1	**4.20e+02**	**2.62e-01**	7148	1
f_8	RPSO	−3.13e+02	1.08e+01	80000	0	−1.90e+02	3.15e+01	120000	0
	LPSO	−3.10e+02	1.06e+01	77337	0.03	−2.07e+02	2.65e+01	120000	0
	NHPSO	**−3.30e+02**	**2.46e-05**	5365	1	**−3.30e+02**	**2.29e-05**	7571	1
f_9	RPSO	−3.13e+02	1.08e+01	80000	0	−1.90e+02	3.15e+01	120000	0
	LPSO	−3.10e+02	1.10e+01	77337	0.03	−2.08e+02	2.60e+01	120000	0
	NHPSO	**−3.30e+02**	**2.74e-05**	5360	1	**−3.30e+02**	**2.71e-05**	7547	1

Table 2. Wilcoxon signed-ranks testsR^+, R^-, z and p-value on the results obtained

PM	Statistical measures	Dimension = 10		Dimension = 30	
		NHPSO vs. RPSO	NHPSO vs. LPSO	NHPSO vs. RPSO	NHPSO vs. LPSO
MF	R^+	45	45	45	45
	R^-	0	0	0	0
	z	−2.668	−2.666	−2.668	−2.666
	p-value	0.008	0.008	0.008	0.008
SD	R^+	44	45	45	45
	R^-	1	0	0	0
	z	−2.547	−2.666	−2.666	−2.666
	p-value	0.011	0.008	0.008	0.008
FE	R^+	42	45	45	45
	R^-	3	0	0	0
	z	−2.310	−2.666	−2.666	−2.666
	p-value	0.021	0.008	0.008	0.008

From Table 1, NHPSO outperforms RPSO and LPSO for all problems tested, across the dimensions. Though, for f_2 and f_3 (see Dimension = 10) RPSO proves to converge faster but eventually got trapped in local optima; whereas NHPSO is more robust, stable and got better solution. For f_3, RPSO proves to be more stable, robust and faster in convergence (see Dimension = 10), but NHPSO obtained better solution. From the statistical analysis presented in Table 2, NHPSO wins over RPSO and LPSO in all the measurements across all dimensions for all the problems. The p-values show

Table 3. Comparison between HTPSO and NHPSO

PB	Variant	MF	SD
f_2	HTPSO	3.55e+01	3.74e+00
	LI-QPSO	2.81e+01	1.78e-01
	NHPSO	**2.78e+01**	**9.06e-01**
f_3	HTPSO	0.00e+00	0.00e+00
	LI-QPSO	1.49e-40	1.39e-40
	NHPSO	0.00e+00	0.00e+00
f_4	HTPSO	−2.00e-06	0.00e+00
	LI-QPSO	-	-
	NHPSO	**0.00e+00**	0.00e+00
f_5	HTPSO	7.55e-03	7.31e-03
	LI-QPSO	5.93e-02	2.42e-02
	NHPSO	**0.00e+00**	**0.00e+00**
f_6	HTPSO	2.96e+01	7.42e+00
	LI-QPSO	7.60e+00	3.49e+00
	NHPSO	**0.00e+00**	**0.00e+00**

that NHPSO performed significantly better than RPSO and LPSO. Therefore, NHPSO is significantly better than RPSO and LPSO in accuracy, stability, robustness, efficiency and reliability.

The success of NHPSO is a result of the combination of the good features of RPSO and LPSO. It was always able to quickly locate promising regions using LPSO; using RPSO, solutions of high quality were quickly discovered with lesser efforts. Its SR was always high because of the ability to escape local optima.

NHPSO is also compared with some recently proposed efficient PSO variants namely, HTPSO [11] and LI-QPSO [3]. These variants are reported to perform well on their studied problems. All the results for HPSO and LI-QPSO were obtained from the respective literature. The parameter settings that were suggested in the original papers are also applied here. Results for dimension = 30 are presented in Table 3 and "–" means results not available. From the results, NHPSO is observed to perform better than HTPSO and LI-PSO in stability, robustness and solution quality.

6 Conclusion and Future Research

Motivated by the common practice of hybridizing PSO technique with non-PSO techniques, a new hybrid PSO (NHPSO) algorithm has been presented in this paper. The algorithm hybridizes linear decreasing inertia PSO (LPSO) and random inertia weight PSO (RPSO), taking advantage of their respective promising properties. It was used to address premature convergence associated with PSO technique. Analysis of the experimental results confirmed that NHPSO is very effective and significantly performed better than RPSO, LPSO and some other efficient PSO variants recorded in literature. Thus, it is practicable and profitable to build optimization algorithm(s) from existing PSO variants. Some of the research directions that can be pursued as future work include comprehensive sensitivity analysis on the effect of the parameter n and application of the proposed algorithm to various real-world problems.

References

1. Eberhart, R.C., Kennedy, J.: A new optimizer using particle swarm theory. In: 6th International Symposium on Micro Machine and Human Science, pp. 39–43. Nagoya, Japan (1995)
2. Arasomwan, M.A., Adewumi, A.O.: Improved particle swarm optimization with a collective local unimodal search for continuous optimization problems. Sci. World J. **2014**, 23 (2013). Special Issue on Bioinspired Computation and Its Applications in Operation Management (BIC)
3. Jiang, S., Yang, S.: An improved quantum-behaved particle swarm optimization algorithm based on linear interpolation. In: IEEE Congress on Evolutionary Computation, pp. 769–775. IEEE Press, New York (2014)
4. Sharifi, A., Kordestani, J.K., Mahdaviani, M.: A novel hybrid adaptive collaborative approach based on particle swarm optimization and local search for dynamic optimization problems. App. Soft Comput. **32**, 432–448 (2015)

5. Samuel, G.G., Asir Rajan, C.C.: Hybrid particle swarm optimization – genetic algorithm and particle swarm optimization – evolutionary programming for long-term generation maintenance scheduling. In: IEEE International Conference on Renewable Energy and Sustainable Energy, pp. 227–232. IEEE Press, New York (2013)
6. Jihong, S., Wensuo, Y.: Improvement of original particle swarm optimization algorithm based on simulated annealing algorithm. In: Eighth International Conference on Natural Computation (ICNC), pp. 777–781 (2012)
7. Sahu, B.K., Pati, S., Panda, S.: Hybrid differential evolution particle swarm optimization optimised fuzzy proportional-integral derivative controller for automatic generation control of interconnected power system. IET Gener. Transm. Dis. 8(11), 1789–1800 (2014)
8. Thangaraj, R., Pant, M., Abraham, A., Bouvry, P.: Particle swarm optimization hybridization perspectives and experimental illustrations. App. Math. Comput. 217(12), 5208–5226 (2011)
9. Sedighizadeh, D., Masehian, E.: Particle swarm optimization methods, taxonomy and applications. Int. J. Comput. Theor. Eng. 1(5), 1793–8201 (2009)
10. Parsopoulos, K.E., Vrahatis, M.N.: UPSO: a unified particle swarm optimization scheme. In: Lecture Series on Computer and Computational Sciences, vol. 1, Proceedings of the International Conference on Computational Methods in Science and Engineering, pp. 868–873. VSP International Science Publishers, Zeist, Netherlands (2004)
11. Hamdan, S.A.: Hybrid particle swarm optimizer using multi-neighborhood topologies. INFOCOMP J. Comput. Sci. 7(1), 36–44 (2008)
12. Qin, Z., Yu, F., Shi, Z., Wang, Yu.: Adaptive inertia weight particle swarm optimization. In: Rutkowski, L., Tadeusiewicz, R., Zadeh, L.A., Żurada, J.M. (eds.) ICAISC 2006. LNCS (LNAI), vol. 4029, pp. 450–459. Springer, Heidelberg (2006)
13. Arasomwan, A.M., Adewumi, A.O.: On the performance of linear decreasing inertia weight particle swarm optimization for global optimization. Sci. World J. 2013, 12 (2013)
14. Zhang, L., Yu, H., Hu, S.: A new approach to improve particle swarm optimization. In: Cantú-Paz, E., et al. (eds.) Genetic and Evolutionary Computation — GECCO 2003. LNCS (LNAI), vol. 2723, pp. 134–139. Springer, Heidelberg (2003)
15. Suganthan, P.N., Hansen, N., Liang, J.J., Deb, K., Chen, Y.-P., Auger, A., Tiwari, S.: Problem definitions and evaluation criteria for the CEC 2005 special session on real-parameter optimization. In: IEEE Congress on Evolutionary Computation, pp. 1–50 (2005)

PSO Applications

An Analysis of Competitive Coevolutionary Particle Swarm Optimizers to Train Neural Network Game Tree Evaluation Functions

Albert Volschenk and Andries Engelbrecht[(✉)]

University of Pretoria, Pretoria, South Africa
engel@cs.up.ac.za

Abstract. Particle swarm optimization (PSO) has been applied in the past to train neural networks (NN) as evaluation functions for zero-sum board games. The NN weights were adjusted using PSO in a competitive coevolutionary approach. Recent analyses of PSO as a NN training algorithm have revealed a serious issue when bounded activation functions are used in the hidden layer of the NNs: Very early during the training process, activation function saturation occurs, at which point weight adjustments stagnate. This paper studies the effect of activation function stagnation on previously used competitive coevolutionary training of NNs using PSO, and shows that the results reported indicates performance similar to making random game moves, and worse than random moves for ply depths larger than one. New results are presented showing more efficient training of NN game tree evaluation functions when unbounded activation functions are used in the hidden layer, and bounded activation functions in the output layer.

Keywords: Competitive coevolution · Particle swarm optimization · Neural network · Game tree

1 Introduction

Competitive coevolutionary training algorithms have recently been used to successfully train the heuristic evaluation functions of game playing agents. Chellapilla and Fogel first introduced the competitive coevolutionary training algorithm in [1]. In their work, the game playing agent did not require any human knowledge to play the game at an expert level. Later, Messerschmidt and Engelbrecht [8] adapted the competitive coevolutionary training algorithm to rather use a particle swarm optimisation (PSO) algorithm to train the neural network (NN). Franken and Engelbrecht [6], and Conradie and Engelbrecht [2] further showed that it is possible to use the approach to train game playing agents for more complex games such as Checkers and Bao.

Although the competitive coevolutionary training algorithm has been proven to be effective, in most cases [2,6] the game tree was limited to a small ply depth due to the large branching factors in games such as Checkers. It was generally

© Springer International Publishing Switzerland 2016
Y. Tan et al. (Eds.): ICSI 2016, Part I, LNCS 9712, pp. 369–380, 2016.
DOI: 10.1007/978-3-319-41000-5_37

accepted that the use of a larger ply depth would increase the performance of the training algorithm. This study shows that, for the PSO algorithms used in these previous studies, that the performance does not improve with increase in ply depth! It is also shown that previous versions of the PSO-based competitive coevolutionary algorithm have performance similar or worse to a random moving player.

Recent analyses of PSO as a NN training algorithm have shown divergence of the swarm very early on in the training process [10,11,13]. The divergent behavior of particles was found to be due to the use of bounded activation functions in the hidden layer of the NNs [14]. Based on this finding, this paper analyzes the performance of the PSO-based competitive coevolutionary training of NNs as game tree evaluation functions using linear activation functions in the hidden layer, while retaining sigmoid activation functions in the output layer. It is shown that performance, considering the game of Checkers, improves significantly, and that performance now improves with increase in ply depth as is expected.

The remainder of the paper is organized as follows: The game of Checkers is briefly described in Sect. 2. The PSO-based competitive coevolutionary algorithm is described in Sect. 3. Section 4 provides an empirical analysis of previous approaches to train NN evaluation functions for Checkers, and provides results for when the hidden unit activation functions are all replaced with linear activation functions.

2   The Game of Checkers

The game of checkers is a well known, and easy to understand board game, yet with its roughly 5×10^{20} possible positions in the search space [9], it is more than complex enough to use as the game of choice for the purposes of this study.

The rules of the world checkers/draughts championship federation [12] are strictly followed with the only modification being the state in which a game is declared as a draw. For this study, a draw is declared if a game exceeds 100 moves (50 per player) without reaching a winning state.

Checkers is a two player game played on an 8×8 board with alternating light and dark cells. One player plays as the *Light* player and the other as the *Dark* player. The board is positioned so that each player has a dark cell in the bottom left corner. The game pieces are placed on the dark cells of the board and can only be moved diagonally onto other dark cells. A game piece may never be on a light cell. There are four different player pieces in the game of Checkers: Light Man, Light King, Dark Man, and Dark King. The Man pieces may only move diagonally forward, whereas the King pieces may move diagonally forward and backwards.

There are two types of moves that a player can make: Normal and Capture/Jump. A normal move is defined as diagonally moving a player's piece onto a neighbouring empty dark cell. A capture move occurs when a player's piece moves diagonally over an opponent's piece and lands on an empty dark cell behind it. A player is forced to make a capture move whenever possible. If there are multiple capture moves available, then it is up to the player to decide which capture move to make. When making a capture move, the player is forced to continue making capture moves with the same piece, until no more capturing moves are available to make. The back row of the opponent is known as the "King Row". When a Man piece reaches the King Row then the piece is "crowned", turning it into a King piece. If a piece is crowned, then the player's turn is immediately ended, even if a capture move presents itself to the newly crowned piece.

Each player starts with 12 Man pieces on the board, filling up the dark cells from the lower left corner. The *Dark* player is always the player to move first and is known as *Player 1*. The *Light* player is referred to as *Player 2*. The game can only have one of two possible outcomes, a win or a draw. A win is defined as when the opponent cannot make a move when it is his/her turn. This occurs when the opponent does not have any more game pieces left, or when the opponent cannot move a piece to an empty cell. Multiple continuous captures/jumps count as one move.

3   PSO Competive Coevolutionary Training Algorithm

The following subsections provide an overview of each individual component within the PSO-based coevolutionary training algorithm and how they interact with one another.

3.1   Game Tree

A standard mini-max game tree of an arbitrary number of lookahead moves (ply depth) is used in the coevolutionary training algorithm. The game tree is constructed by setting the root node as the current board state. Then the child of any node represents a possible move in the game from one board state (represented by the parent) to the next board state (represented by the child). The leaf nodes are given a heuristic value based on the game playing agent's heuristic evaluation function. In the case of the coevolutionary algorithm, the heuristic evaluation function is the neural network. The value of the leaf node's parent is then set based on the level requirement of the parent. For example, a parent on the max level selects the child with the largest heuristic value. The values are then "bubbled" up to the root node in this manner until all of the nodes have been evaluated. The root makes the best move by moving to the child with the best heuristic value. It should be noted that it is assumed that the opponent will always make the best possible move based on the current game playing agent's heuristic evaluation function.

Alpha-beta pruning is used to prune nodes that have no possibility of being selected as the best child.

3.2   Neural Network

The coevolutionary training algorithm uses a feedforward neural network as the heuristic evaluation function of the game tree's leaf nodes. The NN allows for learning to occur in the coevolutionary training algorithm by adjusting the NN's weights after each iteration. The hope is that this will produce a slightly better heuristic evaluation function than before. Thus, over time, a more skilful game playing agent will evolve.

A standard feedforward NN consists of an input layer, a hidden layer and an output layer. The input layer consists of a number of input nodes, each containing a given input to the NN. The hidden layer consists of an arbitrary number of hidden nodes, each node connected through weighted connections to every node in the input layer. The input layer and the hidden layer are both augmented by a bias unit, which has a constant input value of -1 [5]. The output layer also consists of an arbitrary number of nodes, each connected to all the nodes of the hidden layer. The nodes in the hidden and output layer have an activation function which is used to calculate their respective output values. The output of the activation functions in the output layer are used as the output of the NN.

The NN is trained by adjusting the weights of the weighted connections in the hidden and output layers. This study uses one of two types of activation functions, one being the sigmoid activation used in [2,5] and the other being a simple linear activation. The sigmoid activation function is given by [3]:

$$f(net) = \frac{1}{1 + e^{-net}} \tag{1}$$

where net is defined as the sum of the weighted input signals to the neuron:

$$net = \sum_{i=1}^{I+1} w_i z_i \tag{2}$$

where I is the number of input connections, w_i is the weight of the i-th connection and z_i is the output of the i-th input signal. The linear activation function is given by:

$$f(net) = \lambda net \tag{3}$$

where λ is a variable that controls the gradient of the linear function.

3.3   Particle Swarm Optimisation

Messerschmidt and Engelbrecht [8] were the first to adapt the competitive coevolutionary training algorithm, introduced by Chellapilla and Fogel [1], to use a PSO algorithm as a method of training the neural network (NN). In their implementation, each particle's position represents the set of weights in the NN. Thus changing the particle's position, changes the weights of the NN.

The particle's velocity is updated using [7]:

$$v_i(t+1) = wv_i(t) + c_1 r_1(t)[y_i(t) - x_i(t)] + c_2 r_2(t)[\hat{y}_i(t) - x_i(t)] \qquad (4)$$

where w is the inertia weight, c_1 and c_2 are the acceleration coefficients, $x_i(t)$ is the current position of the i-th particle, $y_i(t)$ is the particle's personal best (pbest) position, $\hat{y}_i(t)$ is the neighborhood best (nbest) position, and $r_1(t), r_2(t) \sim U(0,1)^{n_x}$, where n_x is the dimension of the search space [4].

The position update equation is

$$x_i(t+1) = x_i(t) + v_i(t) \qquad (5)$$

A maximum velocity may be applied to prevent particles from exploding in diversity and never being able to converge to an optimal solution [5].

Various types of social network structures exist for the PSO [3], but due to the results provided by Franken and Engelbrecht [6], this study uses only the Von Neumann structure, as it was found to be the most effective social structure to use in a PSO coevolutionary training algorithm [6].

3.4 PSO Coevolutionary Training Algorithm

The PSO coevolutionary training algorithm is described as follows [2,6]:

1. Initialise the swarm of game playing agents to the positions
   represented by the initial weights of their corresponding
   neural networks
2. Repeat until the predefined stopping condition
   (a) Add the pbest of each particle to the competition pool
   (b) For every particle in the competition pool
       i. Set score to zero. This score will be used as the fitness
          value of the particle in the swarm.
       ii. Randomly select n players from the competition pool
           - Play a game as *Player 1* against each of the
             opponents
           - Update the particle's overall score based on the
             outcome of each game
   (c) Update the nbest particles of each neighbourhood by
       selecting the particle with the highest score in each
       neighbourhood
   (d) Update the pbest of each particle in the swarm by setting
       the particle's pbest to the current position, if the current
       score is higher than its pbest's score
   (e) Update the global best (gbest) position of the swarm by
       selecting the particle with the highest pbest score within
       the swarm
   (f) Update the velocities of each particle in the swarm
3. Return the final gbest position in the swarm

The final gbest particle is used as the NN game tree evaluation function.

4   Empirical Analysis

The empirical analysis consists of two steps: The first is to repoduce previous results using the model as proposed in [6] and to illustrate the problem that this model exhibits with increase in ply depth. The second step presents results when linear activations functions are used.

4.1   Experimental Setup

For each experiment, 30 independent runs of the respective algorithm have been executed. Results are presented as the mean and standard deviation over these 30 runs.

The input of the NN is the same as the input used by Franken and Engelbrecht [6]. The player pieces can only be on the dark cells of the board, thus only the states of the 32 dark cells need to be given as input to the NN. Input values per cell are as follows: 1.0 for Player King, 0.75 for Player Man, 0.5 for Empty Cell, 0.25 for Opponent Man, and 0.0 for Opponent King.

The weights of the NN are initialised to random values within the range of $(\frac{-1}{\sqrt{fanin}}, \frac{1}{\sqrt{fanin}})$, where $fanin$ is the number of input weights for the specific neuron [5].

A single hidden layer is used, where the number of nodes in the hidden layer depends on the activation function used. The linear activation function requires twice as many hidden nodes as the sigmoid activation function in order to obtain the same approximation accuracy as the non-linear sigmoid activation function. The number of hidden nodes used within the hidden layer is set to the following:

- Sigmoid activation function: 5 hidden nodes, since Franken and Engelbrecht [6] found that it produced the best results for the Von Neumann structure.
- Linear activation function: 10 hidden nodes, to obtain the same approximation accuracy as the non-linear sigmoid activation function.

A single output node is used in the NN. The output of this node is used as the heuristic value of the leaf nodes in the game tree, thus a larger value represents a better board state for the max player, and a smaller value represents a better board state for the min player.

Throughout the study, the PSO is instantiated with 27 particles using the Von Neumann neighbourhood structure. The structure follows a $3 \times 3 \times 3$ layout. Since a particle's position represents a possible weight configuration of a NN, each particle's initial position is set to its respective NN's initial weight values. The initial velocity of each particle is set to zero based on the findings of Engelbrecht [4], where it was shown that an initial velocity of zero outperforms other velocity initialisation strategies. A predefined maximum number of epochs is used as the stopping condition in each of the experiments. The maximum velocity, velocity clamping method, acceleration coefficients and the inertia value are experiment dependant variables and are listed in the subsections below for each experiment.

The fitness of each particle is calculated within the coevolutionary environment. The competitive coevolutionary algorithm is used as a method of calculating the fitness of a particle after each update step within the PSO algorithm. In this study, the fitness of each particle is calculated by using the same method used by Franken [5]. This is done in *Step 2* of the algorithm given in Sect. 3.4: Each particle's pbest is added to the set of particles that compete against each other during training in order to compare the particle's current performance against the performance of its pbest [6]. The set of particles competing against each other is also known as the competition pool. This is an important step because the fitness value is a relative score based on the performance of the particle in the swarm. If the average swarm performance increases but a particle's pbest position does not change, then the pbest's fitness may still change because it finds itself playing against better opponents. After each game, a score is given as follows [6]: +1 for a win, 0 for a draw, and −2 for a loss. Particles are therefore punished for loosing games, and rewarded for winning games. The fitness of the particle is then set to the sum of the scores obtained for each game played [6]. For the purposes of this study the number of games played by each particle is set to $n = 5$.

The final performance of the algorithm is calculated as the F-measure defined by Franken [5]. The F-measure is calculated by taking the gbest particle in the swarm, after the training is complete, and by playing 10 000 games as player 1 and 10 000 games as player 2 against a random moving player. The performance measure is then calculated for player 1 and player 2 respectively, and the mean value is used as the final F-measure for the game playing agent. The performance measure, μ, is calculated using [6]:

$$\mu = \sum_{i=1}^{l} w_i f(x_i) \tag{6}$$

where x_i is the outcome, w_i is the weight associated with the outcome and $f(x_i)$ represents the probability of the corresponding outcome obtained during the 10 000 games played as that player. There are three types of outcomes, each with the following weights: 3 for a win, 2 for a draw, and 1 for a loss.

Running a simulation of 20 000 games between two random move making agents resulted in the probabilities listed in Table 1. This does slightly differ from the values in [6], but still indicates that there is not a distinct advantage for playing as either player.

Table 1. Probabilities of a Random moving player 1

Outcome	Probability
Win	0.404
Lose	0.401
Draw	0.195

Calculating the mean F-measure of 10 games between two players making random moves resulted in an F-measure of 50.154. This indicates that successful training can only occur if a game playing agent achieves an F-measure greater than 50.154 [6].

4.2 Reproducing Franken and Engelbrecht's [6] Results

Configuring the experiment to be exactly the same as the configuration used by Franken and Engelbrecht [6] resulted in the following configuration: sigmoid activation function, ply depth of one, 5 hidden units, 500 epochs, no velocity clamping, c_1 and c_2 of 1.0, and w of 1.0.

The experiment was run 10 times and resulted in a mean value of 60.7974, which is similar to Franken and Engelbrecht's [6] result of 61.8595. The next step was to run the experiment on a ply depth of two. It is generally excepted that a larger ply depth will increase the performance of the game playing agent. Yet, by analysing the results shown in Table 2, it was clear that a drastic decline in performance occurred when the ply depth was increased by only one. The mean value dropped from 61.8595 to 24.2775.

Since this was a completely unexpected result, an investigation of the particles within the PSO was conducted. By looking at the final position of the gbest particle of an arbitrary run, it was discovered that the particle's position contained extremely large/small values. For the sake of readability, only the first four position values of the particle is given in Table 4, and it can clearly be seen how large/small the position values are. Further investigation yielded the same results when looking at the particle's velocity at the end of training. Table 5, again only containing the first 4 velocity values, shows the same result where the values are either extremely large or extremely small. The effect of the very large velocity values were investigated by running 10 iterations of the same initial experiment, but this time with the number of training epochs set to zero.

Table 2. Franken and Engelbrecht [6] Ply Depth 2

Run	F-Measure
Run 1	32.8575
Run 2	21.8375
Run 3	23.0
Run 4	14.825
Run 5	35.6025
Run 6	14.2325
Run 7	28.2525
Run 8	17.6625
Run 9	19.6125
Run 10	34.7925
Mean	24.2775

Table 3. Ply Depth 1 for Random Moving Agent

Run	F-Measure
Run 1	65.28
Run 2	65.145
Run 3	69.9575
Run 4	68.5375
Run 5	59.9499
Run 6	59.245
Run 7	59.3799
Run 8	59.335
Run 9	48.425
Run 10	79.1875
Mean	63.444

Table 4. gbest's First 4 position values

Particle Position Values
1.436364166794913E28
−4.4464060419901927E27
−2.247930194420741E26
−1.2255266548832014E26

Table 5. gbest's First 4 Velocity values

Particle Velocity Values
−1.840995674142912E26
−5.170844312869841E26
1.1921622636913999E26
1.906136645211881E24

This gives a baseline performance of untrained game playing agents (i.e. random moving agents) which can be used for a comparison against the performance of Franken and Engelbrecht's trained game playing agents [6].

The results obtained are summarised in Table 3, and it is clear that the performance of the untrained gbest, 63.444, was roughly the same as the performance of the trained gbest, 61.8595 (Von Neumann, 25 particles), in Franken and Engelbrecht's paper [6]. This is an interesting result since it was previously accepted that an F-measure greater than 50 shows that training had occurred [6]. Yet, from this result it can be concluded that by just selecting the initial gbest of a swarm, an average F-measure of 63.444 can be obtained. This suggests that training only occurs once an F-measure greater than 63.444 is obtained. Furthermore, it shows that the results obtained by Franken and Engelbrecht [6] was influenced by the large particle velocities to such an extent that no actual training had occurred.

Franken's investigation into the poor performance of the training algorithm found similar results of exploding particle velocities. It was explained that the large values resulted in saturation of the sigmoid activation function used within the NN [5]. Research by van Wyk and Engelbrecht [13] also found similar results where the PSO's swarm drastically diverged when using the sigmoid activation function. They found that the boundaries of bounded activation functions, such as the sigmoid activation function, were responsible for the diverging behaviour of the particles in the swarm. This was due to the flat slope of the activation function near the boundaries, which resulted in a situation where even a big change in a particle's velocity could not change the fitness of the particle. Thus the velocities kept growing in size, resulting in an explosion of divergence in the swarm [13]. Further experiments by Franken [5] found that a maximum velocity (VMax) and a smaller inertia value, reduced divergent behaviour of the swarm. Another possible solution proposed by van Wyk and Engelbrecht [13] is the use of an unbounded activation function in the form of a linear activation function. Van Wyk and Engelbrecht [13] did, however, note that the use of a linear activation function requires more hidden nodes to be added in order to obtain the same approximation accuracy as non-linear activation functions. For the purposes of this study, when using the linear activation function, the number of hidden nodes is doubled to a value of 10.

Another interesting result when using Franken and Engelbrecht's approach [6] with a ply depth of two, was that the average outcomes for playing as Player 1 and Player 2 respectively, showed a slight advantage towards playing as player 1. Table 6 shows the average results obtained for each player playing at a ply depth of 2. There is a clear difference between the two players with Player 1 having a much better average F-measure than Player 2. Since it was shown that with two random moving players neither player had an advantage, it is concluded that there exists a problem with the current training algorithm's method of training Player 1 and Player 2. A possible solution to the problem may be to play as Player 1 and as Player 2 against the n number of randomly selected players in each update step. This is done with the hope of giving a more accurate measure of how the particle performs as Player 1 and Player 2 during training. It should be noted that this was not tested in depth, and this paper leaves the matter open for further investigation since it is not seen as the focus of this study.

Table 6. Player 1 vs Player 2 Ply - Depth 2

	Player 1	Player 2
Win	39030	3559
Lose	46055	88722
Draw	14915	7719
F-measure	46.4875	14.837

4.3   Using Linear Activation Functions

The next experiment shows that the problem illustrated in the previous section is solved when linear activation functions are used in the hidden layer instead of sigmoid activation functions. Note that the output layer still uses sigmoid activiation functions.

A ply depth of four is used since it is large enough to produce a sufficient amount of $\alpha\beta$-pruning, yet it is small enough to run multiple iterations in order to achieve statistical soundness. Franken [5] found that, with the saturated sigmoid activation function, the PSO converges onto a solution extremely early, thus the number of epochs was reduced from 500 to 300. A slightly larger maximum velocity than Franken's [5] maximum velocity is used because the velocity is not clamped to the edge, but rather reset to 0 when the velocity reaches VMax. This is to provide dynamic alternation between exploration and exploitation of the search space, preventing the particles from becoming stuck in a local optimum or not being able to converge onto a global optimum. The inertia value is set to the same value used by Conradie and Engelbrecht [2]. The experiment is run 30 times and is configured as follows: linear activation functions in the hidden layer, a ply depth of 4, 10 hidden units, 300 epochs, VMax of 0.2 with velocities reset to 0 when it exceeds VMax, $c_1 = c_2 = 1.2$, and $w = 0.729844$.

The results are shown in Table 7. The most significant finding is that the final F-measures, with an average of 91.5899, are well above the 63.444 required in order to declare that training had occurred. Another noteworthy result is the fact that Player 1 and Player 2's F-measures are closer than before, yet player 1 still has a slightly better performance. Most notably, when comparing the standard deviations of the two players, Player 2 is found to have almost double the standard deviation value as player 1. Since this is not the focus of this study, it is left as future work to further analyse why this is in fact occurring.

Table 7. Linear Activation Functions Ply Depth 4

	Mean	Minimum	Maximum	Standard Deviation
Player 1	97.192	90.3	99.54	2.11540
Player 2	85.9878	76.785	91.435	4.03593
Average	91.5899	87.81	94.425	2.02789

The results above confirms previous studies that PSO fails to train NNs when sigmoid activation functions are used in the hidden layer, and that using linear activation functions in the hidden layer now provides an efficient competitive coevolutionary approach to train NNs as game tree evaluation functions.

5   Conclusions

This paper showed that previous results obtained by using a PSO-based competitive coevolutionary algorithm to train neural networks (NNs) as evaluation functions for board games, do not reflect good performance as reported in the literature. PSO as NN training algorithm suffers from the problem of activation function saturation when bounded activation functions are used in the hidden layers. Such saturation causes particles to diverge, and no further training of the NN occurs.

It is shown in this paper that saturation results in performance similar to players making random moves at a ply depth of one, and also shows performance significantly worse than making random moves for larger ply depths.

The paper shows that performance significantly improves when linear activation functions are used in the hidden layer, and that performance of the evolved Checker players improves with increase in ply depth, as is expected.

Furture research should investigate in more depth the problem with activation function saturation and the failure of PSO as a NN training algorithm when bounded activation functions are used in the hidden layer.

References

1. Chellapilla, K., Fogel, D.: Evolving neural networks to play checkers without relying on expert knowledge. IEEE Trans. Neural Netw. **10**(6), 1382–1391 (1999)

2. Conradie, J., Engelbrecht, A.: Training Bao game-playing agents using coevolutionary particle swarm optimization. In: IEEE Symposium on Computational Intelligence and Games. pp. 67–74 (2006)
3. Engelbrecht, A.P.: Computational Intelligence: An Introduction, 2nd edn. Wiley Publishing, Hoboken (2007)
4. Engelbrecht, A.: Particle swarm optimization: velocity initialization. In: Proceedings of the IEEE Congress on Evolutionary Computation, pp. 1–8 (2012)
5. Franken, C.J.: PSO-based Coevolutionary Game Learning. Master's thesis (2004)
6. Franken, N., Engelbrecht, A.: Comparing PSO structures to learn the game of checkers from zero knowledge. In: Proceedings of the IEEE Congress on Evolutionary Computation, vol. 1, pp. 234–241 (2003)
7. Kennedy, J., Eberhart, R.: Particle swarm optimization. In: Proceedings of the International Joint Conference on Neural Networks, vol. 4, pp. 1942–1948 (1995)
8. Messerschmidt, L., Engelbrecht, A.: Learning to play games using a PSO-based competitive learning approach. IEEE Trans. Evol. Comput. $8(3)$, 280–288 (2004)
9. Schaeffer, J., Lake, R.: Solving the game of checkers. Games No Chance 29, 119–133 (1996)
10. Scheepers, C., Engelbrecht, A.: Analysis of stagnation behaviour of competitive coevolutionary trained neuro-controllers. In: Proceedings of the IEEE Swarm Intelligence Symposium (2014)
11. Scheepers, C., Engelbrecht, A.: Training multi-agent teams from zero knowledge with the competitive coevolutionary team-based particle swarm optimiser. Soft Comput. $20(2)$, 607–620 (2016)
12. WCDF: Rules of Draughts (Checkers) (2015). http://www.wcdf.net/rules.htm. (Accessed on 19 October 2015)
13. van Wyk, A., Engelbrecht, A.: Overfitting by PSO trained feedforward neural networks. In: Proceedings of the IEEE Congress on Evolutionary Computation (2010)
14. van Wyk, A., Engelbrecht, A.: Lamda-Gamma learning with feedforward neural networks using particle swarm optimization. In: Proceedings of the IEEE Swarm Intelligence Symposium (2011)

Particle Swarm Optimization for Calculating Pressure on Water Distribution Systems

Lala Septem Riza[1](✉), Azhari Fathurachman Azmi[1], Waslaluddin[1],
Eka Fitrajaya Rahman[1], and Kuntjoro Adji Sidarto[2]

[1] Universitas Pendidikan Indonesia, Setiabudhi 229, Bandung, Indonesia
{lala.s.riza,wasluddin,efitrajaya}@upi.edu, azh.udje2@gmail.com
[2] Institut Teknologi Bandung, Ganesha 10, Bandung, Indonesia
sidarto@math.itb.ac.id

Abstract. Flow assurance, aimed to ensure the availability of water flow rate and the sufficiency of pressure on each customer, is one of objectives that should be achieved by water supplying companies. An essential step before dealing with it is to predict pressure distribution on each node. Using the analogy of Kirchoff's Law for the electrical current to the flow of water in pipelines, a non-linear equation system involving fluid dynamics modeling is constructed and used for determining pressure distribution. It is obvious that the system is not a simple one since it contains many non-linear equations expressing the complexity of the network. In this study, we implement Particle Swarm Optimization (PSO) to solve the system by transforming a root-finding task into an optimization problem. Finally, we present a case study using Hanoi network along with a result compared with EPANET, Firefly Algorithm (FA), and a combination of Genetic Algorithm (GA) and Newton's method.

Keywords: Pressure distribution · Swarm intelligence · Fluid dynamics

1 Introduction

In every city, local governments have an obligation to deliver clean water through pipeline networks since it is a principle need for the society. This duty can be accomplished by a company pointed by the governments after signing an official contract. As a common objective in business, the company must increase gain and reduce cost at the same time. Moreover, it has responsibility to the society and government by fulfilling the standard quality of the service as stated in the contract. One of common issues that should be taken into account by the company is related to daily problems happened on water distribution systems.

The water distribution system is a complex pipeline network containing reservoirs (inlets), junctions (intersection points), customers (outlets), pipelines, and other facilities (e.g., pumps, chokes, and valves). In the model, we represent an inlet, outlet, and junction as a node while a line is used to express a pipeline. It should be noted that because of different geographical locations, the network

© Springer International Publishing Switzerland 2016
Y. Tan et al. (Eds.): ICSI 2016, Part I, LNCS 9712, pp. 381–391, 2016.
DOI: 10.1007/978-3-319-41000-5_38

is possible to have elevations on pipelines/nodes. Moreover, the water is usually delivered through a pipeline due to differences of head on both ends of the pipeline.

Based on a review in the literature, basically we can classify some issues on water distribution systems into two following tasks: optimization of networks and prediction of pressure distribution. The first task appears on the design phase of the networks; the following studies are related to this task: [1] and [2] implementing GA for diameter optimization on the networks. An optimal design of water distribution networks can be also achieved by using shuffled complex evolution [3] and tabu search algorithm [4]. Second, on the operational stage where the networks have been established, we need to observe and check pressure distribution on each node. Beside water flow rate into and out from the network, monitoring pressure distribution is an important task since flow assurance, which is a target written the contract, cannot be achieved without it. Some studies aimed to accomplish this task can be found on the following manuscripts: [5–8]. According to the classification of the tasks above, this paper is aimed to carry out the second one.

In order to obtain adequate results, the Hazen-Williams equation is used as a fluid-dynamics model representing water flowing through a pipe [8]. Then, we construct a non-linear equation system involving the formula for expressing the continuity equations at every point on the water distribution system. So, it is obvious that the more complex pipeline networks we have, the harder system of equations to be solved. Finally, PSO, which is a swarm-intelligence method inspired by the movements of particle/birds [9,10], is used to solve the system.

The remainder of this paper is structured as follows. Section 2 introduces to PSO and its pseudo code while Sect. 3 briefly gives an introduction to fluid dynamics used in this research. In Secton 4, we illustrate how to construct a model based on the Hazen-William equation. An example of Hanoi network and its result and discussion are shown in Sects. 5 and 6. Finally, Sect. 7 concludes the paper.

2   Particle Swarm Optimization

PSO was introduced by Kennedy and Eberhart in 1995, inspired by the behavior of the social animals/particles, like a flock of birds in a swarm [9]. The birds dynamically move through space with a velocity according to their historical movements. Moreover, they have a tendency to move to a space area giving better conditions after going through the searching process.

Basically, PSO begins with generating a population of candidate solutions (i.e., particles) randomly. We can represent the population as a two-dimension matrix, where rows represents particles whereas columns contain values of each variable. Each particle will update its better position by adding a certain velocity. The velocity is determined by two values obtained as a solution, which are "global best" (G_{best}) related to the best solution that has ever been achieved by the particles and "local best" (L_{best}) representing the best solution on the population of a certain iteration. For example, in a minimization problem the global best G_{best} will be updated by the local best L_{best} if on a particular iteration we have $L_{best} < G_{best}$. The process is repeated until the given criteria are met.

There are many variant algorithms of PSO in the literature. In this research, we are using PSO with inertia weight (w) as illustrated in Algorithm 1. It can be seen that we need to supply the following parameters: the objective function $F(x)$, numbers of population, initial value of "global best," and two acceleration coefficients. As in evolutionary algorithms, first an initial population is generated by considering the numbers of populations. According to the objective function, we calculate "local best" of the current population, and then compare it with "global best." We update "global best" if "local best" has a better value. Next step is to initialize a velocity. After that, we repeat the following processes according to the number of maximum generation ($MaxGeneration$), populations (N_p), and unknown variables (dim): generating random numbers (on Line 8), updating the velocity (on Line 9) with involving the inertia-weight equation (on Line 10), updating the particle's position (on Line 11), and updating "local best" and "global best" (on Line 12–17). By looking this algorithm, basically the essential part in PSO is on defining the objective/fitness function that represents the problem in hand.

In the literature, we can find some implementations of PSO for dealing with many problems, such as in loading frequency control of energy [12], optimal scheduling of multiple geosynchronous satellites refueling [13], and optimum allocation model for earthquake emergency shelters [14].

3   Fluid Dynamics Modeling

In this section, we present two following subsections: Subsect. 3.1 presenting the Bernoulli formula as the principal knowledge of fluid dynamics modeling and Subsect. 3.2 illustrating the equation of water flow through a pipeline used in this research.

3.1   The Bernoulli Equation

According to the principle of conservation of energy, the sum of all forms of energy in a fluid through a pipeline is the same at all points on that pipeline in a steady-state condition. In order to fulfill this, the sum of all forms of energy must be constant. Three forms of energy that should be considered are kinetic (E_K), potential (E_P), and internal/flow (E_A). So, for incompressible flow the equation, i.e., the Bernoulli equation [15], can be formulated as follows:

$$Constant = E_A + E_P + E_K$$
$$= \frac{p}{\rho} + gz + \frac{v^2}{2}, \tag{1}$$

input : Objective function $F(x)$, where $x = (x^1, \ldots, x^{dim})^T$ and dim is a number of variables/dimension

The number of particles on a population N_p

The initial global best G_{best}^0

The acceleration coefficients: c_l and c_g

output: The best particle x

1 Generate initial population of particles x^i $(i = 1, 2, \ldots, N_P)$;
2 Calculate the initial local best L_{best}^0;
3 Update G_{best}^0 if it is worse than L_{best}^0;
4 Determine the initial velocity v^0;
5 **while** $t < MaxGeneration$ **do**
6   **for** $i = 1 : N_p$ **do**
7   **for** $d = 1 : dim$ **do**
8   Pick random numbers: r_l and r_g;
9   Update the particle's velocity:
$$v^{i,d} = wv^{i,d} + c_l r_l (L_{best}^{i,d} - x^{i,d}) + c_g r_g (G_{best}^d - x^{i,d}),$$
10   where $w = w_{max} - iter \frac{w_{max} - w_{min}}{iter_{max}}$;
11   Update the particle's position: $x^i = x^i + v^i$;
12   **if** $(F(x^i) < F(L_{best}^i))$ **then**
13   Update the local best: $L_{best}^i \leftarrow x^i$;
14   **if** $F(L_{best}^i) < F(G_{best})$ **then**
15   Update the global best: $G_{best} \leftarrow L_{best}^i$;
16   **end**
17   **end**
18   **end**
19   **end**
20 **end**

Algorithm 1. Pseudo code of PSO with inertia weight [11].

where:

- v represents the velocity of fluid on a pipeline,
- ρ is the density of the fluid; in the case of water, $\rho \approx 1$,
- g is the specific weight,
- z is the elevation of the point,
- p is the pressure of the point.

Moreover, each element on Eq. 1 has the following units: Newton-meter ($N.m$) on SI or foot-pounds ($ft - lb$) on U.S. Customary System. The constant in the Bernoulli equation can be normalized in the term of total head or energy head H, as follows:

$$H = z + \frac{p}{\rho g} + \frac{v^2}{2g}. \tag{2}$$

In the case of two different points on a pipeline, we obtain the following equation:

$$H_1 = H_2$$

$$z_1 + \frac{p_1}{\rho g} + \frac{v_1^2}{2g} = z_2 + \frac{p_2}{\rho g} + \frac{v_2^2}{2g}. \tag{3}$$

3.2 The Hazen-William Equation

In the case of water flowing through a pipe, a well-known equation representing head loss is based on Hazen-William [8]. It can be used only for the range of diameter between 1/6 ft (2 inch) and 6.0 ft (72 inch) and flow rate not exceeding 10 ft/s. The formula (in SI) can be expressed by

$$Q = 0.2787 C_k D^{0.63} \left(\frac{h_L}{L}\right)^{0.54}. \tag{4}$$

By considering head loss (h) due to friction (L) on two different points, h_L can be expressed by

$$h_L = \frac{p_i - p_j}{g} + z_i - z_j. \tag{5}$$

Therefore, we have the following equation in U.S. Customary System [8]:

$$Q_{i-j} = 0.4329 C_k D^{0.63} \left(\frac{1}{L}\left(\frac{p_i - p_j}{g} + z_i - z_j\right)\right)^{0.54}. \tag{6}$$

As illustrated in Eq. 6, two indexes of nodes (i.e., i and j) are connected by a pipeline with the length L and the diameter D. Then, Q_{i-j} declares the water flow rate in the pipeline from i to j in ft^3/s, while p_i and p_j denote the pressure at the nodes i and j in lb/ft^2. The elevations of two nodes, z_i and z_j, are in ft. Moreover, the equation uses the specific weight g, which is equal to 62.4 lb/ft^3 on the temperature of $60^o F$.

4 The Model Construction

4.1 Water Distribution Networks

As mentioned previously, water distribution networks consist of reservoirs/inlets, junctions, customers/outlets, pipelines, pumps, chokes, and other facilities, that are represented in a graph containing lines and several symbols of nodes. While a line is used for depicting a pipe, a facility such as a junction, an inlet, and an outlet can be represented as a node. Then, in the computation, we can build an adjacency matrix for representing the graph.

In this research, a constructed model is aimed to determine pressure distribution on water distribution systems. In order to get a valid solution, the following requirements should be fulfilled:

- At least one node has a given pressure, while the rest will be calculate. Usually, reservoir's pressure is known or measured.
- For simplicity, here it is assumed the system was in a steady state, isothermal conditions, and no pumps and control valves on the network.

We build a model by considering Kirchoff's Law for the electrical current in an electrical circuit as the analogy for the water flow in pipeline networks [5]. So, obviously the sum of the amount of water into and out of a node is equal to zero. For example, we have a simple network containing six nodes with one inlet and five outlets as shown in Fig. 1. By considering Kirchoff's Law, we obtain a non-linear equation system consisting of six equations (i.e., f) corresponding to six nodes, as follows:

$$
\begin{aligned}
f_1 &= Q_{N1} - Q_{1-2} = 0, \\
f_2 &= Q_{1-2} - Q_{2-3} - Q_{2-4} - Q_{2N} = 0, \\
f_3 &= Q_{2-3} - Q_{3-6} - Q_{3-5} - Q_{3N} = 0, \\
f_4 &= Q_{2-4} - Q_{4N} = 0, \\
f_5 &= \pm Q_{3-5} \pm Q_{5-6} - Q_{5N} = 0, \\
f_6 &= \pm Q_{3-6} \pm Q_{5-6} - Q_{6N} = 0,
\end{aligned}
\tag{7}
$$

where Q_{N1} is a flow rate of the reservoir/supplier (i.e., Node 1) while Q_{2N}, Q_{3N}, Q_{4N}, Q_{5N}, and Q_{6N} are given flow rates on the outlets 2, 3, 4, 5, and 6, respectively. The minus sign on the equations means that the flow rate goes out from the system. It should be noted that we have the sign $\pm$ on f_5 and f_6, which means the direction of the water flow is unknown on these nodes. It happens when a loop is exist on the water distribution systems. The computation will be looking for the suitable direction based on pressure distribution so that all equations are met. Moreover, Q_{i-j}, which is a flow rate on the pipeline connecting the following nodes: i and j, is substituted by Eq. 6. Usually, the flow rate Q_{N1} and pressure on reservoir p_1 are known along with diameter and length of all pipelines, and we need to determine pressure/head on other nodes so that the values f is closed to 0. After determining pressure/head distribution, we can also define the flow direction on the system.

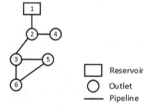

Fig. 1. A simple network containing 6 nodes.

4.2   Solving with Particle Swarm Optimization

As we mentioned, in this research we attempt to solve the non-linear equation system by performing PSO for determining pressure distribution on each node. Basically, three following steps need to be considered to do in this task: defining a fitness function, initializing PSO parameters, and following Algorithm 1.

First, we can define the following objective function: $F(x) = |f_1(x) + f_2(x) + \ldots + f_N(x)|$, where f_i is the continuity equation as the example written on Eq. 7. In this case, obviously PSO is aimed to obtain the minimum value of $F(x)$. Therefore, we basically transform the finding roots on a non-linear equation system to be an optimization task by minimizing the value F, which is closed to 0. Then, we set the following parameters: N_p, G_{best}^0, and the acceleration coefficients c_l, c_g. Lastly, we just need to follow the steps as shown in Algorithm 1.

5   A Case Study: Hanoi Network

As mentioned in the previous study [5], Hanoi network [16] is a model that is often used by many researchers as their experiments. For example, Tabu search algorithm was used for water network optimization on Hanoi network [4].

Hanoi network has one reservoir, 31 customers, and no elevation on each node as shown in Fig. 2. In this experiment, the following data need to be assigned: pressure/head on reservoir, flow rates on each node, diameter, and length. While the flow rate on the reservoir (i.e., Node 1) are 195.5 ft^3/s, rates on other nodes have been illustrated in [5]. Regarding the pressure/head on reservoir (i.e., Node 1), we assign it to 20472.4 lb/ft^2, whereas pressures on other nodes are unknown and will be calculated. Moreover, sizes of diameter and length of each pipeline can be also seen in [5]. Regarding the PSO parameters, in this experiment we set 100 particles, 500 iterations, and 2 for both values of c_l and c_g.

6   Results and Discussion

After running a simulation, a non-linear equation system can be built by considering the schema of Hanoi network and the value of the objective function. Since we have 32 nodes in Hanoi network, we obtain a system consisting of 32 non-linear equations as shown in Table 1. Because Hanoi network contains looping pipelines, basically there are many possibilities of flow directions. So, before a simulation has finished, actually we do not know which the flow direction on the circle is correct and met with the equations. On the other words, the signs of Q_{i-j} can be positive or negative. Therefore, PSO determines the correct one as represented by signs on the equations shown in Table 1 and by arrows on Fig. 2.

Besides constructing the model, we also compare the results with EPANET, which is a software used for developing model of water distribution system. It was developed by the United States Environmental Protection agency's (EPA) Water Supply and Water Resources Division in 1993 [17]. Moreover, results on two previous studies, which are based on FA [5] and a combination of GA and

Table 1. The model of Hanoi network.

Node	Equations	Node	Equations
1	$f_1 = Q_{N1} - Q_{1-2} = 0$	17	$f_{17} = Q_{18-17} - Q_{17-16} - Q_{17N} = 0$
2	$f_2 = Q_{1-2} - Q_{2-3} - Q_{2N} = 0$	18	$f_{18} = Q_{19-18} - Q_{18-17} - Q_{18N} = 0$
3	$f_3 = Q_{2-3} - Q_{3-4} - Q_{3-19} - Q_{3-20} - Q_{3N} = 0$	19	$f_{19} = Q_{3-19} - Q_{19-18} - Q_{19N} = 0$
4	$f_4 = Q_{3-4} - Q_{4-5} - Q_{4N} = 0$	20	$f_{20} = Q_{3-20} - Q_{20-21} - Q_{20-23} - Q_{20N} = 0$
5	$f_5 = Q_{4-5} - Q_{5-6} - Q_{5N} = 0$	21	$f_{21} = Q_{20-21} - Q_{21-22} - Q_{21N} = 0$
6	$f_6 = Q_{5-6} - Q_{6-7} - Q_{6N} = 0$	22	$f_{22} = Q_{21-22} - Q_{22N} = 0$
7	$f_7 = Q_{6-7} - Q_{7-8} - Q_{7N} = 0$	23	$f_{23} = Q_{20-23} - Q_{23-24} - Q_{23-28} - Q_{23N} = 0$
8	$f_8 = Q_{7-8} - Q_{8-9} - Q_{8N} = 0$	24	$f_{24} = Q_{23-24} - Q_{24-25} - Q_{24N} = 0$
9	$f_9 = Q_{8-9} - Q_{9-10} - Q_{9N} = 0$	25	$f_{25} = Q_{24-25} + Q_{26-25} - Q_{25-32} - Q_{25N} = 0$
10	$f_{10} = Q_{9-10} - Q_{10-11} - Q_{10-14} - Q_{10N} = 0$	26	$f_{26} = Q_{27-26} - Q_{26-25} - Q_{26N} = 0$
11	$f_{11} = Q_{10-11} - Q_{11-12} - Q_{11N} = 0$	27	$f_{27} = Q_{16-27} - Q_{27-26} - Q_{27N} = 0$
12	$f_{12} = Q_{11-12} - Q_{12-13} - Q_{12N} = 0$	28	$f_{28} = Q_{23-28} - Q_{28-29} - Q_{28N} = 0$
13	$f_{13} = Q_{12-13} - Q_{13N} = 0$	29	$f_{29} = Q_{28-29} + Q_{30-29} - Q_{29N} = 0$
14	$f_{14} = Q_{10-14} - Q_{14-15} - Q_{14N} = 0$	30	$f_{30} = Q_{31-30} - Q_{30-29} - Q_{30N} = 0$
15	$f_{15} = Q_{14-15} + Q_{16-15} - Q_{15N} = 0$	31	$f_{31} = Q_{32-31} - Q_{31-30} - Q_{31N} = 0$
16	$f_{16} = Q_{17-16} - Q_{16-15} - Q_{16-27} - Q_{16N} = 0$	32	$f_{32} = Q_{25-32} - Q_{32-31} - Q_{32N} = 0$

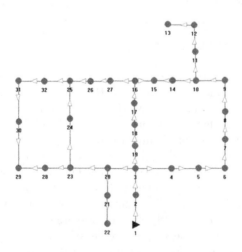

Fig. 2. The flow direction of Hanoi network.

Newton's method [18], are used for comparison. FA is an approach included in swarm-intelligence methods with two essential components, inspired by the flashing pattern of tropical fireflies, for seeking the best solution: attractiveness and movement. In the second one, we conducted GA in order to obtain the best initial values for Newton's method, which is known as the sophisticated and accurate technique in mathematics but needs good initial values.

Pressure distribution along with a comparison with EPANET and the previous work is illustrated in Table 2. It should be noted that the obtained results are the best ones after running several simulations. Moreover, we only compare with EPANET and the other methods because the actual pressures/heads obtained from a measurement are not available. According to Table 2, we obtain that the root mean square deviation (RMSD) between PSO and FA is 1.5464, while 1.3798 and 1.0059 are RMSD of PSO–EPANET and PSO–GA, respectively. The biggest margins of PSO–FA, PSO–EPANET and PSO–GA are 3.43977 m on Node 23, 3.14923 m on Node 13, and 2.5573 m on Node 18. Otherwise, the lowest ones are 0.13018 m on Node 24 for PSO–FA, 0.071 m on Node 15 for PSO–EPANET, and 0.026 m on Node 19. Therefore, it can be seen that the result of PSO is reasonable since it is closed to the results of the other methods.

Table 2. Head distribution (m): Result and comparison with EPANET, FA [5], and the combination GA and Newton's method [18]

Node	PSO	FA	GA	EPANET	Node	PSO	FA	GA	EPANET
2	96.0903	96.7790	97.0910	97.14	18	55.7552	54.6555	53.1979	54
3	61.5956	63.0189	61.0001	61.67	19	58.3271	58.5727	58.3531	59.07
4	58.4974	56.3849	56.7973	57.54	20	51.4179	53.0722	52.8120	53.62
5	52.8662	51.4976	51.6011	52.43	21	43.0236	43.4023	43.3003	44.27
6	44.5737	47.5391	46.2030	47.13	22	37.8451	38.8719	38.0411	39.11
7	44.7719	45.3709	44.9713	45.92	23	38.0110	41.4508	37.7197	38.79
8	43.3563	45.7469	43.5804	44.55	24	34.6335	34.7637	35.2548	36.37
9	39.3250	42.0959	39.2231	40.27	25	32.0195	33.1317	31.9859	33.16
10	37.7160	38.4237	36.1433	37.24	26	31.9569	32.1022	32.2732	33.44
11	35.5557	36.1409	34.5576	35.68	27	32.9929	32.5681	33.2295	34.38
12	33.9386	33.4625	33.3816	34.52	28	30.5366	31.9066	31.472	32.64
13	27.1708	29.2492	29.0999	30.32	29	28.4723	29.8794	28.8264	30.05
14	32.0123	33.3489	32.9273	34.08	30	28.0179	30.4325	28.8826	30.1
15	34.1510	33.7520	32.9251	34.08	31	29.2442	30.9299	29.1323	30.35
16	35.3730	34.5684	35.0136	36.13	32	31.2367	29.4533	29.8895	31.09
17	48.7442	50.4658	47.7420	48.64					

7   Conclusions and Future Work

In this research, a model involving a non-linear equation system for calculating pressure distribution is presented. It can be stated as an initial study of flow assurance on water distribution systems. To solve the system, we have implemented PSO by transforming root finding of non-linear equations into an optimization problem. According to a simulation, PSO can be used as an alternative method to solve the problem since it provides a reasonable result when being compared with EPANET, FA, and the combination of GA and Newton's method. Based on the obtained pressure distribution, the water flow direction is generated as well.

As future work, we plan to determine optimal design of networks by considering behavior pressure on nodes and pipelines. In order to obtain an optimal configuration, we might transform the optimization into regression task, and then solve it by using several machine-learning methods, such as fuzzy rule based systems [19] and rough set theory and fuzzy rough set theory [20]. Moreover, developing an application used for determining pressure distribution that involves various methods based on swarm intelligence can also be a next target.

References

1. Prasad, T.D., Park, N.S.: Multiobjective genetic algorithms for design of water distribution networks. J. Water Res. Planning Manage. (2003)
2. Iglesias, P.L., Mora, D., Martínez, F.J., Fuertes, V.S.: Study of sensitivity of the parameters of a genetic algorithm for design of water distribution networks. J. Urban Environ. Eng. (JUEE) 1(2) (2008)
3. Liong, S.Y., Atiquzzaman, M.: Optimal design of water distribution network using shuffled complex evolution. J. Inst. Eng. 44(1), 93–107 (2004). Singapore
4. da Conceicao Cunha, M., Ribeiro, L.: Tabu search algorithms for water network optimization. Eur. J. Oper. Res. 157(3), 746–758 (2004)
5. Riza, L.S., Kusnendar, J., Munir, H., R.N., Sidarto, K.A.: Determining the pressure distribution on water pipeline networks using the firefly algorithm. In: 2016 7th International Conference on Intelligent Systems, Modelling, and Simulation. IEEE (2016, to appear)
6. Savic, D.A., Walters, G.A.: Genetic algorithms for least-cost design of water distribution networks. J. Water Resour. Plan. Manage. 123(2), 67–77 (1997)
7. Simpson, A., Elhay, S.: Jacobian matrix for solving water distribution system equations with the Darcy-Weisbach head-loss model. J. Hydraul. Eng. 137(6), 696–700 (2010)
8. Walski, T.M., Chase, D.V., Savic, D.A., Grayman, W.M., Beckwith, S., Koelle, E., et al.: Advanced Water Distribution Modeling and Management. Haestad press, Waterbury (2003)
9. Kennedy, J., Eberhart, R.: Particle swarm optimization. In: IEEE International Conference on Neural Networks, Proceedings, vol. 4, pp. 1942–1948 (1995)
10. Kennedy, J.: Particle swarm optimization. In: Sammut, C., Webb, G.I. (eds.) Encyclopedia of Machine Learning, pp. 760–766. Springer, NewYork (2010)

11. Eberhart, R.C., Shi, Y.: Particle swarm optimization: developments, applications and resources. In: Proceedings of the 2001 Congress on Evolutionary Computation, vol. 1, pp. 81–86. IEEE (2001)
12. Dhillon, S., Lather, J., Marwaha, S.: Multi objective load frequency control using hybrid bacterial foraging and particle swarm optimized pi controller. Int. J. Electr. Power Energy Syst. **79**, 196–209 (2016)
13. Zhou, Y., Yan, Y., Huang, X., Kong, L.: Optimal scheduling of multiple geosynchronous satellites refueling based on a hybrid particle swarm optimizer. Aerosp. Sci. Technol. **47**, 125–134 (2015)
14. Zhao, X., Xu, W., Ma, Y., Hu, F.: Scenario-based multi-objective optimum allocation model for earthquake emergency shelters using a modified particle swarm optimization algorithm: a case study in chaoyang district. PloS one **10**(12), e0144455 (2015). Beijing, China
15. Bernoulli, D.: Hydrodynamica, sive De viribus et motibus fluidorum commentarii. Opus academicum ab auctore, dum Petropoli ageret, congestum. Sumptibus Johannis Reinholdi Dulseckeri (1738)
16. Fujiwara, O., Khang, D.B.: A two-phase decomposition method for optimal design of looped water distribution networks. Water Resour. Res. **26**(4), 539–549 (1990)
17. Rossman, L.: The EPANET water quality model. Technical report, Environmental Protection Agency, Cincinnati, OH (United States) (1995)
18. Sidarto, K.A., Siregar, S., Amoranto, T., Riza, L.S., Darmadi, M., Dewi, S.: Predicting pressure distribution in a complex water pipeline network system using combination of genetic algorithm and newtons method. In: SEAMS - GMU International Conference on Mathematics and its Applications (2007)
19. Riza, L.S., Bergmeir, C., Herrera, F., Benítez, J.M.: FRBS: fuzzy rule-based systems for classification and regression in R. J. Stat. Softw. **65**(1), 1–30 (2015)
20. Riza, L.S., Janusz, A., Bergmeir, C., Cornelis, C., Herrera, F., Ślęzak, D., Benítez, J.M.: Implementing algorithms of rough set theory and fuzzy rough set theory in the R package RoughSets. Inf. Sci. **287**, 68–89 (2014)

Content-Based Image Retrieval Based on Quantum-Behaved Particle Swarm Optimization Algorithm

Wei Fang[(✉)] and Xiaobin Liu

School of Internet of Things Engineering,
Jiangnan University, Wuxi 214122, China
fangwei@jiangnan.edu.cn

Abstract. The performance of content-based image retrieval (CBIR) is usually limited since only single visual feature and single similarity measurement are used. In order to solve this problem, the color and texture visual features of an image are analyzed firstly. And then 12 kinds of similarity measurement are used to evaluate similarity between the image being checked and the images in the retrieval library. The CBIR problem is therefore transferred to an optimization problem with the precision ratio as its objective function. Quantum-behaved Particle Swarm Optimization (QPSO) algorithm is used to solve the CBIR optimization problem in order to find the optimal weight and the optimal combination of visual features and similarity measurements. Experimental results show that the proposed method based on QPSO algorithm has better performance on the retrieval effect.

Keywords: Content based image retrieval · Quantum-behaved Particle Swarm Optimization · Feature extraction

1 Introduction

With the rapid development of the Internet, the image information is increased rapidly. To extract valuable knowledge quickly from the mass of digital images has become an urgent need. Since 1990s, the content-based image retrieval (CBIR) technology [1] came into being, and became a research focus in the intelligent information processing field. CBIR technology takes use of human' visual features and makes extraction, processing, quantification, calculation, analysis and matching of images, so as to achieve the retrieval process of images [2]. The technology is based on visual features and semantic information of images to achieve the retrieval process. It is widely used in

W. Fang—This work was partially supported by the National Natural Science foundation of China (Grant Nos. 61105128, 61170119, 61373055), the Natural Science Foundation of Jiangsu Province, China (Grant No. BK20131106), the Postdoctoral Science Foundation of China (Grant No. 2014M560390), the Fundamental Research Funds for the Central Universities, China (Grant No. JUSRP51410B), Six Talent Peaks Project of Jiangsu Province (Grant No. DZXX-025), the PAPD of Jiangsu Higher Education Institutions, China.

© Springer International Publishing Switzerland 2016
Y. Tan et al. (Eds.): ICSI 2016, Part I, LNCS 9712, pp. 392–400, 2016.
DOI: 10.1007/978-3-319-41000-5_39

various fields, such as engine search, pattern recognition, digital-image processing, statistics and computer vision.

Su *et al.* used the HSV color space to extract and quantify the color feature of images, and conducted similarity matching retrieval by using Minkowski matrix distance [3]. Liu *et al.* used color histogram to extract color features and conducted similarity matching retrieval by using Canberra formula to calculate the distance between images in the query image database [4]. Qazi *et al.* used the multi-dimensional texture histogram to extract the texture feature of sample images, and conducted similarity matching retrieval of quantized features by using standard Euclidean distance [5]. Agarwal *et al.* extracted the RGB and YCbCr space features by using method of discrete wavelet transform and color edge detection, extracted features by using Canny operator to conduct color edge detection, and conducted similarity matching retrieval by using Manhattan distance formula [6]. Imran *et al.* implied Color moment method to quantify color features of images and matched the sample images with images in image database one by one by using Euclidean distance formula [7]. Liyu used the method of Gabor filter to extract and quantify texture features of images [8].

The above methods conducted retrieval by designing specified image features and similarity measurement formula artificially, and matching the weight ratio between features artificially. The retrieval effect of images is therefore limited. This paper adopts two kinds of visual features, the color and the texture. And 12 kinds of commonly used similarity measurements are used. Random combinations of visual features are made after quantization, and then optimal matching with similarity measurement and weights between features is done by using Quantum-behaved Particle Swarm Optimization (QPSO) [8]. The best precision rate was taken as the objective function. The CBIR problem is solved through the optimization way to select the best feature combination, similar measurement formula and the approximate optimal weights between features, so as to achieve better retrieval effect. The algorithm of QPSO introduced the random exponential distribution of particle position. And it could cover the entire feasible solution space and increase the search ability of the whole situation, which leads to the good performance of the algorithm.

The remainder of this paper is arranged as follows, Sect. 2 introduces the feature extraction and the CBIR as optimization problems. Section 3 introduces the QPSO algorithm. Section 4 states the CBIR problem based on QPSO algorithm. Section 5 exhibits experimental results. The concluding remarks are given in Sect. 6.

2 Optimization of CBIR

2.1 Problem Statement

CBIR technology first extracted and quantified basic visual features of images. The image content was expressed by the quantized data information. Then, it conducted similarity distance metric between example images provided by user and images in image database to be detected one by one, so as to achieve the retrieval purpose of image database [2]. It can be known that the extraction and quantization of image features and the similar performance measurement algorithm are the basis and core of

CBIR technology. In the stage of visual feature extraction, we use two kinds of commonly used visual features which are the color and the texture. As color feature is concerned, four color spaces are adopted, including RGB, lab, HSV and gray. The color histogram and color moment features are extracted and quantified. For texture feature, the features are extracted and quantified by using the methods of gray co-occurrence matrix and Gabor filter.

In the stage of similarity measurement, twelve kinds of similarity measurements are adopted and listed in Table 1.

Where $F_Q = (f_{Q_1}, f_{Q_2}, \ldots, f_{Q_n})$ represents feature vector of example image; $F_D = (f_{D_1}, f_{D_2}, \ldots, f_{D_n})$ represents feature vector of image in the image database to be detected; D_n and Q_n respectively represent the length of feature vector; $\sigma(i)$ represents the variance in the ith dimension of feature vector; $\|\cdot\|$ represents L_2 normalization of feature vector; $\overline{F_D}$ and $\overline{F_Q}$ respectively represent the mean of F_D and F_Q; σ represents standard deviation.

At the stage of optimization and feature matching, the paper selects the corresponding distance formula of each visual feature is selected. The color feature extraction process is to extract two kinds of features, which are color histogram and color moment, from four kinds of color space. Gabor filter and gray-level co-occurrence matrix method is used in the process of texture feature extraction. The color extracted and texture feature are randomly matched weights as

$$\Phi = \{\Phi_0 | 1 \leq k \leq (n_{Cg} \times n_d) \times (n_{Cm} \times n_d) \times (n_T \times n_d)\} \tag{13}$$

where $n_{C_g} = 4$, $n_{C_m} = 4$, $n_T = 2$ and $n_D = 12$ represent the number of color histogram features, the number of color moments, the number of texture features and the number of distance formula respectively.

In the stage of similarity comparison between example image and query image, we randomly select one formula from twelve commonly used distance formulas to calculate each feature, and randomly match weights to them as

$$\Phi_k = \{(F_{Cg_{i_k}}, d_{C_k}, w_{Cg}), (F_{Cm_k}, d_{C_k}, w_{Cm}), (F_{Tj_k}, d_{T_k}, w_T)\} \tag{14}$$

where $i_k \in \{1, 2, \ldots, n_{C_g}\}$, $m_k \in \{1, 2, \ldots, n_{C_m}\}$ and $j_k \in \{1, 2, \ldots, n_T\}$. $(F_{C_{g_{i_k}}}, d_{C_g})$ represents the $C_{g_{i_k}}$ color histogram feature and the C_k distance formula; $(F_{C_{m_{i_k}}}, d_{C_k})$ represents the $C_{m_{i_k}}$ color histogram feature and the C_k distance formula; $(F_{T_{j_k}}, d_{C_k})$ represents the T_{j_k} texture feature and the C_k distance formula; w_{cg}, w_{cm} and w_T represents the randomly matching weights between different features.

In summary, it will produce a huge amount of combinations in the process, and it will increase the search space redundancy. Therefore, the CBIR problem is proposed to be solved by the optimization algorithm to select the best feature combination, that is, the similarity measurement and the approximately optimal weight of each feature in order to achieve a better retrieval result.

Table 1. Formulation list

Distance	Measurements name	Similarity measurements
	L_1 Reference	$L_1(F_D, F_Q) = \sum_{i=1}^{Q_n} |f_{D_i} - f_{Q_i}|$ (1)
d_2	L_2 Reference	$L_2(F_D, F_Q) = \sqrt{\sum_{i=1}^{Q_n} (f_{D_i} - f_{Q_i})^2}$ (2)
d_3	X^2	$X^2(F_D, F_Q) = \sum_{i=1}^{Q_n} \frac{(f_{D_i} - f_{Q_i})^2}{(f_{D_i} + f_{Q_i})^2}$ (3)
d_4	Standard Euclidean	$SE(F_D, F_Q) = \sum_{i=1}^{Q_n} (\frac{f_{D_i} - f_{Q_i}}{\sigma(i)})^2$ (4)
d_5	Cosine	$Cos(F_D, F_Q) = \sum_{i=1}^{Q_n} \frac{f_{D_i} f_{Q_i}}{\sqrt{\|F_D\|} \cdot \sqrt{\|F_Q\|}} \in [-1, 1]$ (5)
d_6	Correlation	$Cor(F_D, F_Q) = \sum_{i=1}^{Q_n} \frac{(f_{D_i} - \overline{F_D})(f_{Q_i} - \overline{F_Q})}{\sqrt{\|F_D - \overline{F_D}\|} \cdot \sqrt{\|F_Q - \overline{F_Q}\|}}$ (6) where $\overline{F_D} = \frac{\sum_{i=1}^{Q_n} f_{D_i}}{Q_n}, \overline{F_Q} = \frac{\sum_{i=1}^{Q_n} f_{Q_i}}{Q_n}$
d_7	Canberra	$Can(F_D, F_Q) = \sum_{i=1}^{Q_n} \frac{|f_{D_i} - f_{Q_i}|}{|f_{D_i} + f_{Q_i}|}$ (7)
d_8	Minkowski Matrix	$Minkowski(F_D, F_Q) = \left(\sum_{i=1}^{Q_n} |f_{D_i} - f_{Q_i}|^M\right)^{1/M}, M \notin \{1, 2\}$ (8)
d_9	Fu	$Fu(F_D, F_Q) = 1 - \frac{\|F_D - F_Q\|}{\|F_D + F_Q\|}$ (9) where $\|F_D - F_Q\| = \sqrt{\sum_{i=1}^{Q_n} (f_{D_i} - f_{Q_i})^2}$, $\|F_D\| = \sqrt{\sum_{i=1}^{Q_n} (f_{D_i})^2}, \|F_Q\| = \sqrt{\sum_{i=1}^{Q_n} (f_{Q_i})^2}$
d_{10}	Weighted-Mean-Variance	$WMV(F_D, F_Q) = \frac{|\overline{F_D} - \overline{F_Q}|}{|\sigma(\mu)|} + \frac{|\sigma(F_D) - \sigma(F_Q)|}{|\sigma(\sigma)|}$ (10)
d_{11}	Block Metric	$WMV(F_D, F_Q) = \frac{|\overline{F_D} - \overline{F_Q}|}{|\sigma(\mu)|} + \frac{|\sigma(F_D) - \sigma(F_Q)|}{|\sigma(\sigma)|}$ (11)
d_{12}	Chebyshev	$CD(F_D, F_Q) = \max_{1 \leq i \leq Q_n} \{|f_{D_i} - f_{Q_i}|\}$ (12)

2.2 Objective Function

The evaluation about retrieval result can be considered from the two aspects, one is through peoples' subjective feelings and another is the quantitative evaluation criteria. Artificial subjective feeling is affected by subjectivity, and it is difficult to grasp appropriately. Therefore, the quantitative evaluation criteria should adopt an evaluation method which is more intuitive and universal. We adopt Precision Rate and Recall Rate [1] as the criteria. And the Precision Rate is taken as an objective function value of the optimization problem.

Precision Rate is used to detect the accuracy of the retrieval system, which is defined as the ratio of the number of relevant images retrieved and the total number of images retrieved. Recall Rate reflects the comprehensiveness of retrieval system, which is defined as the ratio of the number of relevant images retrieved and the total number of relevant images. The formulations of Precision Rate and Recall Rate are

$$Precision\ Rate = \frac{A}{A + B} \qquad (15)$$

$$Recall\,Rate = \frac{A}{A+C} \qquad (16)$$

where A represents the number of correctly relevant images automatically retrieved by retrieval system, B represents the number of un-relevant images automatically retrieved by retrieval system, and C represents the number of relevant images undetected by retrieval system and can be found out subjectively by human in the database.

From (15), we can see that Precision Rate can not only reflect the accuracy of searching, but also has the ability to reject the number of unrelated images. And Recall Rate mainly reflects the ability of retrieval system to automatically retrieve relevant images. Obviously, the bigger values of these two indicators both reflect that the retrieval effect is better. The paper measures the retrieval algorithm and judges the stability by the statistics of multiple sample images, which consists the mean of Precision Rate and Recall Rate as well as the variance of Precision Rate.

The CBIR retrieval process in this paper is shown in Fig. 1. In the first step, the visual features of sample images and image database to be retrieved are extracted and quantified, and these features are randomly matched weights. Then, the feature similarity matching between sample image and query image is conducted by using similarity search function which is randomly selected, and the retrieval efficiency is also calculated. Finally, the retrieval process is optimally solved by using QPSO. The final output is the combination of the best feature and similarity as well as the feature weight ratio, so as to achieve the better retrieval result in the retrieval process.

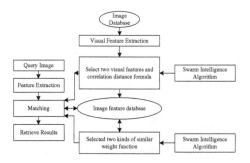

Fig. 1. The flowchart of CBIR problem solved by swarm intelligence algorithm

3   QPSO Algorithm

Compared with PSO algorithm [11], QPSO algorithm [9] is a global optimization algorithm based on PSO and the motion law of particle in potential well δ in quantum mechanics. The difference between QPSO algorithm and PSO algorithm is that particle only has position information in QPSO algorithm, and the position updating is determined by the following three equations:

$$p_{i,j}(t) = \varphi_j(t) \cdot P_{i,j}(t) + [1 - \varphi_j(t)] \cdot G_j(t), \varphi_j(t) \sim U(0,1) \tag{17}$$

$$G_j(t) = \frac{1}{M} \sum_{i=1}^{M} P_{i,j}(t) \tag{18}$$

$$X_{i,j}(t+1) = p_{i,j}(t) \pm \alpha \cdot \left| G_j(t) - X_{i,j}(t) \right| \cdot \ln\left[1/u_{i,j}(t) \right], u_{i,j}(t) \sim U(0,1) \tag{19}$$

where $\varphi_j(t)$ in (17) is a random number in $[0,1]$; $p_{i,j}(t)$ is a random position between the personal best position of the particle and the global best position of the swarm. M is the population size. We can get that *mbest* is the average position of the personal best position of all the individuals. In (19), the parameter α is called the expansion contraction factor which is used to control the convergence speed of the algorithm, and it is the only parameter to be controlled in QPSO algorithm. Generally, a linear decreasing approach is used to get the values of as

$$\alpha = (\alpha_{max} - \alpha_{min}) \cdot (iteration - iter)/iteration + \alpha_{min} \tag{20}$$

where *iter* is current iteration and *iteration* is the maximum number of iteration. α_{max} and α_{min} are two constants and taken the values of 1.0 and 0.5 respectively.

From the position updating equation shown in (19), it can be seen that QPSO algorithm introduces the exponential distribution of position which can cover the whole feasible solution space increase the ability of searching the optimal solution. In addition, QPSO algorithm introduces the mean best position *mbest*, which greatly improves the convergence performance of the algorithm. The reason is that the synergy between particles is enhanced and thus increases the stability of the optimization.

4   CBIR Problem Based on QPSO Algorithm

From the analysis in the Sect. 2, we know that in the stage of feature extraction and similarity measurement, the feature random combination after extraction and quantification, the similarity measure formula and approximately optimal weights between features are all randomly matched. Any one matching result is taken as one feasible solution in QPSO algorithm, which is a particle in the swarm. The dimension of each particle is 9 and can be expressed as $(F_{C_{g_{i_k}}}, d_{C_k}, w_{C_k}, F_{C_{m_{i_k}}}, d_{C_k}, w_{Cm}, F_{T_{j_k}}, d_{T_k}, w_T)$. The paper takes the Precision Rate as the objective function of optimization problem, which is shown in (15). After making clear the representation of each particle and the objective function, we can use the QPSO algorithm to solve the CBIR problem as the following steps:

Step 1: Initializing particle position. In the problem space, we randomly select Φ_k in set Φ and take it as the particle position in the initial particle swarm.
Step 2: Calculate the current average optimal position of all individuals in particle swarm according to formula (18).

Step 3: Update the optimal position of individual particles as to the adaptive value of each optimal $X_i(t)$ in particle swarm. The formula is as follows

$$P_i(t) = \begin{cases} X_i(t) & if \quad f[X_i(t)] < f[P_i(t-1)] \\ P_i(t-1) & if \quad f[X_i(t)] \geq f[P_i(t-1)] \end{cases}.$$

Step 4: Update position of particles according to formulas (17), (18) and (19).

Step 5: If the termination condition of the algorithm is not reached, t is set as t + 1, and then it returns to step two, otherwise the algorithm is finished.

5   Experimental Results

5.1   Parameter Setting

The image database "Corel Image" [1] is used in this experiment. To fairly evaluate retrieval results, we select ten kinds of pictures and select one thousands of pictures sensually similar in each kind. Forty images, which are the most similar in each retrieval, are extracted. In order to further verify the performance of QPSO algorithm, PSO algorithm, CLPSO algorithm [10], and SLPSO algorithm [10] are used to run the same experiments and compare with QPSO. The parameters setting used in this are listed in Table 2.

Table 2. Parameter setting

Algorithm	Parameter setting
QPSO	population size: P = 30, compression coefficient: a = 1
PSO	acceleration factor: c1 = c2: 0–1, inertia weight w: 0.2–0.9
CLPSO	acceleration factor: c1 = c2 = 1.49445, inertia weight w = 1
SLPSO	acceleration factor: c1 = c2 = 1.49445, inertia weight w: 0.2–0.9

Table 3. The average result and the variance of Precision Rate

Group	Average retrieval result (100 %) (variance)			
	QPSO	PSO	CLPSO	SLPSO
1	100.00 (0.0000)	100.00 (0.0000)	100.00 (0.0000)	100.00 (0.0000)
2	99.91 (0.0000)	94.82 (0.0017)	69.93 (0.0742)	98.78 (0.0002)
3	87.52 (0.0056)	81.01 (0.0175)	48.03 (0.0844)	86.48 (0.0074)
4	87.56 (0.0104)	79.66 (0.0248)	58.99 (0.0828)	85.14 (0.0158)
5	83.99 (0.0093)	74.03 (0.0139)	44.49 (0.0497)	85.46 (0.0059)
6	82.58 (0.0095)	73.86 (0.0128)	44.96 (0.0606)	82.49 (0.0095)
7	80.65 (0.0157)	65.11 (0.0217)	40.44 (0.0487)	77.18 (0.0165)
8	78.85 (0.0134)	62.10 (0.0233)	38.17 (0.0756)	75.31 (0.0167)
9	76.13 (0.0083)	61.23 (0.0295)	37.25 (0.0533)	70.07 (0.0125)
10	74.40 (0.0101)	58.47 (0.0305)	35.60 (0.0820)	68.72 (0.0183)
Mean	85.16 (0.0082)	75.03 (0.0176)	51.79 (0.0611)	82.97 (0.0103)

Table 4. The worst Precision Rate

Group	Worst retrieval result (100 %)			
	QPSO	PSO	CLPSO	SLPSO
1	100.00	100.00	100.00	100.00
2	99.60	89.00	31.80	95.60
3	80.50	65.50	20.10	78.20
4	73.20	58.60	16.40	70.70
5	72.50	63.20	20.80	77.60
6	65.30	56.00	16.10	70.30
7	61.20	48.40	24.60	59.70
8	53.70	42.70	18.50	51.20
9	49.20	38.30	20.70	45.30
10	47.30	36.20	16.90	40.10
Mean	70.25	59.80	28.59	68.90

Table 5. The average Recall Rate

Group	Average retrieval result (100 %)			
	QPSO	PSO	CLPSO	SLPSO
1	20.00	20.00	20.00	20.00
2	19.98	18.96	13.99	19.76
3	17.50	16.20	9.61	17.30
4	17.51	15.93	11.80	17.03
5	16.80	14.81	8.90	17.09
6	16.52	14.73	9.00	16.50
7	16.13	13.02	8.81	15.46
8	15.77	12.42	7.63	15.06
9	15.23	12.25	7.45	14.01
10	14.88	11.69	7.12	13.74
Mean	17.03	15.00	10.36	16.59

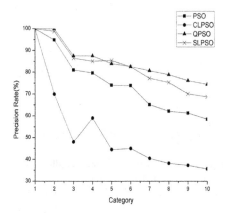

Fig. 2. Comparison of Precision Rate by four algorithms

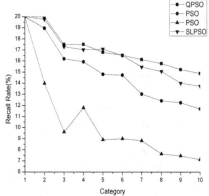

Fig. 3. Comparison of Recall Rate by four algorithms

5.2 Experimental Results

Ten groups of experiments are carried out in this paper. The first group adopts a kind of animal images, which includes 1000 images. Based on the first group, the second group adds another kind of animal images, which also includes 1000 images, and the total is 2000 images. By analogy, based on the sixth group, the tenth experimental group adds another kind of animal images, and the total is 10000 images.

Each group of experiments runs 100 independent times and the average value of Precision Rate and Recall Rate are recorded. The experiment results are given in Tables 3, 4 and 5. Figures 2 and 3 draw the comparison results of precision rate and recall rate by four algorithms.

From results in Table 3, it is clear that the average Precision Rate with QPSO algorithm is better than the other three algorithms and is more robust. The results in Table 4 show that the worst Precision Rate with QPSO algorithm is also better than the other three algorithms. And the experimental results in Table 5 shows that QPSO algorithm has better performance on the Recall Rate.

6 Conclusions

In this paper, the problem of image retrieval based on content is formulated as an optimization problem, which is proposed to solve QPSO algorithm. By using QPSO algorithm, an optimal combination, including the weights and the similarity among visual features can be achieved in order to obtain the optimal retrieval effect. The experimental results showed that the QPSO algorithm can solve the CBIR problem with better performance than PSO algorithm, CLPSO algorithm and SLPSO algorithm.

References

1. Chang, B.-M., Tsai, H.-H., Chou, W.-L.: Using visual features to design a content-based image retrieval method optimized by particle swarm optimization algorithm. J. Eng. Appl. Artif. Intell. 26(10), 2372–2382 (2013)
2. Su, C.H., Chiu, H.-S., Hsieh, T.-M.: An efficient image retrieval based on HSV color space. In: Electrical and Control Engineering (ICECE), pp. 5746–5749 (2011)
3. Liu, G.-H., Yang, J.-Y.: Content-based image retrieval using color difference histogram. J. Pattern Recogn. 46(1), 188–198 (2013)
4. Qazi, M.Y., Farid, M.S.: Content based image retrieval using localized multi-texton histogram. In: Frontiers of Information Technology (FIT), pp. 107–112 (2013)
5. Agarwal, S., Verma, A.K., Dixit, N.: Content based image retrieval using color edge detection and discrete wavelet transform. In: Issues and Challenges in Intelligent Computing Techniques (ICICT), pp. 368–372 (2014)
6. Imran, M., Hashim, R., Abd Khalid, N.E.: New approach to image retrieval based on color histogram. In: Tan, Y., Shi, Y., Mo, H. (eds.) ICSI 2013, Part II. LNCS, vol. 7929, pp. 453–462. Springer, Heidelberg (2013)
7. Hsieh, S.-T., Sun, T.-Y., Liu, C.-C., Tsai, S.-J.: Efficient population utilization strategy for particle swarm optimezer. J. IEEE Trans. Syst. Man Cybern. 2(39), 444–456 (2009)
8. Sun, J., Fang, W., Wu, X., et al.: Quantum-behaved particle swarm optimization: analysis of individual particle behavior and parameter selection. J. Evol. Comput. 20(3), 349–393 (2012)
9. Liang, J.J., Qin, A.K., Suganthan, P.N., et al.: Comprehensive learning particle swarm optimizer for global optimization of multimodal functions. J. IEEE Trans. Evol. Comput. 10(3), 281–295 (2006)
10. Wang, Y., Li, B., Weise, T., et al.: Self-adaptive learning based particle swarm optimization. J. Inf. Sci. 181(20), 4515–4538 (2011)
11. Kennedy, J., Eberhart, R.: Particle swarm optimization. In: Proceedings of IEEE International Conference on Neural Networks, pp. 1942–1948 (1995)

An Approach Using Particle Swarm Optimization and Rational Kernel for Variable Length Data Sequence Optimization

Saritha Raveendran[1(✉)] and S.S. Vinodchandra[2]

[1] College of Engineering, Trivandrum, India
saritha@cet.ac.in
[2] Computer Centre, University of Kerala, Trivandrum, India
vinod@keralauniversity.ac.in

Abstract. This paper proposes a novel approach for unsupervised classification of variable length sequence data using a concept inspired from the Particle Swarm Optimization and rational kernel. The approach uses the distance estimated by the rational kernel as a similarity measure used for clustering the particles. It does not require the normalization of the data sequences into fixed size vectors. Each data sequence has a corresponding particle which moves in the parameter space towards other particles with the similar fitness value. Velocity factor which is used in updating particle position is influenced by the rational distance below a specified threshold. Experimental results display the robustness of proposed algorithm. Misclassification error for clustering the particles into different classes is provided in the results section.

Keywords: Particle swarm optimization · Variable length sequences · Rational kernel · Classification

1 Introduction

Data classification and prediction are two popular sub areas of machine learning. Current algorithms in the sub field concentrates on data sets represented by fixed length feature vector to represent each sample data element. Popular applications from text classification and speech recognition to bioinformatics to information security require the classification of variable-length feature vector representing each data observations. Classifications of such variable length sequences normally employ principle of normalization of the variable length sequences into a fixed length vector before classification. This method suffers from serious drawback of creating degradation and aliasing which affects the accuracy of classification process.

Particle Swarm optimization (PSO) is a nature inspired technique used to optimize a given problem iteratively based on fitness function [1]. PSO originated from the natural observation that a group of birds with similar requisite always flocks together to arrive at their objective. Coordinated group movements are achieved by the individual members by following general rules to try fly near the neighbor. Also keep distance with the neighbor to avoid collision with them neighbor.

© Springer International Publishing Switzerland 2016
Y. Tan et al. (Eds.): ICSI 2016, Part I, LNCS 9712, pp. 401–409, 2016.
DOI: 10.1007/978-3-319-41000-5_40

PSO has been popular and many versions of the algorithms have been proposed so as to make it suitable for different kind of optimization problems like unimodal functions, multi modal and multi objective optimization problems [2]. Parameter fine tuning is an important factor of the algorithm to make it suitable for specific applications [3]. The main disadvantage faced by all these versions is premature converge of the algorithm resulting in sub optimal solutions due to diversity in successive iterations. Another factor which affects the performance of the algorithm is the dimensionality of the search space.

A novel concept inspired from PSO and the rational kernel is proposed in this paper which is successfully used in the clustering of variable length data sequences. Natural behavior of birds to flock near a neighbor of similar species is the motivation behind the proposal [4]. The particles which are scattered in the solution space is made to move closer to the particles of similar functionalities over successive iterations. Rational kernel is used in the evaluation of the fitness function which measures the similarity between two particles of variable length by map variable-length observation sequences into a fixed-dimensional feature-space.

Different phases of the problem is defined as (1) Obtaining the variable length sequence data called as particles of variable length. (2) Map the particles into a parameter space using dynamic kernels (Rational kernel). (3) Optimal classification is done by applying the Novel PSO algorithm such that particles with similar objectives will moved in closer to each other.

2 Variable Length Sequence Data

Some popular applications which require variable length data representations are where documents is to be classified into one or more predefined categories according to their contents [5], applications in bioinformatics that require data of variable length, detection of common patterns from a pool of protein genomics [6], identification of motifs from multiple protein sequences, and discovery of homologous genes from within or across species to determine functionally similar motifs from a set of protein sequences. In the mentioned applications feature vector representing a candidate uses different length feature vectors.

3 Kernel Methods

Kernels functions maps data sequences into a higher dimensional feature space where a linear separation of data is not possible in the original feature dimension. In order to map variable length feature vector into a fixed dimensional feature space dynamic kernels are used. Rational kernels are popular among the general family of kernels and it is based on weighted transducers and acceptors for representing the vectors [7]. Rational kernels specify a general framework based on weighted finite state transducers and acceptors to analyze variable length data sequences.

Using rational kernel each element in the input feature space X is mapped to a higher dimensional parameter space D using a nonlinear mapping function Φ. The task of achieving a linear separability of the input sample is achieved by the mapping function which would not have been possible otherwise.

Kernel defined as function K: $X \times X \rightarrow R$. The value it associates to two examples x and y in input space, K(x, y), equates s with the dot product of their images $\Phi(x)$ and $\Phi(y)$ in higher dimensional parameter space:

$$\forall x, \, y \in X, \, K(x, \, y) = \Phi(x) \cdot \Phi(y) \tag{1}$$

K is similarity measure between x and y. Kernel method provide the advantage of not needing to Explicitly define or compute $\Phi(x)$, $\Phi(y)$, and $\Phi(x)\cdot\Phi(y)$. K is chosen arbitrarily as long as the function Φ exists, known as Mercer's condition. Mercer's condition states that kernel K is positive definite and symmetric (PDS), provided the matrix $(K(x_i, x_j))$ $1 \le i, j \le m$ is symmetric and positive semi-definite for any choice of n points $x_1, \ldots, x_m$ in X.

4   Rational Kernels

A general kernel framework based on weighted finite-state transducers or rational relations is defined by rational kernel to extend kernel methods to the analysis of variable-length sequences. Flexibility and simplicity of rational kernel make it suitable for different types of applications like classification, regression, ranking, clustering, and dimensionality reduction. If operate by computing the shortest distance after mapping variable-length sequences of discrete symbols into a fixed-dimensional feature-space.

4.1   Weighted Transducers

Weighted finite-state acceptors and transducers is an approach for representing and manipulating sequences of discrete observations. Acceptor is a special type of transducer which allow sequences and lattices of discrete observations to be represented within the finite-state transducer framework [8].

A weighted finite state Transducer T over a semiring K is an 8 tuple T = {$\Sigma,\Delta,$Q,I, F,E,λ,ρ}, Σ is a finite set of input alphabet, Δ is the finite set of output alphabet, Q is a finite set of states, I $\subseteq$ Q set of initial states, F $\subseteq$ Q set of final states, E $\subseteq$ Q X (E U {ϵ}) X (Δ U {ϵ}) X K X Q, finite set of transitions, λ : IψK the initial weight function and ρ : FψK the final weight function [9].

Finite-state transducers consist of a set of states joined by directed arcs. They are labeled from 1 to N with single start state and one or many final states represented using double circle. Directed arcs labeled in the form δ: γ where δ ϵ { Σ U ϵ } and γ ϵ { Δ U ϵ } denotes input and output symbols, chosen from the input and output alphabets Σ and Δ respectively are used for connecting states. Transducer is defined with a combination of states and arcs forming paths. Each path is a transformation of a single input sequence into a new output sequence.

In addition to input and output symbols, each transducer arc is assigned a weight, w ∈ W. W denotes a set of valid weights denoted using the notation δ: γ /ω where δ, γ and ω are the input symbol, output symbol and weight respectively. Absence of a specific weight makes the transducer assuming a default value defined by transducer semiring. Semiring specifies a set of operations needed to propagate arc weights through a transducer, and is represented as $(K, +, \cdot, 0, 1)$. The symbols + and · denote operations of addition and multiplication operations respectively. The zero value and the identity value, 1, are used to satisfy the identity axioms of addition $(X + 0 = X)$ and multiplication $(X \cdot 1 = X)$.

4.2 Acceptors

Acceptors is a finite state machine defined using a collection of finite states, joined by a set of arcs. Arc labeled with a single symbol, δ and an assigned weight, w ∈ W denoted using the labeling convention, δ/ω. Acceptors are used to represent inputs or outputs depending upon the operation being performed.

Representation of sequences and sequence transformation using transducers and acceptors is that many sequence operations can be performed using a small number of standardized and efficient algorithms. Commonly used transducer operations are inversion, composition and shortest-distance.

The inverse of a transducer U is measured by swapping the input and output symbols on all transducer arcs. The inverse transducer, U^{-1}, undoes the transformation of the original transducer, U. Composition is the process of chaining together transducers so that the output of the first transducer is used as input for second. Given an acceptor A or transducer U, the shortest distance operator calculates the sum of the weights of all possible paths through A or U. The shortest distance of an acceptor or transducer is denoted by [[A]] or [[U]].

Rational kernels is defined by a combination of transducer operators inversion, composition and shortest-distance. This method provides a powerful platform for measuring distances between variable-length sequences of different observations.

Transducer is viewed as a matrix over a countable set $\Sigma^* \times \Delta^* m$ and composition as the corresponding matrix-multiplication. A kernel is rational if there exist a weighted transducer $T = \{\Sigma, \Delta, Q, I, F, E, \lambda, \rho\}$ over the semiring K and a function $\psi: K \to R$ for all $x, y \in \Sigma^*$

$$K(x, y) = \Psi([T](x, y)) \tag{2}$$

Ψ is an arbitrary function mapping from K to R. K is a rational kernel and T, associated weighted transducer. If A and B are two strings of variable length represented using weighted acceptors, by the definition of rational kernels the shortest distance between two strings can be computed by Algorithm 1.

```
Algorithm1: Shortest Distance using Rational Kernel
 Input : Two variable length sequence X and Y
 Output: Shortest distance between X and Y
 Procedure:
 Construct acyclic composed transducer U = F(X)oToF(Y)
 Compute shortest distance from initial states of U
   to the final states of U  S[U]
       Shortest distance from state N to set of all
       terminal states M is sum of all paths from N to M
       Compute ψ(S[U])
 Repeat the above steps for all the variable length
 sequence combinations
```

5 Modified PSO Algorithm for Clustering Variable Length Data Sequence

Particle Swarm Optimization is inspired from the social behavior like bird flocking or fish schooling is a stochastic global optimization method. Particles move through the environment biasing their movement to good areas of the environment. Bird flocks follow the properties of separation, alignment and cohesion. Principle of separation will help particle maintain a distance from their neighbors so that they do not collide. Alignment implies that the particles move towards the average heading of their neighbors such that the particles remain coordinated. Cohesion implies that particles fly closer to the average position of their neighbors.

PSO consists of a swarm of particles that resides at a position in search space. All particles are assigned initial random positions in the feature space and also initial random velocities. The particles fly in the feature space to locate optimal positions in multi dimensional feature space. The algorithm is executed by changing the position of each particle based on its velocity and previous best position. Thus combining the process of exploration and exploitation particles converge together around single or several optima. The representation of the problem and the fitness function are the two key steps while applying PSO to global optimization problems. Exploration and exploitation is maintained by the algorithm by fine tuning the parameters according to the problem at hand. Exploration helps the particle to move with a pace to find out optimal regions in the search space. There is only one optimal region in case of unimodal functions and multiple regions in multi modal functions. Principle of exploitation directs particle to concentrate around optimal regions to detect the exact optimal positions. Performance of algorithm depends on the parameter fine-tuning. He various parameters to be fine tuned according to the problem addressed are size of swarm, dimension of the particles, maximum velocity, learning factor, inertia weight introduced to limit the change in velocity, terminating condition. The values of parameters are assessed based on empirical analysis.

In this paper, basic idea of PSO is extended for the optimal classification of particles of variable length into different groups based on their functional similarities. In the problem to be addressed here is that the dimension of the particle feature is variable.

Rational kernel is applied to evaluate objective function which is the distance between the particles. Particle positions are updated such that particles of similar functions (which have shortest distance between them) move closer to each other in search space. The modified algorithm for unsupervised classification of particles into different groups based on the similarity of sequences is given as algorithm 2.

```
Algorithm 2: Modified PSO for variable length sequence
  Input Assumptions:
  Number of particles in the solution domain = N
  Each particle X is assigned random position in the
  search domain Current (X)
  Each particle has a feature vector of variable size F(X)
  Neighbourhood of particle is defined using star topology
  Output: Clusters of similar particles
  Procedure:
    for each particle in the solution domain X
        for every other particle Y  with star topology for
            determining the neighbourhood
            Compute the rational kernel Transducer K  between
                particle  F(X) and F(Y)
                    K(X,Y)=F(X)oToF(Y)
            Compute the distance from starting state of K to
            final state.
                    if the K(X)  is better than its Best_P(X)
                            Best_P(X) = K(X,Y) ,   P=Y
            if Best_P(X) ≥ threshold
            V(X)=C1 x V(X)+C2 x rand1 x Best_P(X)-Current(X)
            Current(X)=Current(X) + V(X)
        Repeat above steps for  Max number of iterations
        or till the  convergence criteria is attained
```

5.1   Parameter Selection

Each variable sequence in the database is denoted as a particle assigned a random position denoted as Current (X). Each particle holds a feature vector $F(X)$ of variable length. Particle neighborhood is defined using star topology, where every particle is assumed to be neighbor to every other member in the parameter space. The user defined parameters C1,C2 and rand are empirically estimated. The position of each particle in the search space changes depending on the particle's fitness function which is defined as the distance between the neighbors. N-gram rational kernel is defined as shown in Eq. 3

$$K(X, Y) = F(X)oToF(Y) \qquad (3)$$

The distance between particle X and Y is below a specified threshold the position of the particle is updated such that X and Y moves closer to each other over successive iterations following Eq. 4.

$$V(X) = C1 \times V + C2 \times rand1 \times (Best\_P(X) - Current(X)) \qquad (4)$$

$$Current(X) = Current(X) + V(X) \qquad (5)$$

The current position of a particle is updated using an attraction force derived from a neighbor of similar objective function.000

6 Results and Discussion

The particle swarm optimization algorithm with rational kernels is used over variable length sequence from a randomly generate data set of 1500 particles. Empirical analysis shows the value obtained as 700. A sample sequence from database is represented in the Table 1 with their respective class categorization. Alphabet used for designing the sample sequences is defined as {a,b}. The algorithm is evaluated separately over sample subset consisting of sequences consisting of 2, 3, 4, 5, 6,7 and 8 classes. It is found that particles with shorter distances get their position updated to their species neighbor. Particles are assigned a star topology where each particle is compared with every other particle in the search domain. The velocity update is made only when the distance between them falls below a particular threshold.

Weighted transducer associated with N-gram count rational kernel with N = 2 is depicted in Fig. 1.

Table 1. Sample Data Sequence and their corresponding category.

Particles	Variable sequence Particles	Class Category
S1	Ababababababababababababababababab	1
S2	abababababababab	1
S3	Aaaaaaaaaaaaaaaaaaaaaaaaaaaaaaaaaa	2
S4	aaaaaaaaaaaaabbbbbbbbbbbbbb	3
S5	bbbbbbbbbbbbbbbbbbbbbbbbbbbbbb	4
S6	bbbbbbbbbbbbbbbb	4
S7	bbbbbbbbbbbbbbbaaaaaaaaaaaaaaa	5
S8	aaaaaaaaaaaaaaaa	2
S9	bbbbbbaaaaa	5

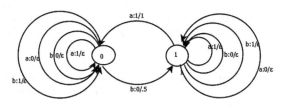

Fig. 1. A sample weighted transducer

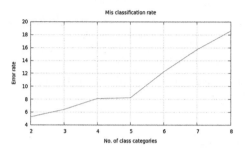

Fig. 2. Misclassification rate

Many variations of PSO is proposed in the literature regarding the clustering of particles. But they all deal with fixed dimensional data sequences. The algorithm is evaluated on a sample dataset for sequences belonging to different classes. The misclassification error rate is specified in Fig. 2. The results denote the promising behavior of the proposed method combining PSO and rational kernel as a prospect to be used in clustering variable length sequence data.

7   Conclusion

A novel method for classification of variable length data sequence into different categories based on the concept of PSO and rational kernel is proposed in the paper. Empirical analysis of the algorithm over a sample data set of multiple categories depicts the robustness of the algorithm. The proposed method can be employed over classification of variable length sequences in different application areas like document classification, bioinformatics, speech classification to information security.

References

1. Poli, R., Kennedy, J., Blackwell, T.: Particle swarm optimization: An overview. Swarm Intell. **1**(1), 33–57 (2007). Springer
2. Coello, C.A.C., Pulido, G.T., Lechuga, M.S.: Handling multiple objectives with particle swarm optimization. IEEE Trans. Evol. Comput. **8**(3), 256–279 (2004)
3. Trelea, I.C.: The particle swarm optimization algorithm: Convergence analysis and parameter selection. Inf. Process. Lett. **85**(6), 317–325 (2003)
4. Jollesa, J.W., King, A.J., Manicab, A., Thornton, A.: Heterogeneous structure in mixed-species corvid flocks in flight. Anim. Behav. **85**(4), 743–750 (2013)
5. Rousu, J., Saunders, C., zedmak, S., Shawe-Taylor, J.: Kernel-based learning of hierarchical multilabel classification models. J. Mach. Learn. Res. **7**, 1601–1626 (2006)
6. Leslie, C.S., Eskin, E., Cohen, A., Weston, J., Noble, W.S.: Mismatch string kernels for discriminative protein classification. Bioinformatics **20**(4), 467–476 (2004)
7. Cortes, C., Haffner, P., Mohri, M.: Rational kernels: Theory and algorithms. J. Mach. Learn. Res. **5**, 1035–1062 (2004)

8. Mohri, M., Pereira, F., Riley, M.: Weighted finite state transducers in speech recognition. Comput. Speech Lang. **16**, 69–88 (2002)
9. Mohri, M.: Semiring frameworks and algorithms for shortest-distance problems. J. Automata Lang. Comb. **7**(3), 321–350 (2002)
10. Rana, S., Jasola, S., Kumar, R.: A review on particle swarm optimization algorithms and their applications to data clustering. Artif. Intell. Rev. **35**(3), 211–222 (2011)

Ant Colony Optimization

A Comparative Approach of Ant Colony System and Mathematical Programming for Task Scheduling in a Mineral Analysis Laboratory

Fabricio Niebles Atencio[1(✉)], Alexander Bustacara Prasca[2],
Dionicio Neira Rodado[1], Daniel Mendoza Casseres[2],
and Miguel Rojas Santiago[3]

[1] Departamento de Ingeniería Industrial, Universidad de la Costa,
Calle 58 # 55–66, Barranquilla, Atlántico, Colombia
{fniebles2, dneira1}@cuc.edu.co

[2] Programa de Ingeniería Industrial, Universidad del Atlántico,
Km 7 Antigua vía Puerto Colombia, Barranquilla, Atlántico, Colombia
alexanderbustacaraa@gmail.com, danielmendoza@mail.
uniatlantico.edu.co

[3] Departamento de Ingeniería Industrial, Universidad del Norte, Km 5 Antigua
vía Puerto, Sincelejo, Colombia
miguel.rojas@uninorte.edu.co

Abstract. This paper considers the problem of scheduling a given set of samples in a mineral laboratory, located in Barranquilla Colombia. Taking into account the natural complexity of the process and the large amount of variables involved, this problem is considered as NP-hard in strong sense. Therefore, it is possible to find an optimal solution in a reasonable computational time only for small instances, which in general, does not reflect the industrial reality. For that reason, it is proposed the use of metaheuristics as an alternative approach in this problem with the aim to determine, with a low computational effort, the best assignation of the analysis in order to minimize the makespan and weighted total tardiness simultaneously. These optimization objectives will allow this laboratory to improve their productivity and the customer service, respectively. A Ant Colony Optimization algorithm (ACO) is proposed. Computational experiments are carried out comparing the proposed approach versus exact methods. Results show the efficiency of our ACO algorithm.

Keywords: Scheduling · Ant Colony Optimization · Multi-objective optimization

1 Introduction

In order to improve competitiveness, manufacturing and service companies require to constantly implementing formal procedures to optimize their processes [1]. In that way, scheduling process is one of the hard optimization problems found in real industrial contexts. Generally speaking, scheduling is a form of decision-making that plays a

© Springer International Publishing Switzerland 2016
Y. Tan et al. (Eds.): ICSI 2016, Part I, LNCS 9712, pp. 413–425, 2016.
DOI: 10.1007/978-3-319-41000-5_41

crucial role in manufacturing and service industries. According to [2], scheduling problems deal with "the allocation of resources to tasks over given time periods and its goal is to optimize one or more objectives". This work focuses on the Scheduling process for a specific configuration of a mineral laboratory located in Barranquilla (Colombia), which is in charge of reception, identification, preparing and analysis on samples of coal and coke according to the customer requirements either for certifying the quality of the material and evaluating the feasibility to open a coal mine, or for selling the coal after mining, or even for knowing the physics and chemical properties of it for a customer utilization, for example in thermoelectric plants and steel companies. Such samples might come from Explorations, where samples are obtained by perforation processes. Exploitations and Development, in which the samples are obtained from exploitation fronts, stationary or during samples transportation in piles, ships, wagons, trucks, conveyors, etc.; also, some samples come from customers directly. These different ways in which the laboratory receives the samples and the variability of the coal market together with external problems related to the mining sector (union strikes, closing of mines for breaking the law, and the bad conditions of colombian roads, etc.), generates that the demand of the laboratory can't be known in advanced and therefore, is quite difficult to plan or schedule the samples for 46 types of analysis that the laboratory carries out.

Therefore, the objective of this paper is to solve a scheduling problem in a real context, which is a mineral laboratory. Objective functions are defined as the minimization of the total completion time of all jobs (or makespan) and total weighted tardiness. Our solution approach is based on a Ant Colony System Optimization Algorithm. The rest of this paper is organised as follows: Sect. 2 is devoted to present the review of literature related to the solution of some particular scheduling problems. Section 3 shows the formulation and mathematical model of the problem under study. Section 4 presents in detail the proposed ACO algorithm, while Sects. 5 and 6 is devoted to computational experiments and the analysis of results. This paper ends in Sect. 7 by presenting some concluding remarks and suggestions for further research.

2 Literature Review

The scientific literature has extensively reported academic works and real-life applications regarding the utilization of metaheuristics for solving scheduling problems. Although most of these works have focused on single objective applications, we have some applications that involve the optimization of two or more objectives simultaneously, especially in shop scheduling problems [3, 4]. On the other hand [5] presented a survey paper of Multiobjective scheduling problems for production planning. In the same way, various intelligent heuristics and meta-heuristics have become popular such as Simulated Annealing (SA), Tabu Search (TS), Multi-Agent System (MAS), Genetic Algorithm (GA) and ACO [6]. For example, authors like: [7, 8, 9, 10, 11], have used those metaheuristics for solving different kind of shop scheduling problems where one of more objectives are optimized and – in general - one of them is the makespan. At the same way, authors like: [8, 9]; have used some heuristics and metaheuristics for different types of scheduling problems like production planning, aircraft schedule or route planning. On the other hand, ACO approaches imitate the behavior of real ants when searching for food. The ACO algorithms use systems formed by several artificial ants. These latter not only simulate the behavior of real ants described above, but also (i) apply additional problem- specific heuristic information, (ii) can manage the deposited quantity of pheromone according to the quality of the solution; moreover it is possible to have various types of pheromone, and (iii) has a memory which is used to store the search history. Each ant uses the collective experience to find a solution to the problem [13, 14, 15].

3 Problem Formulation

As we show later in model formulation, it is necessary to define a decision variable for establish what kind of analysis has to be performed to each sample and also, to know which stages and sub process are necessary to process this sample. The types of analysis are described in Table 1.

Therefore, taking into account the information previously mentioned, the formulation of mathematical model that describes the problem under study is presented below.

3.1 Mathematical Model

Model Parameters:

n: Number of jobs (samples) To be scheduled.

m: Number of process

p_{ik}: $i = 1, \ldots, n; k = 1, \ldots, m$; processing time of job i at process u

d_i: due date of job i

h_i: Priority of job i; 3 if is high, 2 if is middle and 1 if is low

Q_k: $k = 1, \ldots, m$; Capacity of process u

S_{ik}: Binary parameter, i equal to 1 for assigned analysis to job i, if this analysis has to use the process k; 0 Otherwise

Variables:

X_{ijk}: $i, j = 1, \ldots, n; k = 1, \ldots, m$; Binary variable, i equal to 1 if the job i is processed in position j in the process k; 0 Otherwise

r_{ik}: Initialization time of job i in the process k

C_{ik}: Completion time of job i in the process u of stage l

T_i: Tardiness of job i

W_i: weight of job i

C_{max}: Makespan

Table 1. Description of the different types of analysis performed, process and stages.

Type of analysis	Process (u)
Crushing	1
ADL	2
Pulverization	3
Homogenization	4
HGI	5
Coarse series granulometry coal and coke	6
Fine series granulometry	7
Apparent relative density coal	8
Apparent relative density coke	9
Plastometry	10
Dilatometry	11
TGA	12
Ashes	13
Moisture	14
Volatile matter	15
Sulfur	16
Calorific value	17
Ashing	18
SO3	19
Ash elemental analysis	20
AFT	21
Trace elements	22
Equilibrium moisture	23
Forms of Sulfur	24
FSI	25
Australian specific gravity	26
Oxidation index	27
CHN	28
Fluorine	29
Chlorine	30

Objective Function: $Min \sum T_i \times W_i + C_{max}$

Subject to:

(A)

$$\sum_{j=1}^{n} X_{ijk} = S_{ijk}$$

$$\forall i = 1, \ldots, n; \ \forall k = 1, \ldots, m$$

(B)

$$\sum_{i=1}^{n} X_{ijk} = S_{ik}$$

$$\forall j = 1, \ldots, n; \ \forall k = 1, \ldots, m$$

(C)

$$C_{ik} = C_{i(k-1)} + \left[\frac{\sum_{i=1}^{n} S_{ik}}{Q_k} \right] \times p_{ik} + \sum_{k=1}^{m} (X_{ijk} \times p_{ik})$$

Where, $\left[\dfrac{\sum_{i=1}^{n} S_{ik}}{Q_k} \right]$ is the representation of the integer part of this operation.

(D)

$r_{i1} = 0$	$r_{i9} = C_{i8}$	$r_{i17} = C_{i4}$	$r_{i25} = C_{i4}$
$r_{i2} = C_{i1}$	$r_{i10} = C_{i4}$	$r_{i18} = C_{i4}$	$r_{i26} = C_{i4}$
$r_{i3} = C_{i2}$	$r_{i11} = C_{i10}$	$r_{i19} = C_{i18}$	$r_{i27} = C_{i4}$
$r_{i4} = C_{i3}$	$r_{i12} = C_{i4}$	$r_{i20} = C_{i18}$	$r_{i28} = C_{i4}$
$r_{i5} = C_{i4}$	$r_{i13} = C_{i4}$	$r_{i21} = C_{i18}$	$r_{i29} = C_{i4}$
$r_{i6} = C_{i5}$	$r_{i14} = C_{i4}$	$r_{i22} = C_{i18}$	$r_{i30} = C_{i4}$
$r_{i7} = C_{i6}$	$r_{i15} = C_{i4}$	$r_{i23} = C_{i4}$	$\forall i = 1, \ldots n$
$r_{i8} = C_{i7}$	$r_{i6} = C_{i4}$	$r_{i24} = C_{i4}$	

(E)

$$W_i = \frac{h_i}{\sum_{i=1}^{n} h_i}$$

$$T_i = Max(0, C_{max} - d_i) \quad \forall i = 1, \ldots, n$$

(F)

$$C_{max} = Max(C_{ik}) \quad \forall i = 1, \ldots, n; \forall k = 1, \ldots, m$$

(G)

$$C_{ik} \geq 0, \quad \forall i = 1, \ldots, n; \forall k = 1, \ldots, m$$

$$T_i \geq 0, \quad \forall i = 1, \ldots, n$$

The restriction (A) implies that each job has to be processed only once at machines assigned; if the corresponding analysis does not need using some machine, then the machine is not used. The restriction (B) indicates that each job has to be assigned to one machine if the job analysis implies the use of this machine. (C) Refers to termination times of jobs in each machine. In restriction (D), we appreciate the order flow in the samples at the machines. These last two restrictions make sure also, that the jobs is not overlapping in every machines. The sets of constraints (E) and (F) define the criteria C_{max} and $\sum T_i \times W_i$ that are minimized. Finally, the restriction (H) ensures that the value of these criteria are not negatives.

4   Proposed ACO Algorithm

Ant Colony Optimization (ACO) is a meta-heuristic approach proposed by [16] and improved in later research (e.g. see Dorigo et al. [18], Stützle and Hoos [20]). The common behaviour of all variants of ant-based algorithms consists on emulate "real" ants when they find the optimal path between their nest and a food source. Several studies have applied ACO to solve different discrete and continuous optimisation problems [17, 18, 19, 20]. One of these applications involves scheduling problems, as pointed out by [21] in a recently published extended literature review paper.

In this paper, we use the Ant Colony System (ACS) approach to solve the scheduling problem under study. The following elements have to be defined [22]:

- An appropriate model to represent pheromones
- The mechanism to update the pheromone trail
- A heuristic function employed to provide information about the problem under study

These elements are employed to guide the selection of a job to be executed at a given time in the system analysed. This impacts the system behaviour. In order to obtain feasible solutions, job routing sequence has to be respected at each step when building a solution. This is ensured by using a restricted candidate list of all jobs that may be carried out at a given time of the schedule.

4.1 Constructive Procedure

While a feasible solution is constructed, each ant k independently performs a sequence of processing jobs at his step. Hence, each ant k has to make two decisions: On the one hand, it has to select a job from the restricted candidate list L_k, but on the other hand, the ant has to select the position in which this processing job will be carried out. In order to respect processing precedence constraints, our solution approach consist of solving the problem stage by stage (a stage is a specific place of laboratory where particular types of analysis are performed). Solutions are hence constructed by repeating it at each processing stage in the system. Selected jobs are then registered successively in a list that also shows their position in the processing sequence. This avoids the procedure to select a job more than once in the same sequence. The structure of the proposed algorithm is presented in Fig. 1.

Procedure ACO-Metaheuristic
Initialize parameters ( )
   While Stopping condition is not met **do**
      For each ant k
         For each stage
            Select the next processing job applying transition state rule
            Assign a position to selected processing job
            Update pheromone trails locally in order to reduce this quantity for the selected job and position
         End-For
      End-For
      Evaluate solution ( )
      Update pheromone trails globally in order to increase this quantity in current best solution
   End-While
End-Procedure

Fig. 1. Description of the proposed ACO algorithm for the scheduling problem.

4.2 Solving Job Sequence Problem

The procedure uses a state transition rule to select job from the set L_k (referred to the nodes that have not yet visited by ant k). This rule is called random proportional rule. It gives the probability for ant k to select job, based on the pheromone trail τ and the heuristic information η. This rule is formally described by the following expression:

$$
s = \begin{cases} \arg\max\limits_{u \in L_k(r)} \left\{ |\tau(r, u)|^\alpha \, |\eta(r, u)|^\beta \right\} & \text{if } q \leq q_0 \\ S & \text{otherwise} \end{cases} \tag{1}
$$

Where S is a random variable chosen from the following probability distribution (Dorigo and Gambardella [17]):

$$
P_k(O_{ru}, S) = \frac{|\tau(r, u)|\alpha \cdot |\eta(r, u)|}{\sum_{u \in Lk(r)} |\tau(r, u)|\alpha \cdot |\eta(r, u)|} \tag{2}
$$

Where α and β respectively corresponds to relative weights of values $\tau(r, u)$ and $\eta(r, u)$ in the rule; q, with $0 \leq q \leq 1$, is a value randomly chosen from a uniform distribution, and q_0, with $0 \leq q_0 \leq 1$, is a selection parameter that determines the relative importance between intensification and diversification strategies. That is, if $q > q_0$, the system will trend to diversification; but if $q < q_0$, the system will trend to intensification.

4.2.1 Model to Represent Pheromones

As explained before, it is necessary to define an accurate model to represent the level of pheromones. This level of pheromones establishes the "desirability" of having two given jobs next to each other in the sequence. That is, $\tau(i, j)$ represents the convenience of having job j immediately after job i. Hence, the level of pheromones determines the sequencing order of jobs at each machine. It is also used to represent past experiences of ants with regard to the selection of a job from the list of candidates.

4.2.2 Heuristic Information

The heuristic information $\eta(i,j)$ gives specific information about the problem under study. It is used to estimate the convenience of that a job i has to be processed at j position. For the case of this research, the heuristic Information is computed taking into account the relative weight of each job, the time that is received, and the due date. According to this, the heuristic information is obtained by the following equation:

$$
\eta(i,j) = \frac{W_i}{dt_{due\_date(i)} - dt_{arrival(i)}} \tag{3}
$$

Where:

W_i = Relative weight of job i
$dt_{due\_date(i)}$ = Due date of job i
$dt_{arrival(i)}$ = Time of job i is received

4.2.3   Local Updating of Pheromone Trail

The local updating of pheromone trail is executed for each ant once it has built a solution. This rule aims at not to influence the behaviour of other ants. A mechanism is defined to evaporate the pheromone level of the job and position selected by the ant k. This selection becomes less attractive to other ants. This also aims at diversifying the paths that ants are taken and hence avoid convergence to a local optimum. Modification on the pheromone trail is performed using the following expression:

$$\tau(i,j) = (1 - \rho_l) \cdot \tau(i,j) + \rho_l \cdot \tau_0 \tag{4}$$

Where τ_0 is the initial level of pheromone and ρ_l, with $0 \leq \rho_l \leq 1$) is the local evaporation parameter of pheromones.

4.2.4   Global Updating of Pheromone Trail

Once ants have finished their paths, the global rule for updating pheromone trail is applied. This rule intensifies the level of pheromone on paths that allows ants to obtain a better solution. In the next iteration, this path will have a high probability of being chosen. The following expression is employed:

$$\tau(i,j) = (1 - \rho_g).\tau(i,j) + \rho_g.\Delta\tau(i,j) \tag{5}$$

With:

$$\Delta\tau(i,j) = \begin{cases} (ET_b)^{-1} & \text{if } (i,j) \text{ belongs to the best solution} \\ 0 & \text{otherwise} \end{cases} \tag{6}$$

Where ET_b is the best solution that is obtained by multiplying the objective functions (makespan and number of tardy jobs) by their respective weights. In other words, ET_b is the lowest value from over possible schedules obtained by the ants. In addition, ρ_g (with $0 \leq \rho_g \leq 1$) is the pheromone evaporation parameter.

5   Experimental Environment and Parameter Setting

This section first describes the datasets employed for the extended experimental study. Afterwards, the process employed to setup the parameters of the ACO algorithm is described. Finally, the analysis of results is presented, as well as a comparison with an exact method and the laboratory's method.

The algorithm was coded using Visual Basic for Applications (VBA) ®. Experiments were carried out on a PC with processor Intel Pentium ® Dual Core with 2.40 GHz and 4.0 GB of RAM.

5.1   Benchmark Instances

Datasets employed in our experiments were taken from the historical files of laboratory. We considered data from three months with high demand in July, August and

September of 2014. These dataset employed samples (jobs) with different types of priority according to the scale defined in Sect. 3. A total of 30 replications of ACO algorithm were carried out for each dataset in order to compare with the company schedule and with an exact method. For each replication, we compute the value of makespan and total weighted tardiness with its respective weight in the objective function, which was established in 0.5 for each one.

5.2   Parameters Setting and Convergence

Several parameters have to be defined to run the ACO algorithm. Preliminary runs were carried out in order to setup the values of such parameters with a representative instance for each problem size. As performance metrics, the makespan and the total weighted tardiness were considered since these are the objective functions of the problem under study.

For the number of ants, we tested with 15, 20 and 30 ants in the system (these values correspond to the number of jobs to be scheduled according to some datasets employed in the computational experiments). Our performance analysis did not find any significant difference between such values. Hence, the number of ants was set to be 15, just for searching computational efficiency (the higher the number of ants, the higher the computational time).

Regarding the parameters, the values presented in Table 2 were considered in the preliminary experimental design. According to [17] Dorigo and Gambardella, these factors have a great impact on the algorithm's behaviour and, as a consequence, on the quality of solutions. The values of α, β, ρ, pheromones and q_o were defined according to [13], who studied a multi-objective scheduling problem with a configuration relatively similar to the system under consideration in this study with both makespan and total earliness/tardiness as objective functions.

A mixed factorial design was performed. The analysis of results was done using STATGRAPHICS ® software. Results showed that the best combination of values for the different parameters is $\alpha = 2$, $\beta = 2$, $\rho = 0.01$, pheromones = 0.01 and $q_0 = 0.8$. These values were employed further during the full computational experiments. After running several instances, we observed that the algorithm converged after 2000 iterations (see Fig. 2). Hence, this was the number of iterations we have set for running all experiments onwards.

Table 2. Values of parameters for algorithm setup

Parameter	Values
α	1 and 2
β	2 and 3
ρ	0.01 and 0.1
q_o	0.8 and Log(*iter*)/Log(*num_iter*), where *iter* and *num_iter* are, respectively, current iteration and total number of iterations
Quantity of pheromones	0.01 and 0.1

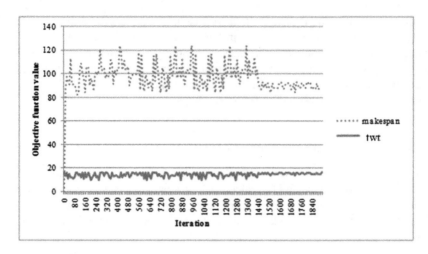

Fig. 2. Convergence of ACO algorithm (Color figure online)

6 Results

This section presents the analysis of results of our computational experiment. All the experiments were run for system configurations among 15 and 45 jobs (that is the maximal capacity of the laboratory for a day trip). As we previously said, the performance of the proposed ACO will be here compared against both real or laboratory schedule, and the exact solution using a Mixed-Integer Linear Programming model. The MILP model was coded and run using AMPL® version 8.0 for MS Windows®. Because of the problem of computational complexity, the model was run for instances with number of jobs less and equal to 45. Hence, optimal solutions were obtained for these small sub problems. For instances with higher number of jobs, to get a solution was not possible in a reasonable time, even for the professional version of AMPL.

The general structure of the set of comparisons is presented in Table 3. We first present the global performance of the proposed ACO algorithm. Table 4 summarizes the obtained values of minimum, maximum and average objective functions values

Table 3. Average values obtained by ACO for the objectives functions

Jobs	Cmax			TWT		
	Min	Max	Avg.	Min	Max	Avg.
15	1089	1198	1516	0.47	0.55	0.51
20	1799	1979	1858	1.58	1.82	1.7
25	1740	1914	2351	2.89	3.32	3.11
30	2738	3012	2043	0.29	0.34	0.32
35	1154	1269	2316	0.05	0.06	0.05
40	3258	3584	2628	3.26	3.75	3.51
45	1748	1923	2035	3.98	4.57	4.27

over the sets of instances. As ACO algorithms are probabilistic in nature, thirty replications were carried out. Hence, a minimum and a maximum value for each objective function correspond to the worst and the best value obtained at a given replication, while the average is computed over the set of thirty replications.

Additionally, in order to maintain certain coherence in the experimental analysis, a relative deviation index, in percentage, was employed, as shown in Eq. (7) and (8), where F_x^{ACO} corresponds to the averages values of the objective function x (i.e., makespan or total weighted tardiness) obtained using proposed algorithm ACO for representative instances (number of jobs). Also, F_x^{MILP} and F_x^{LAB} corresponds to the values of the objective functions using the MILP model, and laboratory approach. These values are shown in Table 5. Note that a negative value of %dev means that the proposed ACO algorithm outperforms the method against it is compared with.

$$\%dev = \frac{F_x^{ACO} - F_x^{MILP}}{F_x^{MILP}} \times 100\% \tag{7}$$

$$\%dev = \frac{F_x^{ACO} - F_x^{LAB}}{F_x^{LAB}} \times 100\% \tag{8}$$

When comparing the proposed ACO algorithm with laboratory's schedule, we observe that ACO is outperformed in only two cases (number of jobs 35 and 43) for makespan criteria. For the rest of instances, ACO always performed better with a significantly difference. On the other hand, for total weighted tardiness criteria, ACO is only for two cases (number of jobs 21 and 34) equal to the laboratory approach. For the rest, ACO outperforms with a great difference (more than 80 % in some cases) to laboratory's schedule. Thus, the ACO approach is widely the best alternative (in comparison for the laboratory approach) for scheduling samples at company's laboratory.

On the other hand, when comparing the proposed ACO algorithm with MILP model, we observe that, although MILP model outperforms to ACO in almost all instances, there is no a statistical difference among these two approach for both objective functions, as we see in Table 5 that shows the z test for mean comparisons was performed using MS-Excel ®. Therefore, we can say, with significantly statistical evidence, that ACO algorithm is as effective as the MILP model, with the plus that ACO schedules the total amount of jobs but MILP model does not. Hence, ACO algorithm can be used as an effective decision support tool for scheduling process in the mineral laboratory, in order to minimize both: makespan and total weighted tardiness.

7 Concluding Remarks and Further Research

This paper studied the job scheduling problem in a real and difficult context of a mineral laboratory. Since in the scientific literature ACO algorithms have shown to be good solution procedures for solving complex scheduling problems, an ACO algorithm was proposed in this paper to solve the multi-objective case of minimising makespan

Table 4. Comparison between ACO algorithm, MILP model and laboratory's approach.

Number of jobs	MILP		ACO		Laboratory		Deviation between ACO and MILP model		Deviation between ACO and laboratory approach	
	Cmax	TWT	Cmax	TWT	Cmax	TWT				
18	716.82	2.74773	752.6	2.746	1254.217	3.255	4.99 %	0.08 %	−39,9999	−15.66 %
21	585.7	0	550.3166	0	1086.667	0	6.04 %	0.00 %	−49.36 %	0.00 %
24	1750.64	7.40878	1921.25	5.838	2273.333	6.341	9.75 %	21.2036	−15.49 %	−7.92 %
27	1490.08	3.97539	1914.2	4.01	2586.6	5.15	28,46 %	0.87 %	−26.00	−22.14 %
28	1304.2	3.16926	1512.1666	3.196	5187.617	3.814	15.95 %	0.85 %	−70.85 %	−16.19 %
34	1044.42	0	1044.4166	0	2067.833	0	0.00 %	0.00 %	−49.49 %	0.00 %
35	1959.35	3.99913	1747.5833	3.975	1496.95	4.215	10.81 %	0.60 %	16.74 %	−5.70 %
37	1164.26	0.473643	1088.5833	0.474	2359.367	0.506	6.50 %	0.16 %	−53.86 %	−6.29 %
41	1512.97	0.0635623	1154.2	0.05	2783.833	0.268	23.71 %	21.93 %	−58.54 %	−81.47 %
42	2874.85	0.840928	3052.0333	0.838	12274.4	7.271	6.16 %	0.31 %	−75.13 %	−88.47 %
43	1261.68	1.19757	1126.85	1.195	1800.05	1.248	10.69 %	0.23 %	−37.40 %	−4.25 %
43	1151.28	0.815308	1873.4333	0.829	1800.05	0.83	62.73 %	1.73 %	4.08 %	−0.05 %
44	1816.88	1.02595	1749.9333	1.081	8072.283	4.369	3.68 %	5.32 %	−78.32 %	−75.27 %
45	2132.09	1.56114	1798.7833	1.58	2023.617	1.642	15.63 %	1.22 %	−11.11 %	−3.76 %

and total weighted tardiness. Computational experiments were carried out using historical datasets. Results showed that very good solutions can be found using our ACO algorithm in comparison with the exact method and laboratory's schedule with a considerably less time and computational effort. It is worthwhile to note that the quality of the solution is not affected by an increase in the number of jobs to be scheduled. Therefore, this computational tool is an effective decision support model that allows scheduling the samples in the laboratory in order to increase the customer level and its productivity simultaneously, even for a large number of jobs (samples). Then, this tool can be replicated in the different branches of laboratory for its flexibility and compatibility with laboratory's software. For further research, the ACO algorithm could be hybridised in order to improve much more its performance when solving instances with large number of jobs. Heuristics or meta-heuristics such as Greedy Randomised Adaptive Search Procedure (GRASP), Genetic Algorithms or Simulated Annealing seems to be good options to be used for this hybridisation. Other opportunities for further research consist on adapting our ACO algorithm to solve for this situation, other types of multi-objective scheduling problems with objectives such as number of tardy jobs, total completion time, earliness, work in process etc. In addition, other versions of the problem can be considered. For example, problems with specific time windows for sample arrivals or stochastic processing times.

References

1. Niebles-Atencio, F., Neira-Rodado, D.: A Sule's method initiated genetic algorithm for solving QAP formulation in facility layout design: a real world application. J. Theor. Appl. Inf. Technol. **84**(2), 157–169 (2016)
2. Pinedo, M.L.: Scheduling: Theory, Algorithms, and Systems. Springer, Heidelberg (2008)
3. Hoogeveen, H.: Multicriteria scheduling. Eur. J. Oper. Res. **167**(3), 592–623 (2005)

4. T'kindt, V., Billaut, J.-C.: Multicriteria Scheduling: Theory, Models and Algorithms. Springer, Berlin (2006)
5. Lei, D., Wu, Z.: Multi-objective production scheduling: a survey. Int. J. Adv. Manuf. Technol. 43(9–10), 926–938 (2009)
6. Khalouli, S., Ghedjati, F., Hamzaoui, A.: Hybrid approach using ant colony optimization and fuzzy logic to solve multi-criteria hybrid flow shop scheduling problem. In: Proceedings of the 5th International Conference on Soft Computing as Transdisciplinary Science and Technology (CSTST 2008), pp. 44–50 (2008)
7. Ponnamambalam, S.G., Ramkumar, V., Jawahar, N.: A multiobjective evolutionary algorithm for job shop scheduling. Prod. Plan. Control 12(8), 764–774 (2001)
8. Armentano, V., Claudio, J.: An application of a multi-objective tabu search algorithm to a bicriteria flowshop problem. J. Heuristics 10(5), 463–481 (2005)
9. Jungwattanakit, J., Reodecha, M., Chaovalitwongse, P., Werner, F.: A comparison of scheduling algorithms for flexible flow shop problems with unrelated parallel machines, setup times, and dual criteria. Comput. Oper. Res. 36(2), 358–378 (2009)
10. Chang, J., Ma, G., Ma, X.: A new heuristic for minimal makespan in no-wait hybrid flowshops. In: Proceedings of the 25th Chinese Control Conference, Harbin, Heilongjiang, 7–11 August 2009
11. Niebles Atencio, F., Solano-Charris, E.L., Montoya-Torres, J.R.: Ant colony optimization algorithm to minimize makespan and number of tardy jobs in flexible flowshop systems. In: Proceedings 2012 XXXVIII Conferencia Latinoamericana en Informática (CLEI 2012), Medellin, Colombia, 1–5 October 2012, pp. 1–10 (2012). doi:10.1109/CLEI.2012.6427154
12. Allaoui, H., Artiba, A.: Integrating simulation and optimization to schedule a hybrid flowshop with maintenance constraints. Comput. Ind. Eng. 47(4), 431–450 (2004)
13. Khalouli, S., Ghedjati, F., Hamzaoui, A.: An integrated ant colony optimization algorithm for the hybrid flow shop scheduling problem. In: Proceedings of the International Conference on Computers and Industrial Engineering (CIE 2009), pp. 554–559 (2009)
14. Khalouli, S., Ghedjati, F., Hamzaoui, A.: A meta-heuristic approach to solve a JIT scheduling problem in hybrid flow shop. Eng. Appl. Artif. Intell. 23(5), 765–771 (2010)
15. Khalouli, S., Ghedjati, F., Hamzaoui, A.: An ant colony system algorithm for the hybrid flow-shop (2011). Alaykýran, K., Engin, O., Döyen, A.: Using ant colony optimization to solve hybrid flow shop scheduling problems. Int. J. Adv. Manuf. Technol. 35 (5–6), 541–550 (2007)
16. Colorni, A., Dorigo, M., Maniezzo, V.: Distributed optimization by ant colonies. In: European Conference of Artificial Life, pp. 134–142 (1991)
17. Dorigo, M., Gambardella, L.M.: Ant colony system: a cooperative learning approach to the Traveling Salesman Problem. IEEE Trans. Evol. Comput. 1, 53–66 (1997)
18. Dorigo, M., Maniezzo, V., Colorni, A.: The ant system: optimization by a colony of cooperating agents. IEEE Trans. Syst. Man Cybern. B 26, 29–41 (1996)
19. Dorigo, M., Stützle, T.: Ant Colony Optimization. MIT Press, Cambridge (2004)
20. Stützle, T., Hoos, H.H.: Max–min ant system. Future Gener. Comput. Syst. 16(9), 889–914 (2000)
21. Tavares Neto, R.F., Godinho Filho, M.: Literature review regarding Ant Colony Optimization applied to scheduling problems: guidelines for implementation and directions for future research. Eng. Appl. Artif. Intell. 26(1), 150–161 (2013)
22. Blum, C., Sampels, M.: Ant colony optimization algorithm for FOP shop scheduling: a case study on different pheromones representations. In: Proceedings of the 2002 Congress on Evolutionary Computation (CEC 2002), vol. 2, pp. 1558–1563. IEEE Computer Society Press, Los Alamitos (2002)

Understanding the Information Flow
of ACO-Accelerated Gossip Algorithms

Andreas Janecek$^{(\boxtimes)}$ and Wilfried N. Gansterer

University of Vienna, Vienna, Austria
{andreas.janecek,wilfried.gansterer}@univie.ac.at

Abstract. Gossip algorithms can be used for computing aggregation functions of local values across a distributed system without the need to synchronize participating nodes. Very recently, we have proposed acceleration strategies for gossip-based averaging algorithms based on ant colony optimization, which reduce the message and time complexity of standard gossip algorithms without additional communication cost. In this paper, we extend our latest studies by analyzing in detail how the proposed acceleration strategies influence the node selection of different variants of PushPull gossip algorithms and show that the directions of information dissemination across the network differ strongly according to the type of the underlying "knowledge" of the neighbors (local vs. global knowledge). This analysis leads to a better understanding of how information is spread throughout the network and provides important insights that can be used to further enhance the acceleration strategies.

Keywords: Applications of ant colony optimization · Gossip-based averaging · Epidemic aggregation · Acceleration

1 Introduction and Related Work

In distributed averaging tasks, the goal is to calculate the average $\bar{v}$ of a set of initial values v_i stored locally at the n nodes. Depending on the application, v_i could be sensor measurements, document attributes, media ratings, etc. One possibility for computing $\bar{v}$ in a completely distributed manner is to use gossip (or "*epidemic*") algorithms. Because of their potential robustness and inherently probabilistic nature (randomized communication schedules), these algorithms have the potential for tolerating dynamic network changes, node failures, or data loss and thus for providing a high degree of resilience and fault tolerance [1], allow for gradually trading runtime performance against communication cost, fault tolerance, energy consumption, and even privacy protection by adapting the intensity and regularity of interaction between nodes. Although several (theoretical) studies have proven that gossip averaging algorithms scale well with the number of nodes n, most of these studies are restricted to fully connected networks and based on rather strong assumptions. Applying gossip algorithms on non-fully connected networks significantly increases the number

© Springer International Publishing Switzerland 2016
Y. Tan et al. (Eds.): ICSI 2016, Part I, LNCS 9712, pp. 426–433, 2016.
DOI: 10.1007/978-3-319-41000-5_42

of messages / rounds, especially on weakly connected networks without a regular structure. Distributed aggregation methods can be accelerated, e.g., based on classical convergence acceleration techniques or Chebyhsev acceleration with time-dependent coefficients [2]), via exploiting additional assumptions on the distributed environment, e.g., based on the optimization of communication patterns with respect to a fixed topology [3] or topology itself [4], or by assuming that nodes have additional global knowledge about the topology and their position and by using multi-hop communication, substantial improvements can be achieved [5,11]. However, often a more general setup without such additional assumptions and only nearest-neighbor communication is required. Very recently [6], we have proposed different acceleration strategies based on ant colony optimization (ACO) in order to improve the diffusion speed of single-hop gossip averaging. The pheromone concept of ACO is adapted such that each node maintains a pheromone deposit for each outgoing link, which influences the probability of selecting a neighboring node as communication partner. Moreover, the (inverse) concept of pheromone evaporation is included to increase the attractiveness of nodes which have not been chosen for a long time. Just as in original gossip, our accelerated versions are based on single-hop communication, where every node only communicates with its direct neighbors without any overlay network. The application of SI in distributed environments is motivated by the fact that many natural examples of SI are based on (simple) individuals that communicate to develop collective behavior in a purely decentralized and self-organized fashion. These characteristics make these natural systems robust to loss of members and adaptable to a changing problem domain — all properties which are highly demanded in the context of distributed computing environments. Due to the underlying distributed setting, the pheromone update in [6] has to be performed locally on each node. The learning encoded in the pheromones is not directly based on the pseudo-random proportional ACO update rule but rather on the (dis-)similarity of the estimates of neighbors compared to the local estimate of nodes. Moreover, our "ants" are restricted to local movements only.

Contributions. We extend our study in [6] with an evaluation of the behavior of our acceleration strategies by analyzing how the proposed acceleration strategies influence the node selection of different variants of PushPull gossip algorithms. Additionally, we show that the directions of information dissemination across the network differ strongly according to the type of the underlying "knowledge" of the neighbors (local vs. global knowledge).

2   Gossip-Based PushPull Averaging

PushPull averaging is based on *exchange* of information (cf. [7]). In the active thread, each node i selects a random neighbor p as communication partner and pushes its current local estimate. Node p receives the packet, replies with its own current estimate (*pull-reply*), and stores the average of the received and its own estimate as its new estimate. Finally, the sender receives the answer and updates its own local estimate. There exists another group of gossip algorithms which are based solely on push averaging. The *PushSum* algorithm ([8]) needs more rounds than PushPull to converge but has the benefit that it preserves the mass conservation invariant — at any time the sum of all values (*i. e.*, approximations) in the network remains constant throughout the algorithm. This guarantees the correctness of algorithm even if messages are delayed [9] — a very important property for distributed systems. The mass conservation invariant can be violated in basic PushPull algorithms if an atomic violation happens, *i. e.*, if a node receives a push while it is waiting for a pull-reply. In order to deal with this problem, an enhanced version of PushPull called Symmetric PushSum Protocol (SPSP, [10]) can be used instead whenever mass conservation needs to be preserved. If there are no atomic violations, SPSP is identical to PushPull. Since PushPull is much simpler than SPSP, we focus on PushPull in the rest of this paper. However, whenever mass conservation needs to be preserved, SPSP can be used instead. PushPull/SPSP can be implemented as purely round-based or round *and* event-based implementation. Moreover, the *update process* may have a significant influence on the number of rounds and on the total number of messages necessary to achieve a given accuracy. If all nodes send their messages at the same time — as often stated in the literature ([7]) — the update of received information can only be performed after all nodes have finished sending in each round. If all nodes send their packets at slightly different times, it is possible that the update based on received information is performed *before* a node performs sending in the current round. A detailed evaluation of these different implementation strategies can be found in [6]. Here, we focus on the round-based PushPull strategy with immediate update, where each node actively sends one packet per round (and additionally replies with a pull-reply message if it receives a push message). Whenever a node p (the receiver) receives a (push) message from node i (the sender), the current local value of p is stored in a temporary variable tx_p and the local value of p is updated with received local value of i. Finally, the local value of i is updated with the temporarily stored previous local value of p. Recall that a delayed update will probably lead to an atomic violation.

3   Acceleration Based on ACO

Consider that each node in the network maintains a pheromone deposit for each outgoing link, as typical in ACO. The amount of pheromone along a directed link influences the probability of selecting a neighbor as communication partner. Our goal is to accelerate the diffusion speed of gossip algorithms by selecting links with higher pheromone value with higher probability. We describe how the amount of pheromone along a path is calculated, and how the (inverse) concept of evaporation is included. The variable λ_i refers to the local estimate at node i, i.e., $\lambda_i = x_i$. We discuss the acceleration strategies for PushPull, although the same ACO-based communication partner selection strategy can be applied for SPSP. At all times t, every node i has a current estimate λ_i of the average $\bar{v}$. Beyond that, node i also has (possibly outdated) information about the estimates of its neighbors, stored in the vector $\boldsymbol{y}_i$ of length $deg(i)$. The elements of this vector are ordered according to the IDs of the neighbors of i.

Example. Consider a node a connected to nodes b, c, d, e. Whenever node a communicates with one of its neighbors, it updates not only its own estimate λ_a, but also $\boldsymbol{y}_a$. The absolute difference between $\boldsymbol{y}_a$ and λ_a is denoted as $\boldsymbol{d}_a$ (*i. e.,* $\boldsymbol{d}_a = |\boldsymbol{y}_a - \lambda_a \boldsymbol{1}|$), and serves as the basis for our biased communication partner selection. Motivated by the concept of ACO, $\boldsymbol{d}_a$ represents the intensity of the pheromone trail along the edges between a and its neighbors. Node a will now choose edges with higher pheromone values with higher probability, i.e., it is more likely that node a selects a node with a strongly different local estimate than a node whose local estimate is very similar. The rationale behind is that more progress towards the true average will be made if an information exchange brings more new information. Clearly, since only ***local knowledge*** about neighbors is available, most elements in $\boldsymbol{y}_a$ and therefore also $\boldsymbol{d}_a$ will be outdated since a does not always know the true current estimate of all of its neighbors. E.g., consider that in round t nodes b and a exchange information, and that in round $t+1$ node a exchanges information only with node c while b exchanges information with another node. At the end of round $t+1$ the information of node a about the estimate of b is outdated and probably differs (at least slightly) from $b's$ current estimate. However, as we will see later, even partly outdated estimates are better than basic PushPull without any information about the neighbors. In ACO, the attractiveness of a pheromone trail is reduced as the pheromones evaporate over time. We exploit this strategy in the opposite direction and increase the attractiveness of edges over time in order to increase the attractiveness of nodes which have not been chosen for a long time. Whenever node a communicates with a neighbor — either as active sender *or* as receiver — it stores the number of the current round in the vector $\boldsymbol{t}_a$. The elements of $\boldsymbol{t}_a$ are also ordered according to the IDs of the neighbors of a. In our example, $\boldsymbol{t}_a(1)$ contains the information in which round the latest information exchange between a and b happened, independently of which of the two nodes initiated the information exchange. For all gossip algorithms, ACO-based acceleration can be implemented using a *roulette-wheel selection or a greedy selection strategy:* The vectors $\boldsymbol{d}_a$ and $\boldsymbol{t}_a$ are used to calculate $\boldsymbol{p}_a$ according to $\boldsymbol{p}_a = \boldsymbol{d}_a^\alpha \otimes (t\boldsymbol{1} - \boldsymbol{t}_a)$, where exponentiation is meant element-wise (the parameter α can be used emphasize edges with high pheromone trails), the symbol $\otimes$ represents element-wise multiplication, and t refers to the number of the current round. After normalization, $\boldsymbol{p}_a$ contains probabilities for selecting each neighboring node based on a *roulette-wheel* selection. Additionally, a *greedy* strategy can be used to select the node with the most different estimate. All values of $\boldsymbol{d}_a$ which are smaller than the maximum value of $\boldsymbol{d}_a$ are set to 0 before calculating the vector $\boldsymbol{p}_a$ in the above equation. If there is only one unique maximum value in $\boldsymbol{d}_a$, this node will be selected, otherwise the roulette wheel selection is used (*i. e.,* frequency of iteration is also an issue in this case).

Overhead. There is only a small overhead compared to basic gossip averaging since there is no additional communication. Only local computation and the following local memory space are required: at each node i, two additional vectors

y_i and t_i with an average length of d_{avg} (the average node degree) must be stored.

ACO-based Acceleration Using <u>Global</u> Knowledge (reference algorithm): In order to demonstrate the best possible results that can be achieved with our acceleration strategies, we have simulated our algorithms based on the assumption that *perfect (global) knowledge* of the current estimates of all neighbors is available at all nodes. Technically, this can be simulated by replacing possibly outdated values in vector y_i with the current estimates of all neighboring nodes.

4   Evaluation and Analysis of Acceleration Strategies

All algorithms are implemented in Matlab in a simulation that allows for executing different algorithmic variants while being able to monitor the state of the network at any time from a bird's-eye-view perspective. The error of the estimated average is calculated after each round as $||v(t) - \bar{v}1||_2 / ||v(0)||_2$, where, $v(t)$ is the n-dimensional vector of all estimates after round t and $v(0)$ is a vector consisting of the initial estimates (cf. [11]). All algorithms are terminated once the error is less than 10^{-8} (single precision). A detailed experimental evaluation of the acceleration strategies can be found in [6]. This evaluation includes (i) different network topologies / sizes, (ii) different initialization fields, (iii) four different gossip algorithms, (iv) purely round-based as well as round- and event-based implementations, and (v) acceleration strategies based on local and global knowledge with roulette-wheel and greedy selection strategies. Summarizing the results, we can say that not only the average node degree significantly influences the amount of acceleration, but also the irregularity of the topology. The acceleration strategies work best for weakly connected, irregular networks. In the optimal case an acceleration factor of up to 2.7 can be observed.

Analysis. In the following, we analyze in detail how the proposed acceleration strategies as well as the reference strategies influence the node selection of one representative gossip algorithm. Figure 1 shows the influence of the ACO-based acceleration on the selection of communication partners for a 2D-torus graph with 256 nodes. The number of rounds was set to 400 for all algorithms, and all algorithms were terminated when the error dropped below 10^{-8}. The results are average values over 10 runs. For larger networks, the results are similar but more difficult to visualize. Figure 1(a) and (b) show the initialization fields used for creating the plots in Fig. 1(c–f); large red dots in Fig. 1(b) refer to initialization values close to 1, and small red dots to initialization values close to 0. We used an initialization with a peak shifted towards the upper right corner in order to better visualize the important information. For being able to reference specific nodes, we labeled the grid with numbers from 1 to 16 along the x-axis, and with letters from A to P along the y-axis. The size of the blue arrows between any two nodes in Fig. 1(c–f) indicates how often nodes have communicated with each other. For example, the arrow pointing from node $A1$ to node $A2$ indicates how often node $A2$ has been selected as communication partner by node $A1$, and vice versa. Note

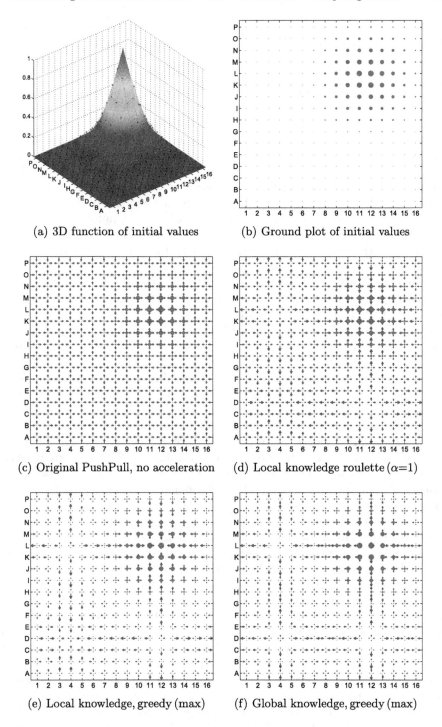

(a) 3D function of initial values

(b) Ground plot of initial values

(c) Original PushPull, no acceleration

(d) Local knowledge roulette ($\alpha=1$)

(e) Local knowledge, greedy (max)

(f) Global knowledge, greedy (max)

Fig. 1. Influence of acceleration strategies on a 2D-torus with 16×16 nodes (Color figure online)

that there is also an arrow from node $A1$ to node $A16$ (the arrow left of node $A1$), and one arrow from node $A1$ to node $P1$ (the arrow underneath node $A1$). This is due to the fact that Fig. 1 shows the results for a 2D-torus graph. The size of the arrows is proportional to the communication frequency along this link — a large arrow indicates that a node has been selected frequently, while a small arrow indicates that a node has been chosen only rarely. Note that each node selects one communication partner per round, although Fig. 1(e) and (f) give the impression that some nodes select more communication partners than others. In fact, the size of the arrows is chosen such that higher-than-average values are represented with large arrows, while the arrow size for average and smaller-than-average values varies only slightly. Moreover, arrows are normalized w.r.t. the largest arrow for each figure. Figure 1(c) refers to the original PushPull algorithm without acceleration. As expected, the selection of nodes is uniformly distributed since each node is connected to four neighboring nodes and all nodes are always selected with the same probability. Figure 1(d–f) illustrate how this distribution changes when the ACO-based acceleration strategy is applied. The results using *local* knowledge are shown in Fig. 1(d) for the roulette-wheel selection strategy with $\alpha = 1$, and in Fig. 1(e) for the greedy selection strategy. The results for the reference algorithm with *global* knowledge using the greedy selection strategy are shown in Fig. 1(f). Since the results in Fig. 1(d) are a hybrid between Fig. 1(c) and (e), we focus on Fig. 1(e) and (f). It is interesting to observe the patterns in Fig. 1(e) and (f): in both cases, there are patterns similar to a "number sign" ("#"), whose lanes intersect at $D4$, $L4$, $D12$, and $L12$. The intersection point at $L12$ is located at the peak of the initialization field; all other intersection points are located at nodes with maximal distance to $L12$: $L4$ and $D12$ are located eight single-hops away from the node at $L12$, and $D4$ is located eight single-hops away from $L4$ and $D12$, respectively. The reason for this special shape is the regular diffusion of information in a rectangular 2D-torus graph (such regular behavior can also be observed for hypercubes, however, this information cannot be displayed in a 2D plot). We point out that the "directions" of these lanes differ between Fig. 1(e) and (f). In Fig. 1(e) (local knowledge), there are two intersections at $L4$ and $D12$ with arrows pointing *away* from them, and two intersections at $D4$ and $L12$ where arrows are pointing *towards* them. In Fig. 1(f) (global knowledge), the arrows tend to point away from *all* intersection points. Indeed, the plot in Fig. 1(f) could be decomposed into four identical squares, which is not the case for Fig. 1(e). One explanation for this difference is the fact that in Fig. 1(e) the neighboring nodes of $D4$ and $L12$ tend to have outdated information about the estimates of $D4$ and $L12$ and chose these nodes more often than others. However, the diffusion of information for these two acceleration strategies is very similar and there are also only small variations in the speed and direction of information diffusion in the network. This analysis provides interesting insights how information diffuses throughout the network. In our current research we aim at applying the observed distribution of node selection from our acceleration algorithms based on global knowledge (Fig. 1(f)) on gossip algorithms with only local knowledge. The goal is to further increase the diffusion speed – however, without the need for exact knowledge of the neighbors' estimates.

5   Conclusions

We have extended our study on ACO-based acceleration strategies for gossip-based averaging algorithms, and evaluated the behavior of our acceleration strategies by analyzing how they influence the node selection of different variants of PushPull gossip algorithms. We have shown that the directions of information dissemination across the network differ strongly according to the type of the underlying "knowledge" of the neighbors (local vs. global knowledge). This analysis leads to a better understanding of how information is spread throughout the network and provides important insights that can be used to further enhance the acceleration strategies. *This research has been partially supported by the Vienna Science and Technology Fund (WWTF) through project ICT15-113.*

References

1. Montresor, A.: Designing extreme distributed systems: challenges and opportunities. In: Proceedings of the 8th ACM SIGSOFT Conference, pp. 1–2. ACM (2012)
2. Golub, G.H., Varga, R.S.: Chebyshev semi-iterative methods, successive overrelaxation iterative methods. Numer. Math. **3**, 157–168 (1961)
3. Boyd, S., Ghosh, A., Prabhakar, B., Shah, D.: Randomized gossip algorithms. IEEE T. Inform. Theory **52**, 2508–2530 (2006)
4. Kar, S., Moura, J.M.F.: Sensor networks with random links: topology design for distributed consensus. IEEE T. Signal Proces. **56**, 3315–3326 (2008)
5. Benezit, F., Dimakis, A., Thiran, P., Vetterli, M.: Order-optimal consensus through randomized path averaging. IEEE Trans. Inform. Theory **56**, 5150–5167 (2010)
6. Janecek, A., Gansterer, W.N.: Aco-based acceleration of gossip averaging. In: GECCO (2016). http://dx.doi.org/10.1145/2908812.2908832
7. Jelasity, M.: Gossip. Self-organising Software. Springer, Heidelberg (2011)
8. Kempe, D., Dobra, A., Gehrke, J.: Gossip-based computation of aggregate information. In: Symposium on Foundations of Computer Science, pp. 482–491. IEEE (2003)
9. Jesus, P., Baquero, C., Almeida, P.S.: Dependability in aggregation by averaging. CoRR abs/1011.6596 (2010)
10. Blasa, F., Cafiero, S., Fortino, G., Di Fatta, G.: Symmetric push-sum protocol for decentralised aggregation. In: AP2PS, IARIA, pp. 27–32 (2011)
11. Dimakis, A., Sarwate, A., Wainwright, M.: Geographic gossip: Efficient averaging for sensor networks. IEEE T. Signal Proces. **56**, 1205–1216 (2008)

Ant Colony Optimization with Neighborhood Search for Dynamic TSP

Yirui Wang[1], Zhe Xu[2], Jian Sun[2,4], Fang Han[1(✉)], Yuki Todo[3], and Shangce Gao[2(✉)]

[1] College of Information Science and Technology,
Donghua University, Shanghai, China
yadiahan@dhu.edu.cn
[2] Faculty of Engineering, University of Toyama, Toyama, Japan
gaosc@eng.u-toyama.ac.jp
[3] School of Electrical and Computer Engineering,
Kanazawa University, Kanazawa, Japan
[4] College of Computer Science and Technology, Taizhou University, Taizhou, China

Abstract. Ant colony optimization (ACO) is one of the best heuristic algorithms for combinatorial optimization problems. Due to its distinctive search mechanism, ACO can perform successfully on the static traveling salesman problem(TSP). Nevertheless, ACO has some trouble in solving the dynamic TSP (DTSP) since the pheromone of the previous optimal trail attracts ants to follow even if the environment changes. Therefore, the quality of the solution is much inferior to that of the static TSP's solution. In this paper, ant colony algorithm with neighborhood search called NS-ACO is proposed to handle the DTSP composed by random traffic factors. ACO utilizes the short-term memory to increase the diversity of solutions and three moving operations containing swap, insertion and 2-opt optimize the solutions found by ants. The experiments are carried out to evaluate the performance of NS-ACO comparing with the conventional ACS and the ACO with random immigrants (RIACO) on the DTSPs of different scales. The experimental results demonstrate our proposed algorithm outperforms the other algorithms and is a competitive and promising approach to DTSP.

Keywords: Ant colony optimization · Dynamic TSP · Neighborhood search

1 Introduction

Ant colony optimization (ACO) proposed by Marco Dorigo et al. was first used to handle the static TSP and performed well [4,6]. The ACO algorithm simulates the foraging behavior of ant colony to search for an optimal route, and ants communicate with each other via the pheromone they deposit on the trail. The pheromone guides ants to find a feasible solution. The more pheromone the previous ants release on the trail, the higher probability of the following

© Springer International Publishing Switzerland 2016
Y. Tan et al. (Eds.): ICSI 2016, Part I, LNCS 9712, pp. 434–442, 2016.
DOI: 10.1007/978-3-319-41000-5_43

ants choosing this trail. After several iterations, ants will acquire the best route finally. Although ACO has been applied successfully to several combinatorial optimization problems [2,5,14], most of the realistic applications are subject to dynamic optimization problems. It means the objective function, the constraints or the problem instances can change with time and the purpose is not to find a static optimal solution but to track the new one to the dynamic problems.

The conventional ACO may not be fit for solving the dynamic problems since the pheromone of the previous route can force ants to follow when the environment changes. This will cause inefficient response to environmental changes [13]. Therefore, several strategies were proposed to enhance the ability to adapt to the changing environment [1]. The simplest assumption is to reset the pheromone on all the trails once the environment changes, however, it is expensive on computing time and not reasonable. The improved method is to adopt the local and global restar strategy proposed to reinitialize the pheromone to response to dynamic environments [8]. Besides, a generating or increasing diversity approach was proposed to resolve dynamic TSP (DTSP) [7]. An effective way to maintain diversity is to set memory to store several solutions to match the previous environments [10,13]. Especially, the P-ACO algorithm used the population-list to store the iteration-best ants for updating the pheromone on the trails [9]. Recently, the ACO algorithm with immigrants schemes was verified to be a good approach to DTSP [13], and the random immigrants ACO (RIACO) performed well in the environment changing quickly and significantly. These approaches can solve the different DTSPs according to the strategies respectively.

In this paper, the ACO with neighborhood search (NS-ACO) is proposed to handle the DTSP composed by random traffic factors. Three moving operations are introduced to adjust the solutions constructed by ants to adapt to the new environments. The experiments implement the comparison among NS-ACO, the conventional ACS and RIACO on the DTSPs of different scales. The experimental results demonstrate the superiority of our proposed algorithm which can track the optimal solution effectively and efficiently.

This paper is organized as follows. Section 2 describes the structure of DTSP. Section 3 presents our proposed NS-ACO algorithm. Section 4 shows the comparative experiments and analyzes the experimental results. Section 5 draws a conclusion.

2   Brief Introduction to DTSP

The conventional TSP is one of the most ordinary and popular NP-hard problems. The several exact and approximation algorithms were proposed to resolve TSP and enhance the solution's quality [3,11,15]. But this kind of TSP that is static and ideal does not accord with the realistic applications. Nowadays, DTSP, the variant of TSP, possesses the dynamic environments such as replacing the cities with time [10,12], changing the cost of the cities' arcs [7] and so on. Addressing the DTSP is not to obtain the global optimal solution completely anymore, but to track the new ones with the changing environments. Therefore, the efficiency and effect of the algorithms are the key factors of solving DTSP.

In this paper, we utilize the DTSP comprised by random traffic factors. This model introduces the cost of the edge between cities i and j. The cost is $D_{ij} \times T_{ij}$, where D_{ij} is the travelled distance between cities i and j, and T_{ij} is the traffic factor indicating the traffic jam between cities i and j. After every f iterations, T_{ij} changes randomly according to a probability value m whose changing range is limited in $[T_L, T_H]$, where T_L and T_H denote the lower and upper bounds of traffic factor respectively. The value of T_{ij} determines the degree of the traffic jam. The larger the T_{ij} is, the severer traffic jam the corresponding edge indicates. Furthermore, $T_{ij} = 1$ represents no traffic jam. In this way, dynamic environments are generated randomly by the frequency f and the magnitude of the change m.

3   Proposed ACO Algorithm

Our NS-ACO algorithm adopts a short-term memory proposed by Michalis Mavrovouniotis et al. [13]. The short-term memory stores K iteration-best ants in the current iteration which are used to update the pheromone in the next iteration, and the corresponding previous ants and pheromone are removed. So no ant can exist more than one iteration in case the current environment changes. The solution construction of our proposed algorithm is the same as the conventional ACO algorithm's. The possibility p_{ij}^k of ant k moving from city i to city j is described as follow:

$$p_{ij}^k = \begin{cases} \dfrac{\tau_{ij}^\alpha \cdot \eta_{ij}^\beta}{\sum_{l \in N_i^k} \tau_{il}^\alpha \cdot \eta_{il}^\beta} & \text{if } j \in N_i^k \\ 0 & \text{otherwise} \end{cases}, \tag{1}$$

where N_i^k indicates the set of the cities which can be visited by ant k at the city i. τ_{ij} denotes the pheromone between city i and city j. $\eta_{ij} = 1/d_{ij}$ is the heuristic information, where d_{ij} is the distance between cities i and j. The parameters α and β determine the relative importance of pheromone τ and heuristic information η respectively.

After executing the solution construction, the neighborhood search is added into the ACO algorithm to optimize the solution. The neighborhood search contains three local search processes: swap operation, insertion operation and 2-opt operation. Each operation is implemented with $1/3$ probability to generate a new solution which may be better than the current solution. The swap operation indicates the positions i and j of a solution are chosen randomly and exchanged each other. The insertion operation is to select randomly position i and position j of a solution $(i \neq j)$ firstly. Then if $i < j$, move position i+1 to position i, and move position i+2 to position i+1, go on until position j, that is to say, move the positions from $i + 1$ to j to the left one step distance respectively. Finally position i replaces position j. Otherwise, move the positions from $i - 1$ to j to the right one step distance respectively and position i replaces position j. Likely, after choosing the positions i and j of a solution randomly the 2-opt operation reverses the segment between i and j. The neighborhood search can effectively optimize the solutions ants construct. Whereafter the pheromone update is implemented by the short-term memory which stores K iteration-best ants in iteration t, which is described as follow:

$$\tau_{ij}(t) = \tau_{ij}(t) + \Delta\tau_{ij}^k(t), \tag{2}$$

where $\Delta\tau_{ij}^k = (\tau_{max} - \tau_0)/K$, τ_{max} and τ_0 indicate the maximum and initial values of pheromone respectively and K is the size of the short-term memory. In this paper, $\tau_{max} = \tau_0 + \sum_{k=1}^{K} \Delta\tau_{ij}^k$. Then in the next iteration $t + 1$, the short-term memory needs to remove the previous ants to make space for the new K iteration-best ants, and accordingly the pheromone deposited previously is subtracted, described as follow:

$$\tau_{ij}(t + 1) = \tau_{ij}(t + 1) - \Delta\tau_{ij}^k(t), \tag{3}$$

this update method not only maintains the pheromone within the definite interval but also enhances the diversity of populations. It should be noted that there is no pheromone evaporation. NS-ACO algorithm is depicted in Algorithm 1, where K_s indicates the short-term memory, N is the number of cities, t is the iteration times and M denotes the number of ants.

Algorithm 1. NS-ACO Algorithm

01: Initialize the parameters $\alpha, \beta, K_s, N, M, t$
02: Initialize pheromone matrix
03: **while** termination criteria **not** be satisfied **do**
04:    Construct dynamic environments
05:    Clear solutions and K_s
06:    Construct solutions by ants
07:    **for** $k = 1$ to M **do**
08:       Execute neighborhood search:
09:          r= generate random number
10:          **if** (r=swap operation) **then**
11:             Generate a swap solution
12:          **end if**
13:          **if** (r=insertion operation) **then**
14:             Generate an insertion solution
15:          **end if**
16:          **if** (r=2-opt operation) **then**
17:             Generate a 2-opt solution
18:          **end if**
19:       Compare and select the best solution
20:    **end for**
21:    Find K iteration-best solutions and add it into K_s
22:    Update K_s using Eq. (2) and Eq. (3)
23:    Compare and record the global optimal solution
24:    t=t+1
25: **end while**

---

4   Experiment and Analysis

4.1   Experimental Settings

In order to demonstrate the performance of our proposed NS-ACO algorithm on the DTSP, we evaluate NS-ACO, the conventional ACS [4] and RIACO [13] on several DTSPs composed by benchmark TSP instances from TSPLIB. eil51 and pr76 indicate small-scale DTSP. kroB100 and pr124 denote medium-scale DTSP. kroB150 and kroB200 represent large-scale DTSP. The ACS is one of the best ACO algorithms for the static TSP since the pheromone evaporation ρ and the pheromone decay coefficient φ change the pheromone of the trails to guide ants to find the optimal solution. RIACO enhances the diversity of population due to its random immigrant scheme, and the short-term memory with the replacement rate r shows efficient performance for DTSP.

In the experiments, all the DTSPs are constructed by random traffic factors set in $[1,5]$, i.e., $T_L = 1$ and $T_H = 5$, respectively. The changing frequency f is set to 10 and 100, which represents the environment changes quickly and slowly respectively. The changing magnitude m is set to 0.1 and 0.9, indicating the small and large degree of environmental changes respectively. Consequently, for each DTSP, 4 dynamic instances are generated to test and analyze the property of three algorithms. Furthermore, for each algorithm on each dynamic instance, the experiment is independently run 30 times and 1000 iterations each time. The experimental data is calculated in terms of the following expression [13]:

$$F^{bs} = \frac{1}{T} \sum_{t=1}^{T} (\frac{1}{N} \sum_{i=1}^{N} f_{ti}^{bs}), \tag{4}$$

where T is the iteration times, N is the number of running and f_{ti}^{bs} is the value of the best solution found in the the i-th running of the t-th iteration. The parameter settings of three algorithms are shown in Table 1. Some parameters are referred in [4,13], where C^{nn} is the length of the tour obtained by the nearest-neighborhood heuristic and q_0 determines the probability of search.

4.2   Experimental Analysis

Table 2 shows the experimental results based on the Eq. (4) among NS-ACO, ACS and RIACO to reveal the average performance of the solutions to each

Table 1. Parameter settings for three algorithms in the experiment.

Algorithm	τ_0	α	β	ρ	φ	q_0	K	r	M
NS-ACO	$1/C^{nn}$	1	5	–	–	0	6	–	30
ACS	$1/C^{nn}$	1	5	0.5	0.5	0.9	–	–	30
RIACO	$1/C^{nn}$	1	5	–	–	0	6	0.4	30

Table 2. Comparison of the experimental results among NS-ACO, ACS and RIACO.

Algorithm	NS-ACO		ACS		RIACO	
	$f = 10$		$f = 10$		$f = 10$	
m	0.1	0.9	0.1	0.9	0.1	0.9
eil51	476.6	1179.7	491.8	1249.2	481.1	1190.9
pr76	130376.5	321490.3	136592.0	336443.9	133712.0	334157.8
kroB100	26704.9	66891.2	26777.2	69894.8	27531.1	69750.6
pr124	71915.8	182928.4	72212.2	190884.8	75817.7	198052.5
kroB150	32553.6	81293.5	32556.3	83854.5	34181.2	85702.8
kroB200	37119.4	93944.6	37127.9	96157.3	39539.5	99797.8
	$f = 100$		$f = 100$		$f = 100$	
m	0.1	0.9	0.1	0.9	0.1	0.9
eil51	452.4	1107.7	477.4	1179.8	454.1	1112.7
pr76	123818.7	299372.9	133243.4	328153.8	125356.0	304831.6
kroB100	25155.0	61748.0	26229.8	67462.6	25438.2	63048.4
pr124	68090.5	168722.2	70658.7	184965.0	69098.0	174331.6
kroB150	30607.7	73670.1	31700.7	79693.8	31436.0	76062.7
kroB200	34628.4	86780.4	36195.1	92780.8	35664.6	89619.6

DTSP. The corresponding statistical results of the Wilcoxon signed ranks test at a level of significance $\alpha = 0.05$ are displayed in Table 3, where the comparison among three algorithms is implemented by the symbols $+$, $-$ and $\sim$ which represent the performance of the former algorithm is better, worse and no significance than that of the latter algorithm respectively. From Tables 2 and 3, several conclusions can be observed. Firstly, our proposed NS-ACO algorithm generally outperforms ACS and RIACO on all the DTSPs, which declares the

Table 3. Statistical results of the Wilcoxon signed ranks test at a level of significance $\alpha = 0.05$ among NS-ACO, ACS and RIACO.

Algorithm	NS-ACO:ACS				NS-ACO:RIACO				RIACO:ACS			
	$f = 10$		$f = 100$		$f = 10$		$f = 100$		$f = 10$		$f = 100$	
m	0.1	0.9	0.1	0.9	0.1	0.9	0.1	0.9	0.1	0.9	0.1	0.9
eil51	+	+	+	+	+	+	+	+	+	+	+	+
pr76	+	+	+	+	+	+	+	+	+	+	+	+
kroB100	+	+	+	+	+	+	+	+	−	+	+	+
pr124	+	+	+	+	+	+	+	+	−	−	+	+
kroB150	∼	+	+	+	+	+	+	+	−	−	+	+
kroB200	∼	+	+	+	+	+	+	+	−	−	+	+

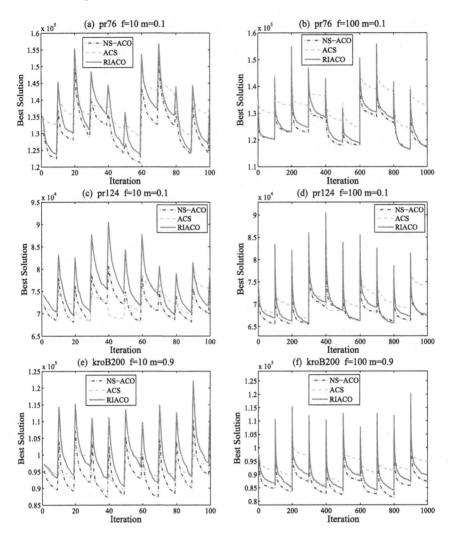

Fig. 1. The convergence graph of the best solution obtained by NS-ACO, ACS and RIACO.

neighborhood search is proper and effective for optimizing solutions to find the optimal solution and three operations randomly selected matching the random changing environments well. The framework of NS-ACO not only maintains the diversity of the solutions but also improves the quality of the optimal solution. The results verify NS-ACO has strong robustness and self-adaptability on DTSP and its ability of tracking the optimal solution is significant. Secondly, except NS-ACO, RIACO performs better on small-scale DTSP such as eil51 and pr76, due to its random immigrant scheme. Nevertheless, on the quickly changing environments of the medium and large scale DTSP, the performance of RIACO declines and is inferior to that of ACS. The reason is that the random immigrants enhance the diversity of the population and destroy the pheromone of the current optimal trail but RIACO has insufficient time to find the optimal solution. On the contrary, ACS can obtain the optimal solution since it utilizes the pheromone evaporation to quickly locate in the optimal trail. Thirdly, when addressing the slowly changing DTSP, RIACO always outperforms ACS since its random immigrants expand the search space to acquire a better solution within enough time whereas ACS is prone to converge to a local optimum due to the slow pheromone evaporation.

To illustrate the convergence characteristics of three algorithms for DTSP intuitively, Fig. 1 is plotted to show the performance of each iteration during three algorithms. Figure 1 depicts the small and medium scale DTSP with $m = 0.1, f = 10$ or 100 and the large-scale DTSP with $m = 0.9, f = 10$ or 100. The horizontal axis and the vertical axis indicate the iteration times and the best-so-far solution respectively. In Fig. 1, each DTSP has 10 environmental changes where Fig. 1(a)(c)(e) and (b)(d)(f) describe 100 and 1000 iterations respectively. From Fig. 1, we can observe that our proposed NS-ACO performs best on each DTSP and RIACO outperforms ACS on slowly changing environments and the small-scale DTSP, and ACS is superior to RIACO on the medium and large scale DTSP whose environment changes quickly.

5   Conclusion

In this paper, we proposed NS-ACO algorithm which adopts neighborhood search to deal with the DTSP comprised by random traffic factors. Swap, insertion and 2-opt operations are randomly utilized to optimize the solutions constructed by ants for tracking the optimal solution efficiently and effectively. The short-term memory is used to store several current iteration-best solutions to enhance the diversity of solutions. To demonstrate the superiority of the proposed algorithm, we compare NS-ACO with the conventional ACS and RIACO on different scales DTSP. The experimental results declare that NS-ACO performs significantly on both convergence and solution quality well. It has the proper ability of convergence that is different from the premature phenomenon of ACS, and it increases the diversity of solutions instead of adding random immigrants which have a high risk in destroying the pheromone of the optimal trail. Therefore, it can be concluded that our proposed NS-ACO is superior to the other two algorithms and shows the promising property for solving the DTSPs.

Acknowledgments. This research was partially supported by the National Natural Science Foundation of China (Grant Nos. 61203325, 11572084, 11472061, and 61472284), the Shanghai Rising-Star Program (No. 14QA1400100) and JSPS KAKENHI Grant No. 15K00332 (Japan).

References

1. Cruz, C., González, J.R., Pelta, D.A.: Optimization in dynamic environments: a survey on problems, methods and measures. Soft Comput. **15**(7), 1427–1448 (2011)
2. Di Caro, G.A., Ducatelle, F., Gambardella, L.M.: AntHocNet: an ant-based hybrid routing algorithm for mobile Ad Hoc networks. In: Yao, X. (ed.) PPSN 2004. LNCS, vol. 3242, pp. 461–470. Springer, Heidelberg (2004)
3. Dorigo, M.: Ant colony optimization. Scholarpedia **2**(3), 1461 (2007)
4. Dorigo, M., Birattari, M., Stützle, T.: Ant colony optimization. Computat. Intell. Mag. IEEE **1**(4), 28–39 (2006)
5. Dorigo, M., Caro, G.D.: The ant colony optimization meta-heuristic. New Ideas Optim. **28**(3), 11–32 (1999)
6. Drigo, M., Maniezzo, V., Colorni, A.: The ant system: optimization by a colony of cooperation agents. IEEE Trans. Syst. Man Cybern. (Part B) **26**, 29–41 (1996)
7. Eyckelhof, C.J., Snoek, M.: Ant systems for a dynamic TSP. In: Dorigo, M., Di Caro, G.A., Sampels, M. (eds.) Ant Algorithms 2002. LNCS, vol. 2463, pp. 88–99. Springer, Heidelberg (2002)
8. Guntsch, M., Middendorf, M.: Pheromone modification strategies for ant algorithms applied to Dynamic TSP. In: Boers, E.J.W., Gottlieb, J., Lanzi, P.L., Smith, R.E., Cagnoni, S., Hart, E., Raidl, G.R., Tijink, H. (eds.) EvoIASP 2001, EvoWorkshops 2001, EvoFlight 2001, EvoSTIM 2001, EvoCOP 2001, and EvoLearn 2001. LNCS, vol. 2037, pp. 213–222. Springer, Heidelberg (2001)
9. Guntsch, M., Middendorf, M.: A population based approach for ACO. In: Cagnoni, S., Gottlieb, J., Hart, E., Middendorf, M., Raidl, G.R. (eds.) EvoIASP 2002, EvoWorkshops 2002, EvoSTIM 2002, EvoCOP 2002, and EvoPlan 2002. LNCS, vol. 2279, pp. 72–81. springer, Heidelberg (2002)
10. Guntsch, M., Middendorf, M., Schmeck, H.: An ant colony optimization approach to dynamic tsp. In: Genetic and Evolutionary Computation Conference (2003)
11. Lin, S., Kernighan, B.W.: An effective heuristic algorithm for the traveling salesman problem. Oper. Res. **21**(3), 498–516 (1973)
12. Mavrovouniotis, M., Yang, S.: A memetic ant colony optimization algorithm for the dynamic travelling salesman problem. Soft Comput. **15**(7), 1405–1425 (2011)
13. Mavrovouniotis, M., Yang, S.: Ant colony optimization with immigrants schemes for the dynamic travelling salesman problem with traffic factors. Appl. Soft Comput. **13**(10), 4023–4037 (2013)
14. Rizzoli, A.E., Montemanni, R., Lucibello, E., Gambardella, L.M.: Ant colony optimization for real-world vehicle routing problems. Swarm Intell. **1**(2), 135–151 (2007)
15. Stutzle, T., Hoos, H.: Max-min ant system and local search for the traveling salesman problem. In: IEEE International Conference on Evolutionary Computation, pp. 309–314 (1997)

A Self-Adaptive Control Strategy of Population Size for Ant Colony Optimization Algorithms

Yuxin Liu[1], Jindan Liu[1], Xianghua Li[1(✉)], and Zili Zhang[1,2(✉)]

[1] School of Computer and Information Science, Southwest University,
Chongqing 400715, China
li_xianghua@163.com, zhangzl@swu.edu.cn
[2] School of Information Technology,
Deakin University, Locked Bag 20000, Geelong, VIC 3220, Australia

Abstract. Ant colony optimization (ACO) algorithms often have a lower search efficiency for solving travelling salesman problems (TSPs). According to this shortcoming, this paper proposes a universal self-adaptive control strategy of population size for ACO algorithms. By decreasing the number of ants dynamically based on the optimal solutions obtained from each interaction, the computational efficiency of ACO algorithms can be improved dramatically. Moreover, the proposed strategy can be easily combined with various ACO algorithms because it's independent of operation details. Two well-known ACO algorithms, i.e., ant colony system (ACS) and max-min ant system (MMAS), are used to estimate the performance of our proposed strategy. Some experiments in both synthetic and benchmark datasets show that the proposed strategy reduces the computational cost under the condition of finding the same approximate solutions.

Keywords: Ant colony optimization · Population size · Self-adaptive control · Travelling salesman problem

1 Introduction

Along with ant system (AS) [1,2] is proposed in 1991, amount of ant colony optimization (ACO) algorithms such as ant colony system (ACS) [3] and max-min ant system (MMAS) [4] have been proposed to pursue strong robustness and excellent solution quality for solving optimization problems, such as travelling salesman problems (TSPs). Existing studies have provided a detailed survey about the historical development of traditional ACO algorithms [5–7].

Although the detailed calculation procedures of ACO algorithms are different, they have a common characteristic: all of them are population-based algorithms. For such kind of algorithms, an appropriate population size is critical to both computational efficiency and global search ability [8,9]. On the one hand, a large population size can improve the sufficiency and precision of a search process. But, it may increase the computational cost and lower the convergence rate of algorithms. On the other hand, although small population size makes

© Springer International Publishing Switzerland 2016
Y. Tan et al. (Eds.): ICSI 2016, Part I, LNCS 9712, pp. 443–450, 2016.
DOI: 10.1007/978-3-319-41000-5_44

algorithms have a higher computation speed, it may weaken the search ability of ant colony and cause the algorithms to fall into the local optimal solutions. Roeva et al. [10] have discussed and provided the optimal population size of ACO algorithms, but the usage of static parameters may not obtain the best results [11]. Hence, how to control the population size dynamically to balance the search ability and the convergence rate is the bottleneck of ACO algorithms.

In order to solve this problem, this paper proposes a universal self-adaptive control strategy of population size for ACO algorithms. The implementation of the strategy is independent of operation details of algorithms, which could be applied to all kinds of ACO algorithms for solving TSPs.

The rest of this paper is organized as follows. Section 2 introduces the definition of TSPs and the principles of ACO algorithms for dealing with TSPs. Section 3 presents the self-adaptive control strategy of population size for ACO algorithms. Section 4 estimates the efficiency of the strategy. Section 5 summarizes the main results.

2   Background

2.1   The Definition of TSPs

The TSP is an *NP*-hard problem, which can be described as follows: There are n cities, defined as $V = \{i|i = 1, 2, ..., n\}$. The distance between cities i and j is defined as d_{ij}, where $i, j \in V$. A salesman leaves a city i and goes back to the city i after he has travelled all other cities. Each city is visited only once and the sequence of cities visited by the salesman is a Hamiltonian circuit Ω. Thus, the solution of the TSP is to find the shortest path S_{min}, which can be defined as:

$$S_{min} = \min(\sum_{i=1}^{n-1} d_{\Omega_i \Omega_{i+1}} + d_{\Omega_n \Omega_1}) \tag{1}$$

where Ω_i denotes the ith city in the Hamiltonian circuit and $\Omega_i \in V$.

2.2   ACO Algorithms for Solving TSPs

As one of typical ACO algorithms, ACS is taken as an example to describe the process of solving TSPs [3]. Supposing there are n cities and s ants. Each ant is first randomly put on a city. At time t, an ant h located at a city i moves to the next city j according to the probability $P_{ij}^h(t)$, as shown in Eqs. (2) and (3).

If $q \leq q_0$:

$$P_{ij}^h(t) = \begin{cases} 1, & \text{if } j = argmax_{j \in N_i^h} \left(\tau_{ij}^\alpha(t) \times \eta_{ij}^\beta(t) \right) \\ 0, & \text{otherwise} \end{cases} \tag{2}$$

else:

$$P_{ij}^h(t) = \begin{cases} \frac{\tau_{ij}^\alpha(t) \times \eta_{ij}^\beta(t)}{\sum \tau_{iu}^\alpha(t) \times \eta_{iu}^\beta(t)}, & \text{if } j, u \in N_i^h \\ 0, & \text{otherwise} \end{cases} \tag{3}$$

where q_0 is a predefined parameter ($q_0 \in [0,1]$), and q is a random number uniformly distributed in $[0,1]$. The pheromone matrix τ_{ij} represents the amount of pheromone in the path connecting cities i and j. The local heuristic value $\eta_{ij} = 1/d_{ij}$ represents the expectation that an ant h moves from city i to city j, which shows that closer cities are more likely to be selected. α and β weight the importance of pheromone matrix τ and the heuristic information η, respectively. u stands for an unvisited city. N_i^h is a set of feasible neighbors of the ant h in the city i, which can make sure that an ant would not visit any cities twice.

Every time an ant travels a city, it releases a definite amount of pheromone on paths. Meanwhile, the pheromone will evaporate at a speed of parameter $\rho(0 < \rho < 1)$. That is, after an ant h visits city j from city i, the pheromone trail on the path (i,j) is updated as Eq. (4).

$$\tau_{ij}(t+1) = (1-\rho)\tau_{ij}(t) + \rho \times \tau_0 \qquad (4)$$

where τ_0 is a constant concerning the initial amount of pheromone in each path.

After all ants have finished their own travels, the best ant that generates the minimum total length of path is allowed to update the global pheromone matrix according to Eq. (5). $S_{best}(t)$ denotes the minimum total length of paths travelled by ants at time t.

$$\tau_{ij}(t+1) = (1-\rho)\tau_{ij}(t) + \rho \times \frac{1}{S_{best}(t)}, \text{if } (i,j) \in global\ best\ tour \qquad (5)$$

By iteratively updating the pheromone matrix, ACS will converge to the shortest Hamiltonian circuit, which is the optimal solution for TSPs. However, during the evolution process of ACO algorithms, the number of ants remains unchanged, which would cost much more computational resources and limit the computational efficiency [12,13]. Hence, a self-adaptive control strategy of population size for ACO algorithms is proposed in the next section.

3   The Self-Adaptive Population Control Strategy

Population size has a great effect on the search ability and computational efficiency of ACO algorithms. Based on numerous experiments, we find that lots of ants have the same travelling tour when the iteration achieves a certain number of steps. It means that the number of ants in guiding the search process of the colony is too much at that time, and in order to improve the computational efficiency, the redundant individuals should be removed. There are two problems as follows for implementing the dynamic change of population size:

– When should the ACO algorithms trigger the change of population size?
– How to change the population size?

The framework of the self-adaptive population control strategy can be described as follows. If S_{min} doesn't change for x iterations and the number of ants s is larger than the minimal value Ant_{min}, the number of ants is reduced

---

Algorithm 1. sdps-ACS for solving TSPs

Input d_{ij}: The distance between cities i and j;
       n: The number of cities;
Output S_{min}: The length of the shortest Hamiltonian circuit;
Begin
1: Initializing parameters α, β, ρ, s, q_0, Ant_{min}, $k1$, $k2$, $m1$ and $m2$
2: Initializing the pheromone trail τ_0 and the total steps of iteration $Steps_{max}$
3: Setting the iteration counter $t := 0$
4: While $t < Steps_{max}$ Do
5:      For $k := 1$ to s Do
6:          Constructing a tour by an ant h
7:          Updating the local pheromone matrix
8:      End For
9:      Updating the global pheromone matrix
10:     $best :=$ the global best ant
11:     $S_{min}(t) :=$ the length of tour generated by the ant $best$ at the iteration step t
12:     If $t \geq x$ Then
13:         If $S_{min}(t-x+1) == S_{min}(t-x+2) == ... == S_{min}(t)$ $\&\&s > Ant_{min}$ Do
14:             Calculate Ant_{same} and S_{slope};
15:             If $k1 < S_{slope} < k2$ Do
16:                 $s = s - Ant_{same} \times m1$;
17:             Else
18:                 $s = s - Ant_{same} \times m2$;
19:             End If
20:         End If
21:     End If
22:     $t := t + 1$
23: End While
24: Outputting the optimal solution S_{min}
End

---

based on the following definition. The minimum number of ants is set to ensure the continuous operation of the colony.

Rule definition: *If the convergence rate of S_{min}, represented by the symbol S_{slope}, belongs to the range of $[k1, k2]$, the number of ants s is changed as $s = s - Ant_{same} \times m1$. Otherwise, s is changed as $s = s - Ant_{same} \times m2$. Ant_{same} represents the number of ants that have the same tour with the last time step, $k1$, $k2$, $m1$ and $m2$ are predefined parameters.*

The framework shows that our strategy has two obvious advantages. First, the new ACO algorithms with the proposed strategy (named as sdps-ACO algorithms) are dynamic, which can change the size of population according to the practical convergence situation of optimal solutions. Second, the strategy has no dependencies on the detailed calculation process of ACO algorithms. Thus, it is universal and can be used to optimize all kinds of ACO algorithms based on swarm computation. Algorithm 1 shows the description of new ACS (named as sdps-ACS) for solving TSPs.

4 Simulation Experiments

4.1 Datasets

Two types of datasets are used in this paper. One is a synthetic dataset. We randomly generate 30 nodes, named as syn30. The coordinates of nodes are shown in Table 1. The other are three benchmark datasets downloaded from the website TSPLIB[1], i.e., att48, kroC100 and gr120. Based on these datasets, we build undirected weighted networks that are fully connected. The weight of the edge is a straight-line distance in the synthetic network, while the weight of edge in the benchmark network is an actual mileage.

4.2 Experimental Analysis

Two typical ACO algorithms, i.e., ACS [3] and MMAS [4], are used to validate the efficiency of the proposed strategy. The new ACO algorithms with the self-adaptive population control strategy are named as the original algorithms with a prefix 'sdps-', i.e., sdps-ACS and sdps-MMAS. All experiments are undertaken in one computer with CPU: Inter(R) Core(TM)2 Duo E4500 2.20 GHz, RAM: 2.00 GB, OS: Windows 7. The application development environment is Microsoft Visual Studio 2010. Through estimating the effect of parameters on the performance of ACO algorithms, some parameters are set as shown in Table 2. Moreover, the parameters involved in our proposed strategy are also analyzed, and

Table 1. The coordinates of 30 synthetic dataset.

No.	Coordinate	No.	Coordinate	No.	Coordinate	No.	Coordinate	No.	Coordinate
1	(2,99)	2	(4,50)	3	(7,64)	4	(13,40)	5	(18,54)
6	(18,40)	7	(22,60)	8	(24,42)	9	(25,62)	10	(25,38)
11	(37,84)	12	(41,94)	13	(41,26)	14	(44,35)	15	(45,21)
16	(54,67)	17	(54,62)	18	(58,35)	19	(58,69)	20	(62,32)
21	(64,60)	22	(68,58)	23	(71,44)	24	(71,71)	25	(74,78)
26	(82,7)	27	(83,46)	28	(83,69)	29	(87,76)	30	(91,38)

Table 2. The parameters values used in this paper.

Dataset	Alg.	α	β	ρ	q_0	$Steps_{max}$
syn30	ACS	2	3	0.1	0.01	1000
	MMAS	1	2	0.95	–	1000
att48	ACS	2	4	0.1	0.1	1000
	MMAS	1	2	0.95	–	1000
kroC100, gr120	ACS	2	5	0.1	0.1	1000
	MMAS	2	5	0.95	–	1000

---

[1] http://www.iwr.uni-heidelberg.de/groups/comopt/software/TSPLIB95/.

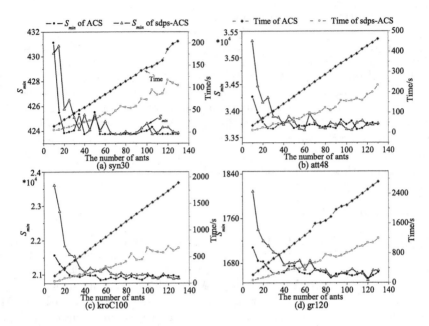

Fig. 1. The comparison of optimal solution and computational cost between ACS and sdps-ACS under the different initial numbers of ants in four datasets.

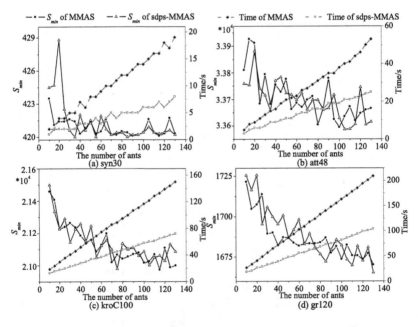

Fig. 2. The comparison of optimal solution and computational cost between MMAS and sdps-MMAS under the different initial numbers of ants in four datasets.

the best values are chosen as $Ant_{min} = 30$, $x = 5$, $k1 = 1/3$, $k2 = 0.5$, $m1 = 0.7$ and $m2 = 0.4$.

Figure 1 shows the comparison results of the computational cost (i.e., Time) and the optimal solution (i.e., S_{min}) between ACS and sdps-ACS under the different initial numbers of ants in four datasets. It can be seen that, S_{min} calculated by sdps-ACS has little difference with that of ACS. These mean that the strategy has no influence on the search ability of traditional ACO algorithms. However, the time spent by sdps-ACS is obviously shorter than that of ACS, which means that the strategy can improve the computational efficiency of the traditional ACO algorithms. The same conclusion can be seen in comparison results between MMAS and sdps-MMAS, as shown in Fig. 2. Therefore, we can conclude that our proposed strategy is independent of the initial scales of population and the types of networks.

5   Conclusion

The number of ants was fixed in the traditional ACO algorithms, which reduced the search efficiency. This paper proposed a self-adaptive control strategy of population size for ACO algorithms. The strategy was independent of the operation details of algorithms, and thus could be used to optimize all kinds of ACO algorithms for solving TSPs. The strategy had been applied to two typical ACO algorithms, i.e., ACS and MACS. Experiments have compared the optimal solution and the computational cost between ACS and sdps-ACS, MACS and sdps-MACS under the different initial numbers of ants. The results in both synthesis and benchmark datasets showed that the optimized ACO algorithms had a higher search efficiency than that of traditional ACO algorithms for solving TSPs.

Acknowledgments. This work was supported by National Natural Science Foundation of China (Nos. 61402379, 61403315), Natural Science Foundation of Chongqing (No. cstc2013jcyjA40022), Fundamental Research Funds for the Central Universities (Nos. XDJK2016D020, XDJK2016A008), Specialized Research Fund for the Doctoral Program of Higher Education (No. 20120182120016), and Chongqing Graduate Student Research Innovation Project. Prof. Zili Zhang and Dr. Xianghua Li are the corresponding authors of this paper.

References

1. Dorigo, M., Maniezzo, V., Colorni, A.: Ant system: optimization by a colony of cooperating agents. IEEE Trans. Syst. Man Cybern. Part B Cybern. **26**(1), 29–41 (1996)
2. Johnson, D.S., McGeoch, L.A.: The traveling salesman problem: a case study in local optimization. In: Aarts, E.H.L., Lenstra, J.K. (eds.) Local Search in Combinatorial Optimization, pp. 215–310. Wiley, Chichester (1997)

3. Dorigo, M., Gambardella, L.M.: Ant colony system: a cooperative learning approach to the traveling salesman problem. IEEE Trans. Evol. Comput. **1**(1), 53–66 (1997)
4. Stützle, T., Hoos, H.: MAX-MIN ant system. Future Gener. Comput. Syst. **16**(8), 889–914 (2000)
5. Dorigo, M., Birattari, M., Stützle, T.: Ant colony optimization. IEEE Comput. Intell. Mag. **1**(4), 28–39 (2006)
6. Dorigo, M., Stützle, T.: Ant Colony Optimization: Overview and Recent Advances. Techreport, IRIDIA, Universite Libre de Bruxelles (2009)
7. López-Ibáñez, M., Stützle, T., Dorigo, M.: Ant Colony Optimization: A Component-Wise Overview. Techreport, IRIDIA, Universite Libre de Bruxelles (2015)
8. Sarker, R., Kazi, M.F.A.: Population size search space and quality of solution: an experimental study. In: Proceedings of the 2003 Congress on Evolutionary Computation, vol. 3, pp. 2011–2018. IEEE (2003)
9. Karafotias, G., Hoogendoorn, M., Eiben, A.E.: Parameter control in evolutionary algorithms: trends and challenges. IEEE Trans. Evol. Comput. **19**(2), 167–187 (2015)
10. Roeva, O., Fidanova, S., Paprzycki, M.: Population size influence on the genetic and ant algorithms performance in case of cultivation process modeling. In: Fidanova, S. (ed.) WCO 2013. Studies in Computational Intelligence, vol. 580, pp. 107–120. Springer, Switzerland (2015)
11. Teo, J.: Exploring dynamic self-adaptive populations in differential evolution. Soft Comput. **10**(8), 673–686 (2006)
12. Eiben, A.E., Hinterding, R., Michalewicz, Z.: Parameter control in evolutionary algorithms. IEEE Trans. Evol. Comput. **3**(2), 124–141 (1999)
13. Castillo, O., Neyoy, H., Soria, J., Melin, P., Valdez, F.: A new approach for dynamic fuzzy logic parameter tuning in ant colony optimization and its application in fuzzy control of a mobile robot. Appl. Soft Comput. **28**, 150–159 (2015)

MPPT of a Partially Shaded Photovoltaic Module by Ant Lion Optimizer

Ekaterina A. Engel[1(✉)] and Igor V. Kovalev[2]

[1] Katanov State University of Khakassia,
Shetinkina, 61, 655017 Abakan, Russian Federation
ekaterina.en@gmail.com
[2] Siberian State Aerospace University, Krasnoyarsky Rabochy Avenue 31,
660014 Krasnoyarsk, Russian Federation
kovalev.fsu@mail.ru

Abstract. This paper is mainly focused on the maximum power point tracking of a photovoltaic module under non uniform solar irradiation level. In such a case multiple local maximum power points are observed but only one global maximum power point exists. In this paper, the Ant Lion Optimizer algorithm is adopted to find the global maximum power point of a partially shaded photovoltaic module. The simulation results show that the performance of the Ant Lion Optimizer is better than the perturbation & observation algorithm in detecting the global maxima of a partially shaded photovoltaic module.

Keywords: Photovoltaic module · Non-Uniform solar irradiation level · Ant Lion Optimizer

1 Introduction

Solar energy is the most inexhaustible and non-polluting among all the clean and renewable energy resources. The performance of a photovoltaic (PV) module is dependent on temperature, solar irradiation level, shading, and module configuration. Under partially shaded conditions, the PV module characteristics become more complex with several peaks [1]. In such situations, it is very important to extract the maximum power. But the conventional methods (such as the perturbation & observation algorithm) fail to optimally track the maximum power point (MPP) of a PV module [1]. This forms the motivation for the use of evolutionary optimization techniques such as the Ant Lion Optimizer (ALO) [2] for detection of the global MPP. The Ant Lion Optimizer is a nature inspired algorithm proposed by Seyedali Mirjalili in 2015 [2]. The Ant Lion Optimizer mimics the hunting mechanism of ant lions in nature. The benefits of the ALO algorithm are global optimization, simplicity, reliability, and effectiveness for real world tasks. The MATLAB toolbox of the ALO algorithm is publicly available at http://www.alimirjalili.com/ALO.html.

The main contributions of the present work are given below.

© Springer International Publishing Switzerland 2016
Y. Tan et al. (Eds.): ICSI 2016, Part I, LNCS 9712, pp. 451–457, 2016.
DOI: 10.1007/978-3-319-41000-5_45

(a) Performance of the ALO algorithm in effectively locating the global MPP for a partially shaded photovoltaic module is tested and the simulation results are discussed.

(b) The simulation results revealed advantages of the ALO algorithm over the conventional perturbation & observation algorithm.

2   Matlab/Simulink Model of the Photovoltaic Module Under Partially Shaded Conditions

The study of the partial shading PV module is carried out based on the MATLAB/Simulink model power_PVArray_PartialShading.slx [3]. This model simulates of a 250-W PV module under partially shaded conditions. The PV module consists of 60 cells connected in series (Fig. 1).

The MATLAB/Simulink model power_PVArray_PartialShading.slx contains a variable DC voltage source for measuring the I-V and P-V characteristics of the PV module under partially shaded conditions. The partially shaded PV module is modeled as three strings of 20 series-connected cells in parallel with bypass diodes that allow current flow when the cells are shaded [3]. A standard irradiance of 1 kW/m^2 is applied on the first string of 20 cells while partial shading is applied on strings 2 (cells 21-40) and string 3 (cells 41-60), resulting in respective insolation of 250 W/m^2 and 700 W/m^2. The control model based on the perturbation & observation algorithm provides only a local MPP of the partially shaded PV module rather than the global

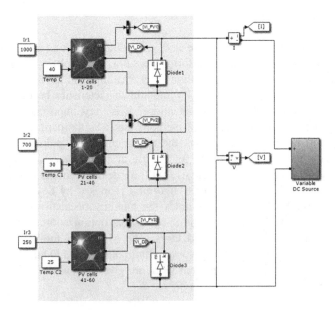

Fig. 1. The MATLAB/Simulink model of PV module under partially shaded conditions.

MPP in its under-utilization. In order to obtain the global MMP of the partially shaded PV module, we adopted the Ant Lion Optimizer [2].

2.1 Maximum Power Point Tracking of the Partially Shaded PV Module Using the Ant Lion Optimizer

The power of a partially shaded PV module is defined as

$$f(X(t)) = X(t) * I, \tag{1}$$

where I represents the current and $X(t)$ represents the voltage. This function f was used as a fitness function for the ALO.

The random voltage values (walks of ants) are all based on the following equation

$$X(t) = [0, \ cumsum(2r(t_1)-1), \ cumsum(2r(t_2)-1), \ \ldots, \ cumsum(2r(t_n) - 1)], \tag{2}$$

where $cumsum$ is the cumulative sum, n is the maximum number of iterations, t shows the step of the random walk, and $r(t)$ is a stochastic function defined as follows:

$$r(t) = \begin{cases} 1 & if \quad rand > 0.5 \\ 0 & if \quad rand \leq 0.5 \end{cases} . \tag{3}$$

The position of the ants is presented with the vector:

$$M_{Ant} = [A_1 \quad A_2 \quad \ldots \quad A_n], \tag{4}$$

where A_i represents the $i - th$ voltage value and n is the number of ants.

The fitness function of each ant is saved in the following vector M_{OA}:

$$M_{OA} = [f(A_1) \, f(A_2) \, \ldots f(A_n)], \tag{5}$$

where f is the objective function.

The vector for each ant lion position is defined as follows:

$$M_{Antlion} = [AL_1 \, AL_2 \ldots AL_n], \tag{6}$$

where $M_{Antlion}$ is the vector for each ant lion position, AL_i represents the value of the voltage of the $i - th$ ant lion, and n is the number of ant lions. The fitness function of each ant lion is saved in the following vector:

$$M_{OAL} = [f(AL_1) f(AL_2) \ldots f(AL_n)]. \tag{7}$$

In order to preserve random walks of ants inside the search space, they are normalized in the following way:

$$X^t = \frac{\left(X^t - a\right) \times \left(b - c^t\right)}{\left(d^t - a\right)} + c, \tag{8}$$

where a is the minimum of the random walk of variable, b is the maximum of the random walk of variable, c^t is the minimum of the variable at the $t - th$ iteration, and d^t is the maximum of the variable at the $t - th$ iteration. Mathematical modeling of ants trapping in the ant lion's pits is given as follows:

$$c_i^t = Antlion_j^t + c^t, \ \ d_i^t = Antlion_j^t + d^t, \tag{9}$$

where c^t is the minimum of variable at the $t - th$ iteration, d^t is the maximum of variable at the $t - th$ iteration, and $Antlion_j^t$ is the position of the selected $j - th$ ant lion at the t-th iteration. The ant lion's hunting capability is modeled by fitness proportional roulette wheel selection. The mathematical model that describes the way the trapped ant slides down towards the ant lion is given as follows:

$$c^t = \frac{c^t \cdot T}{10^W \cdot t}, \ \ d^t = \frac{d^t \cdot T}{10^W \cdot t}, \tag{10}$$

where t is the current iteration, T is the maximum number of iterations, and w is the constant that depends on the current iteration as follows:

$$w = \begin{cases} 2 & if \ \ t > 0.1T \\ 3 & if \ \ t > 0.5T \\ 4 & if \ \ t > 0.75T \\ 5 & if \ \ t > 0.9T \\ 6 & if \ \ t > 0.95T \end{cases} \tag{11}$$

Finally, the best ant lion for iteration is considered as elite. This rule is fulfilled elitism. The elitism indicates that every ant randomly walks near selected the ant lion and has a position $Ant_i^t = (R_A^t + R_E^t)/2$, where R_A^t is the random walk around the ant lion selected by the roulette wheel at the $t - th$ iteration and R_E^t is the random walk around the elite ant lion at the t-th iteration.

To illustrate the benefits of the Ant Lion Optimizer in finding the global maxima of a partially shaded PV module, the numerical examples are revisited.

2.2   Simulation and Results

All the simulations of the 250-W PV module under partially shaded conditions are carried out based on MATLAB/Simulink model power_PVArray_PartialShading.slx [3]. Figure 2 presents the global I-V and P-V curves. It can be seen that the P-V curve exposes three maxima. The results provided by the perturbation & observation algorithm are presented by Fig. 3. In this comparison study of detecting the global maxima of the partially shaded PV module, the performance of the ALO is compared against the perturbation & observation algorithm, under the same conditions.

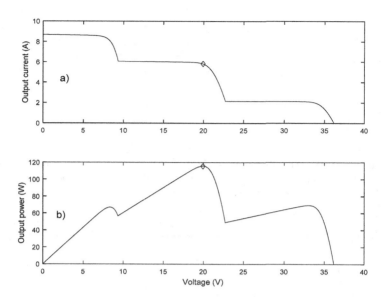

Fig. 2. The global I-V and P-V curves of the PV module.

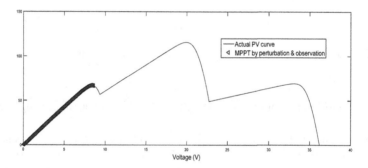

Fig. 3. Plot of the PV module's power provided by the control model based on the perturbation & observation algorithm and the actual P-V curve of the partially shaded PV module respectively.

We use the following initial parameters for ALO: the lower and upper bound of voltage was set to 0 and 40 respectively; the number of search agents was set to 40; the maximum number of iterations was set to 100.

The ALO provided the global MMP after 53 iterations. The voltage at peak power is 20.2 V. Figures 3 to 5 show the simulation results.

It is clear from Fig. 3 that the perturbation & observation algorithm is ineffective. Even after 5000 iterations, the global MPP is still not being detected.

The control model of the partially shaded PV module based on the Ant Lion Optimizer is more robust and provides global MMP trucking in comparison with the control model bases on the perturbation & observation algorithm (Figs. 3–5). Figure 4 shows how the ALO optimized power of the partially shaded PV module (shown by

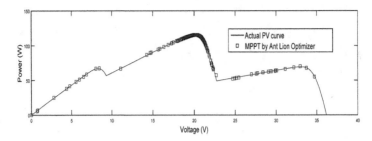

Fig. 4. Plot of the PV module's power provided by the control model based on the Ant Lion Optimizer and the actual P-V curve of the partially shaded PV module respectively.

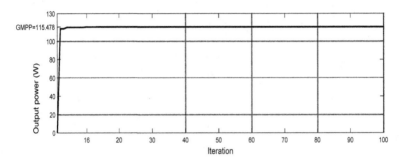

Fig. 5. The convergence profile of the output power against number of fitness function evaluations.

squares) start clustering around the global MPP (GMMP) with an increasing number of iterations.

Figure 5 shows a steady final value of the output power of a partially shaded PV module, as produced by the ALO algorithm. The control model of partially shaded the PV module based on the Ant Lion Optimizer achieves real-time control speed and competitive performance in finding the global MPP of the PV module under a non-uniform solar irradiation level.

3 Conclusions

Simulation comparison results demonstrate the effectiveness of the ALO in finding the global MPP of the PV module under partially shaded conditions as compared with the standard model based on the perturbation & observation algorithm. It is shown that the control model of the PV module based on the Ant Lion Optimizer provided global MMP trucking with real-time control speed and a steady final value of the output power of a partially shaded PV module. These results also suggest that the ALO is a potential candidate for determining the global MPP of a partially shaded PV module with unknown search spaces.

Acknowledgments. The authors wish also to thank Daniel Foty and the reviewers for valuable comments.

References

1. Patel, H., Agarwal, V.: MATLAB-based modeling to study the effects of partial shading on PV array characteristics. IEEE Trans. Energy Convers. **23**(1), 10–302 (2008)
2. Mirjalili, S.: The ant lion optimizer. Adv. Eng. Softw. **83**, 80–98 (2015)
3. The MathWorks Inc. http://www.mathworks.com/examples/simpower/mw/sps_product-pow er_PVArray_PartialShading-partial-shading-of-a-pv-module?s_tid=srchtitle

A Hybrid ACO-ACM Based Approach for Multi-cell Image Segmentation

Dongmei Jiang[1,2], Qinglan Chen[3(✉)], Benlian Xu[1(✉)],
and Mingli Lu[1]

[1] School of Electrical and Automatic Engineering,
Changshu Institute of Technology, Changshu 215500, China
xu_benlian@cslg.edu.cn
[2] School of Information and Electrical Engineering,
China University of Mining and Technology, Xuzhou 221116, China
[3] School of Mechanical Engineering, Changshu Institute of Technology,
Changshu 215500, China
chenql@cslg.edu.cn

Abstract. In this paper, a hybrid multi-cell image segmentation approach is proposed, based on the combination of active contour model (ACM) and ant colony optimization (ACO), for multi-cell image segmentation. This novel image segmentation algorithm integrates the characteristics of ACM model into the ACO with tractable and well defined energy and heuristic functions. Consequently, the problem of cell image segmentation is actually converted to search for the marks of cell contours by group of ants. Experiment results show that our proposed approach is more effective than several existing methods, and it is noted that our proposed approach is developed and implemented in Lab-VIEW as well with performance consistency.

Keywords: Cell segmentation · Active contour · Ant colony optimization

1 Introduction

Image segmentation [1] is a partitioning process of an image into constituent regions or objects, that is the fundamental and important step for image processing. Since Kass et al. introduce the active contour method, known as snake model [2, 3], it has been widely applied in image processing and computer vision for boundary extraction and image segmentation. Active contours are curves defined within an image domain that can move under the influence of internal forces coming from the curve itself and external forces computed from the image data. The internal and external forces are defined so that the snake will conform to an object boundary or other desired features within an image. However, the initial contour should be close to the true boundary or else it will likely converge to the wrong result.

The ACO algorithm [4, 5] is motivated by the natural foraging behavior of ant species, whereby ants deposit a chemical substance, known as pheromone on the ground. The higher the pheromones level along a path, the higher the probability that a

© Springer International Publishing Switzerland 2016
Y. Tan et al. (Eds.): ICSI 2016, Part I, LNCS 9712, pp. 458–466, 2016.
DOI: 10.1007/978-3-319-41000-5_46

given ant will follow the path. Therefore, ants using pheromone can find short paths between their nest and food sources.

Observe that, to locate regions where objects probably appear and adaptively extract objects of interest, some basic elements, such as energy function formulation and minimization in ACM, could be extended into ACO for image segmentation [6–10]. Therefore, we propose a hybrid multi-cell image segmentation approach for multi-cell extraction. In our approach, we integrate the characteristics of ACM model into the ACO with tractable and well defined energy and heuristic functions. Finally, the problem of cell image segmentation is actually converted to search for the marks of cell contours by group of ants.

2 Active Contour Model

The basic idea in active contour models [2, 3] is to evolve a curve, subject to constraints from a given image, to detect objects of interest in an image, which is based on energy minimization. In ACM, the position of the snake is represented parametrically by $v(s) = (x(s), y(s)), s \in (0, 1]$, and the energy function is defined as

$$
\begin{aligned}
E_{snake} &= \int_C E_s(v(s)) ds \\
&= \int_C \left[E_{in}(v(s)) + E_{image}(v(s)) \right] ds
\end{aligned}
\tag{1}
$$

where $E_{in}(v(s))$ is the internal energy, which controls the smoothness of the contour. $E_{image}(v(s))$ is the image energy, which attracts the contour toward the object in the image. The internal energy is defined as

$$
E_{in} = \alpha(s) \cdot |v_s(s)|^2 + \beta(s) \cdot |v_{ss}(s)|^2
\tag{2}
$$

where v_s and v_{ss} are the first and second derivatives of $v(s)$, and parameters $\alpha(s)$ and $\beta(s)$ impose the elasticity and rigidity coefficients of the curve. Adjusting the weights $\alpha(s)$ and $\beta(s)$ could control the relative importance between v_s and v_{ss}. In a simple way, the image energy can be defined as

$$
f(|\nabla I|) = \max(|\nabla I|) + 1 - (|\nabla I|)
\tag{3}
$$

where $|\nabla I|$ is the norm or magnitude of the gradient image. The negative sign results in an attraction of the contour with large values of gradient during energy minimization. So, the term E_{image} can be rewritten as $E_{image} = \omega f(|\nabla I|)$.

In digital image processing, the minimization of the energy functional E_{snake} of ACM is given as

$$E_{snake} = \sum_{i=1}^{n} \left\{ \alpha |v_s(s)|^2 + \beta |v_{ss}(s)|^2 + \omega f(|\nabla I|) \right\} \tag{4}$$

3 Proposed Method

This section gives the principle of the proposed hybrid multi-cell image segmentation approach based on ACM and ACO. Firstly, we obtain the initial contour through the preprocessing multi-cell image, and the searching space is constructed by certain rules on the basis of the initial contour. Secondly, multiple ant systems are built to search for the best cell contour marks through decision model in its own structured searching space, with the aim of minimizing the defined energy function of ACM. Lastly, the optimal segmentation is acquired by restoring the optimal path into the image.

3.1 Searching Space

The searching space where ants work is constructed by certain rules on the basis of a

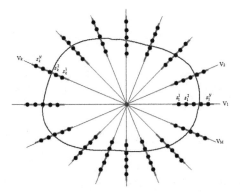

Fig. 1. Searching space

given initial contour. For instance, Fig. 1 shows an initial contour (the black curve), and its initial center is computed accordingly, and plotted as illustrated with a blue point. Then the curve is divided equally into M directions based on the initial center (the red lines). On each direction, the red line is equally divided into N discrete points, as shown with black points, distributed towards or far away from the initial center. These points constitute the so-called one dimensional searching space.

Therefore, we assume that a collection of $v_k = \left\{ s_k^1, s_k^2, \ldots, s_k^N \right\} (k = 1, 2, \ldots, M)$ represents the searching space in the $k - th$ direction, then M collections constitute the ant searching space, denoted by $\Omega = \{v_1, v_2, \ldots, v_M\}$.

3.2 Probability Transition

The aim of our proposed method is to built the multiple ant systems to locate the best points belonging to cell contours through decision model in the structured searching space, namely, minimizing the energy function (Eq. (4)) in the ACM. Specifically, in each searching space, each ant selects the following pixel to be visited according to a stochastic mechanism that belongs to cell contours, with the aim of minimizing the energy function. Assume that an ant is located at $s_k^i \in v_k$, the movement from its location to its neighboring location $s_{k+1}^j \in v_{k+1}$ may happen with the following probability

$$p_{i,j} = \frac{\tau\left(s_k^i, s_{k+1}^j\right)^{\chi} \cdot \eta\left(s_{k+1}^j\right)^{\delta}}{\sum\limits_{x \in v_{k+1}} \tau\left(s_k^i, s_{k+1}^x\right)^{\chi} \cdot \eta\left(s_{k+1}^x\right)^{\delta}} \tag{5}$$

where $\tau\left(s_k^i, s_{k+1}^j\right)$ is the total sum of pheromone amount between s_k^i and s_{k+1}^j; $\eta\left(s_{k+1}^j\right)$ is the heuristic information of location s_{k+1}^j to be defined later; χ and δ are adjustment parameters related to pheromone intensity and importance of heuristic, respectively.

The heuristic information is defined as the gradient of the given image and described as

$$\eta\left(s_{k+1}^j\right) = \frac{1}{f(|\nabla I|)_{s_{k+1}^j}} \tag{6}$$

where the definition of function f follows the same form as Eq. (3). It is easily known that the closer to the cell contour the greater value of the probability.

3.3 Pheromone Update

The pheromone values are updated after each construction process and the ants' decision has been performed. The aim of pheromone update is to increase the pheromone values associated with good solution, and decrease those associated with bad ones. The evolution of pheromone is formulated as follows:

$$\tau_{i,j}(t) = (1 - \rho) \cdot \tau_{i,j}(t) + \Delta\tau \tag{7}$$

where $\rho \in [0, 1)$ is the pheromone evaporation rate; $\Delta\tau$ describes the amount of pheromone external input to trail (i, j).

When ants find the best path, the pheromone update is defined as

$$\tau_{i,j}(t) = (1-\rho) \cdot \tau_{i,j}(t) + \rho \cdot g(t) \qquad (8)$$

where $g(t)$ is defined as a bounded function in the $t-th$ iteration. if $E_{snake}(\hat{t}) < E_{snake}(t)$ holds, then we have $g(\hat{t}) \geq g(t)$. In our approach, the function is given as

$$g(t) = \frac{c_0}{E_{snake}(t)} \qquad (9)$$

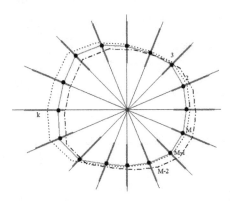

Fig. 2. Several instances of cell contours

where c_0 is a constant, and the function E_{snake} follows the same form as Eq. (4).

Table 1. The pseudo-code of main block of our approach

0.   Initialization
1.   Searching Space
Obtain the initial contour via basic morphology operations.
Build the searching space Ω.
2.   Multi-Ant Systems Decision
While termination condition is not satisfied
For each ant
For collection k in searching space.
Ant movement decision by Eq.(5).
Update pheromone by Eq.(7).
End
End
Compute the energy function of each path.
Find the best set of points with minimum energy value.
Update pheromone by Eq.(8).
End
3.   Output
Acquire contours through sets of points with minimum energy values.

3.4   The Framework of Our Proposed Approach

As shown in Fig. 2, the black points represent the best points of Ω or the marks of cell contours searched by ants. By connecting the best points, we can obtain the contour of the target. For instance, the green curve represents the best path, and the remaining curves represent evolving paths generated by ants during iterations. It is observed that the number of direction is bigger, the yielding curve appears more smooth. However, it suffers from computation burden.

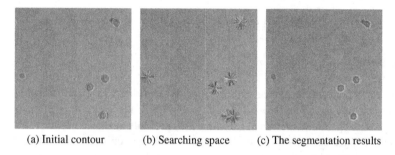

(a) Initial contour   (b) Searching space   (c) The segmentation results

Fig. 3. Initial contour, searching space and segmentation results

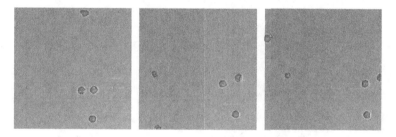

Fig. 4. Segmentation results of other cell images

To visualize our proposed approach in a full view, we summarize the pseudo-code of main block in Table 1.

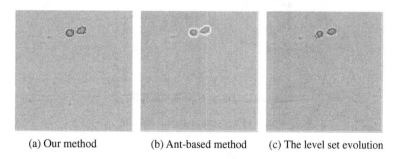

(a) Our method   (b) Ant-based method   (c) The level set evolution

Fig. 5. Cell contour estimate by difference methods

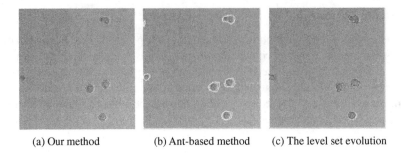

| (a) Our method | (b) Ant-based method | (c) The level set evolution |

Fig. 6. Cell contour estimate by different methods

4   Results

In this section, the performance of our proposed algorithm is tested on two selected multiple cell image sequences. As shown in Fig. 3(a), the yielding cell contours can only give the approximate estimates. Figure 3(b) presents the search space of five cells, each search space consists of 32 collection and each collection consists of 10 point to be visited (denoted by color lines). Figures 3(c) and 4 give the segmentation results of cell image. Note that our algorithm could obtain the accurate cell contours.

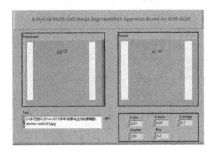

Fig. 7. Front panel of our approach

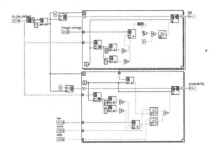

Fig. 8. Ant decision module

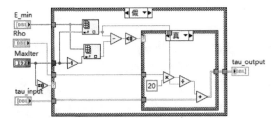

Fig. 9. Pheromone update module

Figures 5 and 6 give the comparisons of multi-cell contour estimates by our proposed approach, the ant-based method [11] and the level set evolution [12], respectively. It can be observes that our method outperforms the other two methods for most of cells in the image.

Our proposed approach is also implemented in LabVIEW and the performance shows the consistency, as shown in Fig. 7. The main function of system consists of five modules: image preprocessing, search space, ant decision, pheromone update and segmentation results. For simplicity, Figs. 8 and 9 illustrate ant decision module and pheromone update module, respectively.

5   Conclusions

This paper proposed a hybrid multi-cell image segmentation approach based on active contour model (ACM) and ant colony optimization (ACO). Through the well defined and tractable energy and heuristic functions, this method converts the problem of image segmentation to a problem of ant searching for the marks of cell contours. Experiment results show that our proposed approach is more effective than several other existing methods. Furthermore, our approach is first implemented in LabVIEW with satisfactory performance consistency.

Acknowledgments. This work is supported by national natural science foundation of China (No.61273312), the natural science fundamental research program of higher education colleges in Jiangsu province (No. 14KJB510001) and the project of talent peak of six industries (DZXX-013).

References

1. Vicent Caselles, F.C., Coil, T., Dibos, F.: A geometric model for active contours. Image Process. Numedsche Math. **66**, 1–31 (1993)
2. Chenyang Xu, a.J.L.P: Snakes, shapes, and gradient vector flow. IEEE Trans. Image Process. **7**, 359–369 (1998)
3. Chenyang Xu, J.L.P.: Gradient vector flow: a new external force for snakes. In: IEEE (1997)
4. Dorigo, M., Stutzle, T.: Ant Colony Optimization. MIT Press, Cambridge (2004)
5. Dorigo, M., Gambardella, L.M.: Ant colony system: A cooperative learning approach to the traveling salesman problem. IEEE Trans. Evolutional Comput. **1**, 53–66 (1997)
6. Rui Li, Y.G., Xing, Y., Li, M.: A novel multi-swarm particle swarm optimization algorithm applied in active contour model. In: IEEE Computer Society (2009)
7. Mahdi Ahmadi Asl, S.A.S.: Active contour optimization using particle swarm optimizer. In: IEEE (2006)
8. Nezamabadi-pour, H., Saryazdi, S., Rashedi, E.: Edge detection using ant algorithms. Soft. Comput. **10**, 623–628 (2005)
9. Tian, J., Yu, W., Xie, S.: An ant colony optimization algorithm for image edge detection. IEEE World Congr. Comput. Intell. **1**, 751–756 (2008)

10. Xu, B., Ren, Y., Zhu, P., Lu, M.: A PSO-based approach for multi-cell multi-parameter estimation. In: The 2014 International Conference on Control, Automation and Information Science (2014)

11. Xu, B., Lu, M., Zhu, P., Shi, J.: An accurate multi cell parameter estimate algorithm with heuristically restrictive ant system. Signal Proces. **101**, 104–120 (2014)

12. Li, C., Xu, C.: Distance regularized level set evolution and its application to image segmentation. IEEE Trans. Image Process. **19**, 3243–3254 (2010)

Brain Storm Optimization

Brain Storm Optimization in Objective Space Algorithm for Multimodal Optimization Problems

Shi Cheng[1](✉), Quande Qin[2], Junfeng Chen[3], Gai-Ge Wang[4,5], and Yuhui Shi[6]

[1] School of Computer Science, Shaanxi Normal University, Xi'an, China
cheng@snnu.edu.cn
[2] Department of Management Science, Shenzhen University, Shenzhen, China
[3] College of IOT Engineering, Hohai University, Changzhou, China
[4] School of Computer Science and Technology,
Jiangsu Normal University, Xuzhou, China
[5] Department of Electrical and Computer Engineering,
University of Alberta, Edmonton, AB, Canada
[6] Department of Electrical and Electronic Engineering,
Xi'an Jiaotong-Liverpool University, Suzhou, China
yuhui.shi@xjtlu.edu.cn

Abstract. The aim of multimodal optimization is to locate multiple peaks/optima in a single run and to maintain these found optima until the end of a run. In this paper, brain storm optimization in objective space (BSO-OS) algorithm is utilized to solve multimodal optimization problems. Our goal is to measure the performance and effectiveness of BSO-OS algorithm. The experimental tests are conducted on eight benchmark functions. Based on the experimental results, the conclusions could be made that the BSO-OS algorithm performs good on solving multimodal optimization problems. To obtain good performances on multimodal optimization problems, an algorithm needs to balance its global search ability and solutions maintenance ability.

Keywords: Brain storm optimization · Brain storm optimization in objective space · Multimodal optimization · Swarm intelligence

1 Introduction

The brain storm optimization (BSO) algorithm is a young and promising algorithm in swarm intelligence. It is based on the collective behavior of human being, that is, the brainstorming process [13,14]. The solutions in BSO are diverging into several clusters. The new solutions are generated based on the mutation of one individual or a combination of two individuals. A comprehensive review on the state-of-the-art BSO research is given in [1]. In the original BSO algorithm, the computational resources are consumed a lot on the clustering strategy at each iteration. To reduce the computational burden, the brain storm optimization in objective space (BSO-OS) algorithm was proposed, and the clustering

© Springer International Publishing Switzerland 2016
Y. Tan et al. (Eds.): ICSI 2016, Part I, LNCS 9712, pp. 469–478, 2016.
DOI: 10.1007/978-3-319-41000-5_47

strategy was replaced by a simple classification strategy based on the fitness values [15]. With modifications, BSO algorithm variants have been utilized to solve different kinds of problems, such as multi-objective optimization problems [19] and multimodal optimization problems [5].

The multimodal optimization is aimed to locate multiple global optima at in a single run and to maintain these found optima until the end of a run [7,9,12]. Two performance criteria can be used to measure the success of search algorithms. One is whether an optimization algorithm could find all desired optima including global and/or local optima, and the other is whether it can maintain multiple candidate solutions stably over a run [7]. Population diversity of swarm intelligence could be a good way to measure the average distance among candidate solutions, which could reflect the algorithm's ability of solutions maintenance.

The remaining of the paper is organized as follows. Section 2 reviews the basic concepts of brain storm optimization algorithms. Section 3 introduces the concepts and performance criteria of multimodal optimization. Experiments of BSO-OS algorithms and four variants of particle swarm optimization (PSO) algorithms on multimodal optimization are conducted in Sect. 4. Finally, Sect. 5 concludes with some remarks and future research directions.

2   Brain Storm Optimization Algorithms

2.1   Original Brain Storm Optimization

The brain storm optimization (BSO) algorithm is based on the collective behavior of human being, that is, the brainstorming process [13,14]. The solutions in BSO are converging into several clusters. The best solution of the population will be kept if the newly generated solution at the same index is not better. New individual can be generated based on the mutation of one or two individuals in clusters. The exploitation ability is enhanced when the new individual is close to the best solution so far. While the exploration ability is enhanced when the new individual is randomly generated, or generated by individuals in two clusters.

The original BSO algorithm is simple in concept and easy in implementation [13,14]. There are three strategies in this algorithm: the solution clustering, new individual generation, and selection [3].

2.2   Brain Storm Optimization in Objective Space

In the original BSO algorithm, the computational resources are spending a lot on the clustering strategy at each iteration. To reduce the computational burden, the brain storm optimization in objective space (BSO-OS) algorithm was proposed, and the clustering strategy was replaced by a simple classification strategy based on the fitness values [15]. The procedure of the BSO in objective space algorithm is given in Algorithm 1.

Algorithm 1. The procedure of the BSO in objective space algorithm

---

1 **Initialization**: Randomly generate n potential solutions (individuals), and evaluate them;

2 **while** *have not found "good enough" solution or not reached the pre-determined maximum number of iterations* **do**

3 | **Classification**: Classify all solutions into two categories: the solutions with better fitness values as elitists and the others as normals;

4 | **New individual generation**: randomly select one or two individuals from elitists or normal to generate new individual;

5 | **Solution disruption**: re-initialize one dimension of a randomly selected individual and update its fitness value accordingly;

6 | **Selection**: The newly generated individual is compared with the existing individual with the same individual index; the better one is kept and recorded as the new individual;

7 | Evaluate all individuals;

2.3 New Individual Generation

The main difference between the original BSO and the BSO-OS is the new individual generation strategy. In the original BSO algorithm, individuals are clustered into several groups. For the BSO-OS algorithm, individuals are classified into two categories according to their fitness values. The procedure of new individual generation strategy is given in Algorithm 2. Two parameters, probability $p_{elitist}$ and probability p_{one}, are used in this strategy. The new individuals are generated according to the functions (1) and (2).

$$x^i_{new} = x^i_{old} + \xi(t) \times \text{rand}() \tag{1}$$

$$\xi(t) = \text{logsig}(\frac{0.5 \times T - t}{k}) \times \text{rand}() \tag{2}$$

where x^i_{new} and x^i_{old} are the ith dimension of $\mathbf{x}_{new}$ and $\mathbf{x}_{old}$; rand() is a random function to generate uniformly distributed random numbers in the range $[0, 1)$; and the value $\mathbf{x}_{old}$ is a copy of one individual or the combination of two individuals. The parameter T is the maximum number of iterations, t is the current iteration number, k is a coefficient to change logsig() function's slope of the step size function $\xi(t)$, which can be utilized to balance the convergence speed of the algorithm.

3 Multimodal Optimization

3.1 Backgrounds

The aim of multimodal optimization is to locate multiple peaks/optima in a single run [10,11] and to maintain these found optima until the end of a run [7,9,12].

Algorithm 2. The **New individual update** Operation

---

1 **New individual generation**: randomly select one or two individual(s) to generate new individual;

2 **if** *random value rand is less than a probability* $p_{elitist}$ **then** /* generate a new individual based on elitists */

3   **if** *random value rand is smaller than a pre-determined probability* p_{one} **then**

4   generate a new individual based on one randomly selected elitist;

5   **else**

6   two individuals from elitists are randomly selected to generate new individual;

7 **else**   /* generate a new individual based on normal */

8   **if** *random value rand is less than a pre-determined probability* p_{one} **then**

9   generate a new individual based on one randomly selected normal;

10   **else**

11   two individuals from normal are randomly selected to generate new individual;

12 The newly generated individual is compared with the existing individual with the same individual index, the better one is kept and recorded as the new individual;

An algorithm on solving multimodal optimization problems should have two kinds of abilities: find global/local optima as many as possible and preserve these found solutions until the end of search.

Several kinds of swarm intelligence algorithms have been utilized to solve multimodal optimization problems, such as niching particle swarm optimization (PSO) with local search [10], variants of PSO algorithms [2], differential evolution algorithm with neighborhood mutation [11], hybrid niching PSO enhanced with recombination-replacement crowding strategy [6], collective animal behavior algorithm [4], sequential niching memetic algorithm [17], and the multiobjective optimization techniques [16,18], *etc.*

3.2 Performance Criteria

Three performance measures are introduced in [8]. The peak ratio (PR) measures the average percentage of all known global optima found over multiple runs, and the success rate (SR) measures the percentage of successful runs (a successful run is defined as a run where all known global optima are found) out of all runs. The equations of PR and SR calculation are given in Eqs. (3) and (4), respectively.

$$PR = \frac{\sum_{run=1}^{NR} NPF_i}{NKP \times NR} = \frac{NPF}{NKP \times NR} \tag{3}$$

$$SR = \frac{NSR}{NR} \tag{4}$$

where NPF_i denotes the number of global optima found in the end of the i-th run, NPF denotes the total number of global optima found in all runs, NKP the number of known global optima, NR the number of runs, and NSR denotes the number of successful runs [8].

The process for determining whether all global optima are found is given in Algorithm 3. The Algorithm 4 gives the process to determine whether a found solution is a new global optimum or not.

Algorithm 3. The algorithm for determining if all global optima are found.

---

1 **Input**: $S_{individuals}$: a set of individuals (candidate solutions) in the population; ϵ: accuracy level; $fit(\mathbf{g}^*)$: the fitness of global optima;
2 **Output** $S_{solutions}$: a set of best-fit individuals identified as unique solutions; *count*: the number of identified global optima found in the end of a run;
3 Initialization: $S_{solutions} = \emptyset$, *count* = 0;
4 **for** *each individual* $\mathbf{x}_i$ *in the candidate solutions set* $S_{individuals}$ **do**
5   **if** $|fit(\mathbf{g}^*) - fit(\mathbf{x}_i)| \leq \epsilon$ **then**
6   **if** *count* == 0 **then**
7   *count* += 1; $S_{solutions} \leftarrow \mathbf{x}_i$;
8   **else if** $\mathbf{x}_i \notin S_{solutions}$ **then**
9   *count* += 1;
10           $S_{solutions} \leftarrow \mathbf{x}_i$;

Algorithm 4. The algorithm for determining if set $S_{solutions}$ contains solution $\mathbf{x}_i$.

---

1 **Input**: r: niche radius;
2 Initialization: *sign* = **false**;
3 **for** *each existing solution* $\mathbf{x}_j$ *in set* $S_{solutions}$ **do**
4   **if** $distance(\mathbf{x}_i, \mathbf{x}_j) < r$ **then**
5   *sign* = **true**;
6   break;
7   **return** *sign*;

The minimum number of function evaluations (FEs) required to locate all known global optima at a specified accuracy level ϵ is utilized to measure the convergence speed of a search algorithm. The Eq. (5) gives the calculation the average FEs ($AveFEs$) over multiple runs.

$$AveFEs = \frac{\sum_{run=1}^{NR} FEs_i}{NR} = \frac{FEs}{NR} \qquad (5)$$

where FEs_i denotes the number of evaluations used in the i-th run, and FEs denotes the total number of evaluations used in all runs. The $MaxFEs$ is used in (5) when the algorithm cannot locate all the global optima until the end of a run [8].

4   Experimental Study

4.1   Benchmark Functions and Parameters Setting

The eight benchmark functions are given in Table 1, and the settings of each function are given in Table 2 [8]. In all experiments, the parameters of BSO-OS algorithms are set as follows: $p_{elitist} = 0.1$; $p_{one} = 0.8$; $p_{disrupt} = 1.0$, slope $k = 500$. Particle swarm optimization algorithm with star, ring, four clusters or von Neumann structure is used to compare the performance of BSO-OS algorithm. The parameters for four variants of PSO algorithms are $w = 0.72984$, $c_1 = c_2 = 1.496172$. The accuracy level ϵ is $1.0E - 03$. The population size and number of iterations are given in Table 3.

Table 1. The benchmark functions used in experimental study, where D is the dimension of each problem.

Func.	Function name	D	Optima (global/local)
f_1	Five-Uneven-Peak Trap	1	2/3
f_2	Equal Maxima	1	5/0
f_3	Uneven Decreasing Maxima	1	1/4
f_4	Himmelblau	2	4/0
f_5	Six-Hump Camel Back	2	2/4
f_6	Shubert	2/3	$D \cdot 3^D$/many
fig f_7	Vincent	2/3	6^D/0
f_8	Modified Rastrigin - All Global Optima	2	$\prod_{i=1}^{D} k_i$/0

4.2   Experimental Results and Analysis

The percentages of global optima found by different algorithms on these functions are listed in Table 4, and the results of success rate are given in Table 5. The results of function f_7, f_6 (3D) are excluded from Table 5. Because these functions have many global optima, all three algorithms in the experiment could not find all global optima at a single run.

In general, PSO with star structure (PSO-Star) will converge to one optimum at the end of a run. The PSO with star structure has no success run on problems with multiple global optima, and it only has some success runs on function

Table 2. The settings of benchmark functions.

Function		r	Maximum	No. of global optima
f_1	(1D)	0.01	200.0	2
f_2	(1D)	0.01	1.0	5
f_3	(1D)	0.01	1.0	1
f_4	(2D)	0.01	200.0	4
f_5	(2D)	0.5	4.126513	2
f_6	(2D)	0.5	186.73090	18
f_6	(3D)	0.5	2709.09350	81
f_7	(2D)	0.2	1.0	36
f_7	(3D)	0.2	1.0	216
f_8	(2D)	0.01	-2.0	12

Table 3. The parameters set in BSO-OS and PSO algorithms.

Function	BSO-OS Algorithm		PSO Algorithm		$MaxFEs$
	Population size	Iteration	Population size	Iteration	
f_1 (1D)	1000	50	100	500	5.0E+04
f_2 (1D)	1000	50	100	500	5.0E+04
f_3 (1D)	500	100	100	500	5.0E+04
f_4 (2D)	500	100	100	500	5.0E+04
f_5 (2D)	500	100	100	500	5.0E+04
f_6 (2D)	1000	200	400	500	2.0E+05
f_6 (3D)	800	500	800	500	4.0E+05
f_7 (2D)	800	250	400	500	2.0E+05
f_7 (3D)	1000	400	800	500	4.0E+05
f_8 (2D)	2000	100	400	500	2.0E+05

f_3. This is because that function f_3 has one global optimum. PSO with ring structure (PSO-Ring) outperforms the other algorithms, which is due to that particles follow different local best positions. The BSO-OS algorithm performs better than PSO-Star algorithm, but worse than PSO-Ring algorithm.

The results of experimental study, it could be concluded that:

- The exploitation ability of BSO-OS algorithm should be enhanced. The solutions found by BSO-OS algorithm are close to the real optima, but the solution accuracy needs to be improved. Combining the BSO-OS algorithm with some local search strategies may be a good approach to improve the performance of BSO-OS algorithm for multimodal optimization problems.
- Large population size could be benefited for BSO-OS algorithm on solving multimodal optimization problems.

Table 4. Peak Ratio (PR) of the BSO-OS and four variants of PSO algorithms (with Accuracy Level $\epsilon = 1.0E - 03$).

Function	$NKP \times NR$	PSO-Star		PSO-Ring		FourClusters		vonNeumann		BSO-OS	
		NPF	PR	NPF	PR	NPF	PR	NPF	PR	NPF	PR
f_1 (1D)	100	50	0.5	100	**1.0**	94	0.94	96	0.96	100	**1.0**
f_2 (1D)	250	50	0.2	249	0.996	147	0.588	233	0.932	250	**1.0**
f_3 (1D)	50	45	0.9	50	**1.0**	49	0.98	50	**1.0**	50	**1.0**
f_4 (2D)	200	50	0.25	199	**0.995**	133	0.665	159	0.795	137	0.685
f_5 (2D)	100	50	0.5	100	**1.0**	92	0.92	100	**1.0**	98	0.98
f_6 (2D)	900	50	0.056	856	**0.951**	180	0.2	705	0.783	73	0.081
f_6 (3D)	4050	50	0.012	2591	**0.640**	193	0.048	1759	0.434	8	0.002
f_7 (2D)	1800	51	0.028	810	**0.45**	157	0.087	365	0.203	103	0.057
f_7 (3D)	10800	52	0.005	2060	**0.191**	182	0.017	887	0.082	79	0.007
f_8 (2D)	600	50	0.083	598	**0.997**	169	0.282	538	0.897	387	0.645

Table 5. Success rate (SR) of the BSO-OS and four variants of PSO algorithms. (with Accuracy Level $\epsilon = 1.0E - 03$).

Function	PSO-Star		PSO-Ring		FourClusters		vonNeumann		BSO-OS	
	NSR	SR	NSR	SR	NSR	SR	NSR	SR	NSR	SR
f_1 (1D)	0	0.0	49	0.98	37	0.74	45	0.9	50	**1.0**
f_2 (1D)	0	0.0	50	**1.0**	0	0.0	31	0.62	50	**1.0**
f_3 (1D)	48	0.96	50	**1.0**	49	0.98	50	**1.0**	50	**1.0**
f_4 (2D)	0	0.0	49	0.98	1	0.02	16	0.32	12	0.24
f_5 (2D)	0	0.0	50	**1.0**	41	0.82	50	**1.0**	48	0.9
f_6 (2D)	0	0.0	22	0.44	0	0.0	1	0.02	0	0.0
f_8 (2D)	0	0.0	49	0.98	0	0.0	16	0.32	0	0.0

5   Conclusions

The aim of multimodal optimization is to locate multiple peaks/optima in a single run and to maintain these found optima until the end of a run. In this paper, brain storm optimization in objective space (BSO-OS) algorithm has been utilized to solve multimodal optimization problems. The performance and effectiveness of BSO-OS algorithm on solving multimodal optimization problems have been validated.

The experimental tests are conducted on eight benchmark functions. Based on the experimental results, the conclusions could be made that the BSO-OS algorithm performs better than PSO-Star algorithm but worse than PSO-Ring algorithm for multimodal optimization problems. Combining the BSO-OS algorithm with some local search strategies may be a good approach to improve the performance of BSO-OS algorithm. In addition, to obtain good performances on multimodal optimization problems, an algorithm needs to balance its global search ability and solutions maintenance ability.

Acknowledgement. The research work reported in this paper was partially supported by the National Natural Science Foundation of China under Grant Number 61273367, 61403121, and 71402103.

References

1. Cheng, S., Qin, Q., Chen, J., Shi, Y.: Brain storm optimization algorithm: a review. Artif. Intell. Rev. (2016, in press)
2. Cheng, S., Qin, Q., Wu, Z., Shi, Y., Zhang, Q.: Multimodal optimization using particle swarm optimization algorithms: CEC 2015 competition on single objective multi-niche optimization. In: Proceedings of 2015 IEEE Congress on Evolutionary Computation (CEC 2015), pp. 1075–1082. IEEE, Sendai, Japan (2015)
3. Cheng, S., Shi, Y., Qin, Q., Gao, S.: Solution clustering analysis in brain storm optimization algorithm. In: Proceedings of The 2013 IEEE Symposium on Swarm Intelligence, SIS 2013, pp. 111–118. IEEE, Singapore (2013)
4. Cuevas, E., González, M.: An optimization algorithm for multimodal functions inspired by collective animal behavior. Soft. Comput. **17**(3), 489–502 (2013)
5. Guo, X., Wu, Y., Xie, L.: Modified brain storm optimization algorithm for multimodal optimization. In: Tan, Y., Shi, Y., Coello, C.A.C. (eds.) ICSI 2014, Part II. LNCS, vol. 8795, pp. 340–351. Springer, Heidelberg (2014)
6. Li, M., Lin, D., Kou, J.: A hybrid niching PSO enhanced with recombination-replacement crowding strategy for multimodal function optimization. Appl. Soft Comput. **12**(3), 975–987 (2012)
7. Li, X.: Niching without niching parameters: particle swarm optimization using a ring topology. IEEE Trans. Evol. Comput. **14**(1), 150–169 (2010)
8. Li, X., Engelbrecht, A., Epitropakis, M.G.: Benchmark functions for CEC'2013 special session and competition on niching methods for multimodal function optimization. Technical report, Evolutionary Computation and Machine Learning Group, RMIT University, Australia (2013)
9. Parrott, D., Li, X.: Locating and tracking multiple dynamic optima by a particle swarm model using speciation. IEEE Trans. Evol. Comput. **10**(4), 440–458 (2006)
10. Qu, B.Y., Liang, J., Suganthan, P.: Niching particle swarm optimization with local search for multi-modal optimization. Inf. Sci. **197**, 131–143 (2012)
11. Qu, B.Y., Suganthan, P., Liang, J.: Differential evolution with neighborhood mutation for multimodal optimization. IEEE Trans. Evol. Comput. **16**(5), 601–614 (2012)
12. Rönkkönen, J.: Continuous Multimodal Global Optimization with Differential Evolution-Based Methods. Ph.D. thesis, Department of information technology, Lappeenranta University of Technology, December 2009
13. Shi, Y.: Brain storm optimization algorithm. In: Tan, Y., Shi, Y., Chai, Y., Wang, G. (eds.) ICSI 2011, Part I. LNCS, vol. 6728, pp. 303–309. Springer, Heidelberg (2011)
14. Shi, Y.: An optimization algorithm based on brainstorming process. Int. J. Swarm Intell. Res. (IJSIR) **2**(4), 35–62 (2011)
15. Shi, Y.: Brain storm optimization algorithm in objective space. In: Proceedings of 2015 IEEE Congress on Evolutionary Computation, CEC 2015, pp. 1227–1234. IEEE, Sendai, Japan (2015)
16. Song, W., Wang, Y., Li, H.X., Cai, Z.: Locating multiple optimal solutions of nonlinear equation systems based on multiobjective optimization. IEEE Trans. Evol. Comput. **19**(3), 414–431 (2015)

17. Vitela, J.E., Castaños, O.: A sequential niching memetic algorithm for continuous multimodal function optimization. Appl. Math. Comput. **218**(17), 8242–8259 (2012)
18. Wang, Y., Li, H.X., Yen, G.G., Song, W.: MOMMOP: multiobjective optimization for locating multiple optimal solutions of multimodal optimization problems. IEEE Trans. Cybern. **45**(4), 830–843 (2015)
19. Xue, J., Wu, Y., Shi, Y., Cheng, S.: Brain storm optimization algorithm for multiobjective optimization problems. In: Tan, Y., Shi, Y., Ji, Z. (eds.) ICSI 2012, Part I. LNCS, vol. 7331, pp. 513–519. Springer, Heidelberg (2012)

Multi-objective Brain Storm Optimization Based on Estimating in Knee Region and Clustering in Objective-Space

Yali Wu[1(✉)], Lixia Xie[1], and Qing Liu[2]

[1] Faculty of Automation and Information Engineering, Xi'an University of Technology, Xi'an 710048, China
yliwu@xaut.edu.cn
[2] Shaanxi Key Laboratory of Complex System Control and Intelligent Information Processing, Xi'an 710048, China

Abstract. The knee region of the Pareto-optimal front is important to decision makers in practical contexts. In this paper, a new multi-objective swarm intelligent optimization algorithm, Multi-objective Brain Storm Optimization based on Estimating in Knee Region and Clustering in Objective-Space (MOBSO-EKCO) algorithm is proposed to get the knee point of Pareto-optimal front. Firstly, the clustering strategy acts directly in the objective space instead of in the solution space, which suggests the potential Pareto-dominance areas in the next iteration more quickly. Secondly, the estimating strategy is used to discover the knee regions, which are the most potential part of the Pareto front. Thirdly, Differential Evolution (DE) mutation is used to improve the performance of MBSO. Experimental results show that MOBSO-EKCO is a very promising algorithm for solving these tested multi-objective problems.

Keywords: Brain storm algorithm · Clustering technique · Multi-objective optimization · Pareto-dominance

1 Introduction

Multi-objective problems have gained much attention in the study of sciences, economic, engineering, etc. The optimum solution for a multi-objective optimization problem is not unique but a set of candidate solutions. In the candidate solution set, no solution is better than any other one with regards to all objectives.

During the last decades, a number of evolutionary algorithms and population-based methods have been successfully used to solve multi-objective optimization problems. For example, there are Multiple Objective Genetic Algorithm [1], Non-dominated Sorting Genetic Algorithm (NSGA II) [2], Strength Pareto Evolutionary Algorithm (SPEA II) [3], Multi-objective Particle Swarm Optimization (MOPSO) [4], to name just a few. Although great improvement has been gained in multi-objective

This paper is supported by National Youth Foundation of China with Grant Number 61503299.

© Springer International Publishing Switzerland 2016
Y. Tan et al. (Eds.): ICSI 2016, Part I, LNCS 9712, pp. 479–490, 2016.
DOI: 10.1007/978-3-319-41000-5_48

optimization algorithm, it is hard to get enough non-dominate solutions to fit real Pareto front in the knee region which presents the maximal trade-offs between objectives.

As we all known, a knee region which is visually a convex bulge in this Pareto-optimal front is important to decision makers in practical contexts. Being inspired by human idea generation process, Shi [5] proposed a novel optimization algorithm - Brain Storm Optimization (BSO) algorithm. The simulation results on two single-objective benchmark functions validated the effectiveness and usefulness of the BSO to solve optimization problems. Two novel component designs [6] were proposed to modify the BSO algorithm. A new Modified Multi-objective Brain Storm Optimization algorithm (MMBSO, SMOBSO) is proposed in [7, 8] for solving multi-objective optimization problems. In this paper, Multi-objective Brain Storm Optimization based on Estimating in Knee Region and Clustering in Objective-Space (MOBSO-EKCO) algorithm is proposed. Two operations of the objective space named Cauchy mutation and clustering can improve the convergence of the proposed algorithm greatly. Circular crowded sorting approach is used to generate a set of well-distributed Pareto-optimal solutions, and the global best individual in multi-objective optimization domain is redefined through a new multi-objective fitness roulette technique. A group of test functions including ZDT series, DO2DK and DEB2DK are simulated to compare with other recently proposed methods. Simulation results show that the algorithm has better uniformity of the Pareto-optimal solutions and high effectiveness for the different kinds of multi-objective optimization problem.

The remaining paper is organized as follows. Section 2 briefly reviews the related works about the BSO and the MOP. In Sect. 3, Multi-Objective Brain Storm Optimization based on Estimating in Knee Region and Clustering in Objective Space (MOBSO-EKCO) is introduced and described in detail. Section 4 contains the simulation results and discussion. Finally, Sect. 5 provides the conclusions and some possible paths for future research.

2   Related Work

2.1   Brainstorm Optimization Algorithm

The BSO algorithm is designed based on the brainstorming process [12]. In the brainstorming process, the generation of the idea obeys the Osborn's original four rules. The people in the brainstorming group will need to be open-minded as much as possible and therefore generate more diverse ideas. Any judgment or criticism must be held back until at least the end of one round of the brainstorming process, which means no idea will be ignored. The algorithm is described as follows. In the initialization, N potential individuals were randomly generated. During the evolutionary process, BSO generally uses the clustering technique, mutation operator and selection operator to create new ideas based on the current ideas, so as to improve the ideas generation by generation to approach the problem solution. Clustering is the process of grouping similar objects together. From the perspective of machine learning, the clustering analysis is sometimes termed as unsupervised learning [11]. In the mutation operator, BSO creates N new individuals one by one based on the current ideas. To create a new individual, BSO first determines whether to create the new individual based on one selected cluster or based on two selected clusters. After the cluster(s) have been selected, BSO then determines whether create the new idea based on the cluster center (s) or random idea(s) of the cluster(s). No matter to use the cluster center or to use random idea of the cluster, we can regard the selected based idea as $X_{selected}$ which can be expressed as $X_{selected} = (x_{selected}^1, x_{selected}^2, \ldots, x_{selected}^d)$, then applying a mutation of the $X_{selected}$ to get the new idea X_{new} which can be expressed as $X_{new} = (x_{new}^1, x_{new}^2, \ldots, x_{new}^d)$. After the new idea X_{new} has been created, BSO evaluates X_{new} and replaces $X_{selected}$ if X_{new} has a better fitness than $X_{selected}$. The procedure of the BSO algorithm is shown in [4].

There have been several improved BSO algorithms to improve its performance. For instance, Brain Storm Optimization Algorithms with K-medians Clustering Algorithms in [14], Predator–Prey Brain Storm Optimization for DC Brushless Motor in [15], Brain Storm Optimization Algorithm for Multi-objective Optimization Problems in [9], and Multi-objective Optimization Based on Brain Storm Optimization Algorithm (MBSO) [10].

2.2 Multi-objective Optimization Problem (MOP)

Without loss of generality, all of the multi-objective optimization problems can be formulated as minimization optimization problems. Let us consider a multi-objective optimization problem:

$$\text{Minimize } \mathbf{F}(\mathbf{X}) = (f_1(\mathbf{X}), f_2(\mathbf{X}), \ldots, f_M(\mathbf{X})). \tag{1}$$

where $\mathbf{X} = (x_1, \ldots, x_D) \in \Re^D$ is called the decision vector in the D dimensional search space and $\mathbf{F} \in \Omega^M$ is the objective vector with M objectives in the M dimensional objective space. The basic concepts of a minimization MOP can be described in [11]. Two goals of a multi-objective optimization are the convergence to the true Pareto-optimal set, and the maintenance of diversity of solutions in the Pareto front set. Many performance metrics have been suggested to measure the performance of multi-objective optimization algorithms. In this paper, we use the metric Υ and metric Δ defined by Deb et al. in [2] to measure the performance of the MBSO algorithm.

2.3 Performance of the Multi-objective Optimization Problem

Unlike the single objective optimization, the optimal solution of multi-objective optimization is a set of Pareto optimal solutions. Whether it can obtain the solutions which are close to the true Pareto optimal or not, the solutions should be determined and evenly distributed on the Pareto optimal.

The metric Υ measures the closeness of solutions to the true Pareto front.

$$\Upsilon = \frac{\sum_{i=1}^{|P|} d(\mathbf{P}_i, \mathbf{TF})}{|\mathbf{P}|}. \tag{2}$$

where 500 uniformly spaced solutions on the true Pareto-optimal front are selected from the true Pareto

TF; P is the Pareto front that has been found by MOBSO-EKCO; d is the minimum Euclidean distance of $\mathbf{P}_i$ and to all of the points in the set TF; and |P| is the size of the set **P**.

The metric Δ measures the non-uniformity in the distribution.

$$\Delta = \frac{d_f + d_l + \sum_{i=1}^{|P|-1} |d_i - \overline{d}|}{d_f + d_l + (|\mathbf{P}| - 1)\overline{d}}. \tag{3}$$

where, the Euclidean distances d_f and d_l are calculated the same as in [2]; d_i is the Euclidean distance between the i-th pair of consecutive solutions in the set P; $\overline{d}$ is the average of all distances d_i.

3 Multi-objective Brain Storm Optimization Based on Estimating in Knee Region and Clustering in Objective-Space (MOBSO-EKCO)

In this paper, Estimating in Knee Region and Clustering in Objective-Space (MOBSO-EKCO) is incorporated into BSO to construct a hybrid multi-objective optimization algorithm. The proposed algorithm intends to increase the convergence and rapidity. Clustering in objective space is used to evaluate potential regions for a better compromise solution. Mutation is used to generate new solutions. And knee solutions are constituted with solutions with the highest values in terms of the marginal rate of return which is the bulges out of Pareto surface. As it is hard to guarantee convergence at a turning point or region with large curvature, a high concentration of solutions on the knee regions is used to produce enough non-dominated solutions to fit the true Pareto-front.

3.1 Clustering Technique

Clustering, as a kind of unsupervised classification, is an effective and useful strategy to help people learn a new object or understand a new phenomenon by seeking the features that can describe it and comparing it with other known objects or phenomena based on the similarity or dissimilarity according to some certain standards or rules. In the operation of the clustering strategy, data are partitioned into a certain number of clusters which have similar patterns in the same cluster and dissimilar patterns between the different clusters [16].

In this paper, clusters are generated in the objective space with the purpose of evaluating the distribution of fitness in the objective space, distinguishing the area whether it belongs non-dominate solutions or not. And the evaluation standard is distance between the fitness of the particle and non-dominate solutions. As we all know, K-means is an iterative scheme attempting to minimize the sum of squared Euclidean distances between data object and cluster centers. It is a great chance. So K-means is used to cluster the objective vectors into k clusters in our proposed algorithm. Firstly, an initial distribution of k cluster centers in the objective space is generated. Then each individual is assigned to the cluster with the closet center according to its fitness. Finally, the centers themselves are updated as the center of mass of all particles belonging to that particular cluster. The procedure is repeated until all the individuals are assigned. The clusters without non-dominate solutions are then mapped into the decision variable space and constitute new population $Q1$.

3.2 Generation Progress

- Mutation Operator

The mutation is used or not is depend on that if there is cluster without non-dominate solutions. If exist, then Mutation operation is used in new population to generate new solutions instead of individuals whose position away from non-dominate solutions. This operation can enhance the search capability. In this paper Differential evolution (DE) mutation, a relatively simple and efficient form of self-adaptive mutation operator, serves as the tool to achieve the purpose. Mutation operation is the key ingredient of DE. To a certain extent, differential evolution mutation considering the correlation between the multivariate has certain advantages in variable coupling. A new trial vectors Z_i is generated by adding weighted differential vectors which is obtained from two or more parameter vectors selected randomly from the current population to another parameter vectors which is called as target vector selected from the same population. The mutation operation can be described as follows [17]:

$$Z_i = P_i + F \cdot (P_m + P_n). \tag{4}$$

where P_m and P_n are generated randomly in the non-dominate solutions, which are mutually different. The mutation parameter $F \in [0.5, 1]$ is a real, constant, user-supplied parameter which controls the amplification of the differential variation. At the same time, the mutation probability is adapted along with the number of iteration for improving the convergence of the proposed algorithm.

- Selection Operator

It is also quite important to decide whether any newly generated solution should survive to the next generation. The selection based on Pareto dominance is utilized in this paper.

3.3   Knee Region-Based Local Optimization

- Estimate Knee region

A non-dominated front that is characterized by a tradeoff between objectives is viewed as the optimal solutions of multi-objective optimization problem. A knee region which is visually a convex bulge in this Pareto-optimal front is important to decision makers in practical contexts, as it often constitutes the optimum in tradeoff, i.e., substitution of a given Pareto-optimal solution with another solution on the knee region yields the largest improvement per unit degradation. The commonly used transformation function used in MOEAs' community maps the objective functions to the interval [0, 1] regardless of their original ranges and is stated in [19].

During the evolution process, preliminary Pareto front is generated as the iterations of population. It is hard to produce enough non-dominated solutions in knee region for the features of Pareto front. Moreover, the position of the knee point will be changed after multiple iterations. And knee point area needs to be estimated after a certain number of iterations owing to the change of the knee solutions. So, mutation operator is adopted to strengthen the local search ability of the decision space mapped by knee regions. Due to the solutions in the mutated regions may not be non-dominated solutions, the performance improvement is becoming gradual process.

On the premise of unknown number of knee points, it is difficult to estimate the knee regions completely and accurately. In this subsection, a new ε- estimation of the knee region algorithm is proposed. The algorithm chooses some top solutions of μ matrix to create the candidate solutions. If the Euclidean distance of two potential knee points is less than ε, these two potential knee points are located in the same real knee point and this knee point is estimated by mean value method. Else if the Euclidean distance of two potential knee points is larger than ε, the two potential knee points are two different knee points. The ε-estimation knee region algorithm is presented in Fig. 1.

---

1.   Input
        CKN :Potential Knee Solutions
        FF :the non-dominate front
        ε *:the coefficients of Knee point estimates(After repeatedly experiment, ε is 0.02)*
2.   Output
        KN : Estimation of the knee region
3.   *FF* ← evaluate trade-off metric
4.   matrix μ ← Calculate the trade-off relationship of each solution on non-dominated set
5.   *CKN* ← choose the top 20 value in matrix μ as potential solutions (*Potential Knees number is 20 which is choose by test*)
    KN ← If Euclidean distance between two potential knee point is less than ε, these two solution is represented to same knee point, average the solutions same knee point, If Euclidean distance between two potential knee point is more than ε, these two solution is represented to different knee point.

Fig. 1. ε-estimation of the knee region algorithm

- Local Mutation

Mutation operator is used to avoid getting stuck in local optima on knee regions. This mechanism can enhance the capability of exploring the vast search space in special regions and improving the convergence performance of the algorithm. The results in [20] show that Cauchy mutation is an efficient search operator for a large class of multimodal function optimization problems. Cauchy mutation has a higher probability of making longer jumps due to its longer flat tails. In this paper, Cauchy Mutation is utilized to generate new individual on decision space mapped by knee region. The detail is as follows:

$$\mathbf{x}_i(t+1) = \mathbf{x}_i(t) + \xi * C(\mu', \sigma'). \tag{5}$$

where $C(\mu', \sigma')$ is the Cauchy random function with mean μ' and variance σ'. Probability density function of Cauchy distribution is defined as (5):

$$g(x) = \frac{a}{\pi \times (x^2 + a^2)}. \tag{6}$$

where a is 0.2.

3.4 Update the Pareto Set

The non-dominated solutions of the archive are composed of two parts. Some of them are new non-dominated solutions in population space; the others are new non-dominated solutions in objective space created by Cauchy mutation. In this step, each new non-dominated solution obtained in the current iteration will be compared with all members in the External archive. If the size of the External archive exceeds the maximum size limit, it is truncated using the diversity consideration. In this paper, the circular crowded sorting operator [21] is adopted to guide the points toward a uniformly spread-out Pareto-optimal front.

3.5 The Procedure of the MOBSO-EKCO

It can be clearly seen that the proposed algorithm makes such improvements as clustering in the objective space, estimating knee regions on Pareto front, mutation and updating external archive. These four parts are described in detail in the foregoing parts. The algorithm of MOBSO-EKCO can be described use the flowchart in Fig. 2.

4 Experiments and Discussions

In this section the MOBSO-EKCO will be tested. Without loss of generality, all the multi-objective optimization problems tested in this paper are minimization problems.

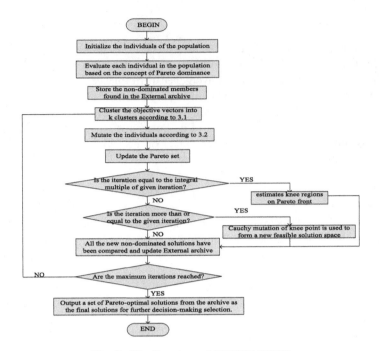

Fig. 2. The flowchart of MOBSO-EKCO

4.1 Test Problems

In order to evaluate the performance of MOBSO-EKCO, the DO2DK and DEB2DK and ZDT test functions [2] are used in this paper.

4.2 Parameter Settings

During all the simulation runs, the population size is set to be 100, the number of clusters is $k = 8$, the maximum iterative time $T_{\max}$ is 250 and the maximum size of the Pareto set is fixed at 100. All algorithms produced final Pareto fronts of the fixed size population. All of the algorithms are implemented in MATLAB using a real-number representation for decision variables. For each experiment, 30 independent runs were conducted to collect statistical results.

4.3 Results

To illustrate the effectiveness of the proposed algorithm, the following two kinds of test functions are tested in the paper to show the effectiveness of the knee region estimation algorithm. The Fig. 3 is forefront profiles under different value in the comparison of MOBSO-CO and MOBSO-EKCO.

	MOBSO-CO	MOBSO-EKCO
DO2DK,k=2,s=1		
DO2DK,k=3,s=1		
DEB2DK,k=2,s=1		
DEB2DK,k=3,s=1		
DEB2DK,k=4,s=1		

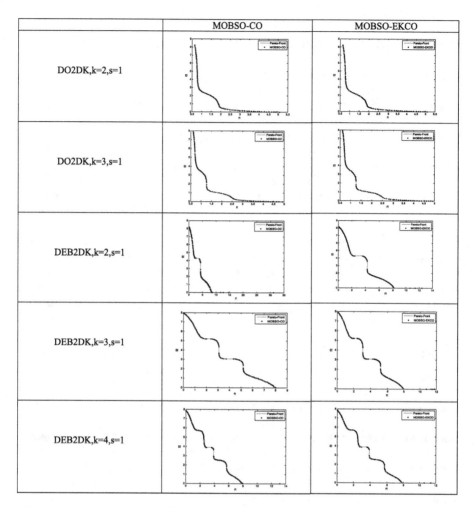

Fig. 3. Pareto fronts produced for MOBSO-CO and MOBSO-EKCO (Color figure online)

In the Fig. 3, red line represents ideal Pareto-front and black points are frontier that has calculated by the MOBSO-CO and MOBSO-EKCO. Moreover, blue stars are the keen points of function. It can be concluded clearly that the proposed algorithm can effectively estimate position of knee region.

For qualitative comparison to other algorithms, Table 1 shows the mean (M) and variance (Var) values of the convergence metric Υ. The diversity metric Δ about the test problems are listed in Table 2. The best results are marked with bold. The line in the Tables 1 and 2 represent that Υ and Δ are not given in the references.

Table 1. Mean (M) and Variance (Var) value of the convergence metric Υ for algorithms

Algorithm	ZDT1		ZDT2		ZDT3		ZDT4		ZDT6	
	M	Var	M	Var	M	Var	M	Var	M	Var
MOBSO-CO	2.4312e-04	8.2752e-07	**1.3142e-04**	**2.4178e-07**	3.2255e-04	1.4565e-06	**1.0258e-04**	**1.4733e-07**	0.0023	7.2542e-05
MOBSO-EKCO	**1.9502e-04**	**5.3244e-07**	2.3504e-04	7.7344e-07	**1.0616e-04**	**1.5778e-07**	1.9825e-04	5.5023e-07	**1.3638e-04**	**2.6040e-07**
MOPSO [4]	*0.00133*	0.00000	0.00089	0.00000	0.00418	0.00000	—	—	—	—
AEPSO [22]	0.00100	0.00000	0.00078	0.00000	0.00462	0.00000	—	—	—	—
NSGA-II [2]	0.033482	0.004750	0.072391	0.031689	0.114500	0.007940	0.296556	0.013135	0.296556	0.013135
SPEA [3]	0.010800	0.002800	0.012700	0.002900	0.095860	0.006200	0.426510	0.026500	0.280551	0.004945
CDMOPSO [23]	0.006900	0.000550	0.006820	0.000548	0.007200	0.000992	0.263000	0.024880	0.245400	0.000130
AIPSO [24]	0.004480	0.000000	0.003860	0.000000	0.006120	0.000000	0.144620	0.00709	0.012630	0.00000

Table 2. Mean (M) and variance (Var) value of the diversity metric Δ for algorithms

Algorithm	ZDT1		ZDT2		ZDT3		ZDT4		ZDT6	
	M	Var	M	Var	M	Var	M	Var	M	Var
MOBSO-CO	0.0018	4.3781e-05	**0.0021**	**6.2652e-05**	0.0012	2.1582e-05	0.0017	4.2113e-05	0.0206	0.0059
MOBSO-EKCO	**0.0016**	**3.7525e-05**	0.0024	8.1750e-05	**6.7667e-04**	**6.4103e-06**	**0.0013**	**2.5147e-05**	**0.0015**	**3.0806e-05**
MOPSO [4]	0.68132	0.01335	0.63922	0.00114	0.83195	0.00892	—	—	—	—
AEPSO [22]	0.55743	0.00142	0.51905	0.00092	0.55187	0.00087	—	—	—	—
NSGA-II [2]	0.390377	0.001876	0.430776	0.004721	0.738540	0.019706	0.668025	0.009923	0.668025	0.009923
SPEA [3]	0.362200	0.039600	0.416700	0.038100	0.521920	0.026980	0.623900	0.002883	0.489800	0.002510
CDMOPSO [23]	0.274000	0.040800	0.270900	0.018200	0.330700	0.011600	0.217800	0.018500	0.457800	0.006890
AIPSO [24]	0.514855	0.00051	0.499099	0.00052	0.505558	0.00028	0.507355	0.00034	0.50349	0.00025

The results in Tables 1 and 2 indicates that the convergence and diversity criterion of MOBSO-EKCO is better than MOBSO-CO for most of benchmark functions. Both MOBSO-EKCO and MOBSO-CO show the better performance than other algorithms in references, especially for ZDT3 which has knee region in the Pareto front.

5 Conclusions and Discussion

In this paper, we proposed the MOBSO-EKCO, an algorithm based on Estimating in Knee Region and Clustering in Objective-Space to enhance the performance of the algorithm. The clustering strategy acts directly in the objective space instead of in the solution space. This mechanism suggests potential areas which far away from Pareto front in the next iteration. The estimating strategy is used to discover knee regions so as to optimize the local position in the next iteration. The simulation results based on a series of test functions show that the performance of the algorithm has great superiority.

References

1. Dutta, D., Dutta, P., Sil, J.: Clustering by multi objective genetic algorithm. In: Recent Advances in Information Technology (RAIT), pp. 548–553 (2012)
2. Deb, K., Pratap, A., Agarwal, S., Meyarivan, T.: A fast and elitist multiobjective genetic algorithm: NSGA–II. IEEE Trans. Evol. Comput. 6(2), 182–197 (2002)
3. Zitzler, E., Laumanns, M., Thiele, L.: SPEA2: improving the strength pareto evolutionary algorithm. In: Evolutionary Methods for Design, Optimization and Control with Applications to Industrial Problems, pp. 95–100 (2001)
4. Coello, C.A.C., Pulido, G., Lechuga, M.: Handling multi-objective with particle swarm optimization. IEEE Trans. Evol. Comput. 8(3), 256–279 (2004)
5. Shi, Y.: Brain storm optimization algorithm. In: Tan, Y., Shi, Y., Chai, Y., Wang, G. (eds.) ICSI 2011, Part I. LNCS, vol. 6728, pp. 303–309. Springer, Heidelberg (2011)
6. Zhan, Z., Zhang, J., Shi, Y., Liu, H.: A modified brain storm optimization. In: IEEE World Congress on Computational Intelligence, pp. 10–15 (2012)
7. Guo, X., Wu, Y., Xie, L.: An adaptive brain storm optimization algorithm for multiobjective optimization problems. Control Decis. 27(4), 598–602 (2012)
8. Xie, L., Wu, Y.: A modified multi-objective optimization based on brain storm optimization algorithm. In: Tan, Y., Shi, Y., Coello, C.A. (eds.) ICSI 2014, Part II. LNCS, vol. 8795, pp. 328–339. Springer, Heidelberg (2014)
9. Xue, J., Wu, Y., Shi, Y., Cheng, S.: Brain storm optimization algorithm for multi-objective optimization problems. In: Tan, Y., Shi, Y., Ji, Z. (eds.) ICSI 2012, Part I. LNCS, vol. 7331, pp. 513–519. Springer, Heidelberg (2012)
10. Shi, Y., Xue, J., Wu, Y.: Multi-objective optimization based on brain storm optimization algorithm. In: Swarm Intelligence Research (IJSIR) 4(3) (2013)
11. Cheng, S., Shi, Y., Qin, Q., et al.: Solution clustering analysis in brain storm optimization algorithm In: 2013 IEEE Symposium on Swarm Intelligence (SIS), pp. 1391–1399 (2013)
12. Shi, Y.: Brain storm optimization algorithm in objective space In: IEEE Evolutionary Computation (2015)
13. Zhan, Z., Chen, W., Lin, Y., Gong, Y., Li, Y., Zhang, J.: Parameter investigation in brain storm optimization. In: IEEE Symposium on Swarm Intelligence (SIS), pp. 103–110 (2013)
14. Zhu, H., Shi, Y.: Brain storm optimization algorithms with k-medians clustering algorithms In: IEEE Seventh International Conference on Advanced Computational Intelligence (2015)
15. Duan, H., Li, S., Shi, Y.: Predator-prey brain storm optimization for DC brushless motor. IEEE Trans. Magn. 49(10), 5336–5340 (2013)
16. Xu, D., Wunsch II, D.: Survey of clustering algorithms. IEEE Trans. Neural Netw. 16(3), 645–677 (2005)

17. Wang, Y., Wu, L.H., Yuan, X.: Multi-objective self-adaptive differential evolution with elitist archive and crowding entropy-based diversity measure. Soft. Comput. **14**(3), 193–209 (2010)
18. Bai, Q.: Analysis of particle swarm optimization algorithm. Comput. Inf. Sci. **3**(1) (2010)
19. Slim, B., Lamjed, B., Khaled, G.: Searching for knee regions of the Pareto front using mobile reference points. Soft. Comput. **15**, 1807–1823 (2011)
20. Zhang, L.B., Zhou, C.G., Ma, M., Sun, C.: A multi-objective differential evolution algorithm based on max-min distance density. J. Comput. Res. Dev. **44**(1), 177–184 (2007)
21. Luo, C., Chen, M., Zhang, C.: Improved NSGA-II algorithm with circular crowded sorting. Control Decis. **25**(2), 227–232 (2010)
22. Chen, M., Zhang, C., Luo, C.: Adaptive evolutionary multi-objective particle swarm optimization algorithm. Control Decis. **24**(12), 1850–1855 (2009)
23. Feng, Y.X., Zheng, B., Li, Z.K.: Exploratory study of sorting particle swarm optimizer for multi-objective design. Math. Comput. Model. **52**, 1966–1975 (2010)
24. Huang, P., Yu, J.Y., Yuan, Y.Q.: Improved niching multi-objective particle swarm optimization algorithm. Comput. Eng. **37**, 1–3 (2011)

Optimal Impulsive Thrust Trajectories for Satellite Formation via Improved Brainstorm Optimization

Olukunle Kolawole Soyinka[1,2(✉)] and Haibin Duan[1]

[1] School of Automation Science and Electrical Engineering,
Beihang University, Beijing 100191, People's Republic of China
kunlesoyinka@yahoo.com
[2] Engineering and Space Systems Department,
National Space Research and Development Agency, Abuja, Nigeria

Abstract. The optimization of the satellite trajectory is a product of the control thrust optimization. Using the Lambert theory formulation this paper considers the problem of a minimum fuel solution of the Lambert formulation by making use of Brainstorm Optimization (BSO). The traditional use of the Lambert formulation for generating control inputs typically pursues time dependent solutions. This paper adapted this formulation to satellite formation control to achieve fuel minimization. The numerical simulations show the feasibility of our method in obtaining minimum fuel solutions of the lambert formulation in contrast to the usual minimum time solution.

Keywords: Brainstorm Optimization (BSO) · Satellite formation · Optimization · Lambert formulation

1 Introduction

Various approaches have been pursued in literature to solve such optimal control problems such as the analytical methods in [1–6]. The primer vector theory in [7] derives a set of necessary conditions for optimal trajectory generation. Recently, bio-inspired intelligence algorithms such as Genetic Algorithm (GA), Particle Swarm Optimization (PSO), and Hybrid PSO and GA [8–11] have been implemented in trajectory optimization. This paper also implements Brainstorm Optimization (BSO) algorithm to solve a two-impulse minimum fuel formation control problem based on Lambert formulation. BSO algorithm mimics the technique used by humans for solving complex problems. Motivation for this study is based on the arguments in [12] and [13] that since man is the most intelligent of all biological species, bio-inspired optimization algorithms developed based on human behavioural pattern should have a better performance. Several scholars have sought some form of optimal solution to the application of the Lambert formulation. [14] introduced an auxiliary transfer phase that minimizes the overall multiple revolution fixed endpoint lambert problem while [15] constrained perigee and apogee altitude leading to the formation of an eight degree polynomial that can be solved by numerical integration. This paper considers fuel minimization in a satellite formation control problem.

© Springer International Publishing Switzerland 2016
Y. Tan et al. (Eds.): ICSI 2016, Part I, LNCS 9712, pp. 491–499, 2016.
DOI: 10.1007/978-3-319-41000-5_49

2 Mathematical Model for Relative Motion

The relative position of a leader and follower satellite can be described using the mean orbital elements [16]. In the mean orbital elements description, the positions of the satellites can be defined by their mean orbital elements vector as

$$\sigma_L = [a_L, e_L, i_L, \Omega_L, \omega_L, M_L]^T, \tag{1}$$

$$\sigma_F = [a_F, e_F, i_F, \Omega_F, \omega_F, M_F]^T, \tag{2}$$

where σ_L and σ_F are the orbital elements vectors of the leader and follower satellite respectively. The parameters a, e, i, Ω, ω, and M are the semi major axis, eccentricity, inclination, Right Ascension of Ascending Node (RAAN), argument of periapsis, and mean anomaly respectively. Following the development in [16, 17], the mean orbital element difference vector can be described as

$$\delta\sigma = \sigma_L - \sigma_F = [\delta a, , \delta i, \delta\Omega, \delta\omega, \delta M]^T, \tag{3}$$

and the relative motion for the follower satellite can be found from the expression

$$\begin{cases} x(fs) = (1 - e\cos(f))\delta a + \frac{ae\sin f}{\Upsilon}\delta M - a\cos f\delta e \\ y(f) = \frac{a}{\Upsilon}(1 - e\cos f)\delta M + a(1 - e\cos f)\delta\omega \\ \qquad + a\sin f(2 - e\cos f)\delta e + a(1 - e\cos f)\delta\Omega \\ z(f) = a(1 - e)(\sin(\omega + f)\delta i - \cos(\omega + f)\sin i\delta\Omega) \end{cases} . \tag{4}$$

The eccentricity parameter Υ is calculated as

$$\Upsilon = (1 - e^2)^{0.5}, \tag{5}$$

and $\delta a, \delta e, \delta i, \delta\Omega, \delta\omega, \delta M$ are the differences in the orbital elements between the leader and a follower satellite.

2.1 Lambert Theorem

Consider the transfer geometry in Fig. 1 between the two fixed points P_1 and P_2 in space having position vectors $\vec{r}_1$ and $\vec{r}_2$ respectively.

 Since an infinite number of trajectories exist that connect the two points, the goal is to determine the elliptical trajectory that connects the points P_1 and P_2 optimally in a specified time. Lambert theorem states that the time of flight (t_{tof}) between P_1 and P_2 depends only on the semi-major axis a, the magnitude of the vectors ($\vec{r}_1 + \vec{r}_2$) and the chord length c. The transfer time for the elliptical transfer orbit can be defined from Kepler's time equation as a function of the mean anomaly (M) at $\vec{r}_1$ and $\vec{r}_2$ as [18]

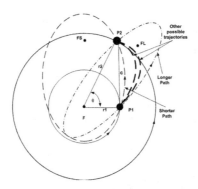

Fig. 1. Geometry for Lambert Problem.

$$t_{tof} = a\sqrt{\frac{a}{\mu}}(2\pi N + (M_2 - M_1)), \tag{6}$$

where N is the number of revolutions, and μ is the gravitational parameter. For our two-impulse control case, (6) reduces to

$$t_{tof} = a\sqrt{\frac{a}{\mu}}(M_2 - M_1). \tag{7}$$

The minimum time solution can be obtained by substituting in (7)

$$a = a_{min} = \frac{s}{2} = \frac{c + r_1 + r_2}{4}. \tag{8}$$

The transfer chord c is determined to be

$$c = \sqrt{r_1^2 + r_2^2 - 2r_1 r_2 \cos(\theta)}, \tag{9}$$

and the transfer angle as

$$\theta = \cos^{-1}\left(\frac{r_1 \cdot r_2}{r_1 r_2}\right). \tag{10}$$

Since the initial position vector $\bar{r}_1$ and the final position vector $\bar{r}_2$ are defined, the two-impulse Δv_1 and Δv_2 required to obtain a desired transfer trajectory is determined following [18] as

$$r_2 = f r_1 + g v_1, \tag{11}$$

and

$$v_2 = fr_1 + \dot{g}v_1. \tag{12}$$

With the foregoing, the velocity required at P_1 to put the satellite on the transfer trajectory is determined as

$$v_1 = \frac{r_2 - fr_1}{g}, \tag{13}$$

and at P_2 as

$$v_2 = \frac{\dot{g}r_2 - r_1}{g}. \tag{14}$$

Assuming a single thruster implementation, the impulsive control thrust ΔV from the Lambert formulation can be found from the expression

$$\Delta v_1 = \begin{bmatrix} \dot{x}_1^+ - \dot{x}_1^- \\ \dot{y}_1^+ - \dot{y}_1^- \\ \dot{z}_1^+ - \dot{z}_1^- \end{bmatrix}, \tag{15}$$

$$\Delta v_2 = \begin{bmatrix} \dot{x}_2^- - \dot{x}_2^+ \\ \dot{y}_2^- - \dot{y}_2^+ \\ \dot{z}_2^- - \dot{z}_2^+ \end{bmatrix}, \tag{16}$$

$$\Delta V = \|\Delta v_1\|_2 + \|\Delta v_2\|_2. \tag{17}$$

3 Brainstorm Optimization

The brainstorming process brings together a group of people with different background, to interactively collaborate to generate great ideas for problem solving.

This study implements a modification to the basic Brainstorm Optimization (BSO) algorithm proposed by [19]. A uniqueness of the BSO algorithm is the way it generates new ideas (individuals) during the iterative process. This idea generation process depends so much on a weighting function. This paper suggests a modification to the weighting function in basic BSO and refers to it as BSO with modified weighting function (BSOMWF). The weighting function controls the search behavior of the algorithm in the individual generation process, it is expressed as

$$\xi(t) = logsig\left(\frac{\frac{T}{2} - t}{k}\right), \tag{18}$$

where $logsig()$ is a logarithmic sigmoid transfer function, T is the maximum iteration, t is the current iteration and k is a predefined parameter that determines the slope of the $logsig$ function. A characteristic of $\xi(t)$ is that it generates large values at the start of the search process and small values towards the end. To improve the performance of BSO an alternative function is implemented and a parameter $\alpha \in (0, 1)$ is introduced as index to the k parameter, this is given as

$$\xi(t)_{new} = \frac{1}{(1 + \exp(-H))}, \tag{19}$$

$$H = \frac{0.5 * T - t}{k^\alpha}. \tag{20}$$

The following describes the stages as implemented in BSOMWF.

(1) **Initialization**

Initialization of the population in the BSOMWF algorithm is done by randomly initializing individuals over the range of the solution space. This is similar to gathering a group of persons having diverse background to generate ideas. For every round of idea generation in the process, a fixed number of n ideas will be generated then problem owners pick up good ideas (individuals) that represent solutions within the solution space.

(2) **Clustering**

This process involves grouping together (clustering) the population of ideas generated within the solution space. During each generation, the population of ideas is grouped into a specified number of clusters using k-means clustering method. The best one in each cluster is chosen as the cluster center this is similar to problem owners' picking up better ideas during each idea generation.

(3) **Individual generation**

In BSOMWF, idea generation is either done partially or by piggybacking. Ideas can be generated partially using the expression

$$x_{new}^d = x_{old}^d + \frac{1}{(1 + \exp(-\frac{0.5 * T - t}{k^\alpha}))} * random(t), \tag{21}$$

while ideas can be generated by piggybacking using the expression

$$x_{new}^d = x_{old}^d + \frac{1}{(1 + \exp(-\frac{0.5 * T - t}{k^\alpha}))} * random(t), \tag{22}$$

$$x_{old}^d = \omega_1 * x_{old1}^d + \omega_2 * x_{old2}^d. \tag{23}$$

where x_{new}, x_{old} are the dimensions of the new and old idea for partial generation and x_{old1}^d, x_{old2}^d are for piggybacking. $random(t)$ is a random function, ω_1 and ω_2 are weighted contributions of the two existing individuals.

(4) **Updating Cluster Center**

To ensure that the algorithm is not trapped in a local minimum the cluster center is updated. To perform the update a cluster center is randomly selected and used as a new focus about which the group will now generate new ideas similar to the functions and appearance of the randomly selected center. This stage is modelled as

$$x_c^d = \begin{cases} x_c^d & if\,(p_1 < p_{5a}) \\ x^d = x_{low}^d + rand\,(x_{up}^d - x_{low}^d), & otherwise \end{cases}, \qquad (24)$$

where p_1 is a random number $(0,1)$, $p_{5a} = 0.2$ and x_c^d is the dimension of the cluster center to be updated.

(a) **Selection**

This process involves determining the individuals moving to the next iteration. The new idea formed is evaluated and compared with the current individual having the same index. The current idea is replaced if the new individual has a better fitness. The next iteration then follows until a terminal constraint is met.

4   Comparative Experiments

Consider a non-planar, assignment problem. The problem seeks to find the optimal times of flight that minimizes the overall amount of fuel required to maneuver each *kth* satellite from an initial state i to a target state j. The optimization problem can be written as

$$min \sum_{k=1}^{n} \Delta V_{ij}. \qquad (25)$$

where n is the total number of deputy satellites.

The assignment of the target is defined under constraints (1) and (2) below:

(1) $\delta a = 0$, $0 < i < 90^0, \delta e \neq 0$, and $\delta i \neq 0$.
(2) $0 < t_k < t_{k+1} \ldots < t_{k+n} \leq T_l$.

Our problem consists of $n = 3$ deputy satellites initially collocated and reconfigured to another position to achieve formation with a leader satellite. The reconfiguration process aims at raising the orbit of the deputy satellites and achieving a fixed geometry with bounded relative motion. The reconfiguration maneuver is expected to change the initial orbit to another having a different inclination and eccentricity. The magnitude of the results is expected since an inclination change is involved in the maneuver. The results indicate that BSOMWF outperforms the other algorithms and its better solution searching ability is obvious from the fitness curves in Fig. 2

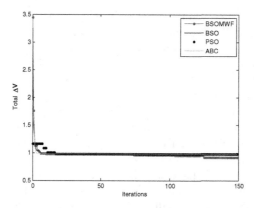

Fig. 2. Optimal thrust fitness curves (Color figure online)

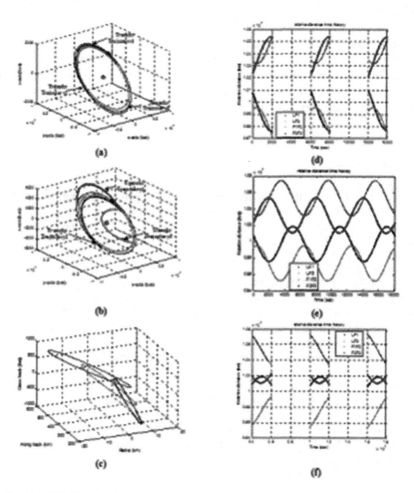

Fig. 3. (a) BSOMWF solution (b) Minimum Time Solution (c) Satellite Formation (d) LF1-LF2 and F1F2-F2F3 Pattern (e) Relative Distance Evolution (f) F1F2-LF1 Pattern

Figure 3(a) is the fuel optimal solution using BSOMWF; Fig. 3(b) shows the transfer trajectory using the minimum time solution of the Lambert formulation. From Fig. 3(c) the bounded relative motion is observed, this indicates that the satellites are in formation. Figure 3(e) shows the time history of the distance between the satellites over three orbits; within each orbit, the nominal distances are not constant but are bounded between a maximum and minimum value as shown in Fig. 3(e). The leader-follower1 (LF1) pair is bounded between −120 m and +307 m of the initial distance; the leader-follower3 (LF3) pair is bounded between −18 m and +408 m of the initial distance; the follower1-follower2 (F1F2) pair is bounded between −410 m and +20 m of the initial value and the follower2-follower3 (F2F3) pair is bounded between −300 m and +130 m of the initial distance. A pattern is observed in Figs. 3(d) and (f). It is observed that in the first third of the orbit LF1 equals LF3 and F1F2 equals F2F3. This occurs twice within an orbit. Further, LF1 equals F2F3 in the last third of an orbit.

5 Conclusion

In this paper, comparative performance analysis of BSOMWF was demonstrated using several benchmark functions. The results indicate satisfactory search ability and convergence. In particular, optimal control thrusts were sought for a non-planar reconfiguration maneuver. The comparative results obtained via the illustrative example are satisfactory, thus demonstrating the viability of the proposed approach.

References

1. Enright, P.J., Conway, B.A.: Discrete approximations to optimal trajectories using direct transcription and nonlinear programming. J. Guid. Control Dyn. **15**(4), 994–1002 (1992)
2. Kiek, D. E.: Optimal Control Theory: An Introduction. Dover Publications (1998)
3. Breakwell, J.V., Redding, D.C.: Optimal Low Thrust Transfers to Synchronous Orbit. J. Guid. Control Dyn. **7**(2), 148–155 (1984)
4. Hargraves, C.R., Paris, S.W.: Direct trajectory optimization using non-linear programming and collocation. J. Guid. Control Dyn. **10**(4), 338–342 (1987)
5. Herman, A.L., Conway, B.A.: Direct optimization using collocation based on high-order Gauss-Lobatto quadrature rules. J. Guid. Control Dyn. **19**(3), 592–599 (1996)
6. Kiusalaas, J.: Numerical Methods in Engineering with Matlab, 2nd edn. Cambridge University Press, Cambridge (2010)
7. Lawden, D.F.: Optimal Trajectories for Space Navigation. Butterworths Publishers, London (1963)
8. Wall, B.J., Conway, B.A.: Near-optimal low-thrust earth-mars trajectories found via a genetic algorithm. J. Guid. Control Dyn. **28**(5), 1027–1031 (2005)
9. Mauro, P., Conway, B.A.: Optimal finite-thrust rendezvous trajectories found via particle swarm optimization. J. Spacecr. Rockets **50**(6), 1222–1233 (2013)
10. Jones, D.R., Schaub, H.: Optimal reconfigurations of two-craft coulomb formation along manifolds. Acta Astronaut. **83**, 108–118 (2013)

11. Duan, H.B., Luo, Q.N., Ma, G.J., et al.: Hybrid particle swarm optimization and genetic algorithm for multi-uavs formation reconfiguration. IEEE Comput. Intell. Mag. **8**(3), 16–27 (2013)

12. Duan, H.B., Shao, S., Su, B.W., et al.: New development thoughts on the bio-inspired intelligence based control for unmanned combat aerial vehicle. Sci. China Technol. Ser. **53**(8), 2025–2031 (2013)

13. Zhan, Z., Zhang, J., Shi, Y.H., et al.: A modified brain storm optimization. In: Proceedings of IEEE World Congress on Computational Intelligence, Brisbane Australia, pp. 10–15 (2012)

14. Shen, H.J., Tsiotras, P.: Optimal two-impulse rendezvous using multiple revolution lambert solutions. J. Guidance Control Dyn. **6**(1), 50–61 (2003)

15. Zhang, G., Zhou, D., Mortari, D.: Optimal two-impulse rendezvous using constrained multiple revolution lambert solutions. Celest. Mech. Dyn. Astr. **110**, 305–317 (2011)

16. Schaub, H., Srinivas, R., Junkins, L., et al.: Satellite formation flying control using mean orbit elements. J. Astronaut. Sci. **48**(1), 69–87 (2000)

17. Schaub, H., Junkins, L.: Analytical Mechanics of Space Systems, 2nd edn., AIAA Education Series (2003)

18. Vallado, D.A.: Fundamentals of Astrodynamics and Applications. Space Technology Series. McGraw-Hill, New York (1997)

19. Shi, Y.H.: An optimization algorithm based on brainstorming process. Int. J. Swarm Intell. Res. **2**(4), 35–62 (2011)

Parameter Estimation of Vertical Two-Layer Soil Model via Brain Storm Optimization Algorithm

Tiew On Ting[✉] and Yuhui Shi

Xi'an Jiaotong-Liverpool University, Suzhou Industrial Park,
Suzhou, Jiangsu, People's Republic of China
{toting,yuhui.shi}@xjtlu.edu.cn

Abstract. A practical soil model is derived mathematically based on the measurement principles of Wenner's method. The Wenner's method is a conventional approach to measuring the apparent soil resistivity. This soil model constitutes two-soil layer with different properties vertically. Thus this model is called the vertical two-layer soil model. The motivation for the mathematical model is to estimate relevant parameters accurately from the data obtained from site measurements. This parameter estimation is in fact a challenging optimization problem. From the plotted graphs, this problem features a continuous but non-smooth landscape with a steep alley. This poses a great challenge to any optimization tool. Two prominent algorithms are applied, namely Gauss-Newton (GN) and Brain Storm Optimization (BSO). Results obtained conclude that the GN is fast but diverges due to bad starting points. On the contrary, the BSO is slow but it never diverges and is more stable.

Keywords: Brain storm optimization · Grounding · Parameter estimation · Soil model

1 Introduction

When referring to the power supply and transmission equipment, it is crucially important to investigate the soil resistivity and grounding system resistance [1]. This is due to the reason that an excellent grounding system can guarantee safety to surroundings. This grounding system not only protects substations but also prevents dangers such as extreme ground potential and other ground faults [2].

Most previous works contributed in studying grounding characteristics [3], soil parameter estimation [4,5], safety design [6], and ground potential rise (GPR) [7] or grounding systems in two- and multi-layer soil. The real soil shows a resistivity change in vertical and horizontal directions and practical modeling of grounding systems based on proposed soil models lead to inevitable errors due to improper models. A vertical-layer soil model was used previously to analyze the current distribution, leakage current density, grounding impedance, and surface potential [8]. Numerical methods for measuring the grounding resistance was carried out in [9]. Simple and accurate equations had been deduced for grounding grid resistance in vertical two-layer soil [10].

© Springer International Publishing Switzerland 2016
Y. Tan et al. (Eds.): ICSI 2016, Part I, LNCS 9712, pp. 500–511, 2016.
DOI: 10.1007/978-3-319-41000-5_50

In terms of parameter estimation, the Gauss-Newton method [11] is one of the widely used conventional methods to estimate soil parameters due to its simplicity, effectiveness, and ease for programming. Further, work in [5] discussed and implemented variants of improved Gauss-Newton methods. In this work, we use the image method to build an accurate and equivalent model for vertical two-layer soil. Based on this model, an equation is proposed to study the apparent resistivity expression via Wenner's method. Having had a solid mathematical model, two algorithms are applied to estimate relevant soil parameters: Gauss-Newton (GN) and Brain Storm Optimization (BSO). Thus it is possible for a result comparison in this work.

The rest of the paper is organized as follows. Section 2 describes the principle of Wenner's method, providing the essense of the mathematical model. Section 3 presents the mathematical model as an optimization problem. In this section, the three algorithms applied for parameter estimation are elaborated in this section. Then the experimental design and results are presented and analyzed in Sect. 5, and finally the conclusions are drawn in Sect. 6.

2   Wenner's Method

Wenner's method is a well-known method applied in measuring soil resistivity with a relatively high accuracy. It consists of four electrodes that are buried into the soil in a straight line at identical spacing, as shown in Fig. 1. In this figure, a source is connected to the outer electrodes and the potential is measured between inner electrodes. By dividing this potential by the injected current, it results as an apparent resistance ρ_a [11], expressed as follows:

$$\rho_a = 2 \cdot \pi \cdot a \cdot \frac{V}{I} \tag{1}$$

whereby a is the electrode spacing [m], V is the measured voltage between inner pair electrodes [V], and I is the source current between outer pair electrodes [A].

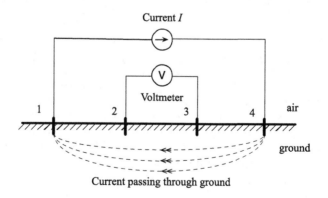

Fig. 1. Electrode arrangement in Wenner's method

In the case of vertical-layer soil, which is shown in Fig. 2(a). It features two vertical stratified layers with different resistivity ρ_1 and ρ_2 respectively. The four electrodes of Wenner's method are buried in the first layer. The first electrode is far from the second layer by a distance d. The line where four electrodes lied on makes an angle β with the perpendicular direction to the interface between the two layers. This image arrangement is shown in Fig. 2(a), depicting the current source located close to the ground surface. This is to satisfy the boundary conditions between the air and the vertical two-layer soil [12].

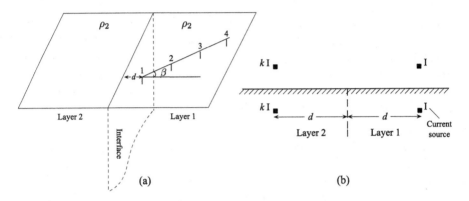

Fig. 2. Vertical-layer soil model. (a) Stratified layers (b) Image distribution around two-layer soil and the air.

The reflection ratio k in Fig. 2(b) is given in [6] as:

$$k = \frac{\rho_2 - \rho_1}{\rho_1 + \rho_2} \tag{2}$$

The voltage at electrode 2 is the summation of the potential due to current injection (current input) in electrode 1 and current sinking (current output) at electrode 4. The potential at electrode 2 due to the current injected at electrode 1 can be expressed as:

$$F_1(x) = \frac{I\rho_1}{2\pi} \left(\frac{1}{x} + \frac{k}{\sqrt{(2d + x\cos\beta)^2 + (x\sin\beta)^2}} \right) \tag{3}$$

where x is the distance between the current source and voltage measuring point, d is the distance between the first electrode and intersection line, and β is the direction of electrodes array. The potential at electrode 2 due to the sinking current at electrode 4 can be obtained as:

$$F_2(x) = -\frac{I\rho_1}{2\pi} \left(\frac{1}{x} + \frac{k}{\sqrt{(2d + (6a - x)\cos\beta)^2 + (x\sin\beta)^2}} \right) \tag{4}$$

where a is the spacing between electrodes. In the same manner, the potential of electrode 3 can be calculated easily. The potential difference between the inner electrodes 2 and 3 is derived from:

$$V = (F_1(a) + F_2(2a)) - (F_1(2a) + F_2(a)) \tag{5}$$

and thus,

$$F_3(x) = \frac{I\rho_1}{2\pi}\left(\frac{1}{x} + \frac{k}{\sqrt{(2d + x\cos\beta)^2 + (x\sin\beta)^2}}\right) \tag{6}$$

By substituting (5) into (1), the apparent resistivity, ρ_a for vertical layer soil shown in (1) becomes

$$\rho_a = a\rho_1 \left(\frac{\frac{1}{a} + \frac{k}{\sqrt{4d^2 + 4da\cos\beta + a^2}} + \frac{k}{\sqrt{4(d+3a\cos\beta)(d+2a\cos\beta)+a^2}}}{-\frac{k}{\sqrt{4d^2+8da\cos\beta+4a^2}} - \frac{k}{\sqrt{4(d+3a\cos\beta)(d+a\cos\beta)+4a^2}}} \right) \tag{7}$$

For the case of $\beta = 0°$, (7) is simplified as

$$\rho_a = a\rho_1 \left(\frac{1}{a} + \frac{k}{\sqrt{4d^2+4da+a^2}} + \frac{k}{2d+5a} - \frac{k}{\sqrt{4d^2+8da+4a^2}} - \frac{k}{2d+4a} \right) \tag{8}$$

whereas when $\beta = 90°$, (7) will be

$$\rho_a = a\rho_1 \left(\frac{1}{a} + \frac{2k}{\sqrt{4d^2 + a^2}} - \frac{2k}{\sqrt{4d^2 + 4a^2}} \right) \tag{9}$$

3   Vertical Soil Parameter Estimation

As explained in the previous section, the perpendicular measurements of Wenner's method are not the same in the vertical layer soil. Obviously, due to the apparent resistivity expression of the vertical layer model, the parameter estimation problem is highly non-linear, which is quite challenging for most optimization algorithms. The objective function for two-layer soil parameter estimation can be formulated as the minimization of the following fitness function:

$$f = \sum_{i=1}^{n} \left[\rho_m^i - \rho_a^i\left(\rho_1, \rho_2, \beta, d, a_i\right)\right]^2 \tag{10}$$

whereby

n is the number of measured data from working site,
ρ_m is the ith measured soil apparent resistivity,
ρ_a is the ith soil apparent resistivity calculated from (6)
ρ_1 is the resistivity of the first layer of the vertically stratified soil layers structure,

ρ_2 is the resistivity of the second layer of the vertically stratified soil layers structure,

β is the angle between the line where four electrodes lied on with the perpendicular direction to the interface between the two vertically stratified layers,

d is the distance between the first electrode to the interface between two layers,

a is the distance between electrodes as illustrated in Fig. 1.

The objective function shown in (10), a is the only variable with known values, and ρ_1, ρ_2, β, and d are unknown variables (parameters). The objective of the optimization process is to estimate these unknown parameters accurately.

3.1   Gauss-Newton Method

There exist different approaches to solving this unconstrained optimization problem [4–6]. Gauss-Newton method is one of the most popular methods for solving these problems. However, it is a gradient-based method and thus sensitive to initial values. Based on the Taylor series, this method attempts to approximate the non-linear equation locally, which is accurate to a certain order of approximations. The general Gauss-Newton program is described in [10], with the basic equation given as

$$x_{k+1} = x_k + (B_k^T B_k)^{-1} B_k f(x_k) \tag{11}$$

where

$$x = [\rho_1, \rho_2, \beta, d]^T \tag{12}$$

$$B_k = - \begin{bmatrix} \frac{\partial f(x_1)}{\partial \rho_1} & \cdots & \frac{\partial f(x_1)}{\partial d} \\ \vdots & \ddots & \vdots \\ \frac{\partial f(x_m)}{\partial \rho_1} & \cdots & \frac{\partial f(x_m)}{\partial d} \end{bmatrix} = \begin{bmatrix} \frac{\partial \rho_c^1}{\partial \rho_1} & \cdots & \frac{\partial \rho_c^1}{\partial d} \\ \vdots & \ddots & \vdots \\ \frac{\partial \rho_c^m}{\partial \rho_1} & \cdots & \frac{\partial \rho_c^m}{\partial d} \end{bmatrix} \tag{13}$$

$$f(x_k) = \begin{bmatrix} \rho_m^1 - \rho_a^1 \\ \rho_m^2 - \rho_a^2 \\ \vdots \\ \rho_m^n - \rho_a^n \end{bmatrix} \tag{14}$$

To obtain a more accurate result for this heteroskedasticity model, an $n \times n$ diagonal matrix W (weight) based on variances is added to (11).

$$W = \begin{bmatrix} \frac{1}{(\rho_M^1)^2} & \cdots & 0 \\ \vdots & \ddots & \vdots \\ 0 & \cdots & \frac{1}{(\rho_M^m)^2} \end{bmatrix} \tag{15}$$

Equation (11) does not result in any satisfactory solution due to divergence problem. Therefore, a scalar value λ is added to the expression in order to overcome the limitation of sensitive initial values and to improve the convergence property, given as

$$x_{k+1} = x_k + \lambda (B_k^T W B_k)^{-1} B_k^T W f(x_k) \tag{16}$$

The Matlab language is used for the implementation so as to model and estimate soil parameters by using Gauss-Newton method. The selection of the value of λ is very sensitive. A fixed λ may contribute to a very slow speed of convergence whereas a large value often causes divergence and badly estimated results. When the estimated results become closer to the true values, the fixed λ causes the results to fluctuate around the true values and thus this leads to a very slow convergence. Therefore, λ should be adjusted during the iterations. Other studies attempted to propose one-dimensional search method to find λ. However, there are other issues arising from this approach. For simplicity, we apply a fixed value for λ in this work.

3.2   Landscape of Optimization Model

In the optimization model given by (10), there exist four variables, which are ρ_1, ρ_2, β, and d. Figure 3(a)–(d) are plotted by setting all non-participating variables at their optimal values. The same approach is implemented for other graphs. From Fig. 3(a) and (b), it is observed that the optimal point is at $(\rho_1, \rho_2) =$

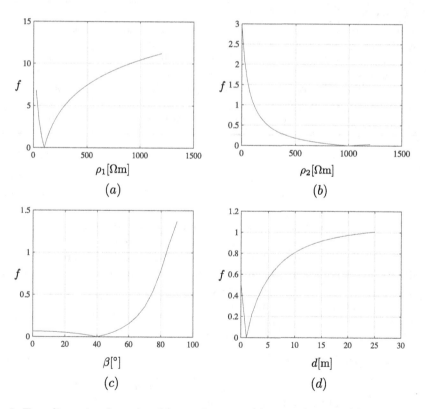

Fig. 3. Two-dimensional graphs of fitness function. (a) ρ_1 vs. fitness f (b) ρ_2 vs. fitness f (c) β vs. fitness f (d) d vs. fitness f

(100,1000). Subsequently, Fig. 3(c)–(d) are the 2D graphs from the perspective of parameters β and d respectively. From these observations, the optimization model has a discontinuous landscape at $\rho_1 = 100\,\Omega\text{m}$. At this point, a steep alley mimicking drain flow is observed. The optimal values of ρ_1 and ρ_2 lie right on this alley. This unique landscape poses a great challenge to many optimization algorithms. The steep alley causes the divergent scenario in case of derivative-based methods, which occurs occasionally, especially when the starting points are not in proper ranges.

4   Brain Storm Optimization

The BSO algorithm was introduced by Shi in 2011 [13]. The idea was based upon the brainstorming process in a group's discussion. The BSO algorithm has a simple concept and thus can be deployed easily in any programming language. The main flow is given here as the Algorithm 1. As from this pseudocode, there are three essential strategies, described in the following subsections.

Algorithm 1. Procedure of the BSO algorithm

---

1 **Initialization**: Initialization and evaluation of n candidate solutions;
2 **while** *not stopping criterion* **do**
3   **Clustering**: Use a clustering algorithm to divide n candidates into m clusters;
4   **Candidate Generation**: Generate new candidate from the ones chosen randomly within different clusters;
5   **Selection**: Evaluate and compare, only retain the better candidate;

4.1   Clustering

This strategy separates the n candidate solutions into m clusters. The popular clustering technique k-means is adopted in the BSO [13]. This clustering technique refines a search area and in a long run, solutions are clustered into a very small independent regions. A cluster probability denoted as p_c acts to control the probability of replacing a cluster's center (the best solution of a cluster) by a randomly generated solution. This encourages diversity, an essential element in avoiding premature convergence.

4.2   Generating New Candidate

A graphical illustration of new candidate generation is given in Fig. 4. New candidates are generated via a simple operator, that is

$$x^i_{\text{new}} = x^i_{\text{old}} + \xi(t) \times N(\mu, \sigma) \qquad (17)$$

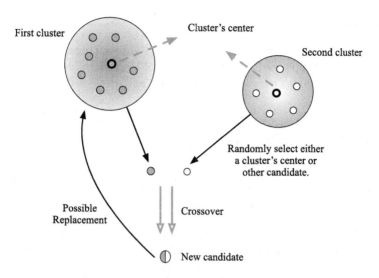

Fig. 4. New candidate generation in the BSO's framework

where x^i_{new} and x^i_{old} are the ith dimension of the new and old values, $\mathbf{x}_{new}$ and $\mathbf{x}_{old}$ respectively. The $N(\mu, \sigma)$ is the Gaussian distribution with μ mean and σ variance, and the coefficient ξ is given as,

$$\xi(t) = \text{sig}\left[\frac{0.5 \times (G - g)}{c}\right] \times \text{rand}() \tag{18}$$

where

sig is the logarithmic sigmoid transfer function,
g is the current generation,
G is the maximum generation,
c is the coefficient to change logsig() function's slope in $\xi(t)$, set to 20 in the original work [13],

Mathematically, the sigmoid function is an "S" curve operator which is popularly used in many neural networks, defined as:

$$S(x) = \frac{1}{1 + \exp(-x)} \tag{19}$$

The aforementioned function $S(x)$ squashes any large x values, either negative or positive, into either 0 or 1. For instance, $S(-1000) = -1$ and $S(1000) = 1$. The x values from -5 to 5 are well distributed within 0 to 1, with almost a linear region in between. In Matlab programming language, the sigmoid function is given as *logsig*. Some readers may think that the *logsig* is the natural logarithm of Eq. (18), which is not the case. The *logsig* function in Matlab merely replicates the sigmoid function, similar to (19).

4.3 BSO Settings

In this work, a similar set of settings found in [13] is adopted. A population size of $n = 100$ is employed, with the number of clusters $m = 5$, and maximum generation $G = 2000$. Other parameters such as $(\mu, \sigma) = (0, 1)$, in (17) and $c = 20$, in (18).

5 Experimental Results

In this experiment, four cases are designed to test the efficiency of each algorithm. Each of these cases involves five measurements. The details are given in Table 1. Each case contains different parameters settings which are reasonable and typical to real measured cases. Results from both GN and BSO methods are tabulated in Table 2. For the case of the GN method, the respective initial values are given in the 4th column, with estimated values in 5th and 7th columns respectively. The percentage of errors (6th and 7th columns) are calculated based on this formula:

$$\text{Err} = \left| \frac{E_v - T_v}{T_v} \right| \times 100\,\% \tag{20}$$

where E_v and T_v are estimated and true values of relevant parameters. The total error E_T will be the summation of errors for n parameters:

$$E_T = \sum_{i=1}^{n} \text{Err}_i \tag{21}$$

Table 2 depicts results obtained via GN and BSO methods. For the case of the GN method, all initial values are given in the 4th column in this table. From this column, the GN method seems to be able to solve all the four cases successfully. In this case, note that the initial values are relatively close to their true values, given in 3rd column. From our empirical observation, this solution is prompt to diverge when initial values are not in a correct range. This is due to the sensitivity of the GN in regards to its initial values. Thus is the major disadvantage of the algorithm that depends on the differentiability of the problem landscape.

The BSO approach, described in Sect. 4, overcomes the limitation of Gauss-Newton in the sense that it is not sensitive to any initial values. The solution obtained is the best obtained over the course of improvements in a number of generations. The initial values are randomly generated within the ranges depicted on the bottom left of the screenshot. The total error recorded for all four variables is 0.2003.

5.1 Result Comparisons

For comparison purposes, we take the summation of all errors of last column in Table 2 and record this total value in Table 3. From this table, the BSO method is superior in terms of its accuracy for cases 1, 2 and 4. Nevertheless, it should be

Table 1. Measured apparent resistivity of the four cases

Case	ρ_m	Measurements				
		1	2	3	4	5
1	ρ_m [Ωm]	105.3339	123.1724	129.3067	131.7586	133.0691
2	ρ_m [Ωm]	200.0042	200.3789	202.0521	204.7726	207.9793
3	ρ_m [Ωm]	999.9388	996.1706	984.0823	968.0910	951.0698
4	ρ_m [Ωm]	599.7958	589.6226	564.5033	538.7287	516.5341
1–4	a [m]	1	5	10	15	20

Table 2. Parameter estimation results via Gauss-Newton and BSO methods

Case	Param	True val	Gauss-Newton			BSO	
			Init val	Est val	Err(%)	Est val	Err(%)
1	ρ_1 [Ωm]	100	105.33	100.0011	0.0011	99.9980	0.00200
	ρ_2 [Ωm]	1000	133.07	999.6913	0.0309	1000.6120	0.06120
	β [°]	40	10	39.7900	0.5250	39.9758	0.06050
	d [m]	1	5	1.0001	0.0100	0.9997	0.02800
2	ρ_1 [Ωm]	200	200.00	200.0000	0.0000	200.000	0.00000
	ρ_2 [Ωm]	500	207.98	540.1353	8.0271	499.4038	0.11924
	β [°]	70	20	69.6083	0.5596	70.0263	0.03757
	d [m]	20	5	20.0018	0.0090	19.9933	0.03350
3	ρ_1 [Ωm]	1000	999.94	1000.0000	0.0000	999.9997	0.00003
	ρ_2 [Ωm]	100	951.07	100.0331	0.0331	100.3153	0.31530
	β [°]	10	40	10.0765	0.7650	10.1942	1.94200
	d [m]	20	5	19.9993	0.0035	19.9936	0.03200
4	ρ_1 [Ωm]	600	599.80	600.0001	0.0000	599.9992	0.00013
	ρ_2 [Ωm]	50	516.53	49.9952	0.0096	50.1388	0.27760
	β [°]	50	10	49.7218	0.5564	50.0293	0.05860
	d [m]	10	5	10.0000	0.0000	9.9986	0.01400

noted that the BSO requires far more iterations compared to the GN method. The solution of the GN will stop after 2000 iterations. However, in the case of the BSO method, there are 100 candidates employed to solve this problem within 2000 iterations, resulted in higher computational cost. This is the major drawback of any stochastic optimization techniques.

Table 3. Results comparison of GN and BSO

Case	Total error (%)	
	GN	BSO
1	0.5670	0.1517
2	8.5957	0.1903
3	0.8016	2.2893
4	0.5660	0.3503

6   Conclusions

In this work, we model the soil apparent resistivity of vertical two-layer soil structure. This modeling is based on the principle of Wenner's method in utilizing four electrodes in measuring the apparent resistivity of the horizontal soil layer. Perpendicular measurements of apparent soil resistivity are different in the vertical two-layer soil for the same spacing between electrodes. The model is then converted to an optimization model with the aim of estimating unknown parameters with several data obtained from site measurements. Two methods are tested to solve this challenging problem. The Gauss-Newton (GN) method implemented in this research shows a good estimation of parameters with great speed. However, this method lacks accuracy and often diverges due to the improper set of initial values.

The limitation of the GN method is overcome by population-based strategy, namely Brain Storm Optimization (BSO) algorithm. Results show better accuracy when utilizing the BSO in search of the unknown parameters in the mathematical optimization model derived. Generally, the BSO algorithm is capable of achieving higher accuracy than the GN. However, this requires relatively higher computational costs. Hence, if one wants to get a crude and quicker estimation, then the GN could be a good choice. However, when accuracy matters more than computational costs, the BSO is absolutely the right tool.

Acknowledgments. This work is supported by the National Natural Science Foundation of China under grant No. 61273367.

References

1. Slaoui, F.H., Erchiqui, F.: Evaluation of grounding resistance and inversion method to estimate soil electrical grounding parameters. Int. J. Multiphys. **4**(3), 201–215 (2010)
2. Berberovic, S., Haznadar, Z., Stih, Z.: Method of moments in analysis of grounding systems. Eng. Anal. Boundary Elem. **27**(4), 351–360 (2003)
3. Yamamoto, K., Yanagawa, S., Yamabuki, K., Sekioka, S., Yokoyama, S.: Analytical surveys of transient and frequency-dependent grounding characteristics of a wind turbine generator system on the basis of field tests. IEEE Trans. Power Delivery **25**(4), 3035–3043 (2010)

4. Althoefer, K., Tan, C.P., Zweiri, Y.H., Seneviratne, L.D.: Hybrid soil parameter measurement and estimation scheme for excavation automation. IEEE Trans. Instrum. Meas. **58**(10), 3633–3641 (2009)
5. Calixto, W.P., Neto, L.M., Wu, M., Yamanaka, K., da Paz Moreira, E., Moreira, E.P.: Parameters estimation of a horizontal multilayer soil using genetic algorithm. IEEE Trans. Power Delivery **25**(3), 1250–1257 (2010)
6. Lagace, P.J., Hoa, M.V., Lefebvre, M., Fortin, J.: Multilayer resistivity interpretation and error estimation using electrostatic images. IEEE Trans. Power Delivery **21**(4), 1954–1960 (2006)
7. Woodhouse, D.J., Middleton, R.H.: Consistency in ground potential rise estimation utilizing fall of potential method data. IEEE Trans. Power Delivery **20**(2), 1226–1234 (2005)
8. Rancic, P.D., Stajic, Z.P., Tosic, B.S., Djordjevic, D.R.: Analysis of linear ground electrodes placed in vertical three-layer earth. IEEE Trans. Magn. **32**(3), 1505–1508 (1996)
9. Zeng, R., He, J., Gao, Y., Zou, J., Guan, Z.: Grounding resistance measurement analysis of grounding system in vertical-layered soil. IEEE Trans. Power Delivery **19**(4), 1553–1559 (2004)
10. Cao, X., Wu, G., Li, S., Zhou, W., Li, R.: A simple formula of grounding grid resistance in vertical two-layer soil. In: Transmission and Distribution Conference and Exposition, Trans. and Dist. IEEE/PES, pp. 1–5 (2008)
11. Seedher, H.R., Arora, J.K.: Estimation of two layer soil parameters using finite wenner resistivity expressions. IEEE Trans. Power Delivery **7**(3), 1213–1217 (1992)
12. Dawalibi, F., Barbeito, N.: Measurements and computations of the performance of grounding systems buried in multilayer soils. IEEE Trans. Power Delivery **6**(4), 1483–1490 (1991)
13. Shi, Y.: Brain storm optimization algorithm. In: Tan, Y., Shi, Y., Chai, Y., Wang, G. (eds.) ICSI 2011, Part I. LNCS, vol. 6728, pp. 303–309. Springer, Heidelberg (2011)

Fireworks Algorithms

Chaotic Adaptive Fireworks Algorithm

Chibing Gong[(✉)]

Department of Computer Information Engineering,
Guangdong Technical College of Water Resource
and Electrical Engineering, Guangzhou 510635, China
spencergong@yahoo.com

Abstract. Fireworks algorithm (FWA) is considered a novel algorithm that reacts the fireworks explosion process. An adaptive fireworks algorithm (AFWA) proposes additional adaptive amplitudes in regards to enhancing performance of the enhanced fireworks algorithm (EFWA). The purpose of this paper is to add chaos to the AFWA with the goal of boosting performance and achieving global optimization. The parameter λ is tuned using ten chaotic maps, and twelve benchmark functions will be tested in regards to chaotic adaptive fireworks algorithm (CAFWA). The final results conclude the CAFWA is able to outperform the FWA, EFWA, and AFWA. Additionally, the CAFWA is compared with the bat algorithm (BA), standard particle swarm optimization 2011 (SPSO2011), harmony search (HS), and firefly algorithm (FA). The research results indicated that the highest performance presented itself when CAFWA is used with Circle maps.

Keywords: Chaotic maps · Fireworks algorithm · Swarm intelligence algorithms

1 Introduction

Fireworks algorithm (FWA) [1] is considered a novel swarm intelligent algorithm. It was introduced in 2010 and reacts the fireworks explosion process. The FWA provides an optimized solution for searching a fireworks location. In the event of a firework randomly exploding, there are explosion and Gaussian sparks produced in addition to the initial explosion. To locate the fireworks local space, a calculation of explosion amplitude and number of explosion sparks, this is then based on other fireworks and fitness functions. Fireworks and sparks are then filtered based on fitness and diversity. Using repetition, FWA focuses on smaller areas for optimized solutions.

Various types of optimized problems have been solved by applying FWA, like factorization of non-negative matrix [2], the design of digital filters [3], parameter optimization regarding detection of spam [4], reconfiguration of networks [5], mass minimization of trusses [6], and scheduling of multi-satellite control resources [7].

However, there are disadvantages to the FWA approach. Although the original algorithm worked well on functions in which the optimum is located at the origin of the search space, when the optimum of origin is more distant, it becomes more challenging to locate the correct solution. Thus, the quality of the results of the original FWA deteriorates severely with increasing distance between the function optimum and the

© Springer International Publishing Switzerland 2016
Y. Tan et al. (Eds.): ICSI 2016, Part I, LNCS 9712, pp. 515–525, 2016.
DOI: 10.1007/978-3-319-41000-5_51

origin. Additionally, the computational cost per iteration is high for FWA compared to other optimization algorithms. For these reasons, the enhanced fireworks algorithm (EFWA) [8] was introduced to enhance FWA.

The explosion amplitude is a significant variable and affects performances of both FWA and EFWA. In EFWA, the amplitude is near zero with the best fireworks, thus employing an amplitude check with a minimum. However, an amplitude like this leads to poor performance. Thus, the adaptive fireworks algorithm (AFWA) introduced a new amplitude [9], where a calculation of distance between filtered sparks and best fireworks as the adaptive amplitude of explosion.

The parameter λ in AFWA is suggested to indicate a fixed value of 1.3 empirically, but our in-depth experiments indicate the reasonable range of value is between 1 and 1.4. For improving the AFWA performance, chaos was added and used to tune the parameter λ. It is called chaotic adaptive fireworks algorithm (CAFWA).

The theory of chaos refers to chaotic dynamical systems studies, they are sensitive to initial states and are non-linear [10]. Swarm intelligence algorithm parameters have recently been tuned using chaotic sequences for boosting performances, like genetic algorithms (GA) [11], harmony search (HS) [12], bee colony optimization [13], particle swarm optimization [14], firefly algorithms (FA) [15], and bat algorithms (BA) [16]. The chaotic integration has indicated promise once the correct chaotic map sets are utilized. These studies have inspired the parameter λ being tuned by chaos within the CAFWA.

The remainder of this paper is organized in the following manner. Both AFWA and CAFWA were shown in Sect. 2. In the third section, ten chaotic maps, and twelve benchmark functions had been utilized for the CAFWA and alternative algorithms. Section 4 covered simulations that had been conducted, while Sect. 5 presents our conclusion.

2   AFWA and CAFWA

2.1   AFWA

Suppose that N denotes the quantity of fireworks while d stands for the number of dimensions, then x_i stands for each firework in AFWA, the explosion amplitude A_i and the number of explosion sparks S_i can be defined according to the following expressions:

$$A_i = \hat{A} \cdot \frac{f(x_i) - y_{min} + \in}{\sum_{i=1}^{N}(f(x_i) - y_{min}) + \in} \tag{1}$$

$$S_i = M_e \cdot \frac{y_{max} - f(x_i) + \in}{\sum_{i=1}^{N}(y_{max} - f(x_i)) + \in} \tag{2}$$

Where $y_{max} = f(x_i))$, $y_{min} = minf(x_i))$, $\hat{A}$ and M_e are two constants. And ε denotes the machine epsilon, $i = (1, 2, \ldots, d)$.

In addition, the number of sparks S_i is defined as follows:

$$S_i = \begin{cases} S_{min} & \text{if } S_i < S_{min} \\ S_{max} & \text{if } S_i > S_{max} \\ S_i & \text{otherwise} \end{cases} \tag{3}$$

Where S_{min} and S_{max} are the lower bound and upper bound of S_i, respectively.

Based on the above A_i and S_i, algorithm 1 is performed by generating explosion sparks for x_i as follows:

```
for j=1 to S_i do
   for each dimension k=1,2,…d do
      obtain r from U(0,1)
      if r<0.5 then
        obtain r from U(-1,1)
        s_ij^(k) ← x_i^(k) + r · A_i
        if s_ij^(k) < LB V s_ij^(k) > UB then
           obtain r again from U(0,1)
           s_ij^(k) ← LB + r · (UB − LB)
        end if
      end if
   end for
end for
return s_ij
```

Where $x \in [LB, UB]$. U (a, b) denotes a uniform distribution between a and b.

After the explosion sparks, algorithm 2 is performed for generating the Gaussian sparks as follows:

```
for j=1 to NG do
   Randomly choose i from 1,2..m
   obtain r from N(0,1)
   for each dimension k=1,2,…d do
      G_j^(k) ← x_i^(k) + r · (x*^(k) − x_i^(k))
      if G_j^(k) < LB V G_j^(k) > UB then
         obtain r from U(0,1)
         G_j^(k) ← LB + r · (UB − LB)
      end if
   end for
end for
return G_j
```

Where NG is the quantity of Gaussian sparks, m stands for the quantity of fireworks, x^* denotes the best firework, $N(0,1)$ denotes normal distribution with an average of 0 and standard deviation of 1.

For the best sparks among above explosion sparks and the Gaussian sparks, algorithm 3 is performed to obtain adaptive amplitude of fireworks in generation $g+1$ as follows [9]:

$A^*(g + 1) \leftarrow UB - LB$
for i=1 **to** n **do**
    if $||s_i - s^*||_\infty > A^*(g + 1) \wedge f(s_i) > f(x)$ **then**
    $A^*(g + 1) \leftarrow ||s_i - s^*||_\infty$
    end if
end for
$A^*(g + 1) \leftarrow \lambda \cdot A^*(g + 1)$
$A^*(g + 1) \leftarrow 0.5 \cdot (A^*(g) + A^*(g + 1))$
return $A^*(g + 1)$

Where $s_{1...}s_n$ denotes all sparks generated in generation g, s^* denotes the best spark in generation g, and x stands for fireworks in generation g.

The above parameter λ is suggested to be a fixed value of 1.3 empirically.

Algorithm 4 demonstrates the complete version of the AFWA.

```
randomly choosing m fireworks
assess their fitness
A* ← UB − LB
repeat
    obtain Aᵢ(except for A*) based on Eq.(1)
    obtain Sᵢ based on Eq.(2) and Eq.(3)
    produce explosion sparks based on Algorithm 1
    produce Gaussian sparks based on Algorithm 2
    assess all sparks` fitness
    obtain A* based on Algorithm 3
    retain the best spark as a firework
    randomly select other m-1 fireworks
until termination condition is satisfied,
returning the best fitness and a firework location
```

Where A^* denotes the adaptive amplitude of the best firework.

2.2   CAFWA

For enhanced EFWA performance, an adaptive amplitude must be calculated based on the best firework within AFWA. Although, the parameter λ is a fixed value of 1.3 when using this adaptive amplitude technique, which cannot be utilized for complex optimization issues. The parameter λ actually has a large impact on AFWA performance as it controls the balance of exploitation and exploration. While exploration helps encourage global search solutions, exploitation targets local search solutions.

Parameter λ should have a specific scope as it relies on a range of fitness functions. A small λ causes the adaptive amplitude to converge too quick, and a large λ causes the adaptive amplitude to be challenging to converge. In-depth experiments indicated the value should range between 1 and 1.4 in CAFWA.

There are many chaotic maps that have ergodicity, stochastic property and certainty. Inspiration developed from [16] that parameter λ_c was replaced using chaotic maps that had a value ranging between 1 and 1.4.

3   Chaotic Maps and Benchmark Functions

3.1   Chaotic Maps

Ten chaotic maps have been used for tuning parameter λ_c within CAFWA to enhance AFWA performance, then the best chaotic map from CAFWA was determined.

Table 1 below presents the ten various chaotic maps [16].

Table 1. Chaotic maps

No.	Name	Chaotic map	Range
1	Chebyshev	$x_{i+1} = \cos(i\cos^{-1}(x_i))$	$(-1,1)$
2	Circle	$x_{i+1} = mod\left(x_i + b - \left(\frac{a}{2\pi}\right)\sin(2\pi x_i), 1\right), a = 0.5, b = 0.2$	$(0,1)$
3	Gauss/mouse	$x_{i+1} = \begin{cases} 1 & x_i = 0 \\ mod\left(\dfrac{1}{x_i}, 1\right) & otherwise \end{cases}$	$(0,1)$
4	Iterative	$x_{i+1} = \sin\left(\frac{a\pi}{x_i}\right), a = 0.7$	$(-1,1)$
5	Logistic	$x_{i+1} = ax_i(1 - x_i), a = 4$	$(0,1)$
6	Piecewise	$x_{i+1} = \begin{cases} \frac{x_i}{p} & 0 \le x_i < p \\ \frac{x_i - p}{0.5 - p} & p \le x_i < 0.5 \\ \frac{1 - p - x_i}{0.5 - p} & 0.5 \le x_i < 1 - p \\ \frac{1 - x_i}{p} & 1 - p \le x_i < 1 \end{cases}, p = 0.4$	$(0,1)$
7	Sine	$x_{i+1} = \frac{a}{4}\sin(\pi x_i), a = 4$	$(0,1)$
8	Singer	$x_{i+1} = \mu\left(7.86x_i - 23.31x_i^2 + 28.75x_i^3 - 13.302875x_i^4\right), \mu = 1.07$	$(0,1)$
9	Sinusoidal	$x_{i+1} = ax_i^2 \sin(\pi x), a = 2.3$	$(0.48,0.92)$
10	Tent	$x_{i+1} = \begin{cases} \dfrac{x_i}{0.7} & x_i < 0.7 \\ \dfrac{10}{3}(1 - x_i) & x_i \ge 0.7 \end{cases}$	$(0,1)$

The parameter λ_c is between 1 and 1.4. Thus, a linear mapping between the chaotic variables of x_i in the range of (*CLB, CUB*) and the parameter λ_c in CAFWA is defined as follows:

$$\lambda_c = 1.0 + 0.4 \cdot \frac{x_i - CLB}{CUB - CLB} \tag{4}$$

Where $x_i \in (CLB, CUB)$, for the Chebyshev map, $CLB=-1$ and $CUB=1$.

Hence, Eq. (4) can produce λ_c using the chaotic variables of x_i derived from chaotic maps.

3.2 Benchmark Functions

In order to assess the performance of CAFWA, we employ twelve standardized benchmark functions [17]. The functions are multi-modal and uni-modal. The global minimum is zero.

Table 2 presents a list of uni-modal functions and features.

Table 2. Uni-modal functions

Function	Dimension	Range	Shift position
$F1(x) = \sum_{i=1}^{n} x_i^2$	40	$[-10,10]$	$[3,3,\ldots,3]$
$F2(x) = \sum_{i=1}^{n}|x_i| + \prod_{i=1}^{n}|x_i|$	40	$[-10,10]$	$[7,7,\ldots,7]$
$F3(x) = \sum_{i=1}^{n}(\sum_{j=1}^{i} x_j)^2$	40	$[-10,10]$	$[-3,-3,\ldots,-3]$
$F4(x) = \sum_{i=1}^{n} i x_i^2$	40	$[-30,30]$	$[-21,-21,\ldots,21]$
$F5(x) = \max\{|x_i|, 1 \leq x_i \leq n\}$	40	$[-100,100]$	$[30,30,\ldots,30]$
$F6(x) = 10^6 x_1^2 + \sum_{i=2}^{n} x_i^2$	40	$[-100,100]$	$[70,70,\ldots,70]$

Table 3 presents a list of multi-modal functions and their features.

4   Simulation Studies

4.1 Success Criterion

In this section we utilize the success rates S_r for comparing the 10 chaotic maps with the CAFWA to determine the best chaotic map. S_r can be defined as follows [16]:

$$S_r = 100 \cdot \frac{N_{successful}}{N_{all}} \tag{5}$$

Table 3. Mul-timodal functions

Function	Dimension	Range	Shift position
$F7(x) = \sum_{i=1}^{n} \left(x_i^2 - 10\cos(2\pi x_i) + 10 \right)$	40	$[-1,1]$	$[0,0,...,0]$
$F8(x) = \frac{1}{4000}\sum_{i=1}^{n} x_i^2 - \prod_{i=1}^{n}\cos(\frac{x_i}{\sqrt{i}}) + 1$	40	$[-100,100]$	$[70,70,$ $...,70]$
$F9(x) = -20exp\left(-0.2\sqrt{\frac{1}{n}\sum_{i=1}^{n} x_i^2} \right)$ $- exp\left(\frac{1}{n}\sum_{i=1}^{n}\cos(2\pi x_i) \right) + 20 + e$	40	$[-10,10]$	$[-3,-3,$ $...,-3]$
$F10(x) = \sum_{i=1}^{n/4}\left((x_{4i-3} + 10x_{4i-2})^2 + 5(x_{4i-1} - x_{4i})^2 + \right.$ $\left. (x_{4i-2} - 2x_{4i-1})^4 + 10(x_{4i-3} - x_{4i})^4 \right)$	40	$[-10,10]$	$[3,3,...,3]$
$F11(x) = 0.1(sin^2(3\pi x_1) + \sum_{i=1}^{n-1}(x_i - 1)^2(1 + sin^2(3\pi x_i + 1))(x_n - 1)^2(1 + sin^2(2\pi x_n)) + \sum_{i=1}^{n} u(x_i, 5, 100, 4)$ $u(x_i, a, k, m) = \begin{cases} k(x_i - a)^m & x_i > a \\ 0 & -a \le x_i \le a \\ k(-x_i - a)^m & x_i < -a \end{cases}$	40	$[-10,10]$	$[7,7,...,7]$
$F12(x) = \frac{\pi}{n}(10\sin(\pi y_1) + \sum_{i=1}^{n-1}(y_i - 1)^2(1 + 10sin^2(\pi y_{i+1})) + (y_n - 1)^2) + \sum_{i=1}^{n} u(x_i, 10, 100, 4)$ $y_i = 1 + \frac{x_i + 1}{4}$	40	$[-10,10]$	$[-7,-7,$ $...,-7]$

Where $N_{successful}$ is the number of trials that were successful, and N_{all} stands for the number of trials. When an experiment locates a solution that is close in range to the global optimum, it is found to be a success. A successful trail is defined as:

$$\sum_{i=1}^{D} \left(X_i^{gb} - X_i^* \right) \le (UB - LB) \times 10^{-4} \qquad (6)$$

Where D denotes dimensions of test function, X_i^{gb} denotes the dimension of the best result by the algorithm.

4.2 Test Studies and Initialization

This paper covered testing the benchmark functions using 100 independent algorithms based on various chaotic maps. In order to fully evaluate the CAFWA performance, statistical measures were used, such as the worse, best, median and mean objective values were used. Standard deviations were also collected.

In addition, the initial point across all of the chaotic maps was 0.7. In the algorithms based on FWA and CAFWA, m=5, $\hat{A} = 100$, $M_e = 200$, $S_{min} = 2$, $S_{max} = 100$ and NG=5. Evaluation times from 65000 to 150000 according to different functions.

Finally, we use Matlab 7.0 software on a notebook PC with a 2.3GHZ CPU, 4 GB RAM, Intel Core i3-2350 and Windows 7 (64 bit).

4.3 CAFWA Results on Benchmark Functions

Table 4 presents the results of CAFWA with the ten chaotic maps and twelve benchmark functions.

Table 4. Success rates of CAFWA for benchmark functions with ten chaotic maps

Chaotic map name	F1	F2	F3	F4	F5	F6	F7	F8	F9	F10	F11	F12	Total
Chebyshev map	44	77	29	49	46	45	48	46	37	5	19	47	492
Circle map	74	87	83	55	72	87	51	45	35	1	38	52	680
Gauss/mouse map	100	37	47	100	0	100	87	48	9	10	53	35	626
Iterative map	42	78	41	34	42	36	46	40	34	6	25	36	460
Logistic map	50	76	41	37	56	47	48	42	35	6	23	42	503
Piecewise map	35	78	36	36	46	40	44	36	32	2	21	41	447
Sine map	66	85	63	51	67	76	44	49	25	8	35	43	612
Singer map	0	3	0	1	1	0	11	30	31	1	1	12	91
Sinusoidal map	7	53	2	17	21	1	28	51	33	2	5	34	254
Tent map	40	72	18	35	45	48	55	57	35	6	32	45	488
Standard AFWA	0	0	0	0	0	0	0	0	26	0	0	0	26

These experimental results indicate the CAFWA is able to increase the performance of AFWA when used with various chaotic maps. Furthermore, the success rates show that the CAFWA chaotic maps are superior to the standard of AFWA. The Circle maps tend to present the best results out of all of them with a 680 success rate. Thus, the Circle map was adopted in CAFWA for the following comparisons.

4.4 Comparison with FWA Based Methods

Table 5 presents the mean error and rank of FWA, AFWA, EFWA and CAFWA using the Circle map.

The results from Table 5 indicate the CAFWA using the Circle map has a greater performance than the AFWA: with the exception of F7 and F9. Additionally, CAFWA also ranks higher than the three FWA based algorithms.

For evaluating whether the CAFWA results was significantly higher than AFWA or not, the Wilcoxon Signed-rank test was utilized at the 5% level, as shown in Table 5.

Here P is the p-value of test and H is the result of the hypothesis test. A value of H=1 indicates the rejection of the null hypothesis at the 5% significance level.

Table 6 indicates that CAFWA had a large improvement over AFWA.

4.5 Comparison with Other Swarm Intelligence Algorithms

Additionally, alternative swarm intelligent algorithms were compared with CAFWA, such as the SPSO2011 [18], HS [19], FA [20] and BA [21]. The resulting evaluation times was the same as the CAFWA.

Table 5. Mean error and rank on twelve functions

Func.	FWA	Rank	EFWA	Rank	AFWA	Rank	CAFWA	Rank
F1	1.16E+00	3	2.36E-03	2	9.34E+00	4	1.63E-03	1
F2	8.37E+00	3	6.59E+00	2	1.20E+01	4	7.94E-02	1
F3	3.21E+01	3	1.56E-02	2	9.60E+01	4	3.96E-04	1
F4	4.37E+02	3	5.77E-01	2	3.21E+03	4	1.82E-02	1
F5	7.95E+00	4	1.01E+00	2	1.46E+00	3	3.13E-02	1
F6	3.11E+02	3	2.64E-01	2	5.26E+03	4	1.18E-02	1
F7	0	1	4.34E-03	2	1.01E+01	4	3.64E+00	3
F8	9.99E-01	4	1.52E-02	2	6.98E-02	3	8.10E-03	1
F9	1.33E+00	3	5.17E+00	4	1.21E+00	1	1.28E+00	2
F10	1.54E+01	3	2.37E-01	2	6.63E+02	4	2.89E-03	1
F11	6.87E-01	2	8.78E+01	4	6.80E+00	3	2.98E-01	1
F12	4.98E-01	3	1.30E+01	4	3.04E-01	2	2.10E-01	1
Mean rank		2.92		2.5		3.33		1.25

Table 6. Wilcoxon Signed-rank test results for CAFWA and AFWA

	F1	F2	F3	F4	F5	F6
H	1	1	1	1	1	1
P	7.80E-086	3.11E-089	0	5.28E-247	8.01E-082	3.40E-155
	F7	F8	F9	F10	F11	F12
H	1	1	1	1	1	1
P	4.14E-019	7.41E-105	1.08E-024	0	7.41E-109	8.36E-042

Table 7 presents the mean error with the five algorithms.

Table 7. Mean error on twelve functions for the five algorithms

Func.	SPSO2011	BA	FA	HS	CAFWA
F1	1.22E-28	2.54E-03	1.25E-06	1.41E-01	1.63E-03
F2	2.26E+01	3.52E+16	2.27E-02	1.29E+00	7.94E-02
F3	5.69E-08	8.75E-03	2.04E+05	3.50E+01	3.96E-04
F4	1.01E-04	8.87E-02	3.65E+05	1.83E+00	1.82E-02
F5	1.34E+01	4.62E+01	2.10E-02	4.45E+00	3.13E-02
F6	4.79E+04	2.96E+05	3.48E+03	3.15E-01	1.18E-02
F7	5.07E+01	2.09E+01	1.94E+01	2.14E+01	3.64E+00
F8	6.53E-01	7.92E+00	6.71E-04	1.03E-02	8.10E-03
F9	1.76E+00	1.01E+01	9.77E+00	3.04E-01	1.28E+00
F10	1.56E-03	3.78E-01	1.14E-02	8.06E+00	2.89E-03
F11	8.21E+00	1.46E+02	2.20E-04	1.48E-02	2.98E-01
F12	3.07E+00	2.06E+01	1.38E+01	5.18E-04	2.10E-01

Table 7 indicates that certain algorithms are good for some functions, but bad for others. Overall, the CAFWA performance was shown to have more stability than the others.

Table 8 presents the rank of the twelve benchmark functions and five algorithms.

Table 8. Rank on twelev benchmark functions for the five algorithms

Func	F1	F2	F3	F4	F5	F6	F7	F8	F9	F10	F11	F12	Mean rank
SPSO2011	1	4	1	1	4	4	5	4	3	1	4	3	2.92
BA	4	5	3	3	5	5	3	5	5	4	5	5	4.33
FA	2	1	5	5	1	3	2	1	4	3	1	4	2.67
HS	5	3	4	4	3	2	4	3	1	5	2	1	3.08
CAFWA	3	2	2	2	2	1	1	2	2	2	3	2	2.00

Table 8 indicates that the CAFWA rank is the best out of the five algorithms.

5   Conclusion

CAFWA was developed by applying chaos to AFWA. Ten various chaos maps were investigated and used for tuning the parameter λ_c in CAFWA. Results indicated a large boost in performance in CAFWA over AFWA when using chaotic maps. Although, the Circle map was shown to be the best map to apply to CAFWA.

The experiments clearly indicated that CAFWA is able to perform better than FWA, EFWA, and AFWA. Additionally, CAFWA is comparable with BA, HS, FA, and SPSO2011. Overall, the research demonstrates CAFWA using the Circle map performed the best.

References

1. Tan, Y., Zhu, Y.: Fireworks algorithm for optimization. In: Tan, Y., Shi, Y., Tan, K.C. (eds.) ICSI 2010, Part I. LNCS, vol. 6145, pp. 355–364. Springer, Heidelberg (2010)
2. Janecek, A., Tan, Y.: Using population based algorithms for initializing nonnegative matrix factorization. In: Tan, Y., Shi, Y., Chai, Y., Wang, G. (eds.) ICSI 2011, Part II. LNCS, vol. 6729, pp. 307–316. Springer, Heidelberg (2011)
3. Gao, H., Diao, M.: Cultural firework algorithm and its application for digital filters design. Int. J. Model. Ident. Control 14(4), 324–331 (2011)
4. He, W., Mi, G., Tan, Y.: Parameter optimization of local-concentration model for spam detection by using fireworks algorithm. In: Tan, Y., Shi, Y., Mo, H. (eds.) ICSI 2013, Part I. LNCS, vol. 7928, pp. 439–450. Springer, Heidelberg (2013)
5. Imran, A., Kowsalya, M., Kothari, D.: A novel integration technique for optimal network reconfiguration and distributed generation placement in power distribution networks. Int. J. Electr. Power Energy Syst. **63**, 461–472 (2014)

6. Pholdee, N., Sujin, B.: Comparative performance of meta-heuristic algorithms for mass minimisation of trusses with dynamic constraints. Adv. Eng. Soft. **75**, 1–13 (2014)
7. Liu, Z., Feng, Z., Ke L.: Fireworks algorithm for the multi-satellite control. In: IEEE Congress on Evolutionary Computation (CEC 2015), pp. 1280–1286 (2015)
8. Zheng, S., Janecek, A., Tan, Y.: Enhanced fireworks algorithm. In: IEEE Congress on Evolutionary Computation (CEC 2013), pp. 2069–2077 (2013)
9. Li, J., Zheng, S., Tan, Y.: Adaptive fireworks algorithm. In: IEEE Congress on Evolutionary Computation (CEC 2014), pp. 3214–3221 (2014)
10. Saremi, S., Mirjalili, S., Lewis, A.: Biogeography-based optimisation with chaos. Neural Comput. Appl. **25**, 1077–1097 (2014)
11. Gharoonifard, G., Moein, F., Deldari, H., Morvaridi, A.: Scheduling of scientific workflows using a chaos-genetic algorithm. Proc. Comput. Sci. **1**, 1445–1454 (2010)
12. Alatas, B.: Chaotic harmony search algorithms. Appl. Math. Comput. **216**, 2687–2699 (2010)
13. Alatas, B.: Chaotic bee colony algorithms for global numerical optimization. Expert Syst. Appl. **37**, 5682–5687 (2010)
14. Gandomi, A., Yun, G., Yang, X., Talatahari, S.: Chaos-enhanced accelerated particle swarm algorithm. Commun. Nonlinear. Sci. **18**(2), 327–340 (2013)
15. Gandomi, A., Yang, X., Talatahari, S., Alavi, A.: Firefly algorithm with chaos. Commun. Nonlinear. Sci. **18**(1), 89–98 (2013)
16. Gandomi, A., Yang, X.: Chaotic bat algorithm. J. Comput. Sci. **5**(2), 224–232 (2014)
17. Mitić, M., Vuković, N., Petrović, M., Miljković, Z.: Chaotic fruit fly optimization algorithm. Knowl. Based Syst. **89**, 446–458 (2015)
18. Zambrano, M., Bigiarini, M., Rojas, R.: Standard particle swarm optimisation 2011 at CEC-2013: A baseline for future PSO improvements. In: IEEE Congress on Evolutionary Computation (CEC 2013), pp. 2337–2344 (2013)
19. Geem, Z., Kim, J., Loganathan, G.: A new heuristic optimization algorithm: harmony search. Simulation **76**(2), 60–68 (2001)
20. Yang, X.: Firefly algorithms for multimodal optimization. Lect. Notes Comput. Sci. **5792**, 169–178 (2009)
21. Yang, X.-S.: A new metaheuristic bat-inspired algorithm. In: González, J.R., Pelta, D.A., Cruz, C., Terrazas, G., Krasnogor, N. (eds.) NICSO 2010. SCI, vol. 284, pp. 65–74. Springer, Heidelberg (2010)

Support Vector Machine Parameters Optimization by Enhanced Fireworks Algorithm

Eva Tuba[1], Milan Tuba[1(✉)], and Marko Beko[2]

[1] Faculty of Computer Science, John Naisbitt University, Bulevar Umetnosti 29,
11070 Belgrade, Serbia
etuba@acm.org, tuba@ieee.org
[2] Computer Engineering Department, Universidade Lusòfona de
Humanidades e Tecnologias, Lisbon, Portugal
marko@isr.ist.utl.pt

Abstract. Support vector machines are widely used as superior classifiers for many different applications. Accuracy of the constructed support vector machine classifier depends on the proper parameter tuning. One of the most common used techniques for parameter determination is grid search. This optimization can be done more precisely and computationally more efficiently by using stochastic search metaheuristics. In this paper we propose using enhanced fireworks algorithm for support vector machine parameter optimization. We tested our approach on standard benchmark datasets from the UCI Machine Learning Repository and compared the results with grid search and with results obtained by other swarm intelligence approaches from the literature. Enhanced fireworks algorithm proved to be very successful, but most importantly it significantly outperformed other algorithms for more realistic cases for which there were separate test sets, rather than doing only cross validation.

Keywords: Support vector machine · SVM parameter tuning · Fireworks algorithm · Swarm intelligence · Metaheuristics

1 Introduction

One of the very important and active research topics in computer science, especially in data mining and machine learning, is classification problem. It addresses the problem of determination of the class that a given data instance belongs to. Classification problem is present in many different science fields such as medicine for classification of tumors and cancers [1,2], image processing for face recognition [3] or optical character recognition for symbol classification [4], economy for risk of financial distress classification [5], etc.

Support vector machine (SVM) was proposed by Vapnik in [6] as a supervised machine learning algorithm. Nowadays SVM is successfully and widely used for

M. Tuba–This research is supported by Ministry of Education, Science and Technogical Development of Republic of Serbia, Grant No. III-44006.

Y. Tan et al. (Eds.): ICSI 2016, Part I, LNCS 9712, pp. 526–534, 2016.
DOI: 10.1007/978-3-319-41000-5_52

many different classification problems in numerous science areas. In [4] support vector machines was used to improve the classification for the OCR of mathematical documents. Based on texture features support vector machine model was built with aim to discriminate between brain metastasis and radiation necrosis [7]. SVM was used as a classificator of electroencephalogram signals [8].

To build a SVM model, few parameters need to be tuned. One parameter is soft margin constant C which affects the trade-off between complexity of the model and proportion of non-separable samples. Another parameter that needs to be tuned is parameter of kernel function. One of the most common used kernel function is Gaussian radial basis function with free parameter γ. Parameter tuning for the SVM is a hard optimization problem which requires a lot of computation.

Stochastic population algorithms, particularly swarm intelligence algorithms, were intensively studied and used for solving hard optimization problems during the past two decades. For complex optimization problems with huge number of local optima or discrete combinatorial exponential problems, deterministic algorithms cannot find solution in reasonable time while stochastic population algorithms proved to be very successful. The idea of swarm intelligence algorithms is to simulate collectives of simple agents with simple behavior. While individually without intelligence, together they exhibit significant intelligence and capability to progress to better solution.

There are in literature numerous papers that proposed swarm intelligence algorithms for parameter tuning and feature selection problems, however most of them concentrated on PSO. PSO based method was introduced for parameter determination and feature selection of the SVM [9], while in [10] for optimization of the values of parameter memetic algorithm based on particle swarm optimization and pattern search was used. Based on the chaotic system theory, new PSO method that uses chaotic mappings for parameter adaptation of variant of SVM, wavelet v-support vector machine was proposed in [11]. PSO algorithm was introduced into least square support vector machine (LS-SVM) model in order to propose an optimized LS-SVM model in [12]. In [13] two modifications to the original artificial bee colony (ABC) were introduced for optimization of parameters of least squares support vector machines.

In this paper we propose enhanced fireworks algorithm for parameters tuning of the support vector machine. We tested the proposed algorithm on standard benchmark datasets from UCI Machine Learning Repository [14] and performed a comparative analysis using [15] as reference.

2   Support Vector Machine

Support vector machine (SVM) is one of the latest and widely used binary classifiers. The main idea of the SVM is to find hyperplane that separates instances from two different classes. Instances are represented as points in space and each one is labeled with corresponding class. Based on instances from the training set, SVM model is built and new examples can be classified using that model.

With the assumption that training data are linearly separable, hyperplane that separates examples from different classes represented as vectors $x_i \in R^d$ labeled with $y_i \in \{-1, 1\}$, $i = 1, 2, ..., n$, could be written as:

$$y_i(w \cdot x_i + b) \geq 1 \quad \textbf{for} \quad 1 \leq i \leq n, \tag{1}$$

where w is normal vector to the hyperplane. Hyperplane defined as described is determined by the instances nearest to the hyperplane. Points that represent these instances are named support vectors.

In the described method all instances from one class must be on the correct side of the hyperplane which means that model built like that will be highly sensitive to the noise in input data. If the training set contains any errors in labels or outliers (instances that are very different from other instances of the same class) hyperplane will not be found. Since the data collected from real world usually have some number of outliers instead of hard margin defined by Eq. (1) the soft margin is introduced with:

$$y_i(w \cdot x_i + b) \geq 1 - \epsilon_i, \quad \epsilon_i \geq 0, \quad 1 \leq i \leq n \tag{2}$$

where ϵ_i are slack variables. Slack variables allow some instances to fall off the margin and these instances cannot be used as support vectors. For finding optimal soft margin the following quadric programming problem need to be solved:

$$\frac{1}{2}||w||^2 + C \sum_{i=1}^{n} \epsilon_i, \tag{3}$$

where C is the parameter for soft margin cost function. If a larger value for C is selected, the model will be similar to the model built with hard margin. By adjusting the value of parameter C, classification accuracy can be drastically improved and vice verse, poor choice of parameter C can lead to the model with unsatisfactory classification accuracy [16].

The second problem with basic SVM classifier is that input data must be linearly separable which in most examples is not the case. The solution is to use non-linear classifier instead of the liner one. Non-linear classifier uses kernel function instead of applying dot product. The purpose is that kernel function project input data into higher dimensional space with the aim to make them linearly separable. The most common used kernel function is Gauss radial basis function (RBF) defined by the following expression:

$$K(x_i, x_j) = \exp(-\gamma ||x_i - x_j||^2). \tag{4}$$

Parameter γ plays a major role in achieving a better classification accuracy. It defines the influence of a single training example to the model. With larger value of parameter γ advantages obtained from the higher dimensional projection are reduced and with too small value of parameter the decision boundary would become extremely sensitive to the noise in training data.

For finding a good SVM model, the first task is to determinate the optimal pair of values for parameters C and γ. One of the most common technique used for this purpose is grid search. Grid search basically builds a SVM model for different pairs (C, γ) and checks the accuracy of the obtained model. Usually, the accuracy is determined by cross validation. Problem with grid search is that it is a local search and it is highly possible to get trapped in local optima. Setting the search interval is also a challenge. For too wide search interval huge amount of unnecessary computations will be induced, while if the search interval too small it is likely that the optimal values will be left out of that interval.

Instead of grid search, for this optimization problem stochastic population based algorithms can be used. Swarm intelligent algorithms were widely used in the past few years for optimization parameters of SVM [10,13,15]. In this paper we propose using enhanced fireworks algorithm for determining the best pair of parameters (C, γ).

3   Enhanced Fireworks Algorithm

Inspired by fireworks explosions at night, fireworks algorithm (FWA) was developed in 2010 by Tan and Zhu [17]. FWAs creators considered two different behaviors of the fireworks explosion, one when fireworks were well manufactured and one when they were not. In the case of well manufactured fireworks numerous sparks centralized around explosion center will be generated, while for a bad fireworks there will be only a few sparks scattered in the space [17]. This behavior was used for search algorithm where in the first case, when a firework is well manufactured, a firework is located in the promising area of the search space and generate more sparks to search around that space. The second case, when sparks are scattered, is used for exploration. In the FWA, more sparks are generated and the explosion amplitude is smaller for a good firework, compared to a bad one [17].

Authors of the classical FWA proposed five major improvements to the FWA introducing enhanced FWA (EFWA) in [18]. They reported that FWA had high computation cost per iteration, problems if the optimal point is not at zero, etc.

The first modification refers to the equation for calculation of the explosion amplitude. Problem with this equation occurs when fitness value is equal or close to zero because the explosion amplitude is very small in that case and the sparks will be located at almost the same position as the firework which results in poor exploitation. Improved equation for explosion amplitude is defined by [18]:

$$A_i^k = \begin{cases} A_{min}^k & if \quad A_i^k < A_{min}^k, \\ A_i^k & otherwise \end{cases} \tag{5}$$

$$A_{min}^k(t) = A_{init} - \frac{A_{init} - A_{final}}{evals_{max}} * t \tag{6}$$

where t is the number of function evaluation at the begining of the current iteration. Values A_{init} and A_{final} are the initial and final minimum explosion amplitudes, while $evals_{max}$ represents maximum number of evaluations.

In order to improve the search for global minimum, authors in [18] suggested modification of operator for generating explosion sparks. In FWA the same value was added to the location of selected dimensions for all sparks explosion, while in the EFWA spark will be mapped to the new location according to the equation:

$$\Delta X_k = A_i \times rand(-1, 1) \tag{7}$$

where A_i represents explosion amplitude.

The third modification proposes new mapping operator. In many cases, a spark will exceed the allowed search space only by a rather small value. Moreover, as the search space is often equally distributed, the adjusted position X_i^k will be usually very close to the origin. As solution for this problem mapping operator was changed to [18]:

$$x_i^k = X_{min}^k + rand * (X_{max}^k - X_{min}^k). \tag{8}$$

In order to avoid the problems of the conventional Gaussian mutation operator, where some Gaussian sparks were located very close to the origin of the function, independent of the initialization range Zheng, Janecek and Tan proposed a new Gaussian mutation operator which is computed as [18]:

$$X_i^k = X_i^k + (X_B^k - X_i^k) * e. \tag{9}$$

where X_B is the location of the currently best firework/explosion spark found so far, and $e = N(0, 1)$.

In [18] authors applied selection method which is referred to as elitism-random selection (ERP) method with aim to speed up the selection process of the population for the next generation.

4   Experimental Results

Quality of our proposed method was tested on 20 different well known benchmark datasets from the UCI Machine Learning Repository [14]. Detailed information about used datasets are listed in Table 1. Missing values were set to the mode and mean of the attributes for categorical and continuous attributes, respectively. All data were scaled to the range $[-1, 1]$. Test sets were scaled with the same values of minimum and maximum for the reasons explained in [19].

The proposed algorithm was implemented using Matlab R2015a and LIB-SVM (Version 3.21) [20]. Experiments were performed on the following platform: Intel ® Core™ i7-3770K CPU at 4 GHz, 8 GB RAM, Windows 10 Professional OS.

Accuracy of the proposed method was determined by performing 10-fold cross validation, except for *Hill-Valley* and *Monks1* datasets which have separate test sets. For this datasets SVM models were trained with the training sets and accuracy was determined using the test sets.

Table 1. Datasets from the UCI machine learning repository

No	Dataset	No. of classes	No. of instan.	No. of featur.	Missing values
1	Australian credit (Australian)	2	690	14	Yes
2	Blood transfusion (Blood)	2	748	4	No
3	Bupa liver disorders (Bupa)	2	345	6	No
4	Contraceptive method choice (CMC)	3	1473	9	No
5	Germen credit (German)	2	1000	24	No
6	Glass	6	214	9	No
7	Haberman survival (Haberman)	2	306	3	No
8	Statlog heart (Heart)	2	270	13	No
9	Hill-Valley	2	1212	100	No
10	Ionosphere	2	351	34	No
11	Iris	3	150	4	No
12	Monks1	2	556	7	No
13	Pima Indians diabetes (Pima)	2	768	8	No
14	Sonar	2	208	60	No
15	Teaching assistant evaluation (Teaching)	3	151	5	No
16	Wisconsin diagnostic breast cancer (WDBC)	2	569	30	No
17	Wine	3	178	13	No
18	Wisconsin breast cancer (Wisconsin)	2	699	9	Yes
19	Wisconsin prognostic breast cancer (WPBC)	2	198	34	Yes
20	Zoo	7	101	17	No

The goal of this paper was to prove that enhanced fireworks algorithm can be successfully used for support vector machine parameter tuning. For this purpose we compared our results with results from literature. Our algorithm was compared with approach proposed by Chen et al. in [15] named PTSPSO-SVM. This method was based on particle swarm optimization algorithm. Chen et al. proposed parallel time variant of PSO with aim to simultaneously perform the optimization of support vector machine parameters. In [15] Chen et al. implemented standard grid search and regular PSO algorithm and compared results of the proposed method with these results. Table 2 shows results from [15] in the first three column while the last column represents results of our proposed algorithm. Better results are printed in bold.

As can be seen from the Table 2, our algorithm shows better results for 10 out of 20 tested datasets. For most cases there is a small improvement by our algorithm or there is small advantage of the PTVPSO-SVM. If that were the only results, we could conclude that enhanced fireworks algorithm is capable of successfully tuning SVM parameters, but it is not a significant improvement. However, for datasets *Hill-Valley* and *Monks1* which were tested on separate test

Table 2. Comparison between our EFWA-SVM and approaches proposed in [15] (in % of successful classification)

Data	Grid-SVM	PSO-SVM	PTVSPSO-SVM	EFWA-SVM
Australian	84.91	86.93	**88.05**	87.10
Blood	78.32	79.98	**80.53**	78.88
Bupa	70.95	72.03	**74.76**	73.91
CMC	53.85	56.67	**58.66**	57.98
German	75.33	76.34	78.02	**79.20**
Glass	69.98	72.45	74.66	**75.23**
Haberman	72.29	75.77	75.77	**75.82**
Heart	82.81	85.24	**86.75**	85.56
Hill-Valley	69.80	71.92	73.21	**89.93**
Ionosphere	93.90	94.34	95.21	**96.58**
Iris	95.67	97.33	98.26	**98.67**
Monks1	80.56	83.64	84.98	**95.37**
Pima	76.65	77.58	**78.14**	77.34
Sonar	88.98	91.27	**92.66**	91.35
Teaching	59.68	60.77	62.86	**62.91**
WDBC	97.45	98.01	**98.44**	96.31
Wine	98.20	98.64	**98.99**	97.19
Winsconsin	96.62	97.55	**98.62**	96.85
WPBC	77.48	79.33	81.22	**82.83**
Zoo	93.55	94.55	96.67	**98.02**

sets, our proposed method achieved significant improvement compared to results from [15]. Accuracy of classification for *Hill-Valley* dataset with our approach was 89.93 % while with PTVPSO-SVM accuracy was only 73.21 %. For dataset *Monks1* our method preformed also very well with accuracy of 95.37 % which is 10.39 % more then the PTVPSO-SVM accuracy of 84.98 %. The most likely explanation for these improvements is that parameters selected by PTVPSO-SVM lead to over-fitting, which was not visible on cross-validation, but the accuracy of the model applied to separate test set was poor.

Our proposed method did not reach better accuracy compared to [15] for some datasets that have large number of instances. In our experiments we limited execution time and for large datasets more time is needed to achieve a better accuracy. Another possible solution could be to increase the number of subsets in cross validation.

5   Conclusion

In this paper we tested the quality of the support vector machine parameter tuning by enhanced fireworks algorithm which was not used for that purpose before. For some datasets this approach increased accuracy of classification compared to the existing methods. The biggest improvement was for datasets tested on separated test sets which is very important because it indicates that the proposed method was superior since it did not lead to over fitting. That establishes the enhanced fireworks algorithm as very appropriate for support vector machine parameter tuning. Besides parameter optimization, support vector machine model highly depends on feature extraction so it can be included in the future research.

References

1. Xian, G.: An identification method of malignant and benign liver tumors from ultrasonography based on GLCM texture features and fuzzy SVM. Expert Syst. Appl. **37**, 6737–6741 (2010)
2. Liu, H., Liu, L., Zhang, H.: Ensemble gene selection for cancer classification. Pattern Recogn. **43**, 2763–2772 (2010)
3. Gumus, E., Kilic, N., Sertbas, A., Ucan, O.N.: Evaluation of face recognition techniques using PCA, wavelets and SVM. Expert Syst. Appl. **37**, 6404–6408 (2010)
4. Malon, C., Uchida, S., Suzuki, M.: Mathematical symbol recognition with support vector machines. Pattern Recogn. Lett. **29**, 1326–1332 (2008)
5. Pai, P.F., Hsu, M.F., Lin, L.: Enhancing decisions with life cycle analysis for risk management. Neural Comput. Appl. **24**, 1717–1724 (2014)
6. Cortes, C., Vapnik, V.: Support-vector networks. Mach. Learn. **20**, 273–297 (1995)
7. Larroza, A., Moratal, D., Paredes-Sanchez, A., Soria-Olivas, E., Chust, M.L., Arribas, L.A., Arana, E.: Support vector machine classification of brain metastasis and radiation necrosis based on texture analysis in MRI. J. Magn. Reson. Imaging **42**, 1362–1368 (2015)
8. Subasi, A., Gursoy, M.I.: EEG signal classification using PCA, ICA, LDA and support vector machines. Expert Syst. Appl. **37**, 8659–8666 (2010)
9. Lin, S., Ying, K., Chen, S., Lee, Z.: Particle swarm optimization for parameter determination and feature selection of support vector machines. Expert Syst. Appl. **35**, 1817–1824 (2008)
10. Bao, Y., Hu, Z., Xiong, T.: A PSO and pattern search based memetic algorithm for SVMs parameters optimization. Neurocomputing **117**, 98–106 (2013)
11. Wu, Q.: A self-adaptive embedded chaotic particle swarm optimization for parameters selection of Wv-SVM. Expert Syst. Appl. **38**, 184–192 (2011)
12. Liu, F., Zhou, Z.: A new data classification method based on chaotic particle swarm optimization and least square-support vector machine. Chemometr. Intell. Lab. Syst. **147**, 147–156 (2015)
13. Mustaffa, Z., Yusof, Y., Kamaruddin, S.S.: Enhanced artificial bee colony for training least squares support vector machines in commodity price forecasting. J. Comput. Sci. **5**, 196–205 (2014)
14. Lichman, M.: UCI machine learning repository (2013)

15. Chen, H.J., Yang, B., Wang, S.J., Wang, G., Liu, D.Y., Li, H.Z., Liu, W.B.: Towards an optimal support vector machine classifier using a parallel particle swarm optimization strategy. Appl. Math. Comput. **239**, 180–197 (2014)

16. Wang, L., Chu, F., Jin, G.: Cancer diagnosis and protein secondary structure prediction using support vector machines. In: Wang, L. (ed.) Support Vector Machines: Theory and Applications, pp. 343–364. Springer-Verlag, Berlin Heidelberg (2005)

17. Tan, Y., Zhu, Y.: Fireworks algorithm for optimization. In: Tan, Y., Shi, Y., Tan, K.C. (eds.) ICSI 2010, Part I. LNCS, vol. 6145, pp. 355–364. Springer, Heidelberg (2010)

18. Zheng, S., Janecek, A., Tan, Y.: Enhanced fireworks algorithm. In: IEEE Congress on Evolutionary Computation (CEC), pp. 2069–2077 (2013)

19. Hsu, C.W., Chang, C.C., Lin, C.J.: A practical guide to support vector classification. Technical report, National Taiwan University (2010)

20. Chang, C.C., Lin, C.J.: LIBSVM: a library for support vector machines. ACM Trans. Intell. Syst. Technol. **2**, 27:1–27:27 (2011)

A Modified Fireworks Algorithm for the Multi-resource Range Scheduling Problem

Zhenbao Liu[1,2], Zuren Feng[1,2], and Liangjun Ke[1,2(✉)]

[1] Systems Engineering Institute, Xi'an Jiaotong University, Xi'an, China
kelj163@163.com
[2] State Key Laboratory for Manufacturing Systems Engineering,
Xi'an Jiaotong University, Xi'an, China

Abstract. In this paper, we describe a modified fireworks algorithm (MFWA) for the multi-resource range scheduling problem which is a highly constrained combinatorial optimization problem. Fireworks algorithm (FWA) is a meta-heuristic method inspired by fireworks explosion at night. The basic components of FWA consist of a local search phase and a selection phase. In the local search phase, explosion sparks are generated with genetic strategy, and Gaussian sparks are produced through interchange operator. In the selection phase, a disparity metric is introduced so as to select representative solutions for the next generation. The experimental results demonstrate this MFWA is more competitive than the original FWA as well as the other two commonly used methods.

Keywords: Range scheduling · FWA · Crossover operator · Disparity metric

1 Introduction

Multi-resource range scheduling problem (MRRSP) is a kind of large-scale combinatorial optimization problem which involves assigning a number of resources to tasks during a scheduling horizon. The objective of the scheduling is to maximize the scheduled task number.

Traditional optimization techniques are unsuitable for MRRSP since it is a NP-complete problem. Many researchers have turned their attentions to the intelligent optimization field. A constraint-based iterative-repair (IR) method was proposed for the GERRY scheduling and rescheduling system [1]. Several algorithms on satellite range scheduling problem were compared in [2]. An genetic method, called Genitor, was presented, and it performed very well on a broad range of instances studied in the paper in [3]. A guidance-solution based ant colony optimization was proposed to avoid premature convergence with the aim of minimizing the working span [4]. An invariant ant colony optimization with two-stage pheromone updating was presented, and it was applied to solve

© Springer International Publishing Switzerland 2016
Y. Tan et al. (Eds.): ICSI 2016, Part I, LNCS 9712, pp. 535–543, 2016.
DOI: 10.1007/978-3-319-41000-5_53

this problem [5]. The original fireworks algorithm was applied to MRRSP in [6], and different neighborhoods are analyzed.

Fireworks algorithm is inspired by fireworks explosion at night and is quite effective at finding global optimal value for continuous optimization problems [7]. The applications of FWA include: non-negative matrix factorization [8], image identification [9] etc. Nevertheless, the neighborhood structure and disparity metric in combinatorial space are difficult to define compared with continuous space. In this paper, neighborhood is constructed by crossover operator and interchange operator, and how to measure the distance between two permutations is also introduced.

The rest of this paper is organized as follows. In Sect. 2, the optimization model is established, and FWA is introduced. In Sect. 3, neighborhood construction strategy and disparity metric are presented. Experiment results are given in Sect. 4. The conclusions of this paper are summarized finally.

2   Preliminaries

The LEO satellites need to contact with ground stations several times each day to transmit data or receive instructions, and these communications are established through visible arcs generated when satellites fly over ground stations.

2.1   Optimization Model

Scheduling Object. In this paper, we define visible arcs as the scheduling object which include all the information needed in a scheduling. An visible arc a_i can be depicted as a five-tuple, $a_i = \{sa_i, st_i, ts_i, te_i, j_i\}$, where sa_i is its satellite; st_i is the ground station; ts_i is the start time of a_i and te_i is the end time of a_i; its corresponding task is j_i. It is worth reminding that different arcs may correspond to the same task.

In order to indicate whether a_i is utilized, a decision variable x_i is introduced:

$$x_i = \begin{cases} 1 \text{ if } a_i \text{ is utilized} \\ 0 \text{ otherwise} \end{cases} \tag{1}$$

A scheduling result is depicted as a state of arcs which is stored in a decision vector x, denoted as $x = [x_1, x_2, \ldots, x_n]$, and n is the number of arcs.

Constraint Analysis. For each task, it should be accomplished between its release time and due time with the relevant satellite and ground station. On top of these, some intrinsic constraints should be satisfied as well.

Antenna unit capacity constraint. An antenna can only serve one satellite until the task is finished. The decision vector satisfying this constraint is stored in set C_1, and

$$C_1 = \{x | \forall a_i, a_j, x_i + x_j \leq 1 \ if \ \{st_i = st_j\} \wedge \{[ts_i, te_i] \cap [ts_j, te_j] \neq \varnothing\} \tag{2}$$

Satellite unit capacity constraint. If a satellite has a possibility to communicate with several different stations, it can exchange data with only one of them. The decision vector satisfying this constraint is stored in set C_2, and

$$C_2 = \{x| \ \forall \ a_i, a_j, \ x_i + x_j \leq 1 \ if \ \{sa_i = sa_j\} \wedge \{[ts_i, te_i] \cap [ts_j, te_j] \neq \varnothing\} \quad (3)$$

Model Construction. The objective of the scheduling is to maximize the scheduled task number under the constraints aforementioned. Since a scheduling result is described as a state vector x, the formulation of MRRSP is given as follows:

$$max \ f(x) = \sum_{i=1}^{n} x_i$$

$$s.t. \ x = [x_1, x_2, \ldots, x_n]$$

$$\forall x_i \in x, \ x_i \in \{0, 1\}$$

$$x \in C_1 \cap C_2$$

where $f(x)$ is the fitness value and it stands for the scheduled tasks.

2.2   Fireworks Algorithm

Fireworks algorithm was proposed for global optimization of complex functions in 2010 [7]. The basic components of FWA includes three parts: initialization, local search and selection, which can be described as follows:

Initialization. N solutions are selected randomly from the search space as fireworks are let off in different positions overhead.

Local Search. FWA contains two kinds of fireworks: Good firework and bad firework. Good firework can generate a big population of sparks within a small range while bad firework generates a small population of sparks within a big range. In FWA, sparks and range correspond to the number of sampled solutions and the distance from the central solution respectively.

For a minimization problem, the explosion radius and the number of explosion sparks of each firework are calculated as follows:

$$A_i = \hat{A} \times \frac{f(x_i) - y_{min} + \varepsilon}{\sum_{j=1}^{N} (f(x_j) - y_{min}) + \varepsilon} \quad (4)$$

$$N_i = \hat{N} \times \frac{y_{max} - f(x_i) + \varepsilon}{\sum_{j=1}^{N} (y_{max} - f(x_j)) + \varepsilon} \quad (5)$$

where A_i is the explosion radius; $\hat{A}$ is the maximum radius; N_i is the number of sparks generated by the ith firework; $\hat{N}$ is the total number of explosion sparks; $f(x_i)$ is the fitness of x_i; $y_{max} = \max(f(x_i))$; $y_{min} = \min(f(x_i))$; ε is an extremely low number used to prevent the denominator from becoming zero.

There are two kinds of sparks been generated: explosion sparks and mutation sparks. Explosion sparks are responsible for the neighborhood search and mutation sparks are mainly intended to increase the diversity of the population.

For a d dimensional solution x_i, $z(z < d)$ dimensions are chosen randomly and the value of the kth dimension is revised as follows:

$$x_{ik} = x_{ik} + deviation \tag{6}$$

For explosion sparks, $deviation = A_i \times U(a, b)$, and $U(a, b)$ stands for a number sampled from $[a, b]$ uniformly. For mutation sparks, $deviation = A_i \times N(\mu, \sigma^2)$, and $N(\mu, \sigma^2)$ represents a number sampled from a Gaussian distribution with mean μ and variance σ^2.

Selection. FWA allocates its reproductive trials with the disparity between one solution and the others, and the best solution is reserved to the next generation due to the holding fittest individual strategy. The individuals with larger distance from others have a higher probability to be chosen regardless of its fitness. Supposing the candidate set is K, roulette wheel method is adopted to choose the rest $n - 1$ solutions. The probability of selecting a candidate solution x_i is given as follows:

$$p(x_i) = \frac{R(x_i)}{\sum\limits_{x_j \in K} R(x_j)} \tag{7}$$

and $R(x_i) = \sum\limits_{x_j \in K} ||x_i - x_j||$.

3   MFWA for MRRSP

In this section, a new neighborhood construction strategy is presented, and the disparity metric is introduced.

3.1   Local Search

The reason that Genitor achieves satisfying results can be attributed to its neighborhood construction strategy. This strategy is also adopted in MFWA, and the mechanism to generate explosion sparks is modified.

Crossover Operator. Two solutions are chosen from the solution space, recorded as Parent 1 and Parent 2, and the new solution is denoted as Child. The crossover operator selects K random positions in Parent 1 firstly. Then, the corresponding elements in Parent 2 are located. Elements in Parent 1 that do not belong to the selected elements are passed to Child directly with the same positions. Finally, the free slots are assigned to the located elements in Parent 2 in the same order.

An illustration is given to show the acting mechanism of the crossover operator in Fig. 1. In this example, $k = 3$. The selected elements in Parent 1 are 5, 7, 2 respectively, and the other elements are passed to Child directly. The free slots are assigned to 2, 7, 5, which is the same order they appear in Parent 2.

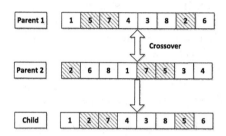

Fig. 1. The mechanism of crossover operator

Explosion Sparks. In the original FWA, for a certain solution S_i, when its spark number and explosion radius are determined, all solutions are generated in its neighborhood and no other solution is involved. However, for the crossover operator, a child solution is generated by two parents, which means, besides S_i, another solution should be involved.

Supposing S_{best} is the current best solution, and S_i is the current solution. Explosion sparks are generated as follows: if $S_i = S_{best}$, S_{best} is chosen as Parent 1, and a different solution is chosen randomly as Parent 2. If $S_i \neq S_{best}$, S_i is chosen as Parent 1, and S_{best} is chosen as Parent 2. Parameter k is equal to the explosion radius. So, when k elements are selected, a child solution will be obtained when crossover operator is applied to Parent 1 and Parent 2.

Mutation Sparks. Mutation sparks are used to maintain the diversity of solutions and enhance the local search ability. So, its acting mechanism should be quite different from crossover operator. In [6], three operators are presented, and interchange operator turns out to be the best. In this paper, mutation sparks are generated with interchange operator.

Supposing Π is the set of all permutations of the number $\{1, 2, \ldots, n\}$. π is one of the permutations, $\pi \in \Pi$. It can be written as follows:

$$\pi = \{\pi_1, \pi_2, \ldots, \pi_n\} \tag{8}$$

π_i is the ith element in π, and $\pi_i \neq \pi_j$ $(i \neq j)$.

The interchange operator chooses two elements π_i and π_j randomly, and their positions are exchanged. If $i > j$, we will get π' as follows:

$$\pi' = \{\pi_1, \ldots, \pi_{j-1}, \pi_i, \pi_{j+1}, \ldots, \pi_{i-1}, \pi_j, \pi_{i+1}, \ldots, \pi_n\} \tag{9}$$

3.2 Selection

FWA allocates its reproductive trials with the Euclidean distance between one solution and the others, however, this kind of metric is not defined explicitly in combinatorial problem. In [6], the distance is related to the scheduled tasks. For MRRSP, solutions differ little in scheduled task number. The fitness of a solution is closely related to the permutation of the arcs. So, the disparities of solutions are measured directly in this paper.

A good distance measure between two permutations is given by the number of different arcs [10]. Supposing $\pi = \{\pi_1, \pi_2, \ldots, \pi_n\}$, and $\sigma = \{\sigma_1, \sigma_2, \ldots, \sigma_n\}$. In order to calculate the distance, π and σ are transformed to the following:

$$\pi_P = \{(\pi_1, \pi_2), (\pi_2, \pi_3), \cdots, (\pi_{(n-2)}, \pi_{(n-1)}), (\pi_{(n-1)}, \pi_n)\} \tag{10}$$

$$\sigma_P = \{(\sigma_1, \sigma_2), (\sigma_2, \sigma_3), \cdots, (\sigma_{(n-2)}, \sigma_{(n-1)}), (\sigma_{(n-1)}, \sigma_n)\} \tag{11}$$

The distance between π and σ are calculated as follows:

$$D(\pi, \sigma) = n - |\pi_P \bigcap \sigma_P| \tag{12}$$

4   Experimental Results and Analysis

In this section, the relevant results are presented. The algorithms were compiled in MATLAB R2012a, the experimental results were obtained on a PC with an Intel core i7-3770 3.4 GHz processor and 8.0 GB of memory.

4.1   The Problem Instances

Four instances are tested. The basic information is given in Table 1. The satellites, ground stations and scheduled tasks are given by Xi'an Satellite Control Center. The visible arcs are calculated by the Satellite Tool Kit (STK)[1], and scheduling horizon is a week.

4.2   Comparison on FWA and MFWA

To study the performance of MFWA, A comparison between FWA and MFWA is carried out. Four instances are tested, and the result of instance 3 is presented. The maximum iteration number is set to 100, and each algorithm is run 20 times. The mean value of each iteration is presented in Fig. 2 (the results on the other three instances are similar to this one).

From two curves in the graph, we can see the convergence speed of MFWA is similar to FWA, but its solution quality is better, which means the combination of crossover operator and interchange operator is much more efficient than that of swap, insert and interchange operator [6].

---

[1] http://www.agi.com/products.

Table 1. The basic information of the scheduling

Instances	Stations num.	Satellites num.	Tasks num.	Arcs num.
Instance1	3	16	532	913
Instance2	3	16	548	917
Instance3	3	24	801	1274
Instance4	3	24	764	1281

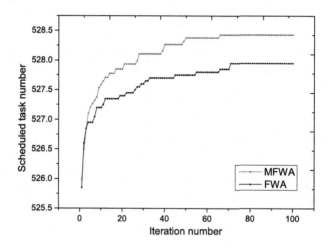

Fig. 2. Comparison between FWA and MFWA (Color figure online)

4.3 Computational Results and Comparisons

MFWA is applied to four instances aforementioned. Since the algorithmic mechanisms are quite different, it is meaningless to set the same number of iterations to compare the performances of algorithms. In this paper, the runtime in each experiment is set to a constant value, and each algorithm is run 20 times for different instances independently.

In this paper, time limit is set to 90 s. Three conclusions can be drawn from Table 2. Firstly, IR is less competitive compared with the other three algorithms,

Table 2. Comparison on four instances in 90 s

Instance	IR		Genitor		FWA		**MFWA**	
	best	mean	best	mean	best	mean	best	mean
Instance1	524	521.37	528	526.91	528	527.48	**529**	**528.16**
Instance2	538	534.50	**540**	538.49	539	538.56	**540**	**539.02**
Instance3	770	759.80	**790**	787.37	**790**	789.45	**790**	**789.77**
Instance4	729	718.82	747	744.10	**754**	750.01	**754**	**751.97**

especially on large scale instances. Secondly, although MFWA is some kind of combination of Genitor and FWA, it is better than Genitor and FWA. Thirdly, MFWA always produce the best solution and solutions with better average values.

5   Conclusion

In this paper, a modified fireworks algorithm is presented for MRRSP. Learning from the neighborhood structure of Genitor algorithm, The neighborhood structure of MFWA is generated with the current solution and the current best solution, which is different from the original FWA. Since the distance in scheduled tasks can not reflect the difference between the permutations of arcs directly, a new kind of disparity metric is introduced. Compared with the other three intelligent algorithms, MFWA is more competitive.

Acknowledgments. This work was supported by the Fundamental Research Funds for the Central Universities, the Open Research Fund of the State Key Laboratory of Astronautic Dynamics under Grant 2014ADL-DW402, the Scientific Research Foundation for the Returned Overseas Chinese Scholars, State Education Ministry, and State Key Laboratory of Intelligent Control and Decision of Complex Systems. We are also thankful to the anonymous referees.

References

1. Zweben, M., Davis, E., Daun, B., Deale, M.J.: Scheduling and rescheduling with iterative repair. IEEE Trans. Syst. Man Cybern. **23**(6), 1588–1596 (1993)
2. Barbulescu, L., Howe, A.E., Watson, J.-P., Whitley, L.D.: Satellite range scheduling: a comparison of genetic, heuristic and local search. In: Guervós, J.J.M., Adamidis, P.A., Beyer, H.-G., Fernández-Villacañas, J.-L., Schwefel, H.-P. (eds.) PPSN 2002. LNCS, vol. 2439, p. 611. Springer, Heidelberg (2002)
3. Barbulescu, L., Watson, J.P., Whitley, L.D., Howe, A.E.: Scheduling spaceCground communications for the air force satellite control network. J. Sched. **7**(1), 7–34 (2004)
4. Zhang, N., Feng, Z.R., Ke, L.J.: Guidance-solution based ant colony optimization for satellite control resource scheduling problem. Appl. Intell. **35**(3), 436–444 (2011)
5. Zhang, Z., Zhang, N., Feng, Z.: Multi-satellite control resource scheduling based on ant colony optimization. Expert Syst. Appl. **41**(6), 2816–2823 (2014)
6. Liu, Z., Feng, Z., Ke, L.: Fireworks algorithm for the multi-satellite control resource scheduling problem. In: 2015 IEEE Congress on Evolutionary Computation (CEC), pp. 1280–1286. IEEE, May 2015
7. Tan, Y., Zhu, Y.: Fireworks algorithm for optimization. In: Tan, Y., Shi, Y., Tan, K.C. (eds.) ICSI 2010, Part I. LNCS, vol. 6145, pp. 355–364. Springer, Heidelberg (2010)
8. Janecek, A., Tan, Y.: Swarm intelligence for non-negative matrix factorization. Recent Algorithms and Applications in Swarm Intelligence Research, 168 (2012)

9. Zheng, S., Tan, Y.: A unified distance measure scheme for orientation coding in identification. In: 2013 International Conference on Information Science and Technology (ICIST), pp. 979–985. IEEE, March 2013
10. Jones, T., Forrest, S.: Fitness distance correlation as a measure of problem difficulty for genetic algorithms. In: ICGA, vol. 95, pp. 184–192, July 1995

Discrete Fireworks Algorithm for Aircraft Mission Planning

Jun-Jie Xue[1]([✉]), Ying Wang[1], Hao Li[1,2], and Ji-yang Xiao[1]

[1] Air Force Engineering University, Xi'an 710051, China
{poot-cupic-xue,sn.poison}@163.com, {6948570,905132738}@qq.com
[2] Air Force Early-Warning Academy, Wuhan 430019, China

Abstract. In order to get the optimal aircraft mission planning path, a new solution approach and mathematical formulation of aircraft mission planning is proposed. Firstly, generalized distance is provided to convert aircraft mission planning problem into travelling salesman problem. Secondly, discrete fireworks algorithm (DFW) is proposed to obtain optimal mission path efficiently. At last, 3 typical benchmarks are simulated by DFW and contrast algorithms. Results show that new solution approach with DFW is effective and efficient to aircraft mission planning.

Keywords: Mission planning · Discrete fireworks algorithm · TSP

1 Introduction

Aircraft mission planning refers to find time efficient flight path under the precondition of minimum possibility of being shot down by enemy fire [1]. And battlefield threat and terrain is the two key factors influencing the safety of aircraft flight [2]. Recently, more and more researchers pay attention to digital terrain simulation beyond multi-objective traveling salesman problem, which relies on weight coefficient of different objective [3]. In order to transform complex battlefield environment into data form [4], threat equivalent simulation is adopted in this paper. Within sets of target coordinates obtained earlier, mission planning is easily mapped to the TSP.

In order to improve computational efficiency of TSP, evolutionary algorithm is applied in the past decades, such as PSO (Particle Swarm Optimization) [5], GA (The Genetic Algorithm) [6], ACO (Ant Colony Optimization) [7], ABC (Artificial Bee Colony Algorithm) [8,9]. 2010, A novel algorithm named fireworks algorithm (FWA) is proposed [10], and then, enhanced [11] and applied to different types of problems. However, its application to control systems is scarce and rarely found in the published literature.

As a consequence, this paper focus on aircraft mission planning based on DFW. Mathematical formulation of aircraft mission planning is built based on the generalized distance. And DFW is provided to improve computational efficiency. Then, 3 typical benchmark symmetric TSPs are simulated to obtain optimal mission path by DFW, compared with PSO, GA, ACO and ABC. Results show that DFW is effective and easily implemented methods to aircraft mission planning.

© Springer International Publishing Switzerland 2016
Y. Tan et al. (Eds.): ICSI 2016, Part I, LNCS 9712, pp. 544–551, 2016.
DOI: 10.1007/978-3-319-41000-5_54

2 Description and Mathematical Formulation of Aircraft Mission Planning

Aircraft mission planning is to choose the optimal path which complete the task of each target safety and efficiently. The path must meet three necessary factors. Firstly, the aircraft must arrive every each target. Secondly, the optimal path must be safe and feasible to avoid being destroyed or being detected. Lastly, the flight path should be the minimal one. Traditional, aircraft mission planning is transformed into multi-objective TSP, but it is unsuitable to combine distance and battlefield threat in a changeless weight factor, because every target pair are difference with others.

2.1 Generalized Distance of Aircraft Mission Planning

When aircraft performs various tasks in the complex warfare environment, it may face the threat of enemy fire, such as surface-to-air missile and anti-aircraft artillery. Up to now, the radar is still the main equipment to detect and track aircraft at a distance, so the enemy's threats can be simplified as radar threat area. Aircraft encountered battlefield threat in the operational area, also can be regarded as a peak.

In order to fuse distance and battlefield threat of every point, we transform threat into mountain by using threat equivalent terrain simulation method. It is an equivalent processing that fuse complex environment of the threat and obstacle into mountain terrain. The enemy threats becomes special terrain, the scope and position of threat superimpose on the digital map. That is to say, the threat is equivalent to raise the range of terrain. Terrain obstacles and enemy threat in flying area are fused into integrated terrain information, and the threat avoidance equivalent to terrain avoidance. The formulation of integrated terrain information model can be presented as follows,

$$z(x,y) = h(x,y) + \sum_{j=1}^{N} h_j^{max}.$$
$$exp\left\{ -\left[k_j^x \cdot \frac{(x - x_j^{max})}{x_j^{max}} \right]^2 - \left[k_j^y \cdot \frac{(y - y_j^{max})}{y_j^{max}} \right]^2 \right\}, \tag{1}$$

where x and y are the abscissa and ordinate of aircraft, $h(x,y)$ is the initial terrain height of position (x,y), N is the number of equivalent threat or the number of peaks, h_j^{max} is the peak height of mountain $j(j = 1, 2, ..., N)$, k_j^x and k_j^y are the slope coefficient of mountain j in the x axis and y axis, x_j^{max} and y_j^{max} are the the the abscissa and ordinate of mountain j peaks respectively.

Obviously, the aircraft navigation distance is a function of terrain height. The higher integrated terrain elevation lead to longer aircraft flight distance, and the distance between target A and target B should be calculated as,

$$d_0(A, B) = \int\limits_{path(A,B)} [z(x, y) - h_0] \, dL, \tag{2}$$

where h_0 is the benchmark height of operational area, and $d_0(A, B)$ is defined as $+\infty$ if it is too dangerous to across. Then, generalized distance between every two targets can be calculated and make up the target pair distance matrix D,

$$D = \begin{bmatrix} d_{11} & d_{12} & \cdots & R_{1N} \\ d_{21} & \cdots & \cdots & d_{2N} \\ \vdots & \vdots & \ddots & \ddots \\ R_{N1} & R_{N2} & \cdots & R_{NN} \end{bmatrix}. \tag{3}$$

2.2 Mathematical Formulation of Aircraft Mission Planning

With the definition of the generalized distance, aircraft mission planning is reduced to classical travelling salesman problem. In order to model Aircraft Mission Planning with TSP, complete edge-weighted graph $G = (V, E)$ is used, in which V is a set of $m = |V|$ nodes representing cities or targets and $E \subseteq V \times V$ is a set of weighted edges representing distance of node pairs. The goal of TSP is to find the shortest length in Hamiltonian Circuit of graph, where Hamiltonian Circuit is a closed path which achieve every nodes in G one and only one time. Then, the solution route is the permutation $X = \{x_1, x_2, \ldots, x_m\}$ in which x_i is a code number of the $i - th$ city that travelling man achieve, and every code number x_i is different from the others. The optimal solution route is permutation X which get minimal length $f(X)$,

$$f(X) = \sum_{i=1}^{m} d(x_i, x_j) + d(x_m, x_1). \tag{4}$$

Edges between two cities are unoriented and aircraft mission planning is symmetric TSP, that it to say, $d(x_i, x_j) = d(x_j, x_i)$.

3 Discrete Fireworks Algorithm for Aircraft Mission Planning

3.1 Design of Discrete Fireworks Algorithm

Discrete fireworks space is make up by n firework devices and m-dimensional cities permutation space(arguments space of optimization function), location of element(fireworks, sparks or mutation sparks) i is $X_i = (x_i^1, x_i^2, x_i^m)$ and gorgeous degree (objective function value or feasible route length) of element i is $Y_i = f(X_i)$. Fireworks algorithm is constituted by one distance (permutation distance), two operators (reverse operator and mutation operator) and three steps (fireworks explosion, mutation explosion and selection of explosion locations).

In the beginning, 1st generation location of fireworks is generated randomly.

(1) Permutation Distance

There must be differences between each elements when solving TSP by fireworks algorithm. In order to describe the difference between each element, permutation distance is proposed to define the difference between elements. The permutation distance between x_i and x_j is defined as follows,

$$D(X_i, X_j) = \sum_{k=1}^{m} V_{i,j}^k, \tag{5}$$

where m is the dimensional of cities permutation, and $V_{i,j}^k$ is the diversity of the kth city permutation between permutation x_i and permutation x_j which is expressed as follows,

$$V_{i,j}^k = \begin{cases} 1 & \text{if } x_i^k \neq x_j^k \\ 0 & \text{if } x_i^k = x_j^k \end{cases}. \tag{6}$$

(2)Reverse Operator of Fireworks Algorithm

Setting off fireworks produced a large number of sparks, it is equivalent to do the searching process of sparks location around fireworks. In order to illustrate the searching process of fireworks explosion, reverse operator is introduced and its procedure is described as follows.

Step 1. Generate the reverse point integer r_1 from 1 to m randomly.

Step 2. Generate the reverse interval length l_1 from 1 to A_i randomly.

Step 3. Calculate the downward reverse point $r_d = r_1 - l_1$ and upward reverse point $r_u = r_1 + l_1$. There are three kinds of different situations for setting two reverse points. Firstly, set max_r as r_1 and min_r as r_d if $r_u > m$. Secondly, set max_r as r_u and min_r as r_1 if $r_d < 1$. Lastly, if $1 < r_d < r_u < m$, generate the other reverse point integer r_2 from r_d and r_u randomly. And then set max_r as the larger one between r_1 and r_2, min_r as the smaller one between r_1 and r_2.

Step 4. Reverse the permutation order in location of firework from point min_r to point max_r.

Notes reverse operator as $R(X, N, A)$, which means that generating N sparks from the fireworks X and the reverse interval length is less than explosion amplitude A.

(3)Mutation Operator of Fireworks Algorithm

In order to improve the global searching ability, mutation operator is applied to mutate the current location of elements to discover new fireworks setting off spot (permutation of solution routes). Mutation operator of DFW reorder all the code number of cities in mutation space randomly.

(4)Fireworks Explosion

Sparks generated from the firework is viewed as searching process around specific point. High-quality firework generates numerous sparks and the sparks centralize the explosion center. However, low-quality firework generates quite few sparks and the sparks scatter in the space. Number of sparks generated by firework i is expressed as follows,

$$N_i = round(N_c \cdot \frac{f_{max} - f(X_i) + \varepsilon}{\sum_{i=1}^{N}(y_{max} - f(X_i)) + \varepsilon}), \tag{7}$$

where N_c is number of total sparks, f_{max} is the maximum function value among the n fireworks and ε is the minimum spark distance that can be resolved. N_i is instead by N_{max} if s_i is bigger than N_{max}, because firework with so high-quality can not be manufactured. N_i is instead by N_{min} if N_i is smaller than N_{min}, because it is substandard firework. The amplitude of firework explosion is an integer which is less than or equal to m and more than or equal to 1. The amplitude of firework explosion by firework i is expressed as follows,

$$A_i = ceil \left[A_c \cdot \frac{f(X_i) - y_{min} + \varepsilon}{\sum_{i=1}^{N}(f(X_i) - y_{min}) + \varepsilon} \right], \tag{8}$$

where $ceil(\cdot)$ is a ceiling function, A_c is constant of the maximum explosion amplitude and f_{min} is the worst function value among the n fireworks.

Then firework i is set off, and then generate N_i sparks by $R(X_i, N_i, A_i)$.

(5)Mutation Explosion

Mutation explosion is different from fireworks explosion, because mutation explosion generates mutation sparks which is able to escape from original location to keep the diversity of sparks. Location of mutation spark is related with the location of mutation explosion firework, and escapes from local extremum area as much as possible. So we define the mutation explosion space $M = \{M_1, M_2, \ldots, M_m\}$ of firework i as follows,

$$M_k = \begin{cases} k & \text{if } r_i^k < g_i^k \\ null & \text{if } r_i^k \geq g_i^k \end{cases}, \tag{9}$$

where r_i^k is a random number, g_i^k is a gaussian random number generated by $G(1, 1)$.

(6) Selection of Explosion Locations

After fireworks explosion and mutation explosion, locations of fireworks, sparks and mutation sparks are obtained. At the same time, the gorgeous degree (objective function value or solution route length) are calculated too. Then, n locations are selected to set off fireworks in next explosion generation. The element with best gorgeous degree is kept to next generation and other $n - 1$ locations are selected based on their distance to other elements so as to keep diversity. Distance can be measured by permutation distance. The probability of whether the element i is selected to next explosion generation is expressed as follows,

$$p(i) = \frac{\sum_{j \in K} D(X_i, X_j)}{\sum_{j \in K}(\sum_{i \in K} D(X_i, X_j))}, \tag{10}$$

where K is the set of all current locations of elements.

3.2   Framework of Discrete Fireworks Algorithm

In order to obtain the optimal aircraft mission planning, it is important to get optimal path planning based on generalized distance of aircraft mission planning. Firstly, Setting off fireworks (sets of path permutation) and generating sparks (sets of new path permutation) and mutation sparks. Secondly, checking out gorgeous degree of all elements (fireworks, sparks and mutation sparks). After that, selecting n location to be spots of next explosion generation. Model of DFW for aircraft mission planning is depicted as Fig. 1.

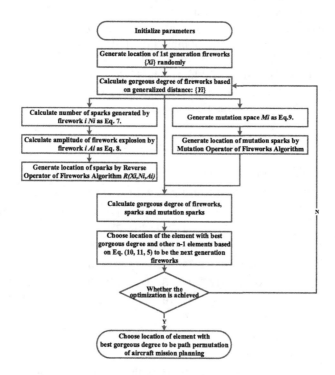

Fig. 1. Flow Chart of Discrete Fireworks Algorithm for TSP

4   Simulations and Analysis

Due to there is no benchmark data library for aircraft mission planning, three benchmark symmetric TSPs (dantzig42.tsp, st70.tsp and gr96.tsp.) from TSPLIB (2012) are used for testing, instead of the aircraft mission planning. At the same time, four distinguished algorithms (PSO, GA, ACO, ABC) for TSP are used for comparison with DFW.

In order to fully compare the performance of different evolutionary algorithms for aircraft mission planning, it is essential to setting up similar simulation environment as soon as possible. Simulations are carried out 40 times per benchmark in MATLAB R2014a on Intel(R) Core(TM) i5-4200U CPU @2.30 GHz under Windows 8.1 environment.

Parameters of contrast algorithms with better performance are chosen to test effectiveness of DFW, and the corresponding parameters are recommended as follows. In PSO, population size $NP_P = 300$ and the maximum iterations $NI_P = 3000$. In GA, $NP_G = 50$, $NI_G = 2000$, selection probability $p_s = 0.9$, crossover probability $p_c = 0.9$, and mutation probability $p_m = 0.05$. In ACO, $NP_A = 50$, $NI_A = 2000$, pheromone trail coefficient $\alpha = 1$, heuristic value coefficient $\beta = 5$, and pheromone trail decay coefficient $\rho = 0.3$. In DFW, population size and the maximum iterations are bote chosen as the minimal as comparison algorithm setting, $NP_D = 50$, $NI_D = 2000$.

Simulating the aircraft mission planning on TSP model by discrete fireworks algorithm with three benchmark symmetric TSPs respectively. Results of Best, Avg., Worst, SD are given as Table. 1, where Best, Avg., Worst, SD are the minimum, average, maximum, and standard deviation of results path length obtained in 40 independent runs respectively.

Table 1. Statistical results

Function	SPLP	Index	PSO	GA	ACO	ABC	DFA
dantzig42	679.2019	Best	679.2019	679.2019	703.8294	679.2019	679.2019
		Avg	699.8715	715.8312	724.6758	695.3776	686.5372
		Worst	722.9854	762.0302	739.5636	715.7774	708.8364
		SD	13.1413	22.0551	10.5018	12.1077	10.2674
st70	678.5975	Best	687.1945	692.4504	699.2357	692.3274	687.1945
		Avg	717.7294	732.0563	710.3917	712.0920	703.9716
		Worst	766.7405	767.2600	715.4182	739.3253	713.1831
		SD	22.7000	21.3766	4.7699	15.0072	8.3675
gr96	512.3094	Best	551.4460	529.3467	539.4167	530.9624	532.4761
		Avg	574.3774	558.2334	546.5933	544.2715	555.4712
		Worst	608.7520	583.1796	552.5655	641.5035	552.5655
		SD	15.5858	16.6326	3.5920	23.6072	16.2541

The average path length describes the quality of the algorithm, SD means the stability of the algorithm convergence. It is shown that discrete fireworks algorithm almost has the best performance of 4 different performance index (Best, Avg., Worst and SD) with typical benchmark symmetric TSPs st70.tsp and dantzig42.tsp, or has little difference with the best performance index but much better than the other evolutionary algorithm with benchmark gr96.tsp according to Table 2. Results show that discrete fireworks algorithm is able to solve TSP much better than PSO, GA, ACO and ABC with greater accuracy and stability. That also means that discrete fireworks algorithm is efficient for the aircraft mission planning problem.

5 Conclusion

In this paper, generalized distance of aircraft mission planning is proposed to translate aircraft mission planning into typical TSP by simulating battlefield threat and terrain peaks in a equivalent distance simulation graph. In order to get high-quality performances aircraft mission planning path, DFW, which is constituted by one distance (permutation distance), two operators (reverse operator and mutation operator) and three steps (fireworks explosion, mutation explosion and selection of explosion locations), is built to optimize aircraft mission planning path. And then3 benchmark symmetric TSPs (dantzig42.tsp, st70.tsp and gr96.tsp.) from TSPLIB (2012) are used for testing. The simulation results show that DFW is effective and easily implemented methods to solve TSP prsoblems and aircraft mission planning.

Acknowledgments. This work is supported by National Natural Science Foundation of China under Grant No. 71171199 and No. 61472443.

References

1. Nikolos, I.K., Valavanis, K.P., et al.: Evolutionary algorithm based offline/online path planner for UAV navigation. IEEE Trans. Syst. Man Cybern. **6**, 898–912 (2003)
2. Shima, T., Rasmussen, S.: UAV Cooperative Decision and Control: Challenges and Practical Approaches (Advances in Design and Control). Society for Industrial and Applied Mathematics, Philadelphia (2008)
3. Cela, E., Deineko, V., Woeginger, G.J.: The x-and-y-axes travelling salesman problem. Eur. J. Oper. Res. **223**, 333–345 (2012)
4. Zhong-hua, H., Min, Z., Min, Y.: Multi-objective and multi-constrained UAV path plan optimum selection based on GRA. J. Grey Syst. **23**, 35–46 (2011)
5. Shi, X.H., Liang, Y.C., Lee, H.P., Lu, C., Wang, Q.X.: Particle swarm optimization-based algorithms for TSP and generalized TSP. Inf. Process. Lett. **103**, 169–176 (2007)
6. Contreras-Bolton, C., Parada, V.: Automatic combination of operators in a genetic algorithm to solve the traveling salesman problem. PLoS ONE **10**, 1–25 (2015)
7. Jianyi, Y., Ruifeng, D., et al.: An improved ant colony optimization (I-ACO) method for the quasi travelling salesman problem (Quasi-TSP). Int. J. Geogr. Inf. Sci. **29**, 1534–1551 (2015)
8. Szeto, W., Wu, Y., Ho, S.C.: An artificial bee colony algorithm for the capacitated vehicle routing problem. Eur. J. Oper. Res. **215**, 126–135 (2011)
9. Zutong, W., Jiansheng, G., Mingfa, Z., Ying, W.: Uncertain multiobjective travelling salesman problem. Eur. J. Oper. Res. **241**, 478–489 (2014)
10. Tan, Y., Zhu, Y.: Fireworks algorithm for optimization. In: Tan, Y., Shi, Y., Tan, K.C. (eds.) ICSI 2010, Part I. LNCS, vol. 6145, pp. 355–364. Springer, Heidelberg (2010)
11. Zheng, S., Janecek, A., Tan, Y.: Enhanced fireworks algorithm. In: 2013 IEEE Congress on Evolutionary Computation, pp. 2069–2077 (2013)

Multi-Objective Optimization

Multi-objective Reconfiguration of Power Distribution System Using an ILS Approach

Abdelkader Dekdouk[1], Hiba Yahyaoui[1(✉)], Saoussen Krichen[2], and Abderezak Touzene[3]

[1] Computer Science Department, College of Arts and Applied Sciences, Dhofar University, Salalah, Oman
hiba.yahyawi@gmail.com
[2] LARODEC, Institut Superieur de Gestion, University of Tunis, Tunis, Tunisia
[3] Department of Computer Science, College of Science Oman, SQU University, Muscat, Oman

Abstract. In this paper, we address a distribution network reconfiguration problem (DNRP) that operates on standard configurations of electrical networks. The main objectives handled by the DNRP are the minimization of power loss, the minimization of the number of switching operations and the minimization of the deviations of bus voltages from their rated values. Due to its multiobjective nature and combinatorial aspects, the DNRP is considered as **NP**-hard. Hence approximate approaches are very promising in generating high quality solutions within a concurrent run time. In this paper, we develop a distribution network reconfiguration approach using an iterated local search (ILS) algorithm, known to be a powerful stochastic local search method. This has been investigated and illustrated on an IEEE 33-bus and IEEE 69-bus radial distribution systems. Indeed, we proposed a novel solution encoding that avoids in a smooth and natural way the creation of isolated components and closed loops, in each generated network configuration.

Keywords: Optimization · Iterated local search · Distribution network reconfiguration problem · Reconfiguration · Artificial intelligence

1 Introduction

Power system distribution is a challenging problem that manages the delivery of energy to final customers in an optimal way. It consists in scheduling the ordering plan while expecting appropriately customers' demands. Power system distribution delivers power utility to numerous final consumers as residential, industrial and commercial customers. Hence, with the recent challenges of power production and consumption, a reliable and economic customer satisfaction is becoming more and more crucial in a modern electric power grid. The continuous increasing of power demand and the high load density in the urban areas make the power distribution management more complex, particularly when operating on

© Springer International Publishing Switzerland 2016
Y. Tan et al. (Eds.): ICSI 2016, Part I, LNCS 9712, pp. 555–563, 2016.
DOI: 10.1007/978-3-319-41000-5_55

the existing power grid. In Power systems, taking into consideration cost efficiency and reliability factors, the power network topology has a significant importance. Thus, it is usually designed as a radial distribution network (RDN). The reconfiguration of a radial distribution system is the mechanism of altering the switch (open/close) operations in order to optimize several objectives (minimize the power line losses, improve the voltage stability profile, well manage the load congestion and enhance system reliability) while maintaining the radial topology of the grid. Over the last, decades, numerous researchers investigated losses minimization in power systems. Merlin and Back [1] studied the minimization of feeder loss for distribution network reconfiguration. The problem therein was formulated as a mixed integer nonlinear optimization problem and solved using a discrete branch-and-bound technique. Imran and Kowsalya [7] proposed a metaheuristic approach termed "Fireworks Algorithm" to solve the network reconfiguration while minimizing power loss and voltage profile of the distribution system. In [7] the proposed algorithm is a recently developed swarm intelligence based optimization metaheuristic which is conceptualized using the fireworks explosion process of searching a best location of sparks. Niknam [10] presented a multiobjective honey bee mating optimization metaheuristic for distributing feeder reconfiguration. To do so, several queens were utilised and considered as an external repository to save non-dominated solutions. Besides, fuzzy clustering techniques were used to control the size of the repository within the limits. Gupta et al. [6] proposed a multi-objective reconfiguration of radial distribution systems in a fuzzy framework using an adaptive genetic algorithm, the genetic operators are designed with the help of graph theory to generate feasible individuals. Andervazh et al. [2] proposed a Pareto-based multi-objective distribution network reconfiguration method using a discrete particle swarm optimisation approach. Probabilistic heuristics and graph theory techniques are employed in [2] to improve the stochastic random search of the algorithm. Nguyen and Truon [9] proposed a reconfiguration methodology based on a cuckoo search algorithm (CSA) to minimize active power loss and the maximize voltage magnitude.

This paper proposes an adaptive Iterated Local Search (Adapt-ILS) that optimizes three objectives for the DNRC, namely the minimisation of power loss, the minimization of the number of switching operations and the minimization of the deviations of bus voltages from their rated values. The Adapt-ILS handles only feasible solutions, due to its encoding. Hence, all generated configurations are radial and there is no isolated nodes. A large stream of experiments shows that Adapt-ILS provides a set of high-quality potentially efficient solutions.

This paper is structured as follows. Section 2 states all related concepts, Sect. 3 enumerate all symbols used throughout this paper followed by their explanations. Section 4 outlines the adaptive ILS metaheuristic. This paper is enclosed with an experimental investigation of the problem and some concluding remarks.

2   Problem Description and Modeling

We state in what follows the mathematical modeling of the DNRC that optimizes concurrently 3 objectives while respecting power flow requirements.

2.1 Objectives

The reconfiguration of radial distribution systems is modeled as a three-objective optimization problem as follows:

Minimize the Power Loss. The minimization of the total power loss, that is determined by the summation of losses in all line sections, which is written as

$$Min \ z_1(x) = \sum_{i=1}^{n} x_i r_i |I_i|^2$$

Minimize the Number of Switch Operations. The second objective aims at minimizing the number of switching operations, the formulation of the objective function is the following [10]:

$$Min \ z_2(x) = \sum_{j=1}^{Ns} |S_j - S_{oj}|$$

Minimize the Deviation of Bus Voltage. The third objective is to minimize the maximum deviation of bus voltages, as stated in [10]:

$$Min \ z_3(x) = \max[(1 - V_{min}), (1 - V_{max})]$$

Where V_{min} and V_{max} are, respectively, the minimum and maximum values of bus voltages.

2.2 System Constraints: Power Flow Equations

Active and reactive power: $P_i = \sum_{i=1}^{N_{bus}} V_i V_j Y_{ij} \cos(\Theta_{ij} - \delta_i + \delta_j)$
$Q_i = \sum_{i=1}^{N_{bus}} V_i V_j Y_{ij} \sin(\Theta_{ij} - \delta_i + \delta_j)$
Radial configuration: $M = N_{bus} - N_f$
Where M is the number of branches, N_{bus} is the number of nodes and N_f is the number of sources.
Limit feeder capacity: $I_{max}.k_b \times |I_{ij}^b| \leq I_{ij,max}^b$
Limit bus voltage: $V_{max}.V_{min} \leq V_i \leq V_{max}$

3 The Adapt-ILS Approach for the DNRC Problem

The ILS algorithm is a challenging metaheuristic able to generate high quality solutions for numerous combinatorial optimization problems, especially minimum spanning tree problems. This neighborhood-based approach, specifically designed for solving the DNRC, outputs a revisited network configuration that fulfills all structural constraints. The Adapt-ILS approach starts by generating an initial feasible solution, as reported in figure [2], then applies multiple k-swap neighborhoods to identify the most promising solution that can be a potential candidate for next iterations. This guarantees that particles with radial topologies can be always generated. The effectiveness of the ILS is strongly correlated with the adaptation of the ILS features, mainly the neighborhood generation [5]. The quality of the generated results is highlighted in the experimental part of this paper.

3.1   Solution Encoding

The solution is coded as a set of binary vectors that express the belongingness of the edges to the loops vectors. Our encoding is reported in Fig. 1.

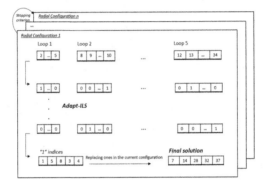

Fig. 1. Solution encoding of the ILS algorithm

3.2   Neighborhood Generation

As the search process evolves, newly explored solutions are generated in such a way to to overcome the infeasibility by applying the following graph theoretic rules:

- *Rule 1: All loop vectors should belong to the solution:* As the search process evolves, the swap moves operate on all loop vectors. Each loop vector

- *Rule 2: Only one edge can be selected from a common branch vector:* All switches of a common branch should belong to one vector. Hence, there is no case in which two switches from a common branch are opened.

- *Rule 3: All common branch vectors of any prohibited group vector cannot be involved simultaneously:* Only one switch from a group of branches is opened as we dispose of only one zero for each loop vector.

The neighborhood structure switches between k swaps according to the iteration counter. The loop vectors, common branches and prohibited groups are reported in Table 1 as follows:

Table 1. Vectors of the 33-bus system [2,6]

Loop vectors
$L_1 = \{2, 3, 4, 5, 6, 7, 33, 20, 19, 18\}$
$L_2 = \{33, 8, 9, 21, 35, 11, 10, 9\}$
$L_3 = \{11, 10, 9, 34, 12, 13, 14\}$
$L_4 = \{6, 7, 8, 34, 15, 16, 17, 25, 26, 27, 28, 29, 30, 31, 32, 36\}$
$L_5 = \{6, 7, 8, 34, 15, 16, 17, 25, 26, 27, 28, 29, 30, 31, 32, 36\}$
Common branch vectors
$L_{12} = \{33\}$, $L_{14} = \{3, 4, 5\}$, $L_{15} = \{6, 7\}$, $L_{23} = \{9, 10, 11\}$
$L_{25} = \{8\}$, $L_{35} = \{34\}$, $L_{45} = \{25, 26, 27, 28\}$
Prohibited group vectors
$R_5 = \{L_{14}, L_{15}, L_{45}\}$, $R_7 = \{L_{12}, L_{15}, L_{25}\}$,
$R_8 = \{L_{23}, L_{25}, L_{35}\}$, $R_{57} = \{L_{12}, L_{14}, L_{25}, L_{45}\}$,
$R_{78} = \{L_{12}, L_{15}, L_{23}, L_{35}\}$, $R_{572} = \{L_{12}, L_{14}, L_{23}, L_{35}, L_{45}\}$

3.3 Stopping Rule

The ILS algorithm stops processing after a predefined number of iterations.

4 Simulation and Results

In order to validate the Adapt-ILS, two radial distribution test feeders are applied: IEEE 33-bus and IEEE 69-bus systems. Technical data can be found in [4] and [28] We coded the proposed Adapt-ILS algorithm with MATLAB and we used MATPOWER.

MATPOWER [13] is an open-source Matlab-based power system simulation package that provides a high-level set of power flow and optimal power flow. Our experiments was executed on a personal computer with $Intel^{®}$ $Core^{TM}$ i7-4610M CPU @ 3.00 GHz 3.00 GHz 16 GB RAM and Windows 8.1 pro, 64-bit operating system, x64-based processor.

4.1 The IEEE 33-bus System

The IEEE 33-bus [3] system includes 32 buses, 37 branches, 5 looping branches, 5 tie switches and 32 sectionalizing switches and 1 main transformer. As shown

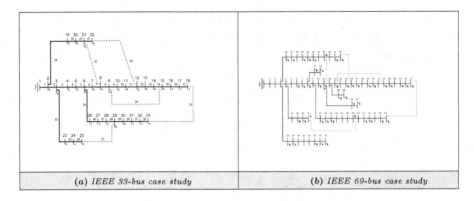

| (a) IEEE 33-bus case study | (b) IEEE 69-bus case study |

Fig. 2. A single line diagram of IEEE 33-bus system and IEEE 69-bus system

in Fig. 3, the five initially open switches (tie switches) are 33, 34, 35, 36 and 37 (dotted lines) and the normally closed switches are 1–32 (solid lines). When all tie switches are open, the power loss is 202.67 kW and minimum voltage per unit is 0.913. Table 2 reports the empirical results obtained by the proposed Adapt-ILS. To validate the presented results, it have been compared with other works from the literature for each objective function. As shown in Table 2 the Adapt-ILS is capable of finding the optimum solutions for each objective function. It converges to the global optimum configuration. Considering the third objective function, the result for minimizing the number of switching operations is the initial configuration. In this case there is no change in the status of all branches (the ties are opened and the sectionalizing switches are closed). The system voltage profile and the power loss at each line are reported in Fig. 2.

Table 2. Comparison of simulation results for 33-node network.

Approach	Open switches	$F1$: Power Loss (KW)	$F2$: Number of switches	$F3$: Voltage deviation ($p.u$)	Minimum voltage
Base case	33-34-35-36-37	202.67	0	0.8691	0.087
Adapt-ILS (this paper)	7-9-14-32-37	138.87	8	0.06217	0.94235
	7-9-14-28-32	146.39	10	0.058754	0.941246
CSA (Nguyen et al. 2015 [9])	7-9-14-32-37	138.87	8	0.06217	0.94235
PSO (2015)	7-9-14-32-37	138.87	8	0.06217	0.94235
MHBMO (Niknam 2013 [10])	7-9-14-32-37	138.87	8	0.06217	0.94235
Heuri. (2011)	7-9-14-32-37	138.87	8	0.06217	0.94235
HSA	7-10-14-28-36	146.39	8	0.0664	0.9336
ITS	7-9-14-36-37	145.11	8	0.0664	0.9336

4.2   The IEEE 69-bus System

The IEEE 69-bus [8] system includes 68 buses, 73 branches, 5 looping branches, 5 tie switches and 68 sectionalizing switches and 1 main transformer. As shown

Table 3. Comparison of simulation results for 69-node network

Approach	Open swit.	F1: Power Loss (KW)	Red. %	F2: Numb. of swit.	F3: Volt. dev. ($p.u$)	Min. volt.
Base case	69-70-71-72-73	224.95	-	0	0.0908	0.9092
Adapt-ILS (this paper)	14-57-61-69-70	98.59517617076	0	6	0.0505	0.9495
	14-55-61-69-70	98.59517617023	5.2e-10	6	0.0505	0.9495
	14-58-61-69-70	98.59517617033	4.2e-10	6	0.0505	0.9495
CSA (Nguyen et al. 2015 [9])	14-57-61-69-70	98.595	0	6	0.0505	0.9495
HSA (2013)	13-18-56-61-69	99.35	-	8	-	0.9428

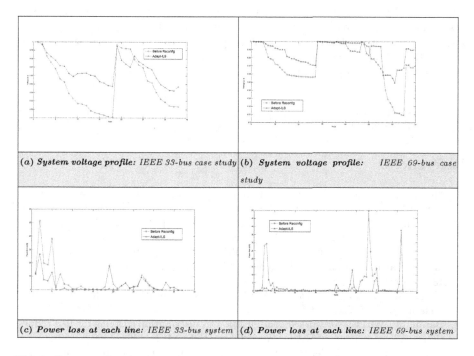

(a) *System voltage profile: IEEE 33-bus case study* (b) *System voltage profile: IEEE 69-bus case study*

(c) *Power loss at each line: IEEE 33-bus system* (d) *Power loss at each line: IEEE 69-bus system*

Fig. 3. Computational experiments in terms of IEEE 33-bus and IEEE 69-bus systems (Color figure online)

in Fig. 3, the five initially open switches (tie switches) are 69, 70, 71, 72 and 73 (dotted lines) and the normally closed switches are 1–68 (solid lines). When all tie switches are open, the power loss is 224.95 kW and minimum voltage per unit is 0.909. Table 3 reports the empirical results obtained by the proposed Adapt-ILS. To validate the presented results, it have been compared with other works from the literature for each objective function. As shown in Table 3 the Adapt-ILS is capable of finding the best solution in the literature. Moreover, it is observed that the Adapt-ILS outperforms other approaches in optimising the first objective function (power loss minimization). The system voltage profile and the power loss at each line are reported in Fig. 2.

5   Conclusion

This paper presents an Iterated Local Search algorithm (Adapt-ILS) to solve the problem of multi-objective power system reconfiguration. The investigated objectives are the minimization of power loss, the number of switching operations and voltage deviations. We proposed a novel solution encoding that maintains, smoothly and elegantly, the feasibility of solutions during the search process of power network reconfiguration. Simulation results illustrate that Adapt-ILS is highly efficient in solving multi-objective DNRP. We addressed in our experiments two standard power systems IEEE 33-bus and 69-bus. Considering the first network, compared to other related works, our approach confirms its performance while it converges to the global optimum solution. Besides that, in the case of IEEE 69-bus systems, we notice that Adapt-ILS outperforms the state-of-the-art approaches for the first objective which is power loss minimization.

Acknowledgments. This research was financially supported by the TRC research project "TFSDPG" ORG/DU/EI/14/020.

References

1. Merlin, A., Back, H.: Search for a minimal loss operating spanning tree configuration in an urban power distribution system. In: Proceeding of 5th Power System Computation Conference (PSCC), Cambridge (UK), pp. 1–18 (1975)
2. Andervazh, M.-R., Olamaei, J., Haghifam, M.-R.: Adaptive multi-objective distribution network reconfiguration using multi-objective discrete particles swarm optimisation algorithm and graph theory. IET Gener. Transm. Distrib. **7**(12), 1367–1382 (2013)
3. Venkatesh, B., Rakesh, R., Gooi, H.: Optimal reconfiguration of radial distribution systems to maximize loadability. IEEE Trans. Power Syst. **19**, 260–266 (2004)
4. Baran, M., Wu, F.: Network reconfiguration in distribution systems for loss reduction and load balancing. IEEE Trans. Power Deliv. **4**(2), 1401–1407 (1989)
5. den Besten, M., Stützle, T., Dorigo, M.: Design of iterated local search algorithms. In: Boers, E.J.W., Gottlieb, J., Lanzi, P.L., Smith, R.E., Cagnoni, S., Hart, E., Raidl, G.R., Tijink, H. (eds.) EvoWorkshop 2001. LNCS, vol. 2037, p. 441. Springer, Heidelberg (2001)
6. Gupta, N., Swarnkar, A., Niazi, K., Bansal, R.: Multi-objective reconfiguration of distribution systems using adaptive genetic algorithm in fuzzy framework. IEE Gener. Transm. Distrib **12**(4), 1288–1298 (2010)
7. Imran, A.M., Kowsalya, M.: A new power system reconfiguration scheme for power loss minimization and voltage profile enhancement using fireworks algorithm. Electr. Power Energy Syst. **62**, 312–322 (2014)
8. Baran, M., Wu, F.F.: Optimal sizing of capacitors placed on a radial distribution system. IEEE Trans. Power Syst. **4**, 735–743 (1989)
9. Nguyen, T.T., Truon, A.V.: Distribution network reconfiguration for power loss minimization and voltage profile improvement using cuckoo search algorithm. Electr. Power Energy Syst. **68**, 233–242 (2015)
10. Niknam, T.: An efficient multi-objective HBMO algorithm for distribution feeder reconfiguration. Expert Syst. Appl. **38**, 2878–2887 (2011)

11. Rajaram, R., Kumar, K.S., Rajasekar, N.: Power system reconfiguration in a radial distribution network for reducing losses and to improve voltage profile using modified plant growth simulation algorithm with distributed generation (dg). Energy Rep. **2015**, 116–122 (2015)

12. Zhu, J.Z.: Optimal reconfiguration of radial distribuion system using artificial intelligence methods. Electr. Power Syst. Res. **62**, 37–42 (2002)

13. Zimmerman, R., Murillo-Sánchez, C., Thomas, R.: Matpower: steadystate operations, planning and analysis tools for power systems research and education. IEEE Trans. Power Syst. **26**(1), 12–19 (2011)

Cooperative Co-evolutionary Algorithm for Dynamic Multi-objective Optimization Based on Environmental Variable Grouping

Biao Xu[1,2], Yong Zhang[1], Dunwei Gong[1(✉)], and Miao Rong[1]

[1] School of Information and Electrical Engineering,
China University of Mining and Technology, Xuzhou 221116, China
dwgong@vip.163.com
[2] School of Mathematics and Science,
Huaibei Normal University, Huaibei 235000, China

Abstract. This paper presents a cooperative co-evolutionary dynamic multi-objective optimization algorithm, i.e., DNSGAII-CO for solving DMOPs based on environmental variable grouping. In this algorithm, a new method of grouping decision variables is first presented, in which all the decision variables are divided into two subcomponents according to whether they are interrelated with or without environment parameters. Then, when cooperatively optimizing the two subcomponents by using two populations, two prediction methods, i.e., differential prediction and Cauchy mutation, are employed to initialize them, respectively. The proposed algorithm is applied to a benchmark DMOPs, and compared with two state-of-the-art algorithms. The experimental results demonstrate that the proposed algorithm outperforms the compared algorithms in terms of convergence and distribution.

Keywords: Dynamic multi-objective optimization · Cooperative co-evolution · Decision variable grouping

1 Introduction

A dynamic multi-objective optimization problem (DMOP) has multiple conflicting objectives and one or more constraints which vary with time, and can be regarded as several consecutive static optimization problems. The optimal solutions of this problem generally change over time. DMOPs have been widely existed in real-world applications, such as dynamic job shop scheduling [1,2,4,7], greenhouse control [11], dynamic airspace re-sectorization [9], vehicle motion planning [6,10]. So it is of considerable significance in theory and applications to research DMOPs [3]. Without loss of generality, a DMOP can be defined as follows [5].

$$\min F(X,t) = (F_1(X,t), F_2(X,t), \cdots, F_M(X,t))$$
$$s.t. \begin{cases} g_i(X,t) \leq 0, i = 1,2,\cdots,q; \\ h_j(X,t) = 0, j = 1,2,\cdots,r; \\ X \in [X_{\min}, X_{\max}] \end{cases} \tag{1}$$

© Springer International Publishing Switzerland 2016
Y. Tan et al. (Eds.): ICSI 2016, Part I, LNCS 9712, pp. 564–570, 2016.
DOI: 10.1007/978-3-319-41000-5_56

In Eq. (1), $X = (x_1, \cdots, x_D)$ is a decision vector of D dimensions, F means the set of objectives, M refers to the number of objectives, g_i and h_j represent the set of inequality and equality constrains, respectively. The definition of $PS(t)$, $PF(t)$ and four possible types of DMOPs according to the changes of $PS(t)$ and $PF(t)$, Type I-Type IV, can also be found in [5].

The organization of this paper is as follows. Section 2 presents the method of decomposing decision variables based on environmental variable grouping and the approach of constructing a complete solution. An improved cooperative co-evolutionary algorithm is proposed. Section 3 provides the experimental results and analysis. Finally, Sect. 4 concludes the whole paper.

2 The Proposed Algorithm

2.1 The Proposed Decomposition Method

In this paper, the above strategy which was proposed by Omidvar et al. [8] is extended to the case of DMOPs. In the improved strategy, the decision variables are divided into ones interrelated with and without environment parameter t, with the environment parameter t being a reference.

Algorithm 1 is the pseudo code of the proposed grouping method, which employs Theorem 1 to group interrelated variables with the environment variable, t.

Algorithm 1. [group1, group2] $\longleftarrow$ group(F, lowb, upb, M, D)

1: group1=ϕ //group1 is a null set, Coordinate set of inseparable variable with t.

2: group2=ϕ //group2 is a null set, Coordinate set of separable variable with t.

3: for m=1:M //M is the number of objectives.

4:     for i=1:D

5:         X1=lowbones(1, D+1)

6:         X2=X1(1:D)

7:         X2(D+1)=lowb+rand*(upb-lowb)

8:         delta1=F(m)(X1) - F(m)(X2)

9:         X1(i)=(lowb+upb)/2

10:        X2(i)=(lowb+upb)/2

11:        delta2=F(m) (X1) - F(m) (X2)

12:        if $|delta1 - delta2| > \delta$

13:        group1=group1∪i

14:        end

15:    end

16: end

17: group2 =(1, 2, $\cdots$, D) - group1

2.2   Construction of the Complete Solutions

Based on the above grouping strategy, the whole decision variable, X, is decomposed into the following two groups. One is $X^1 = (x_1^1, x_2^1, \cdots, x_r^1)$ inseparable with the environment variable, t, and the other is $X^2 = (x_1^2, x_2^2, \cdots, x_{D-r}^2)$ separable with t. In this paper, the two subcomponents are evolved by two sub-populations, P^1 and P^2. Supposing that the number of representative individuals from the other sub-population is 3, i.e., $N_1 = 3$, Fig. 1. shows the process of constructing complete solutions, where $p_i^1 = (p_{i1}^1, p_{i2}^1, \cdots, p_{ir}^1)$ is the i-th individual of P^1. p_j^2, p_k^2, and p_l^2 are representative individuals randomly selected from P^2. p_1, p_2, and p_3 are generated complete solutions.

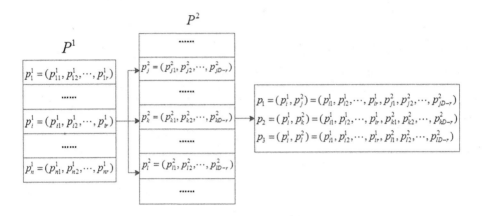

Fig. 1. Construction of complete solutions

2.3   Cooperative Co-evolutionary Dynamic NSGA-II Based on Environment Decomposition

NSGA-II is a state-of-the-art representative in evolutionary algorithms, and it has been widely used in real-world applications. We propose a cooperative co-evolutionary dynamic NSGA-II based on environment decomposition (DNSGAII-CO) by combining the proposed paradigm with NSGA-II. The Steps of DNSGAII-CO is summarized as follows.

Step1: The decision variable in optimization problem is divided into two groups with the strategy of Subsect. 2.1, i.e., $X1 = X(group1)$ and $X2 = X(group2)$. Initialize two sub-populations, P^1, P^2;

Step2: Whether the environment changes or not. If yes, initialize the two sub-populations with differential prediction and Cauchy mutation, respectively;

Step3: Update P_k^1 and P_k^2 by using crossover and mutation operators of NSGA-II, and the new two offspring sets, Q_k^1 and Q_k^2, are generated;

Step4: Evaluate the new offspring in Q_k^1 and Q_k^2 by the strategy in Subsect. 2.2, and update the archive A;

Step5: Generate the new sub-populations, P_{k+1}^1 and P_{k+1}^2, by the nondominated sorting in NSGA-II;

Step6: If the maximal number of iterations or other termination conditions are met, output the nondominated complete solutions from the archive; otherwise, go to Step 2.

3 Experimental Study

In this section, the performance of DNSGAII-CO is evaluated by comparing them with DNSGAII-A [4], DNSGAII-B [4].

3.1 Test Instances

In order to demonstrate the performance of the proposed algorithms, a function F1is obtained by modifying the following benchmark problem, i.e., FDA4 [5]. The function is provided in Table 1.

Table 1. Description of benchmark functions

Definition and the true of PS/PF	Type and separablility
F1 $f_1(x) = (1 + g_1 + g_2)\cos(0.5\pi X_I) f_2(x) = (1 + g_1 + g_2)\sin(0.5\pi X_I)$ $g_1(x) = \sum_{x \in X_{II}} (x_i - 0.5)^2 g_2(x) = \sum_{x \in X_{III}} (x_i - G(t))^2$ $G(t) = \|\sin(0.5\pi t)\|,\ t = \frac{1}{n_t}\lfloor\frac{\tau}{\tau_t}\rfloor$ $where: x_i \in [0,1], i = 1,2,...,20;$ $X_I = (x_1), X_{II} = (x_2,...,x_{10}), X_{III} = (x_{11},...,x_{20})$ $PS(t): x_1 \in [0,1]; x_i = 0.5, x_i \in X_{II}; x_i = G(t), x_i \in X_{III};$ $PF(t): f_1(x) = \cos(0.5\pi x_1), f_2(x) = \sin(0.5\pi x_1)$	Type I, and X_I, X_{III} are inseparable with time t; X_{II} is separable

3.2 Comparison Algorithms and Parameter Settings

The Matlab program of NSGAII and MOPSO can be downloaded from MAT-LAB CENTRAL (http://cn.mathworks.com/matlabcentral/). The parameters of the test functions and these algorithms are listed in Table 2.

Table 2. Parameters setting

Algorithms	Parameters
DNSGAII − A	Crossoverrate $p_c = 0.9$
DNSGAII − B	
DNSGAII − CO	Mutation rate $p_m = 0.1$

Table 3. Mean and standard deviation of MIGD and MHV for each algorithm over 20 runs

Problem	Stats.	MIGD			MHV		
		DNSGAII-A	DNSGAII-B	DNSGAII-CO	DNSGAII-A	DNSGAII-B	DNSGAII-CO
F1	Mean	0.025756*	0.025835*	**0.008705**	0.177632*	0.177542*	**0.2017**
	Std	0.000686	0.000717	0.000218	0.001054	0.000828	0.000707

In this paper, the size of population and archive are 100, the rate of cooperative individual is 10 %, and 20 times independent runs are conducted for each algorithm, and the mean and the standard deviation of the above two performance metrics are calculated. Besides, Mann-Whitney U test is employed to test performance metric value at the significant level of 5 %.

For F1, $n_t = 10, \tau_t = 2$, each algorithm tracks 50 times environment changes, and each function is evaluated 200 times after each environment change.

3.3   Comparison Results of DNSGAII-CO, DNSGAII-A, and DNSGAII-B

We do a group of experiment which is to compare DNSGAII-CO with DNSGAII-A and DNSGAII-B.

The mean and the standard deviation of MIGD and MHV for DNSGAII-A, DNSGAII-B, and DNSGAII-CO are listed in Table 3, where the bold values mean that the algorithm has either the minimal MIGD or the maximal MHV. (*) indicates there is significant difference between comparison algorithm and proposed algorithm at the 5 % significance level under Mann-Whitney U test. From Table 3, DNSGAII-CO significantly outperforms the two compared algorithms for F1. The MIGD values of DNSGAII-A and DNSGAII-B are 0.025756 and 0.025835, respectively, and that of DNSGAII-CO is 0.008705. Therefore, DNSGAII-CO has a better capability in exploration than DNSGAII-A and DNSGAII-B.

The curves of average IGD and HV over 20 runs on F1 is depicted in Fig. 2. Figure 2 reports that,

For F1, the IGD value of DNSGAII-CO has a slighter fluctuation than those of DNSGAII-A and DNSGAII-B, as Fig. 2 shows. The reason is that there is no historical information that can be utilized for DNSGAII-A and DNSGAII-B at the initial time. Similarly, DNSGAII-A and DNSGAII-B are greatly influenced by the environment, and their IGD values are periodically fluctuant. Hence, DNSGAII-CO has good performances in convergence, diversity, and robustness.

4   Conclusion

This paper focuses on DMOPs. A strategy of cooperative co-evolution based on environmental variable grouping is proposed by analyzing the relationship

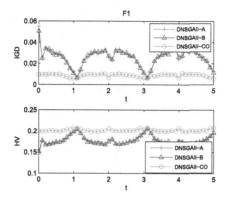

Fig. 2. Average IGD and HV values over 20 runs versus time on F1

between the decision variables and the environment variable. Two subcomponents are first obtained by dividing the decision variables into those interrelated with and without environment. And then, they are optimized by two subpopulations, respectively. So the search space of each sub-population is reduced, and the capability of the algorithm in exploration is improved. The simulation results demonstrate that DNSGAII-CO significantly outperform DNSGAII-A, DNSGAII-B for most benchmark instances whose decision variables are separable with the environment variable.

Acknowledgments. This work was jointly supported by National Natural Science foundation of China (No. 61375067, 61473299, 61573361), National Basic Research Program of China (973 Program) (No. 2014CB046306-2), Natural Science foundation of Anhui Province (No. 1608085QG169), and Natural Science Foundation of Anhui Higher Education Institutions (No. KJ2014B17).

References

1. Abello, M.B., Bui, L.T., Michalewicz, Z.: An adaptive approach for solving dynamic scheduling with time-varying number of tasks-Part II. In: 2011 IEEE Congress on Evolutionary Computation (CEC), pp. 1711–1718. IEEE (2011)
2. Amiri, B., Hossain, L., Crowford, J.: A multiobjective hybrid evolutionary algorithm for clustering in social networks. In: Proceedings of the 14th Annual Conference Companion on Genetic and Evolutionary Computation, pp. 1445–1446. ACM (2012)
3. Deb, K.: Single and multi-objective dynamic optimization: two tales from an evolutionary perspective. Indian Institute of Technology (2011)
4. Deb, K., Rao, U.B.N., Karthik, S.: Dynamic multi-objective optimization and decision-making using modified NSGA-II: a case study on hydro-thermal power scheduling. In: Obayashi, S., Deb, K., Poloni, C., Hiroyasu, T., Murata, T. (eds.) EMO 2007. LNCS, vol. 4403, pp. 803–817. Springer, Heidelberg (2007)

5. Farina, M., Deb, K., Amato, P.: Dynamic multiobjective optimization problems: test cases, approximations, and applications. IEEE Trans. Evol. Comput. **8**(5), 425–442 (2004)

6. Kaiwartya, O., Kumar, S., Lobiyal, D., Tiwari, P.K., Abdullah, A.H., Hassan, A.N.: Multiobjective dynamic vehicle routing problem and time seed based solution using particle swarm optimization. J. Sens. **2015** (2015)

7. Nguyen, S., Zhang, M., Johnston, M., Tan, K.C.: Automatic design of scheduling policies for dynamic multi-objective job shop scheduling via cooperative coevolution genetic programming. IEEE Trans. Evol. Comput. **18**(2), 193–208 (2014)

8. Omidvar, M.N., Li, X., Mei, Y., Yao, X.: Cooperative co-evolution with differential grouping for large scale optimization. IEEE Trans. Evol. Comput. **18**(3), 378–393 (2014)

9. Tang, J., Alam, S., Lokan, C., Abbass, H.A.: A multi-objective evolutionary method for dynamic airspace re-sectorization using sectors clipping and similarities. In: 2012 IEEE Congress on Evolutionary Computation (CEC), pp. 1–8. IEEE (2012)

10. Wu, P.P., Campbell, D., Merz, T.: Multi-objective four-dimensional vehicle motion planning in large dynamic environments. IEEE Trans. Syst. Man Cybern. Part B: Cybern. **41**(3), 621–634 (2011)

11. Zhang, Z.: Multiobjective optimization immune algorithm in dynamic environments and its application to greenhouse control. Appl. Soft Comput. **8**(2), 959–971 (2008)

Novel Local Particle Swarm Optimizer for Multi-modal Optimization

Yuechao Jiao[1], Lei Yang[1], Boyang Qu[1(⊠)], Dingming Liu[1],
J.J. Liang[2], and Junming Xiao[1]

[1] School of Electronic and Information Engineering,
Zhongyuan University of Technology, Zhengzhou 450007, China
qbyl984@hotmail.com
[2] School of Electrical Engineering, Zhengzhou University,
Zhengzhou 450001, China

Abstract. Recently, evolutionary computation has become an active research area. Multi-modal optimization is one of the most important directions in this area. The target of multi-modal optimization is to locate multiple peaks in one single run. Particle swarm is one of the most effective global optimization methods. However, most of the existing PSO-based algorithms suffer from the problems of low accuracy and requirement of prior knowledge of some niching parameters. To tackle these issues, this paper proposed a Novel Local Particle Swarm Optimizer to solve multi-modal optimization problems. To enhance the algorithm's ability of locating multiple peaks, a new local best based velocity updating formula is introduced. With the proposed updating formula, the probability of finding global/local optima is greatly increased. The experimental results reveal that the proposed algorithm is able to generate satisfactory performance over a number of existing state-of-the-art multimodal algorithms.

Keywords: Evolutionary computation · Niching technique · Multi-modal optimization · Particle swarm optimization

1 Introduction

In artificial intelligence, stochastic methods such as evolutionary algorithms (EAs) have shown to be effective and robust in solving difficult optimization problems [1]. Compared with classical optimization method, EAs have a high chance to avoid getting stuck into a local optimum. The original form of most EAs is designed for locating one single global optimum. However, in real-world application, many optimization problems fall into multi-modal optimization category. These problems require locating not only one optimum but also the entire set of global/local optima. In such a problem, if multiple solutions (local and global) are located, the implementation can be quickly changed to another solution while still maintaining an optimal system performance [2]. To handle multi-modal problems, a niching method is generally used to modify the behavior of a classical EA in order to divide the whole population into multiple groups, thereby locating different optima. Many niching techniques have been proposed in literature, such as crowding [3], restricted tournament selection [4], and speciation [5].

© Springer International Publishing Switzerland 2016
Y. Tan et al. (Eds.): ICSI 2016, Part I, LNCS 9712, pp. 571–578, 2016.
DOI: 10.1007/978-3-319-41000-5_57

Various niching methods have also been incorporated into particle swarm optimizer (PSO) to enhance their ability to handle multimodal optimization problems [5, 6]. However, most of these niching methods still have many disadvantages/difficulties in solving multi-modal problems. The disadvantages include low accuracy, requirement of knowledge of niching parameters, ignoring local optima, etc. To overcome the above issues, a novel local particle swarm optimizer (NLPSO) is proposed to solve multi-modal optimization problem. This research introduces a new velocity updating method and aims to develop an algorithm that:

- does not require specification of any niching parameter.
- is able to locate all desired optima with a high accuracy.
- is able to maintain the found optima till the end of the search.

To demonstrate the effectiveness of the proposed algorithm, NLPSO is compare with a number of the state-of-the-art multimodal optimizers on 15 multi-modal benchmark test functions.

The rest of this paper is organized as follows. Section 2 gives a brief overview of some classical niching techniques, PSO algorithm as well as some existing PSO niching methods. In Sect. 3, the NLPSO algorithm is introduced. The experimental setup and experimental results are presented and discussed in Sects. 4 and 5, respectively. Finally, the paper is concluded in Sect. 6.

2 Literature Review

2.1 Reviewing Classical Niching Methods

Similar to many artificial intelligence techniques, *niching* method is also inspired by nature. *Niching* refers to the method of finding and maintaining multiple stable niches, or favorable parts of the solution space possibly around multiple solutions, so as to prevent convergence to a single solution [2].

Crowding is one of earliest *niching* technique that introduced by De Jong in 1975 [7]. It was originally designed to preserve the diversity of the population and it allows competition among similar individuals in the population. In crowding, an offspring is compared with the most similar individual within a small random sample taken from the current population. The size of the sample is determined by the parameter CF (crowding factor). Although crowding is simple and effective for certain problems, the problem of replacement error limits its usage in real-world application.

Clearing [8] is another commonly used approach in EA. To maintain the diversity of the current population, clearing eliminates the bad individuals and keeps only the fittest individual (or a few top individuals) within each niche. Pétrowski has shown that clearing method has a lower complexity than that of the fitness sharing technique.

Beside the above mentioned methods, many other niching techniques have also been developed over the years, including restricted tournament selection [4], speciation [5], NichePSO [9], ring topology-based PSO [10].

2.2   Particle Swarm Optimization

Particle swarm optimization was first introduced by Eberhart and Kennedy in 1995 [11]. Although the operation is a relative simple, it has been proven to be one of the most effective optimization method [12–14]. Particle swarms concept was originally designed for simulating the social behavior commonly observed in animal kingdom and subsequently extended as an intelligent search technique. Each particle of the population represents a potential solution in the D-dimensional search space. Same as evolutionary algorithms, PSO is also a population-based search technique. However, in the search process, instead of using evolutionary operators, PSO modifies each particle's position in the search space based on its velocity, previous best position found by itself and previous best position found by the whole population.

In a canonical PSO, the position X and velocity V of each particle is updated using the following equations:

$$V_i^d = \omega * V_i^d + c_1 * rand1_i^d * (pbest_i^d - X_i^d) + c_2 * rand2_i^d * (gbest_i^d - X_i^d). \qquad (1)$$

$$X_i^d = X_i^d + V_i^d. \qquad (2)$$

Where c_1 and c_2 are the acceleration constants, ω is the inertia weight to balance the global and local search. $Rand1_i^d$ and $rand2_i^d$ are two random numbers within the range of [0, 1]. $Pbest_i$ is the best previous position yielding the best fitness value for the ith particle while $gbest_i$ is the best position found by the entire population.

Beside the above mentioned $gbest$ PSO, another common approach is called $lbest$ PSO. In $lbest$ PSO, the algorithm only allows each particle to be influenced by the best position that found by the particles within its neighborhood. In this version of PSO, the $gbest$ in the velocity update equation (Eq. 1) is replaced by $lbest$. The neighborhood is formed by various topologies [15]. Kennedy and Mendes [16] have shown that different topologies could give superior performance over different optimization problems.

3   Novel Local Particle Swarm Optimizer

3.1   Drawback of Existing Niching Particle Swarm Optimizer

As mentioned in Sect. 1, various niching techniques have been incorporated into PSO to handle multi-modal optimization problems. However, most of these methods have different drawbacks which limit their usage in real-world application. Take Speciation-based PSO (SPSO) as an example; the method divides the population into different species and the dominant (fittest) particle in each species is considered as the species seed. Within each of the species, it can be treat as a separated $gbest$ PSO itself. Multiple peaks can be located by SPSO, as the species are adaptively formed around different regions. However, a niche radius must be specified in order to define the size of the species. Choosing a proper radius is difficult if little prior knowledge is available about the problem. A Large radius may miss some of the peaks while a small radius slows down the converge speed which leads to low accuracy. This problem is demonstrated in Fig. 1. Recently, Li [10] introduced a new PSO based multi-modal optimization algorithm using a ring topology which successfully solve the problem of specifying the radius. However, the accuracy of the algorithm is relatively low because of the indices based neighborhood used in the method. In this method, the particles are only allowed to interact with their immediate neighbors on their left and right according to the indices. Due to the random initialization, it is likely that many of the neighbors belong to different niches which causes low converge speed and low accuracy.

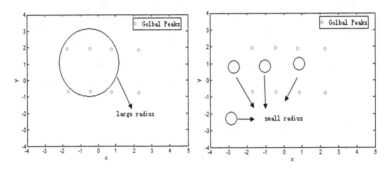

Fig. 1. Problem with niching radius

Motivated by the above observations, a novel local particle swarm optimizer (NLPSO) is proposed to solve the problems of specifying niching radius and low accuracy.

3.2   Novel Local Particle Swarm Optimizer (NLPSO)

To ensure that most of the neighbors belong to the same niche and increase the accuracy, a new neighborhood selection method which based on Euclidean distance is proposed. In each of the iteration, the particles are led by its personal best (*pbest*) and neighborhood best (*gbest*). The neighborhood best is selected as the personal best of other particles which is closest (based on Euclidean distance) to the current particle. The steps of selecting the neighborhood best are presented in Table 1.

Table 1. Steps of choosing neighborhood best

For $i=1$ to *NP* (population size)
Calculate the Euclidean distance between the ith particle and the personal bests of all the
    rest particles.
    Find the personal best with the smallest Euclidean distance $pbest_k$.
Set $pbest_k$ as the neighborhood best of the ith particle and let $nbest_i = pbest_k$.
End for

With the selected neighborhood best, the velocity update of NLPSO uses the formula given below while the position update keeps unchanged:

$$V_i^d = \omega * V_i^d + c_1 * rand1_i^d * (pbest_i^d - X_i^d) + c_2 * rand2_i^d * (nbest_i^d - X_i^d). \tag{3}$$

where $nbest_i^d$ is the dth dimension of the ith particle's neighborhood best and the rest notations are the same as Eq. (1).

Compared with other classical PSO based niching algorithms, there are many advantages of NLPSO. The concept of NLPSO is simple and it is convenient to be used in real world application. Similar to ring topology based PSO, NLPSO does not have any niching parameters and it does not require any prior knowledge of the problems. Moreover, the neighborhood best selection method ensures that the $nbest$ and $pbest$ are from the same niches and it improves converge speed as well as the accuracy. To give a further explanation of NLPSO algorithms, the overview of NLPSO is presented in Table 2.

Table 2. Overview of NLPSO

Step 1	Randomly generate the initial particles.
Step 2	Evaluate the initial particles and initialize the $pbest$
	For $i = 1$ to NP (population size)
Step 3	Indentify the neighborhood best using the method described in Table 1.
Step 4	Update the particles velocity using Eq. (3).
Step 4	Update the particles position using Eq. (2).
Step 5	Evaluate the newly generated particle.
Step 6	Update the $pbest$ for the ith particle.
	End for
Step 7	Stop if a termination criterion is satisfied. Otherwise go to Step 2.

4 Experimental Setup

4.1 Test Functions

In order to access the performance of various multi-modal algorithms, we investigate 10 benchmark test functions with different properties (modality, search space size and number of variables). The test functions, together with the dimension and number of known peaks, are depicted in Table 3.

4.2 Experimental Settings

Ten different multimodal algorithms are examined by our experiments:

- NLPSO: Novel local particle swarm optimizer.
- CDE [3]: The original crowding DE.
- SDE [17]: Speciation-based DE.
- CLDE [18]: Clearing DE.
- FERPSO [19]: Fitness-Euclidean distance ratio PSO.
- SPSO [8]: Speciation-based PSO.

Table 3. Test functions

Function No.	Function Name	No. of Dimension	No. of Optima
F1	Equal Maxima	1	5
F2	Decreasing Maxima	1	5
F3	Uneven Maxima	1	5
F4	Uneven Decreasing Maxima	1	5
F5	Branin RCOS	2	3
F6	Ursem F1	2	2
F7	Himmelblau's function	2	4
F8	Six-Hump Camel Back	2	2
F9	Shekel's foxholes	2	25
F10	Inverted Shubert function	2	18

- r2pso [10]: An *lbest* PSO with a ring topology, each member interacts with only its immediate member to its right.
- r2pso-*lhc* [10]: The same as r2pso, but with no overlapping neighborhoods.
- SGA [5]: Fitness sharing genetic algorithm.
- RTSGA: Restricted tournament selection genetic algorithm.

To access the performance of different multi-modal algorithms, 60 independent runs of each of the methods are executed on each function. The following criteria are used to measure the performance of the compared multi-modal algorithms:

- MPR (Maximum peak ratio. To test the quality of optima without considering the distribution of the population)

The maximum peak ratio is defined as follows: (assuming a maximization problem)

$$MPR = \frac{\sum_{i=1}^{q} f_i}{\sum_{i=1}^{q} F_i} \tag{4}$$

where q is the number of optima, $\{f_i\}_{i=1}^{q}$ are the fitness values of the optima in the final population while $\{f_i\}_{i=1}^{q}$ are the values of real optima of the objective function. Note that the larger the MPR value indicates the better performance of the particular algorithm [20].

5   Experimental Results

The maximum peak ratio (MPR) of the compared algorithms are presented in Table 4. A smaller MPR value means a high accuracy of the compared algorithm. The ranks of each algorithm are presented in the parentheses under the MPR value while the total rank is listed in the last row of the table. NLPSO ranks the best out of the ten compared algorithms which reveals that NLPOS is able to generate stable niching behavior with a high accuracy.

Table 4. MPR

fnc	NLPSO	CDE	SDE	CLDE	FER-PSO [28]	SPSO [28]	r2pso [28]	r2pso-lhc [28]	SGA	RTSGA
F1	4.29e-05 (1)	1.28e-04 (5)	0.0497 (9)	6.14e-04 (7)	1.20e-04 (4)	4.71e-05 (2)	0.01 (8)	5.02e-05 (3)	1.48e-04 (6)	0.71 (10)
F2	5.77e-05 (1)	1.42e-04 (3)	1.3954 (8)	4.60e-04 (4)	1.9504 (9)	7.19e-05 (2)	1.9926 (10)	0.1194 (5)	0.1605 (6)	0.721 (7)
F3	4.52e-05 (1)	1.31e-04 (6)	0.0928 (9)	4.42e-04 (8)	6.17e-05 (4)	5.85e-05 (3)	7.59e-05 (5)	5.13e-05 (2)	1.32e-04 (7)	0.7565 (10)
F4	6.44e-05 (1)	1.26e-04 (3)	1.2574 (8)	3.83e-04 (4)	1.7799 (10)	7.66e-05 (2)	1.6172 (9)	0.0575 (5)	0.1888 (6)	0.7537 (7)
F5	0.0123 (1)	0.1954 (7)	0.0140 (2)	0.3226 (8)	0.0145 (3)	0.0482 (6)	0.0271 (4)	0.0458 (5)	0.7254 (9)	2.5973 (10)
F6	0.0030 (1)	0.0045 (2)	2.9693 (9)	0.0281 (6)	3.1004 (10)	0.0220 (5)	1.0422 (8)	0.0045 (2)	0.0150 (4)	0.1587 (7)
F7	0.0072 (1)	0.0875 (6)	1.5623 (8)	0.1990 (7)	0.0103 (2)	0.0373 (4)	0.0272 (3)	0.0391 (5)	1.9994 (9)	10.4052 (10)
F8	2.17e-04 (2)	0.0033 (6)	2.15e-04 (1)	0.0184 (8)	2.51e-04 (3)	0.0107 (7)	5.06e-04 (5)	4.04e-04 (4)	0.3022 (9)	1.2255 (10)
F9	1.1959 (1)	1.2573 (2)	1091 (10)	1.7234 (6)	99.5909 (8)	1.3955 (4)	1.6784 (5)	1.3380 (3)	72.7716 (7)	557.8573 (9)
F10	0.0051 (1)	0.0327 (3)	5.9916 (8)	7.0792 (9)	0.1925 (7)	0.1240 (6)	0.0686 (5)	0.0647 (4)	0.0096 (2)	50.4437 (10)
F11	6.2272 (1)	6.49 (2)	58.20 (8)	10.7509 (3)	60.32 (10)	28.1270 (7)	21.79 (5)	21.4131 (4)	60.07 (9)	21.84 (6)
F12	11.1542 (1)	12.80 (2)	53.46 (10)	15.4815 (3)	40.36 (8)	26.0253 (7)	21.87 (5)	20.8074 (4)	43.50 (9)	21.87 (5)
F13	5.0421 (2)	4.63 (1)	55.01 (10)	6.6717 (3)	24.68 (8)	21.4353 (7)	15.84 (6)	15.7341 (5)	51.94 (9)	10.58 (4)
F14	5.7243 (1)	5.95 (2)	44.49 (9)	8.2182 (3)	38.73 (8)	21.0532 (7)	16.55 (5)	15.9127 (4)	52.86 (10)	16.55 (5)
F15	5.6700 (2)	4.55 (1)	53.42 (10)	7.7458 (3)	40.31 (8)	21.8981 (7)	15.62 (5)	16.3281 (6)	52.17 (9)	11.68 (4)
Total Rank	**18**	51	119	82	102	76	88	61	111	114

6   Conclusion

In this work, a novel local particle swarm optimizer (NLPSO) is proposed to generate niching behavior for solving multi-modal problems. NLPSO introduces a distance-based new *nbest* (neighborhood best) selection method. Leaded by this *nbest* and their own *pbest*, the particles are able to fly through the search space and identify different global/local optima fast and accurately. Moreover, NLPSO is able to eliminate the need to specify any niching parameters. In the experimental results section, we have demonstrated the effectiveness of NLPSO. It is able to provide a either superior or comparable performance over a number commonly used niching algorithms. Therefore, we can conclude that NLPSO can generate good and stable niching behavior over various multi-modal problems.

Acknowledgement.   This research is partially supported by National Natural Science Foundation of China (61305080, 61473266, 61379113) and Postdoctoral Science Foundation of China (2014M552013) and the Scientific and Technological Project of Henan Province (132102210521).

References

1. Kennedy, J., Eberhart, R.: Swarm Intelligence. Morgan Kaufmann, San Francisco (2001)
2. Das, S., Maity, S., Qu, B.Y., Suganthan, P.N.: Real-parameter evolutionary multimodal optimization-A survey of the state-of-the-art. **1**, 71–88 (2011). Swarm and Evolutionary Computation
3. Thomsen, R.: Multimodal optimization using Crowding-based differential evolution. In: IEEE 2004 Congress on Evolutionary Computation, pp. 1382–1389 (2004)
4. Harik, G.R.: Finding multimodal solutions using restricted tournament selection. In: 6th International Conference on Genetic Algorithms, pp. 24–31. San Francisco (1995)
5. Li, X.: Adaptively choosing neighbourhood bests using species in a particle swarm optimizer for multimodal function optimization. In: Deb, K., Tari, Z. (eds.) GECCO 2004. LNCS, vol. 3102, pp. 105–116. Springer, Heidelberg (2004)
6. Qu, B.Y., Suganthan, P.N., Zhao, S.Z.: Current based fitness euclidean-distance ratio particle swarm optimizer for multi-modal optimization. In: Nature and Biologically Inspired Computing (NaBIC), Japan (2010)
7. De Jong, K.A.: An analysis of the behavior of a class of genetic adaptive systems. In: Doctoral Dissertation, University of Michigan (1975)
8. Pétrowski, A.: A clearing procedure as a niching method for genetic algorithms. In: Proceedings of the IEEE International Conference on Evolutionary Computation, pp. 798–803. New York (1996)
9. Brits, A.E.R., van den Bergh, F.: A niching particle swarm optimizer. In: Proceedings of the 4th Asia-Pacific Conference on Simulated Evolution and Learning 2002(SEAL 2002), pp. 692–696 (2002)
10. Li, X.: Niching without niching parameters: particle swarm optimization using a ring topology. IEEE Trans. Evol. Comput. **14**, 1233–1246 (2010)
11. Eberhart, R.C., Kennedy, J.: A new optimizer using particle swarm theory. In: Proceedings 6th International Symposium Micromachine Human Sci., vol. 1, pp. 39–43 (1995)
12. Liang, J.J., Qin, A.K., Suganthan, P.N., Baskar, S.: Comprehensive learning particle swarm optimizer for global optimization of multimodal functions. IEEE T Evol. Comput. **10**(3), 281–295 (2006)
13. Liu, Y., Qin, Z., Shi, Z., Lu, J.: Center particle swarm optimization. Neurocomputing **70**, 672–679 (2007)
14. Yi, D., Ge, X.: An improved PSO-based ANN with simulated annealing technique. Neurocomputing **63**, 527–533 (2005)
15. Mendes, R., Kennedy, J., Neves, J.: The fully informed particle swarm: simpler, maybe better. IEEE Trans. Evol. Comput. **8**, 204–210 (2004)
16. Kennedy, J., Mendes, R.: Population structure and particle swarm performance. In: Proceedings of the 2002 Congress on Evolutionary Computation, pp. 1671–1675 (2002)
17. Li, X.: Efficient differential evolution using speciation for multimodal function optimization. In: Proceedings of the conference on genetic and evolutionary computation, pp. 873–880. Washington DC (2005)
18. Qu, B., Liang, J., Suganthan, P.N., Chen, T.: Ensemble of clearing differential evolution for multi-modal optimization. In: Tan, Y., Shi, Y., Ji, Z. (eds.) ICSI 2012, Part I. LNCS, vol. 7331, pp. 350–357. Springer, Heidelberg (2012)
19. Li, X.: A multimodal particle swarm optimizer based on fitness Euclidean-distance ration. In: Proceedings of Genetic and Evolutionary Computation Conference, pp. 78–85 (2007)
20. Qu, B.Y., Liang, J.J., Suganthan, P.N.: Niching particle swarm optimization with local search for multi-modal optimization. Information Sciences, doi:10.1016/j.ins.2012.02.011

Interval Cost Feature Selection Using Multi-objective PSO and Linear Interval Programming

Yong Zhang[1(⊠)], Dunwei Gong[1,2], Miao Rong[1], and Yinan Guo[1]

[1] School of Information and Electrical Engineering, China University of Mining
and Technology, Xuzhou 221116, China
yongzh401@126.com
[2] School of Electrical Engineering and Information Engineering,
Lanzhou University of Technology, Lanzhou 730050, China

Abstract. Interval cost feature selection problems (ICFS) are popular
in real-world. However, since the optimized objectives not only are multiple but also contain interval coefficients, there have been few solving
methods. This paper first transforms the ICFS into a multi-objective one
with exact coefficients by the linear interval programming. Second, by
combining a multi-objective particle swarm algorithm (which has a good
performance in exploration) with a powerful problem-specific local search
(which is good at exploitation), we propose a memetic multi-objective
feature selection algorithm (MMFS-PSO). Finally, experimental results
confirmed the advantages of our method.

Keywords: Feature selection · Cost · Interval · Particle swarm ·
Multi-objective

1 Introduction

Feature selection (FS) is one of the most important issues in the field of data
mining and pattern recognition [14]. Traditional FS approaches, such as [10,13],
usually assume that the data are already stored in datasets and available without
charge. However, as we all know, data are not free in real-world applications.
There are various costs, such as money, time, or other resources to obtain feature
values of objects [8,11].

Although there are a few attempts to deal with this cost-based FS problem
[2,6,8,15], all the existing approaches assume that the cost associated to any feature is precise. However, in many practical cases, resulting from many objective
and/or subjective factors, it is not easy or even impossible to exactly specify the
cost for part features. On the contrary, it is easy for us to address the costs with
some intervals. If some coefficients of a FS problem are expressed with intervals,
rather than exact values, the existing methods are hardly to solve it.

Particle swarm optimization (PSO) is inspired from the behavior of bird
flocks [9]. It has recently been used as an effective technique in feature

© Springer International Publishing Switzerland 2016
Y. Tan et al. (Eds.): ICSI 2016, Part I, LNCS 9712, pp. 579–586, 2016.
DOI: 10.1007/978-3-319-41000-5_58

selection [4,12]. However, the use of PSO for cost-based FS is still few. In our previous research [15], we applied successfully PSO to feature selection with precise costs.

The overall goal of this paper is to solve the kind of feature selection problems with interval cost (ICFS). The main contributions are as follows: (1) A transformation based on linear interval programming is introduced, by which the ICFS problem is converted into a multi-objective one with exact coefficients; (2) A powerful problem-specific local search, namely random moving search (RSM) is proposed; (3) a memetic multi-objective feature selection algorithm (MMFS-PSO) is presented by combining effectively PSO with RSM.

2   Problem Description

Using the binary strategy, a solution of the ICFS problem, X, is denoted as

$$X = (x_1, x_2, \cdots, x_D), \quad x_j \in \{0, 1\} \tag{1}$$

where D is the feature number of data set; $x_j = 1$ represents the j-th feature is selected into the feature subset. In contrast, $x_j = 0$ means the feature is not selected.

This paper adopts an interval number $\bar{u} = [u^-, u^+]$ to represent the cost of obtaining data for a feature, where u^- and u^+ are the lower and upper limits of the interval $\bar{u}$ respectively, $0 \le u^- \le u^+ \le 1$. Furthermore, the cost values of all the D features can be represented by the following interval vector:

$$U = [\bar{u}_1, \bar{u}_2, \cdots, \bar{u}_D], \ \bar{u}_j = [u_j^-, u_j^+], \ j = 1, 2, \cdots, D. \tag{2}$$

Then, the first objective of the ICFS problem, i.e., the cost that may be associated to features, can be calculated, as follows:

$$f_1(X) = \sum_{j=1}^{D} \bar{u}_j x_j \tag{3}$$

Due to easy implementation and coefficient-free, the one nearest neighbour (1-NN) method [4], as a classifier, is used to evaluate the classification performance of feature subsets. In this method, the classification error rate of a solution is the proportion of incorrectly predicted samples (S_{ipc}) to all the samples (S_{all}), as follows:

$$f_2(X) = \frac{S_{ipc}}{S_{all}} \tag{4}$$

Based on the above discussion, the ICFS problems are modeled as follows:

$$ICFS1: \quad \min \ F(X) = (f_1(X), f_2(X)) = \left\{ \sum_{j=1}^{D} \bar{u}_j x_j, \ \frac{S_{ipc}}{S_{all}} \right\} \tag{5}$$

3   Problem Transformation

Since the first objective values become interval, the existing PSO algorithms based on precise fitness are no longer suited to the ICFS1. Focused on this, a transformation approach based on linear interval programming is introduced in this section.

The linear interval programming with pessimistic cost was proposed by Lai et al. [7]. In this method, the center of an interval is defined to denote the expected value of that special interval variable, and the right limit of an interval is used to denote its pessimistic cost. By the definition above, the interval function $f_1(X)$ becomes the following bi-objective problem:

$$f_1(X) = \min \left\{ \sum_{j=1}^{D} u_j^+ x_j, \sum_{j=1}^{D} \frac{u_j^- + u_j^+}{2} x_j, \right\} \tag{6}$$

According to the conclusion in [3], we have

$$\lambda \sum_{j=1}^{D} u_j^- x_j + (1-\lambda) \sum_{j=1}^{D} \frac{u_j^- + u_j^+}{2} x_j = \sum_{j=1}^{D} \frac{[(u_j^- + u_j^+) - \lambda(u_j^+ - u_j^-)]}{2} x_j \tag{7}$$

Based on (7), the ICFS1 can be converted into one with precise coefficients.

$$ICFS2: \quad \min \ F(X) = \left\{ \sum_{j=1}^{D} \frac{[(u_j^- + u_j^+) - \lambda(u_j^+ - u_j^-)]}{2} x_j, \frac{S_{ipc}}{S_{all}} \right\} \tag{8}$$

where $\lambda \in [0,1]$ is a weight coefficient, which controls the decision maker's preference on the expected value and the pessimistic cost.

4   The Proposed Memetic Algorithm

4.1   Encoding of Particles

In the MMFS-PSO, a particle refers to a possible solution of the optimized problem, thus it is very important to define a suitable encoding strategy first. This paper adopts the probability-based encoding strategy proposed in [2]. In this strategy, a particle is represented as a vector of probability,

$$P_i = (p_{i,1}, p_{i,2}, \cdots, p_{i,D}), \ p_{i,j} \in [0,1] \tag{9}$$

where the probability $p_{i,j} > 0.5$ means that the j-th feature will be selected into the i-th feature subset. Obviously, for any particle P_i, its corresponding feature subset X_i is constructed by the following equation:

$$x_{i,j} = \begin{cases} 1, & x_{i,j} \geq 0.5 \\ 0, & \text{otherwise} \end{cases} \tag{10}$$

4.2   Problem-Specific Local Search

To improve the performance of MMFS-PSO, this section designs a local search operator, namely random moving search.

The random moving search is designed to search around sparse solutions in the archive by using a stochastic move. First a non-dominated solution with the largest crowding distance in the archive, notified X_{best}, is selected as a base solution. Then, the binary tournament with the return ratio is utilized to select another non-dominated solution from the archive, notified X_k, which has a good return ratio for X_{best}, as the auxiliary solution. Here the return ratio is defined to evaluate the reward which the base solution obtains possibly when a solution is set as its auxiliary solution. For a non-dominated solution in the archive, the larger its return ratio is, the higher the probability of being selected as the auxiliary solution is.

Return ratio (Ereturn): Let X_1 and X_2 be two non-dominated solutions of the feature selection problem, then the return ratio that X_1 learns from X_2 is defined as follows:

$$Ereturn(X_1, X_2) = \left| \frac{1 - 0.5(f_2(X_1) + f_2(X_2))}{Ham(X_1, X_2) + 1} \right| \tag{11}$$

Where the hamming distance $Ham(X_1, X_2)$ is used to calculate the difference between the encoding of X_1 and that of X_2.

On the one hand, the higher the difference $Ham(X_1, X_2)$ between X_1 and X_2 is, the more the neighborhoods located between X_1 and X_2 is. The increase of the size of neighborhoods must decrease the probability of finding a good non-dominated solution from these neighborhoods. Therefore, $Ham(X_1, X_2)$ reflects the difficulty degree of exploiting the neighborhoods. On the other hand, the average $1 - 0.5(f_2(X_1) + f_2(X_2))$ reflects partly the benefit obtained possibly by searching the neighborhoods.

After determined the base solution X_{best} and its auxiliary solution X_k, the following random search is implemented to generate a candidate solution X'_{best}. Firstly, X'_{best} is initialized with the encoding of X_{best}. Then, compared the binary encoding of X_{best} and that of X_k, all D bits are divided into the equal bits and the unequal bits. These unequal bits and an equal bit randomly chosen from the set of the equal bits are considered as candidate bits. Next, the value of X'_{best} on these candidate bits are randomly set to 0 or 1.

For example, as shown in Fig. 1, {2,5} is the unequal bits, and {1,3,4,6} is the unequal bits. In this case, the two unequal bits and a random bit 3, chosen from {1,3,4,6}, form the candidate bits {2,3,5}. Then the value of X'_{best} on the 2,3,5 bits are randomly set to (0,1,0). So we can get the candidate solution $X'_{best} = (1\ 0\ 1\ 1\ 0\ 0)$.

4.3   Implement of the MMFS-PSO

Bare-bones multi-objective PSO algorithm (BMOPSO) was first proposed in [16]. Compared with the traditional MOPSO, this version eliminates the need

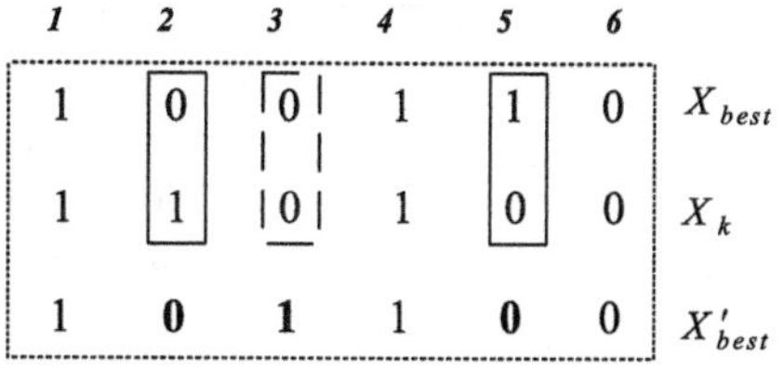

1	2	3	4	5	6	
1	0	0	1	1	0	X_{best}
1	1	0	1	0	0	X_k
1	0	1	1	0	0	X'_{best}

Fig. 1. Example of Random moving search

for tuning control parameters, such as inertia weight and acceleration coefficients. Therefore, this paper set the algorithm as a global exploration operator in the MMFS-PSO.

Combining the random moving search and some established operators introduced by BMOPSO, *Algorithm* 1 shows the detailed steps of the new MMFS-PSO algorithm. In this Algorithm, the Pareto domination-based strategy is used to update the personal best position of each particle (Pbest). If the decoded solution of a particle is dominated by the solution contained in its memory, then we keep the old solution in memory; otherwise, the decoded solution replaces it. The global best position of each particle (Gbest) is selected from the archive based on the diversity of non-dominated solutions. Herein the crowding distance [5] is used to estimate the diversity of non-dominated solutions. The higher crowding distance indicates a better chance to be selected as the Gbest. After determined the Gbest and Pbest, the new update equation proposed in [16] is used to generate a new position for each particle.

5   Experiment and Analysis

The NSGAII-based feature selection algorithm (NSGAFS) [5], and SPEA2-based feature selection algorithm (SPEAFS) [10] are employed to compare the performances of MMFS-PSO. In all these algorithms, the swarm/population size is set to 30, the maximum size of archive is set to 30, and the maximum iteration is set to 100.

Four real-world benchmark datasets are selected, including Vowel, Ionosphere, Wisconsin Diagnosis Breast Cancer (WDBC) and Sonar [1]. Without loss of generality, a synthetic cost vector $\bar{U} = [u_1, u_2, \cdots, u_D]$ is used in this paper, as follows:

U = [0.54 0.60], [0.18 0.22], [0.26 0.30], [0.68 0.78], [0.85 0.95], 0.68, 0.25, 1, 0.66, 0.92, 0.6, 0.72, 0.26, 0.24, 0.96, 0.98, 0.58, 0.67, 0.65, 0.69, 0.15, 0.31, 0.56, 0.88, 0.22, 0.66, 0.77, 0.64, 0.26, 0.58, 0.92, 0.01, 0.17, 0.36, 0.96, 0.91, 0.66, 0.3, 0.28, 0.67, 0.17, 0.53, 0.21, 0.12, 0.47, 0.85, 0.02, 0.94, 0.78, 0.97, 0.64, 0.41, 0.85, 0.64, 0.31, 0.33, 0.37, 0.06, 0.59, 0.92.

Among the vector $\bar{U}$, the costs of the first five features are interval. Note that, the cost vector of each dataset is composed of the elements of U, which begin at the first element 0.92 until the last feature is assigned.

Algorithm 1. the proposed MMFS-PSO algorithm

Parameters: the maximal number of generations, T_{max}, the swarm size, N, the archive size N_a

Step 1) Initialize the positions of a number of particles; Set the archive $A(0) = \emptyset$; Set the Pbest of each particle to be the particle itself;

Step 2) Let t=1. // the number of iterations

Step 3) Iteration

for $i = 1, 2, ..., N$, **do**

Step 3.1) Decode the particle $P_i(t)$ to a feature subset X_i;

Step 3.2) Calculate the objective values of the particle $P_i(t)$ by substituting X_i in Eq. (8);

Step 3.3) Save the feature subset X_i into $A(t)$, and prune the archive $A(t)$ by the method introduced in BMOPSO;

Step 3.4) Update the $Pbest_i(t)$ and $Gbest_i(t)$ of the particle $P_i(t)$;

Step 3.5) Update the particle position by the new update equation proposed in BMOPSO;

Endfor

Step 3.6) Implement the random moving search proposed in subsect. 4.2;

Step 4) Stopping criterion: If t¡Tmax, let t++, and go to **Step 3**; otherwise, stop the algorithm

---

5.1   Results Analysis

Considering that non-dominated solutions may result in estimation, we ran MMFS-PSO, NSGAFS and SPEAFS 20 times respectively for the datasets. The hyper-volume (HV) metric is used to estimate the performance of an algorithm. Table 1 lists their comparative results in terms of HV. Furthermore, the paired t-test at the significance level of 0.05 is employed to demonstrate the statistical significance of the results of three algorithms. Note that, 'Y+' indicates that MMFS-PSO is significantly better than the selected one at the significance level of 0.05, and 'N' means that the difference between two algorithms is not statistically significant.

Table 1. The HV values obtained by the three algorithms for the datasets

HV	MMFS-PSO	NSGAFS		SPEAFS	
	Average/Std.	Average/Std.	t-test	Average/Std.	t-test
Wine	**13.45/0.0998**	13.20/0.2502	Y+	13.43/0.0523	N
Ionosphere	**33.15/0.2168**	31.65/0.5501	Y+	31.31/0.7727	Y+
WDBC	**30.88/0.1541**	29.78/0.7712	Y+	29.60/0.3997	Y+
Sonar	**55.57/0.6760**	52.56/0.7426	Y+	50.40/0.6589	Y+

Furthermore, Fig. 2 shows the best Pareto fronts produced by the three algorithms. We can see that, for the dataset Wine with a small number of features,

MMFS-PSO and SPEAFS have the similar Pareto fronts, and are slightly better than NSGAFS. For the dataset WDBC, MMFS-PSO achieves better Pareto front than SPEAFS, and NSGAFS. Regarding the datasets Ionosphere and Sonar, MMFS-PSO still achieves better results than SPEAFS, and NSGAFS in terms of both the cost (the objective f1) and the error rate (the objective f2). And each solution obtained by the two compared algorithms is dominated by at least one solution obtained by MMFS-PSO.

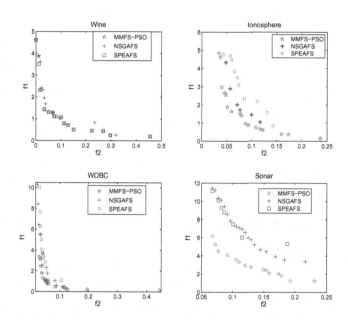

Fig. 2. Results obtained by the three compared algorithms for the four data sets.

6   Conclusion

This paper has conducted the first study on the feature selection problems with interval cost, presented a memetic multi-objective feature selection algorithm by combing multi-objective PSO with a powerful problem-specific local search. The experimental results demonstrate that MMFS-PSO outperforms all the two compared algorithms for most datasets, especially ones with a large number of features.

Acknowledgments. This research is jointly supported by the National Natural Science Foundation of China (No. 61473299, 61573361, 61503384), and China Postdoctoral Science Foundation (No. 2014T70557, 2012M521142).

References

1. Blake, C.: UCI machine learning databases. http://www.ics.uci.edu/?mlearn/MLRepository.html (2015)
2. Bolón-Canedo, V., Porto-Díaz, I., Sánchez-Maroño, N., Alonso-Betanzos, A.: A framework for cost-based feature selection. Pattern Recogn. **47**(7), 2481–2489 (2014)
3. Chankong, V., Haimes, Y.Y., Chankong, V., Haimes, Y.Y.: Multiobjective decision making: theory and methodology. North-Holland series in system science and engineering (1983)
4. Chuang, L.Y., Yang, C.H., Li, J.C.: Chaotic maps based on binary particle swarm optimization for feature selection. Appl. Soft Comput. **11**(1), 239–248 (2011)
5. Hamdani, T.M., Won, J.-M., Alimi, M.A.M., Karray, F.: Multi-objective feature selection with NSGA II. In: Beliczynski, B., Dzielinski, A., Iwanowski, M., Ribeiro, B. (eds.) ICANNGA 2007. LNCS, vol. 4431, pp. 240–247. Springer, Heidelberg (2007)
6. Kenndy, J., Eberhart, R.: Particle swarm optimization. In: Proceedings of IEEE International Conference on Neural Networks, vol. 4, pp. 1942–1948 (1995)
7. Lai, K.K., Wang, S.Y., Xu, J.P., Zhu, S.S., Fang, Y.: A class of linear interval programming problems and its application to portfolio selection. IEEE Trans. Fuzzy Syst. **10**(6), 698–704 (2002)
8. Min, F., Hu, Q., Zhu, W.: Feature selection with test cost constraint. Int. J. Approximate Reasoning **55**(1), 167–179 (2014)
9. Shi, Y., Eberhart, R.: A modified particle swarm optimizer. In: The 1998 IEEE International Conference on Evolutionary Computation Proceedings, IEEE World Congress on Computational Intelligence, pp. 69–73. IEEE (1998)
10. Tabakhi, S., Moradi, P., Akhlaghian, F.: An unsupervised feature selection algorithm based on ant colony optimization. Eng. Appl. Artif. Intell. **32**, 112–123 (2014)
11. Turney, P.D.: Cost-sensitive classification: empirical evaluation of a hybrid genetic decision tree induction algorithm. J. Artif. Intell. Res. **2**, 369–409 (1995)
12. Wang, X., Yang, J., Teng, X., Xia, W., Jensen, R.: Feature selection based on rough sets and particle swarm optimization. Pattern Recogn. Lett. **28**(4), 459–471 (2007)
13. Xue, B., Cervante, L., Shang, L., Browne, W.N., Zhang, M.: Multi-objective evolutionary algorithms for filter based feature selection in classification. Int. J. Artif. Intell. Tools **22**(04), 1–34 (2013)
14. Xue, B., Zhang, M., Browne, W.N.: Particle swarm optimization for feature selection in classification: a multi-objective approach. IEEE Trans. Cybern. **43**(6), 1656–1671 (2013)
15. Zhang, Y., Gong, D.W., Cheng, J.: Multi-objective particle swarm optimization approach for cost-based feature selection in classification (2015)
16. Zhang, Y., Gong, D.W., Ding, Z.: A bare-bones multi-objective particle swarm optimization algorithm for environmental/economic dispatch. Inf. Sci. **192**(6), 213–227 (2012)

Hybrid Differential Evolution-Variable Neighborhood Search to Solve Multiobjective Hybrid Flowshop Scheduling with Job-Sequence Dependent Setup Time

Budi Santosa[1(⊠)] and Ong Andre Wahyu Riyanto[2]

[1] Industrial Engineering Department,
Institut Teknologi Sepuluh Nopember, Surabaya, Indonesia
budi_s@ie.its.ac.id
[2] Universitas Wijaya Putra, Surabaya, Indonesia
ongandre@uwp.ac.id

Abstract. This paper proposes a hybrid algorithm which combines the differential evolution algorithm (DE) with variable neighborhood search (VNS) to solve multi-objective hybrid flexible flowshop with job-sequence dependent setup time (HFFS/SDST). The objective is to minimize makespan and lateness maximization on a hybrid flexible flowshop. Each stage has at least two units operating in parallel machines as well as considering skipping the stage, where not all jobs must be processed at each stage of operation. The model also considers the machine's setup time that depends on the sequence of jobs that are processed directly on the machine. Pareto solution is used as the process of collecting the points of non-dominated solutions. To evaluate the performance of our algorithm, we compare the results with those of DE-Insert and Particle Swarm Optimization (PSO)-VNS. Computational results and comparisons indicate that DE-VNS is more efective than DE-Insert and PSO-VNS.

Keywords: Differential evolution · Variable neighborhood search · Particle swarm optimization · Skipping the stage · Hybrid flowshop scheduling · Pareto solution

1 Introduction

The flow shop scheduling problem is a production planning problem in which n jobs have to be processed in the same sequence on m machines. In flow shop problem, there is only one machine at each stage of the operation. However, real shop floor rarely employs single machine for each stage of the operation. Flow shop model is often modified by employing a number of identical parallel machines at each stage [1]. The purpose of duplicating machines in parallel is to increase the capacity of the shop floor and reduce bottlenecks in the overall production line. Duplication of parallel machines at each stage is categorized as a hybrid flow shop (HFS).

The hybrid flow shop (HFS) is common manufacturing environment in which a set of n jobs are to be processed in a series of m stages. All jobs are processed following

© Springer International Publishing Switzerland 2016
Y. Tan et al. (Eds.): ICSI 2016, Part I, LNCS 9712, pp. 587–598, 2016.
DOI: 10.1007/978-3-319-41000-5_59

the same production flow: stage 1, stage 2,....., stage m. A job might skip some of stages provided. A job is processed in at least one of stages. In the HFS has at least two stages of operation and at least each stage has more than one unit machine in parallel [2]. Another frequent situation faced in real shop floor is setup time. Setup activities are non productive operations carried out on machines in order to prepare the next job in the sequence. Often, there are sequence dependent setup times, something referred to as SDST. Variants of hybrid flow shop (HFS) that considers the two characters mentioned above, is termed with a hybrid flexible flow shop with sequence dependent setup times (HFFS/SDST). Iterated local search (ILS) was used for makespan minimization [1].

Production scheduling needs to consider two or more objectives that might conflict with each other simultaneously. Multi-objective optimization focused on finding the solution set, where there is no single performance criterion that can be improved without sacrificing the quality of the performance of other criteria. This is commonly referred to as non-dominated solutions or pareto solutions [3].

Multi-objective optimization in the scheduling problem using methods based metaheuristic concept of evolution has been done by several researchers. Among them, Ishibuci and Murata [4] used multi-objective genetic local search (MOGLS) to solve multi-objective permutation flowshop scheduling. The PSO-based hybrid algorithm was used to minimize makespan and maximum tardiness simultaneously by using the concept of pareto solutions [5]. GA was used to minimize the makespan and the number of tardy jobs in flexible flow shop scheduling problem [6]. A hybrid differential evolution (DE)-based algorithm was used to minimize makespan and maximum tardiness simultaneously using the pareto solutions in job flow shop scheduling with limited buffers [8]. It was stated that the hybridization of PSO-VNS algorithm provides a multi-objective performance better than the GA algorithm [4].

This paper focuses on utilization of hybrid of differential evolution (DE) algorithm with a variable neigborhood search (VNS) for minimizing makespan and maximum lateness simultaneously in HFFS/SDST. Simultaneous minimization of two objectives uses the concept of non-dominated solutions (pareto solution). Some common performance metrics on the concept of pareto solution are used to determine the effectiveness of the algorithm. To the best of our knowledge, this is the first study on the utilization of DE-VNS algorithm for the hybrid flexible flow shop with sequence dependent setup time (HFFS/SDST) scheduling problem with minimization of the makespan and maximum lateness criteria.

The rest of the paper is organized as follows. In Sect. 2, the hybrid flexible flow shop with sequence dependent setup times (HFFS/SDST) is stated and formulated. Also the concept of pareto optimization is briefly introduced. In Sect. 3, steps of DE algorithm is briefly reviewed. In Sect. 4, the scheme of hybridization of DE-VNS algorithm for minimizing the two-objective in HFFS/ SDST is described in detail. Section 4 explains the experimental results and comparison with other algorithms. Finally, in Sect. 5, we conlude the results of this paper.

2 Literature Review

An effective hybrid differential evolution (HDE) was proposed for the no-wait flow-shop scheduling problem (FSSP) with the makespan criterion [7]. The DE-based parallel evolution mechanism and framework is applied to perform effective exploration, and a simple but efficient local search developed according to the landscape of FSSP is applied to emphasize problem dependent local exploitation.

A hybrid algorithm based on differential evolution (HDE) was proposed to solve multi-objective permutation flow shop scheduling problem (MPFSSP) with limited buffers between consecutive machines [7]. An efficient local search, which is designed based on the landscape of MPFSSP with limited buffers combined with DE is applied to emphasize exploitation. The concept of Pareto dominance is used to handle the updating of solutions in sense of multi-objective optimization. The convergence property of HDE is analyzed by using the theory of finite Markov chain. The proposed HDE demonstrate the effectiveness and efficiency.

In [10], a new self-adaptive mechanisms to improve the previous work, NSDE and SaDE, SaNSDE was proposed. The algorithm is intended to solve continuous optimization problems. The proposed SaNSDE showed the superiority over SaDE and NSDE. A differential evolution algorithm based on a variable neighborhood search algorithm (DE-VNS) is proposed in order to solve the constrained real-parameter optimization problems. The computational results show that the simple DE-VNS algorithm was very competitive to some of the best performing algorithms from the literature [11]. A differential evolution algorithm with a variable neighborhood search was also proposed to solve the multidimensional knapsack problem. Computational results show its efficiency in solving benchmark instances and its superiority to the best performing algorithms from the literature [12].

2.1 Definition of HFFS/SDST

We define hybrid flexible flow shop with sequence dependent setup time (HFFS/SDST) as follows: a set of N jobs, $N = \{1,2, .., n\}$ must be processed on a set of M stages of the operations, where $M=\{1,2,..,m\}$. All jobs visit stage at the same direction. Each stage of the operation i have m_i identical parallel machines, where $M_i = \{1,2,...,m_i\}$. Every job does not always have to be processed at each stage i. There might be some jobs skip beyond the specific stage. We denote by F_j the set of stages that job j has to visit, where $j \in N$ and $1 \leq |F_j| \leq m$. The processing time of job j at stage i is denoted by p_{ij}. The setup time between job j and k denoted by S_{ijk} where $k \in N$ at stage i. Due date of job j is denoted by d_j. The objective function is the minimization of makespan and the maximum lateness simultaneously.

2.2 The Concept of Pareto Optimization

In multicriteria optimization there is no single performance criteria that can be improved without sacrificing the quality of the performance of other criteria. Therefore, one solution can not be considered better or worse than other solutions. Generally, a multiobjective optimization problem with w objectives without constraints can be described as follows:

$$\text{Minimize} : f_1(x), f_2(x), \ldots, f_w(x) \tag{1}$$

where $f_1(x)$, $f_2(x)$, ..., $f_w(x)$ are w objectives to be minimized.

For the multi-objective optimization, the following concepts are of importance:

(1) Pareto dominance: A solution **x1** is said to (Pareto) *dominate* another solution **x2** (denoted **x1** $\succ$ **x2**) if and only if

$$(\forall i \in \{1, 2, \ldots, w\} : f_i(x1) \leq f_i(x2)) \wedge (\exists j \in \{1, 2, \ldots, w\} : f_j(x1) < f_j(x2)) \tag{2}$$

(2) Optimal pareto solution: A solution **x1** is said to be an optimal pareto solution if and only if there is no any solution x_2 that satisfies **x1** $\succ$ **x2**.
(3) Optimal pareto set: The set containing all optimal pareto solutions.
(4) Optimal Pareto front: The set of all objective function values corresponding to the solutions in the optimal pareto set.

2.3 Differential Evolution (dE) Algorithm

The Differential evolution (DE) algorithm introduced by Storn and Price [9] for complex continuous optimization problems. DE can be categorized into a class of floating-point encoded evolutionary algorithms, where each variable's value in the chromosome is represented by a real number. DE starts with the random initialization of a population of individuals in the search space. It finds the global optima by utilizing the distance and direction information according to the differentiations among population. Moreover, the searching behavior of each individual is adjusted by dynamically altering the differentiation's direction and step length. At each generation, the *mutation* and *crossover* operators are applied to individuals to generate a new population. Then, *selection* takes place and the population is updated. The key parameters in DE are size of population (PS), scaling factor (F) and crossover parameter (CR).

3   Proposed Completion Time Procedure in HFFS/SDST

In the flow shop scheduling, it is required only one permutation of jobs, where a single permutation sequence is used for all machines. In the HFFS/SDST problem, a single permutation of jobs, where this permutation is maintained through all stages, results in poor performance [1]. A better method is needed to determine the job sequence at the beginning at each stage. In this paper we used earliest completion time (ECT) rule that we adopted from Naderi et al. [1]. We develop a procedure that does not separate the decisions of job sequencing and machine assignment. An outline of the procedure is shown in Fig. 1 below.

> **The Lowest Completion time Procedure**
> At the first stage all the ready time of jobs at machines are zero
> **For** $i=1$ to m do
> N=set of all n jobs
> Ready time of jobs at stage i=completion time of jobs at stage i-1
> **While** $| N | > 0$ do
> - Calculate completion time of job j after assigning job j to each parallel machine at stage i
> -   Assign job j to parallel machine that resulting lowest completion time.
> -   Extract job j from N
> -   Update the ready time of each parallel machine at stage i
> **Endwhile**
> **Endfor**

Fig. 1. The lowest completion time procedure

3.1   The Scheme of dE-VNS for Two-Objective HFFS/SDST

In this section, we presents the hybridization of DE and VNS for two-objective HFFS/SDST. The procedure of DE-VNS is summarized as folows:

Step 0:

- Let g denote a generation, P_g a target population with size NP in generation g, $X_{i,g}$ the ith individual with dimension N ($N=n$) in P_g. $x_{i,j,g}$ the jth variable of individual $X_{i,g}$. CR the crossover probability, F the fraction of mutation, and random(0,1) the random value in the interval [0,1]. Where NP is the number of population, $i=1,2,...,$ NP. $Xmax$ the upper limit, $Xmin$ the lower limit. $Gmax$ the maximum generation or iteration.
- Input the parameter value N, PS, $CR \in [0,1]$, $F \in [0,1]$, $Gmax$. Let $Xmax=4$, $Xmin=0$.
- Set $S=\emptyset$, where S is the archive of non-dominated solutions.

Step 1a, population initialization:

- Generate population using

$$X_{i,j0} = \text{random } (0, 1) * (Xmax_{i,j} - Xmin_{i,j}) + Xmin_{i,j}. \qquad (3)$$

- Convert $X_{i,j\ 0}$ being an individual job permutation using smallest parameter value (SPV) rule.

Step 1b, population initialization:

   We develop combination of largest processing time (LPT) rule and earliest due date (EDD) rule, and then we named it as LPT-EDD rule. LPT-EDD rule serves to improve initial population. The individual solution produced by LPT-EDD rule was collected in a population that we named it as $P_{LPT\text{-}EDD}$. Procedure of the LPT-EDD rule is explained as follows:

1. Set iteration = 1, and determine the number of job permutations N to be produced (the number of job permutations N is set equal to NP).

2. Generate *rand*1 and *rand*2 using the random value in the interval [0,1] for each individual of job permutation to make the weight vector, $w1$ and $w2$, where $w1 = rand1 / (rand1 + rand2)$, and $w2 = rand2 / (rand1 + rand2)$.
3. Generate a job permutations Π^1 by ordering jobs based on the order of the largest job processing time down to the smallest (LPT rule). Then also generate another job permutations Π^2 by using EDD rule.
4. Compute the weighted sum of jobs position value

$$\Phi_j = w_1 \Phi_j(\Pi^1) + w_2 \Phi_j(\Pi^2), \tag{4}$$

where $j = 1,2,\ldots,n$, and $\Phi_j(\Pi^1)$ denotes the position value of j in permutation Π^1, and $\Phi_j(\Pi^2)$ denotes the position value of j in permutation Π^2.
5. Generate a permutation Π^0 by ordering jobs in ascending weighted sum of their position value Φ_j.
6. If iteration $\le$ N, then iteration = iteration + 1 and proceess go back to step no. 2. Otherwise, stop the procedure.

Step-2, Selection between initial target population (P_0) and $P_{LPT\text{-}EDD}$:

- Calculate the two-objective function of each individual target population, $X_{i,0}$ and individual populations generated by LPT-EDD.
- Calculate the value of fitness function of each individual initial population, $f_{i,0}(P_0)$ as well as each individual populations produced by LPT-EDD, $f_i(P_{LPT\text{-}EDD})$.
- Evaluation method of multi-objective function in this paper adopts the scalar fitnees function [7] the following:

$$f(x) = w1 f_1(x) + w2 f_2(x) + \ldots w_K f_K(x) \tag{5}$$

where, $w_i \ge 0$, $i = 1,2,..,K$, and K is the number of the objective function. To maintain the diversity of non-dominated solutions (pareto solution) obtained as well as to enrich the search direction, the weight w_i randomly generated as follows:
Step-3, Compare the fitness values between $f_{i,0}(P_0)$ and $f_i(P_{LPT\text{-}EDD})$:

- Each individual solution in initial target population (P_0) will be compared with each solution generated by LPT-EDD rule ($P_{LPT\text{-}EDD}$).
- Individuals solution that has the smallest fitness value was selected into the selection population, $P_{selection}$

Step-4, initialization of the archive of pareto solutions, $Sp^{g=0}$:

- Find the non-dominated solution in a selection population, $P_{selection}$. Each point of non-dominated solutions that have been found from the $P_{selection}$ will be entered into the archive of pareto solutions, $Sp^{g=0}$.

Step-5, update the count generation:

- Set $g = g + 1$

Step-6, generate mutant population:

- For each target individual, $X_{i,g}$ at generation g, a mutant individual, $V_{i,g+1}$= is determined by the following:

$$V_{i,g+1} = X_{a,i,g} + F * \left(X_{best,i,g} - X_{a,i,g}\right) + F * \left(X_{b,i,g} - X_{c,i,g}\right) \tag{6}$$

Where a_i b_i, c_i are three individual randomly choosen from the population such that $a_i \neq b_i \neq c_i$. $F > 0$ is a mutant factor which affects the differential variation between two individuals. X_{best} is one of the individual randomly taken from the current archive of pareto solutions.

Step-7, generate trial population:
Following the mutation phase, the crossover (recombination) operator is applied to obtain the trial population.

Step-8, the evaluation of the fitness function:

- Calculate the two-objective function of each individual trial population, $U_{i,g+1}$
- Evaluation method of multi-objective using scalar fitness function

Step-9, selection between target population ($X_{i,g}$) and trial populations ($U_{i,g+1}$) to determine the best solution for the next generation:

Step-10, Applying VNS local search techniques to the 1/10 best individuals in target population (P_{g+1}):
The ten percent of the best individual in population (P_{g+1}) is selected by applying VNS algorithm. It is used to improve the solution obtain in each generation.

Step-11, Updating the archives of pareto solutions, $Sp^{g=g\ +1}$
Update the archive of pareto solutions $Sp^{g=g\ +1}$ by finding the non-dominated in a population of generation $g+1$ (P_{g+1}). Each point of non-dominated solutions that have been discovered, is added to the archive of pareto solutions. If there are points of non-dominated solution in the archives of pareto solutions are dominated by the new non-dominated solutions, then the dominated solutions will be eliminated from the archives of pareto solutions.

Step-12, stopping criterion:
if the maximum number of generations is reached, stop. Otherwise, go to *step*-6.

3.2 Performance Metric

We need a performance metric to measure how well the set of solutions obtained by an algorithm. The performance metrics are explained as follows.

3.2.1 Number of Non-dominated Solutions (NNDS)
This metric counts the total number of those those solution obtained by a algorithm that are not dominated by any other solution obtained by another algorithm. It is formulated as follow:

$$\mathrm{NNDS}(S_j) = |S_j - \{x \in S_j | \exists y \in S : xy\}| \tag{7}$$

Where S_j (j=1,2,3) is the non-dominated solution set obtained by algorithm j, and $S = \cup S_j$ the union of all the non-dominated solution sets. $x \prec y$ means that the solution x is dominates by the solution y. The larger value of NNDS(S_j) is the better solution set S_j.

3.2.2   Average Distance (D1$_R$)

This metric is used to measure the performance of non-dominated solution set S_j relative to a reference set S^* of the *Pareto front*. The average distance (D1$_R$) is the average of those shortest normalized distance from all the reference solution S^* to the set S_j. This metric is formulated as follow:

$$D1_R(S_j) = \frac{1}{|s*|} \sum_{y \in S_*} d_y(S_j) \tag{8}$$

$$d_y(S_j) = \min_{x \in S_*} \left(\sqrt{\sum_{i=1}^{2} \left(\frac{f_i(x) - f_i(y)}{f_i^{\max}(\cdot) - f_i^{\min}} \right)^{-2}} \right) \tag{9}$$

where d_y denotes the shortest normalized distance from a reference solution $y \in S^*$ to solutions set S_j. When comparing the metrics of two non-dominated solution sets S_1 and S_2 if the Pareto front is not known, we will combine the two sets and select all the non-dominated solutions to form set S^*, $f_i(\cdot)$ is the ith objective value, and $f_i^{\max}(\cdot)$ and $f_i^{\min}(\cdot)$ are the maximum and minimum of the ith objective value in the reference set S^*, resepectively. Obviously, a smaller D1R(S_j) value corresponds to a better distribution of Sj and better approximation to the Pareto front. Furthermore, if D1R(S_j)=0, all the reference solution S^* are include in the soluiton set Sj. In other word, the algorithm j can find all the reference solutions [7].

3.2.3   Overall Non-dominated Vector Generation (ONVG)

This metric is used to measure the number of distinct non-dominated solution set S_j. ONVG (S_j) is formulated as follows:

$$\mathrm{ONVG}(S_j) = |S_j| \tag{10}$$

4   Experiment and Analysis

To test the performance of the proposed DE-VNS for HFFS/SDST, we generate test instances with the following combination of n and m, where n=[20,50,120] and m= [2, 4, 8]. We also generate group of instances with two parallel machines at each stage and group using a uniform distribution between [2, 5]. The processing times (p_{ij}) are generated from a uniform distribution [50,100]. S_{ijk} denotes the sequence dependent setup time generated from a uniform distribution [20,40]. T_{ij} denoted the setup time of job j if job j is assigned to parallel machine at the first position at stage i generated from a uniform distribution [20,40]. The probability of skipping a stage for each job (S_p) is set at 0.1. The due date of each job is uniformly distributed at interval [$\mu *m$, $\mu *(n+m-1)$], where μ is the mean of processing times. Therefore, the total number of test instances is 18. A summary of the instances is given in Table 1. below.

Table 1. Data employed for the benchmark generation

Factor	Levels		
N	20	50	120
M	2	4	8
m_i	Constant	2	
	Variable	U(2,5)	
P_{ij}	U(50,100)		
S_{ijk}	U(20,40)		
S_p	0,1		
T_{ij}	U(20,40)		
d_i	U(μm,μ(n+m-1))		

The proposed DE-VNS algorithm is implemented with the parameters CR=0,8; F=0,5. For each instance, each algorithm is run for 10 replications and for each replication the above three performance metrics are computed. The average value of these metrics of 10 replications are calculated as the statistics for the performance.

We compared DE-VNS with DE-Insert and PSO-VNS. The algorithm is executed for solving HFFS/SDST problem. We set the maximum generation of HDE as $Gmax$=50 and 100, respectively for stopping criteria. The statistics of NNDS, $D1_R$ and ONVG produced by the three algorithms are reported in Table 2. It can been seen from

Table 2. Average of NNDS, $D1_R$ and ONVG

	Problem		NNDS			$D1_R$			ONVG		
No	N,m	m_i	DE-VNS	DE-Insert	PSO-VNS	DE-VNS	DE-Insert	PSO-VNS	DE-VNS	DE-Insert	PSO-VNS
1	20,2	2	0,6	1	1	0,67	0,678	0,841	1,000	1,800	1,400
2	20,4	2	1,6	0,1	0,1	,978	8,648	4,391	1,000	1,800	2,700
3	20,8	2	4,9	0	0,0	2,064	6,881	7,809	5,100	2,300	4,300
4	20,2	variable	3,4	0,3	0,4	0,269	0,305	0,452	3,800	4,700	3,400
5	20,4	Variable	1	0,3	0,2	1,619	16,843	5,71	1,000	1,000	1,100
6	20,8	Variable	1,3	0	0	6,306	18,078	26,649	1,300	1,000	1,000
7	50,2	2	3,6	0	0,3	1,892	39,63	6,716	4,100	3,200	4,400
8	50,4	2	2,8	0,3	1,2	4,539	39,067	7,023	3,300	4,300	5,100
9	50,8	2	4,6	0	0	1,557	82,229	10,414	4,600	4,000	4,600
10	50,2	Variable	1,4	0,8	0,1	0,46	0,979	0,948	1,400	2,400	1,500
11	50,4	Variable	4,8	0	0,0	11,401	138,38	48,455	4,800	3,400	4,400
12	50,8	variable	3,6	0	0,0	0,689	19,411	3,81	3,600	4,200	3,250
13	100,2	2	3	0	0,2	0,7	11,871	2,720	3,000	0,000	0,200
14	100,4	2	3,4	0	0	1,835	56,178	11,034	3,400	2,400	2,800
15	100,8	2	3,6	0	0	8,676	1399,72	113,74	3,600	1,400	2,000
16	120,2	Variable	1,6	0,6	0	,546	10,534	2,01	1,600	1,000	1,400
17	120,4	Variable	3	0	0	0,696	44,324	6,605	3,200	2,200	3,200
18	120,8	Variable	1	0	0	4,667	128,764	18,003	1,000	1,00	2,200
		average	2,733	0,189	0,194	2,754	112,362	15,407	2,837	2,494	2,897

that the DE-VNS algorithm performs better than the other two algorithms in terms of NNDS and also $D1_R$. Especially, in term of the $D1_R$ metric, it indicates that DE-VNS algorithm produces solutions which closer to the *Pareto front* than DE-Insert and PSO-VNS. But, PSO-VNS performs better than DE-VNS in terms of the ONVG metric, all non-dominated solution set obtained by PSO-VNS is dominated by non-dominated solution set obtained by DE-VNS.

To better understand the performance of the DE-VNS, typical results of DE-VNS, DE-Insert and PSO-VNS for instances $n=50$ $m=4$ $m_i=2$ and $n=120$ $m=2$ $m_i=2$ are shown in Fig. 2 (a) and (b). The figures show the solution combinations of makespan and maximum lateness for each method. It can be seen that the DE-VNS outperforms the DE-Insert and PSO-VNS based on the values of makespan and maximum lateness

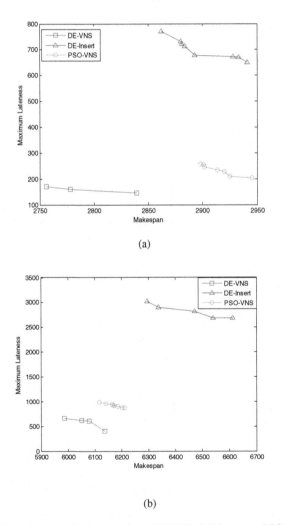

(a)

(b)

Fig. 2. Non-dominated solution of the DE-VNS, DE-Insert and PSO-VNS

which both are smaller compared to those of DE-Insert and PSO-VNS. The similar results are obtained for the other problems with different sizes.

5 Conclusion

We have applied DE-NVS for hybrid flow shop scheduling. We use shortest parameter value (SPV) rule to convert the continuous values of individuals in DE to job permutation. Based on the above comparison, we conclude that the proposed DE-VNS algorithm is more efficient than DE-Insert and PSO-VNS for the hybrid flexible flow shop with sequence dependent setup time (HFFS/SDST) scheduling problem with minimization of the makespan and maximum lateness criteria. The set of non-dominated solutions generated by DE-VNS provides performance metrics of non-dominated solutions (NNDS) which is greater than those generated by DE-Insert and PSO-VNS. The set of non-dominated solutions generated by DE-VNS algorithm gives performance metric average distance (D1R) which is smaller than those of DE-Insert and PSO -VNS. This means that the DE-VNS is able to generate the number of solutions set which is closer to the set of Pareto compared to DE-Insert and PSO -VNS. Although the overall metric of non-dominated vector generation (ONVG) of PSO -VNS has greater value, points of solution produced by the PSO -VNS is generally dominated by a set of solutions of DE-VNS. The compromise solutions through a set of non-dominated solutions generated by DE-VNS provide alternative solutions for decision-making on the production floor. Finally, we realize that we need to extend the application of the proposed algorithm on large size problems in order to strengthen our conclusion.

Acknowledgements. This research was supported by Minsitry of Research and Technology and Higher Education Indonesia.

References

1. Naderi, B., Ruiz, R., Zandieh, M.: Algorithms for a realistic variant of flow shop scheduling. Comput. Oper. Res. **37**, 236–246 (2010)
2. Ruiz, R., Antonio, J., Rodriquez, V.: Invited review: the hybrid flow shop scheduling problem. Eur. J. Oper. Res. **205**, 1–18 (2010)
3. Ruiz, R., Rodrquez, J.A.: The hybrid flow shop scheduling problem. Eur. J. Oper. Res. **16**, 1–18 (2010)
4. Ishibuci, H., Murata, T.A.: Balance between genetic search and local search in memetic algorithm for multiobjective permutation flowshop scheduling. IEEE Trans. Evol. Comput. **7** (2), 204–223 (2003)
5. Tasgetiren, M.F, Liang, Y.C, Sevkli, M, Gencyilmaz, G.: Particle Swarm Optimization Algorithm for Makespan and Maximum Lateness in Permutation Flowshop Sequencing Problem, Dept. of Management, Fatih University, Istambul-Turkey (2004)
6. Li, B.-B., Wang, L., Liu, B.: An effective PSO-based hybrid algorithm for multiobjective permutation flow shop scheduling. IEEE Trans. Syst. Manuf. Cybern. **17**, 314–317 (2008)

7. Jungwattanakit, J., Reodecha, M., Chaovalitwongse, P., Werner, F.: Algorithms for flexible flow shop problems with unrelated parallel machines, setup times, dual criteria. Int. J. Adv. Manuf. Technol. **37**(3), 354–370 (2008)

8. Qian, B., Wang, L., Huang, D., X, W.: An effective hybrid DE-based algorithm for multiobjective flow shop scheduling with limited buffers. Comput. Oper. Res. **36**, 209–233 (2009)

9. Storn, R., Price, K.: Differential evolution: a simple and efficient heuristic for global optimization over continuous spaces. J. Global Optim. **11**(4), 341–359 (1995)

10. Yang, Z., Tang, K., Yao, X.: Self-adaptive differential evolution with neighborhood search, Evolutionary Computation. In: CEC 2008 (IEEE World Congress on Computational Intelligence), pp. 1110–1116 (2008)

11. Tasgetiren, M., Suganthan, P.N., Ozcan, S., Kizilay, D.: A differential evolution algorithm with a variable neighborhood search for constrained function optimization. In: Fister, I., Fister Jr., I. (eds.) Adaptation and Hybridization in Computational Intelligence. ALO, vol. 18, pp. 171–184. Springer, Heidelberg (2015)

12. Tasgetiren, M.F., Pan, Q.-K., Kizilay, D., Suer, G.: A differential evolution algorithm with variable neighborhood search for multidimensional knapsack problem, Evolutionary Computation. In: 2015 IEEE Congress on Evolutionary Computation (CEC), pp. 2797–2804 (2015)

Objective Space Partitioning with a Novel Conflict Information Measure for Many-Objective Optimization

Naili Luo[1,2], Jianping Luo[1,2], and Xia Li[1,2(✉)]

[1] College of Information Engineering, Shenzhen University,
Shenzhen 518060, China
lixia@cuhk.edu.cn
[2] Shenzhen Key Lab of Communication and Information Processing,
Shenzhen 518060, China

Abstract. In this paper, we present a novel conflict information measure used for objective space partitioning in solving many-objective optimization problems. Obtained from the current Pareto front approximation to estimate the degree of conflict between objectives, the conflict information is simply evaluated by counting the occurrence of conflict (improvement vs deterioration) out of all decision making sample pairs. The proposed method is compared with the latest objective space partitioning based on Pearson correlation coefficient conflict information. The results show that the proposed method outperforms the comparison method on identifying the conflicting objectives.

Keywords: Many-objective optimization · Multi-objective evolutionary algorithm · Objective space partitioning · Conflict information

1 Introduction

Since the first Pareto optimality-based multi-objective evolutionary algorithm (MOEA) [1] appeared to solve multi-objective optimization problems (MOPs) in the mid-1980 s, a variety of new effective and efficient MOEAs [2–4] have been proposed and have aroused the interest of many researchers. However, many real-world problems involve simultaneously optimizing many conflicting objectives (in most cases, more than three), which are commonly referred to as many-objective optimization problems (MaOPs) [5, 6]. Compared with low-dimensional MOPs, the large number of conflicting objectives brings many challenges to multi-objective evolutionary algorithms (MOEAs), such as the inefficiency of selection operators, since most solutions in the population of an MaOP are non-dominated [7], and the high computational cost, e.g., if it meets the regularity property, the dimensionality of its Pareto optimal front can be M-1, and for a continuous MaOP with M objectives, the number of points needed for approximating the whole Pareto optimal front increases exponentially [8].

In general, methods used for MaOPs can be roughly classified into two categories [9]: one making use of alternative preference relations and the other transforming the original MaOP into a related one with reduced size.

Y. Tan et al. (Eds.): ICSI 2016, Part I, LNCS 9712, pp. 599–605, 2016.
DOI: 10.1007/978-3-319-41000-5_60

Objective space partitioning [10] falls into the second category and has been proposed to solve the MaOPs. The basic idea is to use objective space partitioning to divide the original high dimensional objective space into non-overlapping subspaces of lower dimensions, such that conventional MOEAs such as the well-known NSGA-II can be efficiently used. There are two primary ways to partition the objective space, random subspace exploration, as the name implies, distributing objectives to different subspaces in an arbitrary manner, or partitioning based on conflict information evaluation between objectives. Recently researchers have proposed to use the Pearson correlation coefficient between objectives for partitioning [11]. To estimate the degree of conflict and the results are very encouraging. This paper presents a new method for measuring conflict information between objectives. The information is derived from the Pareto front approximation set inspired by the dominance of solutions in MOPs. It is used to partition the objective space in the redundant many-objective optimization problem and to identify the conflicting objectives.

The remainder of this paper is organized as follows: Sect. 2 presents problem description and the basic concepts adopted throughout the paper. Section 3 briefly describes objective space partitioning methods and the new conflict information estimation method. Section 4 tests the effectiveness for the new method and the experimental results analysis. Finally, we present our conclusions in Sect. 5.

2   Problem Description and Basic Concepts

Without loss of generality, we consider many-objective optimization problems (MaOPs) from a minimization perspective. The corresponding objective can be defined as follows:

$$\text{Minimize } F(x) = [f_1(x), \cdots, f_M(x)]^T.$$
$$\text{subject to } x \in X. \tag{1}$$

where $M \geq 4$, $X \subseteq \mathbb{R}^V$ is the decision-making set of the problem and the vector function $F : X \to \mathbb{R}^M$ is composed of M scalar objective functions with $f_i : \mathbb{R}^V \to \mathbb{R}$ $(i = 1, \cdots, M)$. $\mathbb{R}^V$ and $\mathbb{R}^M$ are known as the decision variables space and objective function space, respectively. The image of X under the function F is a subset of the objective function set denoted by $\Phi = F(X)$, or $\Phi = \{(f_1(x), \cdots, f_M(x)) | x \in X\}$, in this paper, we call Φ objective space.

Some basic concepts [11] are listed below for later use.

Definition 1 (Pareto dominance relation): A solution x_1 is said to Pareto dominate another solution x_2 in the objective space formed by Φ, denoted by $x_1 \prec x_2$, if and only if: $\forall f_i \in \Phi : f_i(x_1) \leq f_i(x_2) \wedge \exists f_i \in \Phi : f_i(x_1) < f_i(x_2)$.

A solution $x^* \in X$ is Pareto optimal if another solution $x \in X$ does not exist such that $x \prec x^*$. $P_{opt} = \{x \in X | \nexists y \in X : y \prec x\}$ is called the optimal Pareto set, and $PF_{opt} = \{z = f_1(x), \cdots, f_M(x) | x \in P_{opt}\}$ is the Pareto front.

Definition 2 (Pareto front approximation): Let $PF_{approx} \subset \Phi$ be a set of objective function vectors, it is called an approximation set if any element of PF_{approx} does not dominate or is not equal to any other objective function vector in PF_{approx}.

Definition 3 (subspace): The set ψ is an objective subspace of objective space Φ if ψ is a lower-dimensional space that includes some of the objective functions in Φ, i.e., $\psi \subset \Phi$.

If $\Psi = \{\psi_1, \cdots, \psi_{N_s} | [\cup_{i=1}^{N_s} \psi_i = \Phi] \wedge [\cap_{i=1}^{N_s} \psi_i = \phi]\}$ where N_s is the number of the subspaces, then Ψ is a partitioning of objective space Φ.

3 Objective Space Partitioning with Novel Conflict Information

Objective space partitioning [10] splits the M dimensional space $\Phi = \{f_1, \cdots, f_M\}$ into N_s non-overlapping subspaces $\Psi = \{\psi_1, \cdots, \psi_{N_s}\}$. For the MaOPs with conflicting objectives, it is intuitive that conflicting objectives may contribute more in finding the solution to the problem while less conflicting objectives weigh less. Therefore, one recently proposed partitioning strategy makes use of conflict information which is estimated by calculating the Pearson correlation coefficient obtained from the current Pareto front approximation between objectives [11].

In this section, we propose a new conflict information measuring scheme based on an intuitive statement [12] for conflict among objectives, i.e., the conflict between two objectives means that the improvement on one objective would deteriorate on the other.

The conflict information used in this paper is measured as follows:

Given an MaOP with M objectives, suppose there is an N-individual population in the decision-making set $POP = \{x_1, \cdots, x_N\}, x_i \in X, i = 1, \cdots, N$, with C_N^2 different pairs of individuals $(x_k, x_l), k, l = 1, \cdots, N, k \neq l$. For any two objectives $f_i, f_j \in \Phi, i, j = 1, \cdots, M$, if $K(0 \leq K \leq C_N^2)$ different pairs of individuals satisfy the following conditions:

$$[f_i(x_k) \geq f_i(x_l)] \wedge [f_j(x_k) \geq f_j(x_l)]$$
$$\text{or} \tag{2}$$
$$[f_i(x_k) \leq f_i(x_l)] \wedge [f_j(x_k) \leq f_j(x_l)]$$

then the conflict information between the i-th objective f_i and the j-th objective f_j will be simply evaluated as

$$P_{i,j} = 1 - K/C_N^2 \tag{3}$$

Clearly, the new conflict information between the i-th objective f_i and the j-th objective f_j $P_{i,j}$ has the following properties:

(1) Reflexivity: $P_{i,j} = 0, i = j$;
(2) Symmetry: $P_{i,j} = P_{j,i}$.

A symmetric conflict information matrix $C=[P_{i,j}]_{M \times M}$ with zero-diagonal element can thus be constructed. Specifically, $P_{i,j} = 1$ implies that two objectives f_i and f_j conflict with each other.

The focus of this paper is to investigate the efficiency of the proposed conflict information estimation. Therefore, the framework for objective space partitioning remains the same as that described in [11]; the difference lies in the evaluation of the conflict information. It is expected that partitioning is carried out in an ordered manner with the first objective subspace being constructed with the lowest conflict while the last with the highest conflict.

4   Experimental Results and Analysis

To evaluate the effectiveness of the proposed conflict information for objective space partitioning, we use DTLZ5(I,M) [13] as a test problem. In this problem the conflicting objectives are priori known, i.e., DTLZ5(I,M) implies that of the M objectives, only I of them are required to completely generate the Pareto front. So, the corresponding objective space can be classified into two subsets, the redundant subset $F_R = \{f_1, \cdots, f_{M-I+1}\}$ and the necessary subset $F_N=\{f_{M-(I-2)}, f_{M-(I-3)} \cdots, f_M\}$, composed of the last $I - 1$ objectives. The Pareto front can be generated using only one objective from F_R and all of the elements in F_N. There is conflict among every objective in F_N but no conflict among elements in F_R. Nonetheless, there is conflict from elements in F_R to objectives in F_N.

In the experiment, we first investigate the effects of the population size on the correct rate of the partition for DTLZ5($I = M/2,M$) with different number of objectives. The size of the population whose elements are randomly generated is supposed to be from 5 to 500 and the number of objectives varies between $8 \sim 56$. The objective space is partitioned into two subspaces. The experiments are carried out with 1000 independent runs and the results are averaged and shown in Fig. 1.

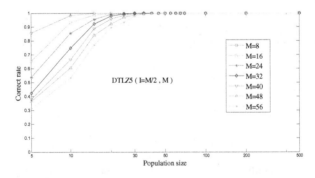

Fig. 1. Correct rate of partition with different population for DTLZ5($I = M/2,M$) (Color figure online)

Figure 1 shows that with the increase of the size of the population, the correct rate of the partition grows gradually up to 100 %. An interesting observation is that when the population size reaches certain value, say 100, or above, the partition is 100 % correct for objectives as many as 56. This implies that in real applications, a suitably chosen population size is sufficient and efficient in solving the complicated optimization problem. In the following experiment we set $N = 200$.

To evaluate the performance of the new conflict information measure, we use the same partitioning framework used in [11] and compare it with the partitioning performance. In this framework, the algorithm repeats G iterations of a two-phase search process that consists of approximation phase and partitioning phase. In the approximation phase the NSGA-II searches G_Φ generations in the original objective space, to update the PF_{approx} and generate new population for partitioning. In the partitioning phase, subspaces are created with NSGA-II applied to search for the optimal solution in each subspace.

In this experiment, we adopt the same parameter values for NSGA-II as those used in [11]. Regarding the setting of the dimensions of the objective subspace, it is assumed that NSGA-II, which claims to be one of the best MOEAs when $2 \sim 6$ objectives are used, is the same as that employed in [11].

We experimented on DTLZ5 (I, M) with 6 to 24 objectives. In addition, we use $I = M/2$ for all M. In the following, we simply show results for DTLZ5($I = 5$, $M = 10$) and DTLZ5($I = 12$, $M = 24$). Apparently, if the size of the subspace is around $2 \sim 6$ and as large as possible, then the former will have two subspaces with $f_7 \sim f_{10}$ conflict, whereas the latter will have six subspaces with $f_{14} \sim f_{24}$ conflict.

Figure 2 shows one realization of the partitioning for DTLZ5($I=5$, $M=10$), in which X axis denotes the number of the iterations for the two-phase search process. The partitioning has two subspaces indicated by either "o" or "☆". It is expected that one of the subspace should contain objectives $7 \sim 10$ and any one of the other objectives. From Fig. 2(a), it is obvious that even though eventually the partitioning is correct, we observe that at the 1st, 3rd, 4th and the 5th iteration, the partitioning is incorrect for the Pearson correlation based conflict information; while for our method, partitioning is always correct, which is shown in Fig. 2(b).

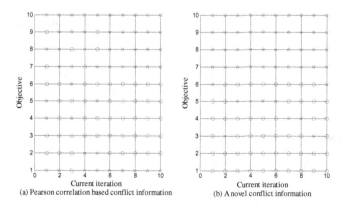

(a) Pearson correlation based conflict information   (b) A novel conflict information

Fig. 2. Objective space partitioning results for DTLZ5 ($I = 5$, $M = 10$). Objectives $7 \sim 10$ and any one of the other objectives are the conflicting objectives.

To further explore the validity of the proposed conflict information measure, we test on the more complicated DTLZ5($I = 12$, $M = 24$) with six subspaces and the results are illustrated in Fig. 3. The partitioning has six subspaces, subspaces 1-3 are indicated by "o", "+" and "*", while subspaces 4-6 are marked with "◇", "◁"and "▷". It is expected that objectives $14 \sim 24$ and any one of the other objectives should be partitioned into subspace 4 to 6 (conflicting objectives in subspaces with larger labels), and the remaining objectives belong to subspaces 1 to 3. From Fig. 3(a), it is clearly seen that, only the 8th and the 9th partitioning contain the correct subspaces. On the contrary, Fig. 3(b) shows that even though the number of objectives is large, the new conflict information based method still provide correct objective space partitioning. What is more interesting is that the subspaces generated at the initial stage of the evolution is also correct, thus making the whole searching process justifiably efficient.

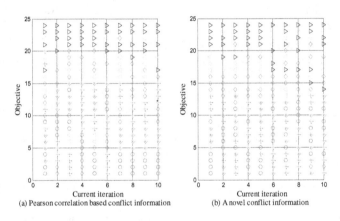

(a) Pearson correlation based conflict information   (b) A novel conflict information

Fig. 3. Objective space partitioning for DTLZ5 ($I = 12$, $M = 24$). Objectives $14 \sim 24$ and any one of the other objectives are the conflicting objectives.

5   Conclusion

In this paper, we propose a new conflict information estimation method which employs statistical method to count the conflict occurrence between objectives in the approximation Pareto optimal set for objective space partitioning in solving many-objective optimization problems. Experimental results show the effectiveness of the proposed method compared with the latest Pearson correlation coefficient based objective space partitioning.

Acknowledgments. This work is supported by the National Natural Science Foundation of China (Grant No. 61171124, 61301298) and Shenzhen Key project for Foundation Research (JC201105170613A).

References

1. Schaffer, J.D.: Multiple objective optimization with vector evaluated genetic algorithms. In: Proceedings of the 1st international Conference on Genetic Algorithms. pp. 93–100. L. Erlbaum Associates Inc. (1985)
2. Deb, K., Pratap, A., Agarwal, S., Meyarivan, T.: A fast and elitist multi-objective genetic algorithm: NSGA-II. IEEE Trans. Evol. Comput. **6**, 182–197 (2002)
3. Coello, C.A.C., Pulido, G.T., Lechuga, M.S.: Handling multiple objectives with particle swarm optimization. IEEE Trans. Evol. Comput. **8**, 256–279 (2004)
4. Xiaohong, C., Xia, L., Na, W.: Objective reduction with sparse feature selection for many objectve optimization problem. Chin. J. Electron. **43**, 1300–1307 (2015)
5. Li, B., Li, J., Tang, K., Yao, X.: Many-objective evolutionary algorithms: a survey. ACM Comput. Surv. (CSUR) **48**, 1–35 (2015). Article no. 13
6. Chand, S., Wagner, M.: Evolutionary many-objective optimization: a quick-start guide. Surv. Oper. Res. Manag. Sci. **20**, 35–42 (2015)
7. Schutze, O., Lara, A., Coello, C.A.C.: On the influence of the number of objectives on the hardness of a multi-objective optimization problem. IEEE Trans. Evol. Comput. **15**, 444–455 (2011)
8. Corne, D.W., Knowles, J.D.: Techniques for highly multi-objective optimization: some non-dominated points are better than others. In: Proceedings of the 9th annual conference on Genetic and evolutionary computation. pp. 773–780. ACM (2007)
9. von Lücken, C., Barán, B., Brizuela, C.: A survey on multi-objective evolutionary algorithms for many-objective problems. Comput. Optim. Appl. **58**, 1–50 (2014)
10. Aguirre, H., Tanaka, K.: Many-objective optimization by space partitioning and adaptive ε-ranking on MNK-landscapes. In: Ehrgott, M., Fonseca, C.M., Gandibleux, X., Hao, J.-K., Sevaux, M. (eds.) EMO 2009. LNCS, vol. 5467, pp. 407–422. Springer, Heidelberg (2009)
11. Jaimes, L., Jaimes, L., Coello, C.A.C., Aguirre, H., anaka, K.: Objective space partitioning using conflict information for solving many-objective problems. Inf. Sci. **268**, 305–327 (2014)
12. Carlsson, C., Fullér, R.: Multiple criteria decision making: the case for interdependence. Comput. Oper. Res. **22**, 251–260 (1995)
13. Deb, K., Saxena, D.: Searching for pareto-optimal solutions through dimensionality reduction for certain large-dimensional multi-objective optimization problems. In: Proceedings of the World Congress on Computational Intelligence (WCCI-2006). pp. 3352–3360 (2006)

Adaptive Multi-level Thresholding Segmentation Based on Multi-objective Evolutionary Algorithm

Yue Zheng[1], Feng Zhao[1($\boxtimes$)], Hanqiang Liu[2], and Jun Wang[1]

[1] School of Telecommunication and Information Engineering,
Xi'an University of Posts and Telecommunication,
Xi'an, People's Republic of China
xiah_yzheng@163.com, fzhao.xupt@gmail.com
[2] School of Computer Science, Shaanxi Normal University,
Xi'an, People's Republic of China

Abstract. In this paper, an adaptive multi-level thresholding segmentation based on multi-objective evolutionary algorithm (AMT_ME) is proposed. Firstly, the between-class variances and the entropy criteria are utilized as the multi-objective fitness functions. Then the threshold-based encoding technique, variable length sub-populations, crowed binary tournament selection, hybrid crossover operation, mutation and non-dominated sorting are produced in this article. Finally, the weight ratio of intra-class variance and between-class variance is used to obtain the optimum number of the thresholds and the optimal thresholds. The experimental results demonstrate good performance of the AMT_ME in solving thresholding segmentation problem by compared with ATMO, Otsu's and Kapur's methods.

Keywords: Image segmentation · Multi-objective evolutionary · Adaptive thresholding · Hybrid crossover

1 Introduction

Image segmentation is one of the most important and difficult task in image processing and pattern recognition. Its main purpose is to classify the input image into more meaningful parts such as foreground and background segments [1, 2]. Among all the segmentation techniques, image thresholding technique has become a widely used tool in image segmentation because of its simplicity and effectiveness [3, 4]. Image thresholding is based on the assumption that the targets can be distinguished by their gray-level. The pixels belonging to the same class have gray-levels within a specific range defined by several thresholds [5]. Among all the thresholding technique, the Otsu's method [6] and the Kapur's method [7] are the most popular ways. Otsu's method is the histogram thresholding method which utilizes the between class variance to choose an optimal threshold, and it can achieve a desire result to the images with obvious bimodal histogram. Kapur's method which is to maximize the entropy of the image histogram can also obtain a good segmentation result with no obvious bimodal. However, with the increasing of thresholds number, computation time grows

© Springer International Publishing Switzerland 2016
Y. Tan et al. (Eds.): ICSI 2016, Part I, LNCS 9712, pp. 606–615, 2016.
DOI: 10.1007/978-3-319-41000-5_61

exponentially owing to its exhaustive searching strategy which limit the application of the multi-level thresholding segmentation algorithm.

To overcome these problems, many optimization methods such as genetic algorithm [8] and particle swarm optimization [9] have been applied to image segmentation to eliminate the computation complexity and time. Recently, researchers proposed many thresholding approaches based on multi-objective optimization [10, 11]. Djerou and Khelil [11] proposed an automatic threshold based on multi-objective optimization (ATMO) in which used three objective functions simultaneously. In the existing multi-objective thresholding optimization, most of them may not obtain the optimum number of the threshold.

In this paper, an adaptive multi-level thresholding segmentation based on multi-objective evolutionary algorithm (AMT_ME) is proposed. The maximization of entropy and the maximization of between-class variances are considered as two objective function. In AMT_ME, we used Non-dominated Sorting Genetic Algorithm II (NSGA-II) [12] as the optimization method. Moreover, the hybrid crossover method between different numbers of thresholds is introduced into the proposed algorithm. After producing the set of non-dominated the above objective functions, square error analysis method [13] is used for eventually to choose the optimum threshold and the optimum threshold number. In the experiment, ATMO, Otsu's and Kapur's methods are chosen as the comparison methods. Experimental results on synthetic and natural images show the effectiveness of the proposed method.

2   Adaptive Multi-level Thresholding Segmentation Based on Multi-objective Evolutionary Algorithm

In this paper, we present an adaptive multi-level thresholding segmentation based on multi-objective evolutionary algorithm (AMT_ME). This method involves some issues are as follows.

2.1   Segmentation Criteria

Let N is the number of pixels in a given image whose gray-level range is from 0 to L-1. If n_i is the occurrence of gray-level i, the probability p_i of the occurrence of gray-level i can be described as:

$$p_i = \frac{n_i}{N}. \tag{1}$$

In our approach, entropy criteria and between-class variance criterion are used as the segmentation criteria.

Entropy Criteria. Kapur [7] proposed a thresholding segmentation method which is based on the entropy theory. The entropy of an image is given as follows:

$$f(t_1, t_2, \ldots, t_n) = H_0 + H_1 + \ldots + H_n. \tag{2}$$

Where :
$$H_0 = -\sum_{i=0}^{t_1-1} \frac{p_i}{\omega_0} ln \frac{p_i}{\omega_0} \quad and \quad \omega_0 = \sum_{i=0}^{t_1-1} p_i$$

$$H_1 = -\sum_{i=t_1}^{t_2-1} \frac{p_i}{\omega_1} ln \frac{p_i}{\omega_1} \quad and \quad \omega_1 = \sum_{i=t_1}^{t_2-1} p_i$$

$$H_n = -\sum_{i=t_n}^{L} \frac{p_i}{\omega_n} ln \frac{p_i}{\omega_n} \quad and \quad \omega_n = \sum_{i=t_n}^{L} p_i.$$

The optimal segmentation threshold vector $(t_1^*, t_2^*, \ldots, t_n^*)$ can be obtained by Eq. (3) that tried to maximize the total entropy.

$$\left(t_1^*, t_2^*, \ldots, t_n^*\right) = Arg \max_{0 < t_1 < t_2 \ldots < L} f(t_1, t_2, \ldots, t_n). \tag{3}$$

Between-Class Variance Criterion. The Otsu's method [6] is based on the discriminant analysis. The between class variance of the image is defined as:

$$\begin{aligned}
f(t_1, t_2, \ldots, t_n) = &\; \omega_0\omega_1(\mu_0 - \mu_1)^2 + \omega_0\omega_2(\mu_0 - \mu_2)^2 + \\
&\; \omega_0\omega_3(\mu_0 - \mu_3)^2 + \ldots + \omega_0\omega_n(\mu_0 - \mu_n)^2 + \\
&\; \omega_1\omega_2(\mu_1 - \mu_2)^2 + \omega_1\omega_3(\mu_1 - \mu_3)^2 + \ldots + \\
&\; \omega_1\omega_n(\mu_1 - \mu_n)^2 + \ldots + \omega_{n-1}\omega_n(\mu_{n-1} - \mu_n)^2
\end{aligned} \tag{4}$$

where $\omega_k = \sum_{i=t_k}^{t_{k+1}-1} p_i$, $\mu_k = \sum_{i=t_k}^{t_{k+1}-1} \frac{i \times p_i}{\omega_k}$ and $0 \le k \le n$. The optimal segmentation threshold vector $(t_1^*, t_2^*, \ldots, t_n^*)$ can be obtained by Eq. (5) that tried to maximize the between class variance.

$$\left(t_1^*, t_2^*, \ldots, t_n^*\right) = Arg \max_{0 < t_1 < t_2 \ldots < L} f(t_1, t_2, \ldots, t_n). \tag{5}$$

2.2   Chromosome Representation and Population Initialization

In AMT_ME, threshold-based encoding is used. The chromosomes are made up of real numbers which represent the gray level of the threshold. If a particular chromosome encodes the thresholds number n, its length is taken to be n. For example, the chromosome {50, 90, 162, 189} encodes four thresholds: 50, 90, 162 and 189. In the population, each threshold is considered to be indivisible.

   In this paper, the population are divided into five sub-populations, each of them is the same size. v_s is the sub-population s, where s is the number of thresholds with the range from 1 to 5. Then, each sub-population is initialized separately. In the initial of each sub-population, for a particular chromosome i, each string is encoded by randomly chosen from the range between g_{min} and g_{max}, where $g_{min}(g_{max})$ is a minimum (maximum) of the gray level value of image pixel. Furthermore, the length of chromosomes in different sub-populations is different from each other.

2.3   Selection, Hybrid Crossover and Mutation

Selection is a genetic operator that chooses a chromosome from the population to be parents to crossover and mutation. In the proposed method, conventional binary tournament method [12] is used in the crowed binary tournament selection method to produce a tournament subset of chromosomes.

In AMT_ME, a new crossover method, called hybrid crossover (Shown in Fig. 1), that combines two chromosomes what are randomly selected from different sub-populations or same sub-populations to produce two new offspring with the crossover probability p_c. The number of thresholds for each sub-population is not same, thus hybrid cross can generate more diverse solutions. The double-point crossover operation is introduced in the hybrid crossover method. In double-point two crossover operation, crossover points are chosen uniformly at random.

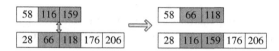

Fig. 1. Hybrid crossover between sub-populations v_3 and v_5.

After the hybrid crossover operation, each threshold encoded in a chromosome should be mutated with the mutation probability p_m. If the kth threshold t_k need to be mutated, its value will become $t_k + 10\alpha$ or $t_k - 10\alpha$ with equal probability, where α is a random number in a given range between 0 and 1 with uniform distribution.

2.4   Elitism and Optimum Solution Selection Criterion

In this paper, NSGA-II is adopted as the underlying multi-objective framework in AMT_ME to develop the thresholding solutions. The elitism operation is the most characteristic part in NSGA-II. Through elitism operation, the non-dominated solutions which are chosen from the parent and child populations are propagated to the next generation. Therefore, the good solutions found so far are retained.

In the final generation of AMT_ME, a set of non-dominated solutions is obtained and all the solutions in final generation are equally important, but the user may want only one solution. The square error analysis method [13] is used for eventually to select the optimum threshold and the optimum number of the thresholds. In this method, the image in each class can obtain a weight ratio of Intra-class variance and between class variance F. By comparing the weight ratio F to achieve the optimum threshold. The maximum of F in the sub-population v_s corresponds to the optimum threshold. F is defined as follows:

$$F = \frac{S_A/(s-1)}{S_E/(N-s)}. \tag{6}$$

Where $S_A = \sum_{j=1}^{s} N_j y_j^2 - N y^2$ and $S_E = \sum_{j=1}^{s} \sum_{i=1}^{N_j} (x_{ij} - y_j)^2$. s is the number of thresholds, N is the total number of pixel, N_j is the total number of pixel in segmented region j, y is the mean of the gray-levels, y_j is the mean of the gray-levels of those pixels in segmented region j. Calculated value of $F_\Delta = F_S - F_{MAX}$ under the current s, where F_{MAX} is the maximum of all F. Select the maximum value from all the F_Δ, which corresponds to the number of s is the optimal number of the threshold.

2.5 AMT_ME Algorithm

The adaptive multi-level thresholding segmentation based on multi-objective evolutionary algorithm (AMT_ME) works as follows:

Step 1. Set the maximum iteration number G.

Step2. Initialize each sub-population v_s by randomly chosen from the range between g_{min} and g_{max}.

Step 3. For each sub-population v_s, calculate the objective function f separately using Eqs. (2) and (4), then non-dominated sorting and selection for individuals to produce mating pool.

Step 4. Hybrid crossover and mutation in a large mating pool which mixed all the mating pool, to produce new offspring and then according the number of thresholds to assign the offspring to the corresponding sub-populations, then obtain the next generation population through elitism operation in each sub-population v_s.

Step 5. Repeat step 3 through 4 until termination criteria are met.

Step 6. Compute the index F using Eq. (6) in each sub-population.

Step7. The maximum of F corresponds to the optimum threshold and the maximum of F_Δ corresponds to the optimum number of the thresholds, then output the result.

3 Experimental Results

In order to evaluate the performance of AMT_ME algorithm, segmentation experiments are performed on natural and synthetic images (Shown in Fig. 2). The actual number of classes $C_{Optimal}$ for synthetic image was known in advance and the optimal range for number of classes $C_{Optimal}$ for natural images were based on a visual analysis survey conducted by many people [11, 14]. In this paper, $C_{Optimal}$ represents the optimal range for number of thresholds. In this experimental section, ATMO [11], Otsu's [6] and Kapur's [7] methods are adopted as the comparison methods. For Otsu's and Kapur's methods, the number of thresholds is selected in advance. To evaluate the algorithm performance, the criterion of Peak Signal to Noise Ratio (PSNR) are used. The parameters used for AMT_ME in our experimental study are set as follows: the

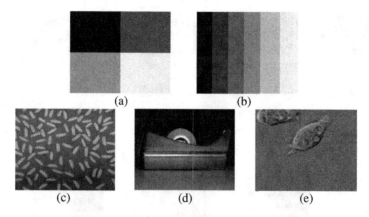

Fig. 2. Synthetic and natural images: (a) Mmi3; (b) Mmi5; (c) Rice; (d) Tape; (e) Cell.

maximum number of thresholds is 5, the population size is 250, thus each of sub-population size is 50, the maximum iteration number is 200, the crossover probability is 0.9, and the mutation probability is 0.1.

Table 1 present the optimal threshold values of the text images. Table 2 shows the PSNR values of different algorithms on those text images. From the results, it can be seen that the AMT_ME algorithm obtains the highest PSNR value among all the methods. In order to more obviously illustrate the visual results, the segmentation results on those text images are shown in Figs. 3, 4, 5, 6 and 7(a)–(d). For Rice, Tape and Mmi3, comparing with ATMO, Otsu and Kapur method, AMT_ME has a better visual effect and more details are displayed in the segmentation results. Although ATMO also contain the idea of automatic threshold, it is worse than our method on selecting the optimal threshold. Therefore, AMT_ME outperform other methods since the adaptive multi-level thresholding and the hybrid crossover between different numbers of thresholds are introduced into the method.

Table 1. Optimal threshold values obtained by four comparison methods

Images	$C_{Optimal}$	Otsu	Kapur	ATMO	AMT_ME
Mmi3	3	20-100-180	21-107-237	63-101-184	55-112-227
Mmi5	5	20-70-120-170-220	27-103-139-185-226	30-101-123-171-224	22-86-139-174-231
Rice	[1, 2]	83-141	86-129	114-180	78-143
Tape	[2, 3]	62-163	85-149	40-90-124	60-192
Cell	[1, 2]	98-138	98-126	98	76-137

Table 2. Comparison of PSNR values for these four methods

Images	Otsu	Kapur	ATMO	AMT_ME
Mmi3	22.65	18.85	27.24	**28.61**
Mmi5	26.57	25.51	26.65	**26.93**
Rice	13.64	12.86	10.42	**13.92**
Tape	19.20	17.68	13.52	**19.31**
Cell	18.89	17.09	8.99	**19.82**

Fig. 3. Segmentation results on the image Mmi3: (a) Otsu; (b) Kapur; (c) ATMO; (d) AMT_ME.

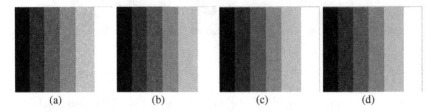

Fig. 4. Segmentation results on the image Mmi5: (a) Otsu; (b) Kapur; (c) ATMO; (d) AMT_ME.

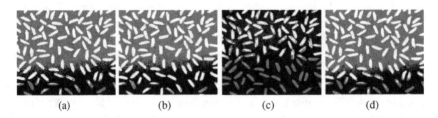

Fig. 5. Segmentation results on the image Rice: (a) Otsu; (b) Kapur; (c) ATMO; (d) AMT_ME.

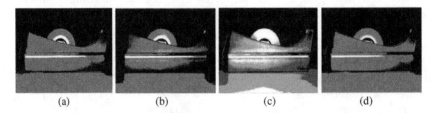

Fig. 6. Segmentation results on the image Tape: (a) Otsu; (b) Kapur; (c) ATMO; (d) AMT_ME.

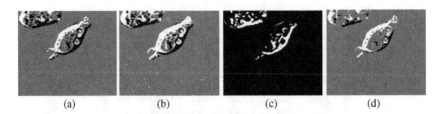

Fig. 7. Segmentation results on the image Cell: (a) Otsu; (b) Kapur; (c) ATMO; (d) AMT_ME.

A Berkeley image #55067 is shown in Fig. 8(a). Figure 8(b)–(d) are the image segmentation results of the corresponding population size taken as 100, 250, and 400. The segmentation accuracy (SA) [15] is used to evaluate the algorithm performance. Figure 9 presents the SA values under different population size obtained by 10 independent trials on Berkeley image. Results reveal that the population size v has a certain effect on the segmentation performance. In the case of v = 100, AMT_ME algorithm will be unstable and has the possibility of serious misclassification (Shown in Figs. 8 (b) and 9). In this paper, the population size v is set to 250 which can obtain a better and stable segmentation results (Shown in Figs. 8(c) and 9). It can also achieve almost the same segmentation results as v = 250 under v = 400 (Shown in Figs. 8(d) and 9). However, algorithms will be longer time with the increasing of v.

(a)   (b)   (c)   (d)

Fig. 8. Segmentation results on the Berkeley image #55067: (a) original image; (b) population v = 100; (c) population v = 250; (d) population v = 400.

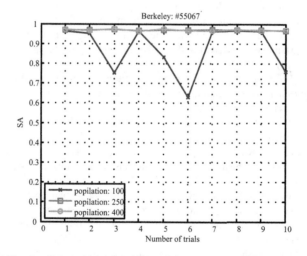

Fig. 9. Comparison of the SA values with v = 100, 250, 400.

4 Conclusions

In this article, an adaptive multi-level thresholding segmentation based on multi-objective evolutionary algorithm (AMT_ME) is proposed. In order to obtain the optimum number of the thresholds and the optimal thresholds, the adaptive multi-level thresholding and the hybrid crossover between different numbers of thresholds are introduced into the proposed method. The experimental results on the natural and synthetic images showed that AMT_ME outperform automatic threshold based on multi-objective optimization (ATMO), Otsu's and Kapur's methods.

In future works, some more effective adaptive multi-level thresholding methods and newly proposed MOEAs can be incorporated into the image segmentation.

Acknowledgements. This work is supported by the National Natural Science Foundation of China (Grant Nos. 61571361, 61102095, and 61202153), the Science and Technology Plan in Shaanxi Province of China (Grant No. 2014KJXX-72), and the Fundamental Research Funds for the Central Universities (Grant No. GK201503063).

References

1. Gonzalez, R.C., Woods, R.E.: Digital Image Processing. Addison-Wesley, Massachusetts (1992)
2. Hosseinzadeh, A., Mozafari, S.: Provide a hybrid method to improve the performance of multilevel thresholding for image segmentation using GA and SA algorithms. In: IKT2015 7th International Conference on Information and Knowledge Technology, pp. 1–6. IEEE Press, Urmia (2015)
3. Al-Amri, S.S., Kalyankar, N.V., Khamitkar, S.: Image segmentation by using threshold techniques. J. Comput. **2**, 83–86 (2010)
4. Mapayi, T., Viriri, S., Tapamo, J.-R.: A new adaptive thresholding technique for retinal vessel segmentation based on local homogeneity information. In: Elmoataz, A., Lezoray, O., Nouboud, F., Mammass, D. (eds.) ICISP 2014. LNCS, vol. 8509, pp. 558–567. Springer, Heidelberg (2014)
5. Zhang, J., Li, H., Tang, Z., Lu, Q., Zheng, X., Zhou, J.: An improved quantum-inspired genetic algorithm for image multilevel thresholding segmentation. J. Math. Probl. Eng. **2014**, 1–12 (2014)
6. Otsu, N.: Threshold selection method from gray-level histograms. J. IEEE Trans. Syst. Man Cybern. **9**, 62–66 (1979)
7. Kapur, J.N., Sahoo, P.K., Wong, A.K.: A new method for gray-level picture thresholding using the entropy of the histogram. J. Comput. Vis. Graph. Image Process. **29**, 273–285 (1985)
8. Muppidi, M., Rad, P., Agaian, S.S., Jamshidi, M.: Image segmentation by multi-level thresholding using genetic algorithm with fuzzy entropy cost functions. In: International Conference on Image Processing Theory, Tools and Applications, pp. 143–148. IEEE Press, Orleans (2015)
9. Ouarda, A.: Image thresholding using type-2 fuzzy c-partition entropy and particle swarm optimization algorithm. In: International Conference on Computer Vision and Image Analysis Applications, pp. 1–7. IEEE Press, Sousse (2015)
10. Nakib, A., Oulhadj, H., Siarry, P.: Image thresholding based on Pareto multi-objective optimization. J. Eng. Appl. Artif. Intell. **23**, 313–320 (2010)
11. Djerou, L., Khelil, N., Dehimi, N.H., Batouche, M.: Automatic threshold based on multi-objective optimization. J. Appl. Comput. Sci. Math. **13**, 24–31 (2012)
12. Deb, K., Agrawal, S., Pratap, A.: A fast elitist non-dominated sorting genetic algorithm for multi-objective optimization: NSGA II. J. IEEE Trans. Evol. Comput. **6**, 180–197 (2002)

13. Yue, Z.J., Qiu, W.L., Liu, C.L.: A self-adaptive approach of multi-objective image segmentation. J. Image Graph. **9**, 674–678 (2004). (in Chinese)
14. Omran, M.G., Salman, A., Engelbrecht, A.P.: Dynamic clustering using particle swarm optimization with application in image segmentation. J. Pattern Anal. Appl. **8**, 332–344 (2006)
15. Zhang, H., Fritts, J.E., Goldman, S.A.: Image segmentation evaluation: a survey of unsupervised methods. J. Comput. Vis. Image Underst. **110**, 260–280 (2008)

Large-Scale Global Optimization

Large-Scale Global Optimization Using a Binary Genetic Algorithm with EDA-Based Decomposition

Evgenii Sopov[(✉)]

Department of Systems Analysis and Operations Research,
Siberian State Aerospace University, Krasnoyarsk, Russia
evgenysopov@gmail.com

Abstract. In recent years many real-world optimization problems have had to deal with growing dimensionality. Optimization problems with many hundreds or thousands of variables are called large-scale global optimization (LGSO) problems. The most advanced algorithms for LSGO are proposed for continuous problems and are based on cooperative coevolution schemes using the problem decomposition. In this paper a novel technique is proposed. A genetic algorithm is used as the core technique. The estimation of distribution algorithm is used for collecting statistical data based on the past search experience to provide the problem decomposition by fixing genes in chromosomes. Such an EDA-based decomposition technique has the benefits of the random grouping methods and the dynamic learning methods. The results of numerical experiments for benchmark problems from the CEC'13 competition are presented. The experiments show that the approach demonstrates efficiency comparable to other advanced algorithms.

Keywords: Estimation of distribution algorithm · Genetic algorithm · Large-scale global optimization · Problem decomposition

1 Introduction

Evolutionary algorithms (EAs) have proved their efficiency at solving many complex real-world optimization problems. However, their performance usually decreases when the dimensionality of the search space increases. This effect is called the "curse of dimensionality". Optimization problems with many hundreds or thousands of objective variables are called large-scale global optimization (LGSO) problems.

There exist some classes of optimization problems that are not hard for either classical mathematical approaches or more advanced search techniques (for example, linear programming). At the same time, black-box LSGO problems have become a great challenge even for EAs as we have no information about the search space to include it into a certain algorithm. Nevertheless, some assumption can be done, and there exist many efficient LSGO techniques for the continuous search space [6].

Many real-world optimization problems encode different complex structures and contain variables of many different types, which cannot be represented only by real values. In this case binary genetic algorithms (GAs) can be used. As we can see from papers, there is a lack of LSGO approaches using the GA as the core technique.

© Springer International Publishing Switzerland 2016
Y. Tan et al. (Eds.): ICSI 2016, Part I, LNCS 9712, pp. 619–626, 2016.
DOI: 10.1007/978-3-319-41000-5_62

In this paper a novel LSGO technique using a GA with a decomposition based on the estimation of distribution algorithm (EDA) is proposed. The binary EDA is used to present a statistic of the past search experience of the GA and to predict the values of problem subcomponents that are being fixed to decrease the problem dimensionality.

The rest of the paper is organized as follows. Section 2 describes related work. Section 3 describes the proposed approach. In Sect. 4 the results of numerical experiments are discussed. In the Conclusion the results and further research are discussed.

2   Related Work

There exist a great variety of different LSGO techniques that can be combined in two main groups: non-decomposition methods and cooperative coevolution (CC) algorithms. The first group of methods are mostly based on improving standard evolutionary and genetic operations. But the best results and the majority of approaches are presented by the second group. The CC methods decompose LSGO problems into low dimensional sub-problems by grouping the problem subcomponents. CC consists of three general steps: problem decomposition, subcomponent optimization and subcomponent coadaptation (merging solutions of all subcomponents to construct the complete solution). The problem decomposition is a critical step. There are many subcomponent grouping methods, including: static grouping [8], random dynamic grouping [13] and learning dynamic grouping [5, 7]. A good survey on LSGO and methods is proposed in [6]. As we can observe in papers, almost all studies are focused on continuous LSGO, and there is a lack of techniques for binary (or other discrete) representations.

The EDA is a stochastic optimization technique that explores a space of potential solutions by building and sampling explicit probabilistic models. The estimated distribution can be used for improving standard search techniques. There exist some hybrid EDA-EA approaches for LSGO [1, 11]. These hybrid EDA-EA techniques are also designed for continuous LSGO.

The most widely known competition on LSGO has been held within the IEEE Congress on Evolutionary Computation (CEC) since 2008. As we can see from the last competition, the majority of proposed methods are based on the random dynamic grouping and continuous search techniques.

3   Proposed Approach

3.1   EDA-Based Decomposition

The main idea of the LSGO problem decomposition methods is based on the divide-and-conquer approach which decomposes the problem into single-variable or multiple-variable low dimensional problems. In this case, only part of the variables are used in the search process; the rest are fixed and their values are defined using some strategy (for example, values from the best-found solution are used).

The finding of an appropriate decomposition is part of the general search process. It is obvious and has been presented in many studies that the best performance is achieved with separable LSGO problems. In the case of non-separable problems, the performance strongly depends on the decomposition strategy.

In this work, we will formulate the following requirements for the proposed decomposition method. The grouping should be dynamic to realize the "exploration and exploitation" strategy. The grouping should be random to avoid the greedy search and the local convergence. The grouping should be based on the past search experience of the whole population (to provide the global search options). The grouping should be adaptively scalable to provide efficient decomposition at every stage of the search process.

As is known, GAs do not collect a statistic of the past generations in an explicit form, but it is contained in the genes of individuals in the population. One of the ways to present the statistic is to evaluate the distribution of binary values as in the binary EDA. The following probability vector can be used (1):

$$P(t) = (p_1(t), p_2(t), \ldots, p_n(t)), p_i(t) = P(x_i = 1) = \frac{1}{N} \sum_{j=1}^{N} x_i^j, i = \overline{1,n} \qquad (1)$$

where t is the number of the current generation, p_i is the probability of a one-value for the i-th position in chromosomes of individuals in the last population, x_i^j is the value of the i-th gene of the j-th individual, n is the chromosome length, and N is the size of the population.

The distribution calculated at the t-th generation describes the generalized statistic collected by the GA in the population. We can also analyse the dynamic of the statistic over a series of generations. In [9] a convergence property of the probability vector components is discussed. Experiments have shown that for a GA that converges to the global optima, the probability vector values converge to one if the corresponding position of the optimal solution contains a one, and converge to zero otherwise.

We will use this convergence property to define the values for fixed genes at the grouping stage. If the i-th position in a chromosome at the t-th generation is fixed, its value is defined by the corresponding value of the probability vector (2):

$$x_i^j(t) = \begin{cases} 0, & \text{if } p_i(t) < (0.5 - \delta) \\ random, & \text{if } (0.5 - \delta) \leq p_i(t) \leq (0.5 + \delta) \\ 1, & \text{if } p_i(t) > (0.5 + \delta) \end{cases} \qquad (2)$$

where δ is a threshold (a confidence level), $\delta \in (0, 0.5)$.

We will explain the proposed approach using Fig. 1. The diagram visualizes an arbitrary component of the probability vector for an arbitrary run of a GA on the Rastrigin function. For the chosen gene the corresponding value of the optimal solution is equal to zero. As we can see from Fig. 1, the GA starts with random initialization, thus the value of the probability vector is equal to 0.5. At the first generations the GA actively explores the search space and number of 1's and 0's genes are almost equal, thus the value of the probability vector is still about 0.5. After that, the GA locates a

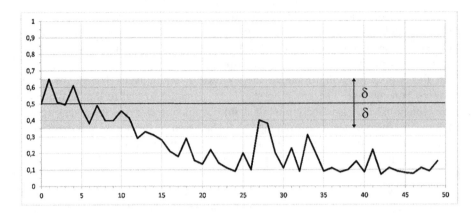

Fig. 1. The dynamic of the probability vector component

promising region in the search space and increases the number of 0's in this position, thus the value of the probability vector decreases towards zero.

The confidence level δ is a parameter that defines a threshold for the probability value around 0.5, when we cannot make a decision about the gene value.

Although a decision about fixed variables is made by stand-alone components, the estimated distribution contains information about the problem solving in general. Thus the method is not focused only on separable LSGO problems.

Next we need to define the number of variables that will be fixed. There exist many strategies. For example, the splitting-in-half method divides an n-dimensional problem into two $n/2$ subcomponents. In general, we will define the number of fixed variables as a percentage of the chromosome length and will denote it as α. The value of α can be constant or can change during the run of the algorithm. The variables and corresponding components of the probability vector are fixed for some predefined number of generations, which is called an adaptation period (denoted as t_{adapt}). The list of fixed components is randomly defined.

In this paper, the straight-forward approach is used, α and t_{adapt} are predefined and constant.

The main advantage of such EDA-based decomposition is that we do not lose the previously collected statistic as we fix components of the probability vector. The GA solves the problem of reduced dimensionality and updates the probability only for active components. After each adaptation period we will randomly fix other components, and the previously fixed components will continue updating their saved values.

3.2 The GA with EDA-Based Decomposition for LSGO

We will describe the proposed LSGO algorithm in detail.

First, we need to encode the initial problem into a binary representation. The standard binary or Grey code can be used. A chromosome length n is defined. Next, specific parameters of the EDA-based decomposition and the chosen GA, maximum number of fitness evaluations (*MaxFE*) or maximum number of generations (*MaxGEN*) are defined.

Finally, the following algorithm is used:

Input: n, N, α, δ, t_{adapt}, *MaxFE*, the GA operators' parameters.
Initialization. Randomly generate a population of N individuals of the length n. Calculate $P(0)$ using formula (1).
Main loop. Until *MaxFE* is reached:

1. **Problem decomposition stage.** Start new adaptation period. Fix random α components in chromosomes and in the probability vector.
2. **Subcomponent optimization stage.** Run the GA for t_{adapt} generations:
   (a) Fitness evaluation. Set values in fixed positions of chromosomes according to $P(t)$ using formula (2).
   (b) Perform selection, crossover and mutation operations.
   (c) Create next generation, update the probability vector $P(t)$ for active components.

Output: the best-found solution.

3.3   Parallel (The Island Model) Modification

Many proposed LSGO approaches with the subcomponent grouping are based on cooperative coevolution. In this case, many populations are used, which evolve different groups of subcomponents. The cooperation is used on the fitness evaluation step to define the values of components that were fixed during the algorithm run. Usually, components of the best individuals from other populations are used.

We will introduce many parallel populations using the following scheme. At the main loop, the total population of size N is divided into K populations of size M, where $N = K \cdot M$. For each population the problem decomposition and the subcomponent optimization steps are independently performed. Thus each population can be viewed as an island with its own decomposition strategy. When the adaptation period is over, all individuals from all populations are collected back into the total population and the summary statistic is updated. This step can be viewed as the cooperation.

As is known, the island model GA can outperform the standard single-population GA for many complex optimization problems. We can also decrease the computational efforts by implementing the GA with a parallel multi-core or multi-processor computer.

4   Experimental Results

To estimate the proposed approach performance, we have used 15 large-scale benchmark problems from the CEC'2013 Special Session and Competition on Large-Scale Global Optimization [3]. These problems represent a wider range of real-world large-scale optimization problems and provide convenience and flexibility for comparing various evolutionary algorithms specifically designed for large-scale global optimization. There are 3 fully-separable problems (denoted as *f1–f3*), 8 partially separable problems (*f4–f7* with a separable subcomponent and *f8–f11* with no separable subcomponents), 3 problems with overlapping subcomponents (*f12–f14*), and 1 non-separable problem (*f15*).

The experiment settings are:

- Dimensions for all problem are $D = 1000$.
- The standard binary encoding is used with accuracies: $\varepsilon = 0.1$ for *f1*, *f4*, *f7*, *f8* and *f11–15*, $\varepsilon = 0.05$ for *f3*, *f6* and *f10*, and $\varepsilon = 0.01$ for *f2*, *f5* and *f9*.
- For each problem the best, mean, and standard deviation of the 25 independent runs are evaluated.
- Maximum number of fitness evaluations is *MaxFE* = 3.0e+6.
- The performance estimation is performed when the number of fitness evaluations is 1.2e+5, 6.0e+5 and 3.0e+6.

The EDA-based decomposition GA settings are:

- Population sizes are $N = 1000$ for the single-population version, $N = 500$ for the island version with 3 islands, and $N = 400$ for 5 islands.
- The adaptation period is $t_{adapt} = 100$.
- The probability threshold is $\delta = 0.15$.
- Numbers of fixed components are $\alpha = 25$ %, 50 % and 75 % of the chromosome length.

All algorithms have been implemented in Visual Studio C++ using the OpenMP for parallel computing with multi-core PC. Free C++ source codes of the benchmark problems are taken from [10].

We have carried out the above-mentioned experiments and have established the following. In the case of single population, the best performance is achieved with 50 % fixed components. In the case of the island model, the best results are obtained by the 5 island model with 75 % fixed components. Almost for every considered value of α, the island model outperforms the single population version of the algorithm. The experimental results for the best found settings are presented in Table 1.

The summary results are compared with other techniques presented at the CEC'13 competition. The algorithms are DECC-G (differential evolution (DE) based cooperative coevolution (CC) with random dynamic grouping) [13], VMO-DE (variable mesh optimization using differential evolution) [4], CC-CMA-ES (Covariance Matrix Adaptation Evolution Strategy using Cooperative Coevolution) [5], MOS (Multiple Offspring Sampling (MOS) based hybrid algorithm) [2], and SACC (smoothing and auxiliary function based cooperative coevolution) [12]. We have averaged the performance estimates of all algorithms over all problems and have ranked algorithms by the Best and the Mean values. The results are in Table 2.

As we can see from Table 2, the proposed approach has taken 4th place by the Best criterion and 5th place by the Mean value. We should note that all algorithms except the proposed are specially designed for continuous LSGO problems. The EDA-based decomposition GA does not use any knowledge about search space. Moreover, the chromosome length in the binary algorithm is greater than in the case of the continuous

Table 1. Experimental results for the EDA-based decomposition GA with 5 islands and $\alpha = 75\ \%$

		f1	f2	f3	f4	f5	f6	f7	f8
1.2e5	Best	1.42E+07	9.63E+03	1.08E+02	1.39E+11	6.11E+14	2.90E+05	7.05E+08	3.97E+15
	Mean	5.50E+07	1.06E+04	4.52E+01	9.15E+11	7.17E+14	7.78E+05	2.76E+09	2.71E+16
	StDev	2.98E+07	1.53E+03	1.44E+01	5.63E+11	7.45E+08	2.61E+05	1.44E+09	6.77E+15
6.0e5	Best	6.89E+03	9.11E+03	3.04E+00	1.95E+10	3.07E+14	5.21E+05	2.01E+08	2.09E+14
	Mean	1.99E+04	1.25E+04	1.30E+01	9.07E+10	5.03E+14	6.05E+05	9.41E+08	2.18E+15
	StDev	1.68E+03	1.18E+03	6.32E−01	6.03E+10	2.50E+07	2.60E+05	7.56E+08	1.52E+15
3.0e6	Best	4.59E−05	1.82E+03	2.94E−05	6.60E+09	7.59E+14	6.25E+04	7.65E+07	4.49E+13
	Mean	5.68E−04	3.34E+03	4.81E−01	2.32E+10	9.75E+14	4.75E+05	2.53E+08	3.64E+14
	StDev	4.29E−04	2.54E+02	2.28E−01	1.14E+10	2.18E+06	3.35E+05	8.35E+07	5.21E+14
		f9	f10	f11	f12	f13	f14	f15	Average
1.2e5	Best	1.08E+09	8.87E+06	1.46E+11	3.87E+06	2.66E+10	1.88E+11	3.61E+07	3.05E+14
	Mean	1.80E+09	7.14E+07	3.47E+11	4.36E+08	2.98E+10	5.78E+11	2.69E+08	1.85E+15
	StDev	4.27E+08	1.57E+07	2.25E+11	7.89E+08	1.12E+10	3.67E+11	9.91E+07	4.52E+14
6.0e5	Best	6.42E+08	7.91E+06	1.34E+10	2.40E+03	6.28E+09	5.68E+10	1.80E+07	3.44E+13
	Mean	1.25E+09	1.38E+07	9.84E+10	6.66E+03	1.47E+10	1.03E+11	2.43E+07	1.79E+14
	StDev	5.21E+08	1.65E+07	1.18E+11	5.53E+03	4.83E+09	6.70E+10	8.68E+06	1.02E+14
3.0e6	Best	4.15E+08	6.18E+06	2.60E+10	7.72E+02	8.02E+09	1.42E+10	2.40E+07	5.36E+13
	Mean	8.06E+08	1.61E+07	7.01E+10	2.30E+03	1.27E+10	1.69E+11	3.05E+07	8.93E+13
	StDev	1.72E+08	7.89E+06	4.29E+10	2.41E+03	2.96E+09	4.81E+10	5.13E+06	3.47E+13

Table 2. LSGO approaches comparison

Average	SACC	MOS	VMO-DE	DECC-G	CC-CMA-ES	EDA-GA
Best	9.80E+12	2.17E+11	4.90E+13	5.80E+13	6.25E+13	5.36E+13
Ranking by best	2	1	3	5	6	4
Mean/StDev	8.0E+13/5.08E +13	5.33E+11/2.04E +11	5.32E+13/4.81E +12	7.7E+13/1.02E +13	8.58E+13/2.39E +13	8.93E+13/3.47E +13
Ranking by mean	4	1	2	3	6	5

space. Nevertheless, the EDA-based decomposition GA outperforms the CC-CMA-ES by two measures and the DECC-G by the Best value on average.

Our hypothesis is that the proposed approach will be a good tool for solving complex real-world LSGO problems, which usually contain not only continuous variables, but can represent arbitrary complex structures. Further investigations of the algorithm structure and parameters can probably improve its performance. In particular, the α value can be adjusted adaptively during the algorithm run using information about the probability vector convergence.

5   Conclusions

In this paper a novel technique for LSGO that uses a binary GA with EDA-based decomposition is proposed. The EDA is used for collecting statistical data based on the past search experience to predict the convergence of subcomponents and to decrease the problem dimensionality by fixing some genes in chromosomes. We have compared a single population and the island model implementations of the algorithm. The best results have been obtained with the island model version. It yields state-of-the-art LSGO techniques, but the performance is comparable. The advantage of the proposed approach is that it can be applied to problems with arbitrary representations and it needs no a priori information about the search space.

In further work, more detailed analysis of the EDA-based decomposition GA parameters will be provided. A self-configuration will be introduced into the algorithm.

Acknowledgements. The research was supported by the President of the Russian Federation grant (MK-3285.2015.9).

References

1. Dong, W., Chen, T., Tino, P., Yao, X.: Scaling up estimation of distribution algorithms for continuous optimization. IEEE Trans. Evol. Comput. **17**(6), 797–822 (2013)
2. LaTorre, A., Muelas, S., Pena, J.-M.: Large scale global optimization: experimental results with MOS-based hybrid algorithms. In: 2013 IEEE Congress on Evolutionary Computation (CEC), pp. 2742–2749 (2013)
3. Li, X., Tang, K., Omidvar, M.N., Yang, Z., Qin, K.: Benchmark functions for the CEC 2013 special session and competition on large-scale global optimization. Evolutionary Computation and Machine Learning Group, RMIT University, Australia (2013)
4. Li, X., Tang, K., Omidvar, M.N., Yang, Z., Qin, K.: Technical report on 2013 IEEE Congress on Evolutionary Computation Competition on Large Scale Global Optimization. http://goanna.cs.rmit.edu.au/ ~ xiaodong/cec13-lsgo/competition/lsgo-competition-sumary-2013.pdf
5. Liu, J., Tang, K.: Scaling up covariance matrix adaptation evolution strategy using cooperative coevolution. In: Yin, H., Tang, K., Gao, Y., Klawonn, F., Lee, M., Weise, T., Li, B., Yao, X. (eds.) IDEAL 2013. LNCS, vol. 8206, pp. 350–357. Springer, Heidelberg (2013)
6. Mahdavi, S., Shiri, M.E., Rahnamayan, S.: Metaheuristics in large-scale global continues optimization: a survey. Inf. Sci. **295**, 407–428 (2015)
7. Omidvar, M.N., Li, X., Mei, Y., Yao, X.: Cooperative co-evolution with differential grouping for large scale optimization. IEEE Trans. Evol. Comput. **18**(3), 378–393 (2014)
8. Potter, M., De Jong, K.A.: Cooperative coevolution: an architecture for evolving coadapted subcomponents. Evol. Comput. **8**(1), 1–29 (2000)
9. Sopov, E., Sopov, S.: The convergence prediction method for genetic and PBIL-like algorithms with binary representation. In: IEEE International Siberian Conference on Control and Communications, SIBCON 2011, pp. 203–206 (2011)
10. Test suite for the IEEE CEC 2013 competition on the LSGO. http://goanna.cs.rmit.edu.au/ ~ xiaodong/cec13-lsgo/competition/lsgo_2013_benchmarks.zip
11. Wang, Y., Li, B.: A restart univariate estimation of distribution algorithm: sampling under mixed gaussian and Lévy probability distribution. In: IEEE Congress on Evolutionary Computation, CEC 2008, pp. 3917–3924 (2008)
12. Wei, F., Wang, Y., Huo, Y.: Smoothing and auxiliary functions based cooperative coevolution for global optimization. In: 2013 IEEE Congress on Evolutionary Computation (CEC), pp. 2736– 2741 (2013)
13. Yang, Z., Tang, K., Yao, X.: Large scale evolutionary optimization using cooperative coevolution. Inform. Sci. **178**(15), 2985–2999 (2008)

Grouping Particle Swarm Optimizer with P_{best}s Guidance for Large Scale Optimization

Weian Guo[1(✉)], Ming Chen[1], Lei Wang[2], and Qidi Wu[2]

[1] Sino-German College of Applied Sciences,
Tongji University, Shanghai 201804, China
guoweian@163.com
[2] Department of Electronics and Information Engineering,
Tongji University, Shanghai 201804, China

Abstract. As a classic Swarm Intelligence (SI), Particle Swarm Optimization (PSO), inspired by the behavior of birds flocking, draws many attentions due to its significant performance in both numerical experiments and practical applications. During the optimization process of PSO, the direction of each particle is guided by its current velocity, its own historical best position (*pbest*) and current global best position (*gbest*). However, once the two positions, especially *gbest*, are local optimum, it is difficult for PSO to achieve a global optimum. To overcome this problem, in this paper, we design a novel swarm optimizer, termed Grouping PSO with *Pbest* Guidance (GPSO-PG), to eliminate the effects from *gbest* in order to enhance the algorithm's global searching ability. By employing the benchmarks in CEC 2008, we apply GPSO-PG to large scale optimization problems (LSOPs). The comparison results exhibit that GPSO-PG is competitive to address LSOPs.

Keywords: Swarm intelligence · Particle swarm optimization · Local optima · Grouping strategy · Large scale optimization

1 Introduction

In current decades, Swarm Intelligence (SI) draws worldwide attentions due to its simplicity and efficiency in optimization [1]. Both researches and applications of PSO demonstrate that the algorithm is competitive to handle optimization problems and plays a very active role to deal with practical cases [2,3]. Nevertheless, there still remain some problems in PSO. An example is given that the canonical PSO is likely to get trapped into local optimum, especially for complex optimizations, such as multi-modal problems, large scale optimization, etc.

In conventional PSO, *gbest* always affects the whole swarm. On one hand, since *gbest* is a global optimal solution, it accumulates the convergence speed. On the other hand, *gbest* will not be updated until a better solution appears, and therefore all particles will move towards *gbest* for a number of generations. Once *gbest* is in local optimal position, it is hard for the whole swarm to get

© Springer International Publishing Switzerland 2016
Y. Tan et al. (Eds.): ICSI 2016, Part I, LNCS 9712, pp. 627–634, 2016.
DOI: 10.1007/978-3-319-41000-5_63

rid of it. In this paper, considering that *gbest* is the best solution in the set of *pbests*, we employ the set of *pbest*, instead *gbest*, to balance convergency and diversity. In each generation, a *pbest* will be selected from the set of *pbests* as an exemplar, but not *gbest*. The design has two advantages. First, since the set of *pbest* contains the good records of the whole swarm, it provide the swarm good information for convergence. Second, this prevents the whole swarm from being affected by any only one position, say *gbest*, for many generations. In addition, we employ grouping strategy to divide the whole swarm into many groups. In each group, the worse solutions will learn from the group best solution.

The rest of this paper is organized as follows. Section 2 will present an introduction of canonical PSO. In Sect. 3, grouping Particle Swarm Optimization with *pbests* guidance (GPSO-PG) will be proposed. The parameter setting will be discussed in this section. In Sect. 4, large scale optimization is considered as benchmarks to compare GPSO-PG with other competitor in CEC'2008 as well as the analysis for the results will be presented. In Sect. 5, we conclude this paper and present our future work.

2   Canonical Particle Swarm Optimization

As a typical swarm intelligence algorithm, PSO was first proposed by Kennedy and Eberhart in 1995 [4], which is given in (1):

$$V_i(t+1) = \omega V_i(t) + c_1 R_1(t)(pbest_i(t) - X_i(t))$$
$$+ c_2 R_2(t)(gbest(t) - X_i(t)).$$
$$X_i(t+1) = X_i(t) + V_i(t+1). \tag{1}$$

where t is the iteration (generation) number, $V_i(t)$ and $X_i(t)$ represent the velocity and position of the ith particle, respectively; ω is termed inertia weight, c_1 and c_2 are the acceleration coefficients, $R_1(t)$ and $R_2(t)$ are tow vectors randomly generated within $[0, 1]^n$; $pbest_i(t)$ is the best solution of the $i - th$ particle found by far, and $gbest(t)$ is the global best particle found by far.

In (1), if *gbest* is a local optimum, it is hard for PSO to get rid of it. In current years, many researches are done to alleviate the influence of *gbest*. In [5], Liang proposes a variant of PSO, which particles learn from *pbest* only, but not *gbest*. However, the probability setting, which affects algorithms' performance, is on the basis of empirical experiments. In [6], the authors employ the competitive mechanism in particle pairs. The loser particle will learn from the winner one. However, since in one generation only half of particles, say loser particles, are involved in updating, the convergency speed is low.

3   Grouping Particle Swarm Optimizer with *Pbests* Guidance (GPSO-PG)

3.1   Design of GPSO-PG

To eliminate the effects of local optimum to PSO's performance, we design a novel variant of PSO, termed Grouping Particle Swarm Optimization with *pbest*

Guidance (GPSO-PG). The motivation is to reduce the effects of any only particle to the whole swarm. First, at the beginning of each generation, we divide the whole swarm into several groups in a random way. In each group, we compare the particles and select a local best particle which will be learned from by the rest particles in the group. In addition, we employ the set of *pbests*, but not *gbest*, to guide the swarm direction. In each group, all particles but the best one will learn from a selected *pbest*. The updating mechanism is shown in (2).

$$V_i(t+1) = \omega V_i(t) + c_1 R_1(t)(X_{lb}(t) - X_i(t))$$
$$+ c_2 R_2(t)(pbest_s(t) - X_i(t)).$$
$$X_i(t+1) = X_i(t) + V_i(t+1). \tag{2}$$

where X_{lb} is the local best position in each group, $pbest_s$ is a selected *pbest* and the selection rule will be presented in the following illustration. In each generation, the local best particle in each group does not update.

In (2), we employ *pbest* to instead *gbest*, which increases the diversity of exemplars for the swarm. The way to select *pbest* is based on the ranking of each *pbest* in the set of *pbests*. The probability to be selected is large if the fitness of *pbest* is good, while a poor *pbest* has a small probability to be selected. The probability to select *pbest* is given in (3).

$$Prob_i = \frac{Ranking_i}{\sum_{j=1}^{N} Ranking_j}. \tag{3}$$

where $Ranking_i$ is the rank of $pbest_i$ in the set of *pbests*. The value of $Ranking_i$ is small when $pbest_i$ has a top ranking, while the value of $Ranking_i$ is large if $pbest_i$ has a behind ranking. With the descriptions and definitions above, the pseudo-codes of GPSO-PG can be summarized in Algorithm 1.

The advantages of this design has two folders: The first one is that the particles are guided by the best solution in each group, which reduces the probability that all particles will learn from local optimum. The second one is that we use *pbest* to instead *gbest*, which helps algorithm get rid of local optimum.

3.2   Group Size

Group size is an important factor to algorithm's performance. It is obvious that the number of X_{lb} is equal to the number of groups. On one hand, a small number of groups means that only a few particles are selected as X_{lb}, which lead to the result that one X_{lb} will affect many other particles. If one of the X_{lb}s is local optimum, the particles in the same group cannot easily break away the local optimum. Meanwhile, in the next generation, the local optimum still has a high probability to be X_{lb}. Hence a small number of groups will increase the risk to get trapped into local optimum. On the other hand, a large number of groups means that only a few particles in one group. If two bad particles or two good particles are in one group, the learning is meaningless. It also slows

Algorithm 1. Pseudo-codes of the GPSO-PG Algorithm

Initialize Population and Parameters including $N,G,c1,c2$
while terminal condition is not satisfied **do**
  for $i = 1$ to N **do**
    Record the best solution $pbest_i$ for Particle i
  end for
  Rank $pbest$ and Assign a Ranking Value for each $pbest$
  Calculate $Prob_i$ Using (3)
  Select pb from $pbests$ with $Prob_i$
  Randomly divide the whole swarm into G groups
  for $i = 1$ to G **do**
    Select a best solution X_{lb} in Group i
    for $j = 1$ to $\frac{N}{G}$ **do**
      if X_j is not X_{lb} **then**
        Update X_j Using (2)
      end if
    end for
  end for
end while

down convergence speed. Therefore, a suitable group size is helpful to get rid of local optimum, and also plays an active role to enhance algorithm's convergency speed.

3.3 *Pbest* Selection Model

In Sect. 3, we employ a linear model for *pbest* selection. According to the design in (2), *Pbest* selection model will affects the diversity of the whole swarm. In addition, *pbest* plays a role to guide the whole swarm. Hence *pbest* selection model is sensitive to algorithm's performance. In this section, we give three different models to select *pbest*, which is given as follows, which are called constant selection model, quadric selection model and cosine selection model. Their performances are investigated and compared in experiments.

$$Prob_i = \frac{1}{N}. \tag{4}$$

$$Prob_i = \left(1 - \frac{Ranking_i}{N}\right)^2. \tag{5}$$

$$Prob_i = 0.5\left(cos\left(\frac{Ranking_i}{N}\pi\right) + 1\right). \tag{6}$$

where N is the population size.

4   Experiments Studies and Discussions

In this section, we do experiments as follows. First, we employ the standard GPSO-PG to compare with other algorithms that are in the CEC 2008. Second, the parameter setting will be investigated. Third, the selection model of *pbest* will be explored to study GPSO-PG's performance with different selection models. In experiments, seven benchmark functions proposed in the CEC08 on large scale optimization problems are employed, which are Shifted Sphere Function, Shifted Schwefel's Problem 2.21, Shifted Rosenbrock' Function, Rastrigin's Function, Shifted Griewank's Function, Shifted Ackley's Function, Shifted Ackley's Function and FastFractal "DoubleDip" Function. The details can be found in [7]. All statistical results are averaged over 25 independent runs. For each independent run, the maximum number of FEs is set to $5000d$, where d is the search dimension of the test functions.

4.1   Comparisons of Results

In (2), there are three parameters, ω, c_1 and c_2 respectively. We employ a linear decreasing value of ω as in [8] from 0.9 to 0.4. The c_1 and c_2 are considered as two weights of group best solution and selected *pbest* respectively. Considering that group best solution is used for local search, while the selected *pbest* attracts the swarm in a global way, c_1 and c_2 should be set to balance global and local searching. In this paper, we set $c_2 = \rho c_1$. In this subsection, we set $\rho = 0.15$. The population size is set as 500 and we set the group number as 50.

In experiments, we employ the competitors in CEC 2008 to present a comparisons with dimension of 1000. In Table 1, we cite the data from [9]. According to the results, we know that GPSO-PG is very competitive in dealing with large scale global optimization problems.

Table 1. Numerical simulation results of CEC'2008 functions with 1000 dimensions.

	CEC08_1	CEC08_2	CEC08_3	CEC08_4	CEC08_5	CEC08_6	CEC08_7
MLCC [10]	8.46E-13	1.09E+02	1.80E+03	**1.37E-10**	4.18E-13	1.06E-12	−**1.47E+04**
EPUS-PSO [11]	5.53e+02	4.66E+01	8.37E+05	7.58E+03	5.89E+00	1.89E+01	−6.62E+03
jDEdynNP-F [12]	1.14E-13	1.95E+01	1.31E+03	2.17e-04	3.98E-14	1.47E-11	−1.35E+04
UEP [13]	5.37E-12	1.05E+02	1.96E+03	1.03E+04	8.87E-04	1.99E+01	−1.18E+04
MTS [14]	**0.00E+00**	4.72E-02	**3.41E-04**	0.00E+00	0.00E+00	1.24E-11	−1.40E+04
DEwSAcc [15]	8.79E-03	9.61E+01	9.15E+03	1.82E+03	3.58E-03	2.30E+00	−06E+04
DMS-PSO [16]	**0.00E+00**	9.15E+01	8.98E+09	3.84E+03	0.00E+00	7.76E+00	−7.51E+03
LSEDA-gl [17]	3.22E-13	**1.04E-05**	1.73E+03	5.45E+02	1.71E-13	4.26E-13	−1.35E+04
ALPSEA [18]	3.58E+04	1.47E+02	2.33E+09	1.89E+02	3.04E+02	1.13E+01	N/A
ALPSEA-100M	1.90E-05	9.30E+01	1.77E+03	1.53E-02	1.17E-06	4.40E-04	N/A
GPSO-PG	2.58E-25	1.07E+02	1.88E+03	1.29E+03	**4.44E-16**	**4.61E-14**	−1.38E+04

4.2   Discussion for Parameter Setting

Group Size. Two extreme cases are given as follows. First, in GPSO-PG, if the number of groups is 1, it means the whole swarm is in one group and all particles will learn from the best solution. This goes against the motivation to reduce the effects from only one particle. Second, if in each group, there are only two particles, only half of swarm are involved in updating in one generation. This slows down the convergence speed. In addition, the credibility of the good one decreases since only two particles are in a group. To obtain a suitable number of particles, we set the number of particles in each group as $0.01N$, $0.02N$, $0.05N$, $0.1N$ respectively, where N, the population size, is set as 500. The results are given in Table 2. According to the results, we know that the group size is influential to algorithm's performance. The population with size of $0.02N$ wins the best for most benchmarks. A large size of group deteriorates algorithm's performance. As mentioned above, a small group size decreases the efficiency of evolutionary process, while a large group size may cause a local optimal solution. Hence, a suitable groups size is crucial to improve algorithms' performance.

Table 2. Numerical simulation results of CEC'2008 functions with 1000 dimensions.

	CEC08_1	CEC08_2	CEC08_3	CEC08_4	CEC08_5	CEC08_6	CEC08_7
GPSO-GP(0.01N)	1.65E-25	9.14E+01	1.32E+03	9.89E+02	3.37E-16	3.19e-14	**−1.36E+04**
GPSO-GP(0.02N)	**9.16E-27**	**8.28E+01**	6.20E+02	4.84E+02	**3.33E-16**	**2.48E-14**	−1.29E+04
GPSO-GP(0.05N)	1.52E+03	1.01E+02	5.05E+03	2.81E+03	2.46E+01	9.42E+00	−1.11E+04
GPSO-GP(0.1N)	1.54E+06	1.38E+02	4.35E+11	1.16E+04	1.33E+04	2.02E+01	−9.74E+03

Setting of ρ. In above sections, we have discussed that c_1 and c_2 are the weight of effects from local group and *pbest* respectively. For the sake of convenience in comparison, we set $c_2 = \rho c_1$ and set the the value of c_1 as 1. The value of ρ will be discussed in the subsection. The reason we set $\rho < 1$ has been explained in above section. The performances of algorithms with different ρ are given as follows. According to results, we know that weight of population diversity and convergence is much influential to algorithms' performance. As shown in Table 3, a small weight of population diversity is helpful to pursue good results. However, c_2 cannot be set as zero. If so, algorithm's performance deteriorates since there is no setting for population diversity.

***Pbest* Selection Model.** In Sect. 3, we introduce three kinds of *pbest* selection model. Their performances are compared in this section. In the experiments, we employ different selection model for *pbest* and use the same parameters for other parameters. The results are given in Table 4. From the table, we know that for CEC08_1, CEC08_2, CEC08_3, CEC08_4 and CEC08_5, the differences among different models are very little, which has same order of magnitude. However, for CEC08_6, Linear Model dramatically outperforms others. These results demonstrate that for different objective, the models have difference performance. Hence the investigation about the effects from parameter setting to models' performance will be studies in our future work.

Table 3. Numerical simulation results of linear model with $\rho = 0.1, 0.2, 0.3, 0.4$ and 0.5 of CEC'2008 functions with 1000 dimensions.

	CEC08_1	CEC08_2	CEC08_3	CEC08_4	CEC08_5	CEC08_6	CEC08_7
$\rho = 0.1$	1.56E-25	**6.84E+01**	**9.73E+02**	**4.93E+02**	**3.33E-16**	3.55E-14	**−1.47E+04**
$\rho = 0.2$	**1.15E-25**	9.03E+01	2.11E+03	1.20E+03	5.55E-16	**2.84E-14**	−1.36E+04
$\rho = 0.3$	2.33E+01	9.77E+01	3.23E+07	2.23E+03	1.20E+00	4.34E+00	−1.34E+04
$\rho = 0.4$	1.08E+04	1.03E+02	2.11E+09	2.77E+03	1.04E+02	9.27E+00	−1.33E+04
$\rho = 0.5$	1.09E+04	1.08E+02	2.27E+09	2.19E+03	1.10E+02	7.93E+00	−1.37E+04

Table 4. Numerical simulation results of Quadric, Cosine, Const and Linear model for CEC'2008 functions with 1000 dimensions.

	CEC08_1	CEC08_2	CEC08_3	CEC08_4	CEC08_5	CEC08_6	CEC08_7
Quadric Model	1.96E-25	1.08E+02	1.70E+03	1.65E+03	4.47E-16	1.44E+00	−1.36E+04
Cosine Model	1.94E-25	**1.01E+02**	1.90E+03	1.68E+03	4.53E-16	1.30E+00	−1.37E+04
Const Model	**1.88E-25**	1.12E+02	**1.67E+03**	1.70E+03	4.71E-16	1.38E+00	−1.36E+04
Linear Model	2.58E-25	1.07E+02	1.88E+03	**1.29E+03**	**4.44E-16**	**4.61E-14**	**−1.38E+04**

5 Conclusions and Future Work

In this paper, we propose a variant of PSO, termed Grouping Particle Swarm Optimization with *Pbests* Guidance (GPSO-PG). For each generation, the proposed algorithm will divide the whole swarm into several groups. In each group, the best solution will be set as an exemplar to learn from. In addition we employ the set of *pbest*, but not *gbest*, to guide swarm direction. The benchmarks in CEC' 2008 are employed in experiments and the results demonstrate that the proposed algorithm is very competitive to address large scale optimization problems. In future, the proposed algorithm will be applied to other complex problems such as dynamic optimization, multi-objective optimization problems and so on.

Acknowledgements. This work was sponsored by the National Natural Science Foundation of China under Grant no. 61503287, Supported by the Fundamental Research Funds for the Central Universities.

References

1. Shi, Y., Eberhart, R.: Fuzzy adaptive particle swarm optimization. In: Proceedings of the Congress on Evolutionary Computation, vol. 1, pp. 101–106 (2001)
2. Eberhart, R., Shi, Y.: Particle swarm optimization: developments, applications and resources, vol. 1, pp. 81–86, Seoul, Republic of Korea (2001)
3. del Valle, Y., Venayagamoorthy, G.K., Mohagheghi, S., Hernandez, J.C., Harley, R.G.: Particle swarm optimization: basic concepts, variants and applications in power systems. IEEE Trans. Evol. Comput. **12**(2), 171–195 (2008)
4. Kennedy, J., Eberhart, R.: Particle swarm optimization, vol. 4, pp. 1942–1948. Perth (1995)

5. Liang, J.J., Qu, B.-Y., Suganthan, P.N.: Problem definitions and evaluation criteria for the CEC 2014 special session and competition on single objective real-parameter numerical optimization, Tech. rep., Zhengzhou University, Zhengzhou China and Technical Report, Nanyang Technological University, Singapore, December 2013

6. Cheng, R., Jin, Y.: A competitive swarm optimizer for large scale optimization. IEEE Trans. Cybern. **45**(2), 191–204 (2015)

7. Tang, K., Yao, X., Suganthan, P.N., MacNish, C., Chen, Y.P., Chen, C.M., Yang, Z.: Benchmark functions for the CEC 2008 special session and competition on large scale global optimization. Tech. rep, Nature Inspired Computation and Applications Laboratory, USTC, China, November 2007

8. Shi, Y., Eberhart, R.: A modified particle swarm optimizer. In: The 1998 IEEE International Conference on Evolutionary Computation Proceedings, IEEE World Congress on Computational Intelligence, pp. 69–73 (1998)

9. Tang, K.: Summary of results on CEC 2008 competition on large scale global optimization. Tech. rep, Nature Inspired Computation and Applications Laboratory, USTC, China, June 2008

10. Yang, Z., Tang, K., Yao, X.: Multilevel cooperative coevolution for large scale optimization. In: IEEE Congress on Evolutionary Computation (CEC 2008), IEEE World Congress on Computational Intelligence, pp. 1663–1670 (2008)

11. Hsieh, S.-T., Sun, T.-Y., Liu, C.-C., Tsai, S.-J.: Solving large scale global optimization using improved particle swarm optimizer. In: IEEE Congress on Evolutionary Computation (CEC 2008), IEEE World Congress on Computational Intelligence, pp. 1777–1784 (2008). doi:10.1109/CEC.4631030

12. Brest, J., Zamuda, A., Boskovic, B., Maucec, M., Zumer, V.: High-dimensional real-parameter optimization using self-adaptive differential evolution algorithm with population size reduction. In: IEEE Congress on Evolutionary Computation (CEC 2008), IEEE World Congress on Computational Intelligence, pp. 2032–2039 (2008)

13. MacNish, C., Yao, X.: Direction matters in high-dimensional optimisation. In: IEEE Congress on Evolutionary Computation (CEC 2008), IEEE World Congress on Computational Intelligence, pp. 2372–2379 (2008). http://dx.doi.org/10.1109/CEC.2008.4631115

14. Tseng, L.-Y., Chen, C.: Multiple trajectory search for large scale global optimization. In: IEEE Congress on Evolutionary Computation (CEC 2008), IEEE World Congress on Computational Intelligence, pp. 3052–3059 (2008)

15. Zamuda, A., Brest, J., Boskovic, B., Zumer, V.: Large scale global optimization using differential evolution with self-adaptation and cooperative co-evolution. In: IEEE Congress on Evolutionary Computation (CEC 2008), IEEE World Congress on Computational Intelligence, pp. 3718–3725 (2008). http://dx.doi.org/10.1109/CEC.2008.4631301

16. Zhao, S., Liang, J., Suganthan, P., Tasgetiren, M.: Dynamic multi-swarm particle swarm optimizer with local search for large scale global optimization. In: IEEE Congress on Evolutionary Computation (CEC 2008), IEEE World Congress on Computational Intelligence, pp. 3845–3852 (2008)

17. Wang, Y., Li, B.: A restart univariate estimation of distribution algorithm: sampling under mixed gaussian and lévy probability distribution. In: IEEE Congress on Evolutionary Computation(CEC 2008), IEEE World Congress on Computational Intelligence, pp. 3917–3924 (2008). http://dx.doi.org/10.1109/CEC.2008.4631330

18. Hornby, G.S.: ALPS: the age-layered population structure for reducing the problem of premature convergence. In: Proceedings of the 8th Annual Conference on Genetic and Evolutionary Computation (GECCO 2006), NY, USA, pp. 815–822. ACM, New York (2006)

Biometrics

Achievement of a Multi DOF Myoelectric Interface for Hand Prosthesis

Sofiane Ibrahim Benchabane[1](✉), Nadia Saadia[1],
and Amar Ramdane-Cherif[2]

[1] Electronic and Computing Faculty (FEI),
Robotics Parallelism and Embedded Systems (LRPE),
University of Science and Technology Houari Boumediene (USTHB),
Bab Ezzouar, Algeria
benchabane.ibrahim@gmail.com,
saadia_nadia@hotmail.com
[2] LISV Laboratory, Université de Versailles Saint-Quentin-en-Yvelines,
10-12 Avenue de l'Europe, 78140 Vélizy, France
rca@lisv.uvsq.fr

Abstract. Nowadays, bionic systems are a real revolution and they are increasingly used by thousands of victims of amputation all over the world. Whether of lower limb or upper limb amputation, bionic systems are a concrete solution that helps people to outdo disability. People with amputated leg can anew walk and person with amputated hand can afresh hold and manipulate objects by dint of myoelectric systems. All bionic systems are built in order to be as much as possible humanoid, regarding their aspect or their functioning, that is why a myoelectric interface should be intuitive and able to analyze and decode by itself the myoelectric excitation to command prosthesis. This paper is about the design and the achievement of a Multi-DOF myoelectric interface used to command an artificial hand that we designed.

Keywords: Myoelectric interface · Humanoid prosthesis · EMG signal classification · Artificial hand · Bionic hands

1   Introduction

Since the eighties, myoelectric systems are built in order to give back some mobility to persons suffering from handicap caused by an amputation of the upper or lower limbs. Nowadays, myoelctric systems are made and marketed, for example the I-LIMB QUANTUM [1]. This hand has a humanoid design, with a carpe and five fingers independently animated by using encoding excitation. The hand grip is more human like and can accomplish a wide range of postures, but the prosthesis is commanded in non intuitive way because of encoding excitation. Other work is also carried out in this field, like the myoelectric arm designed in APL John Hopkins university. This one is controlled by chest myoelctric signal, indeed the team has graft the nerves of the amputee arm on the chest [2–6]. This is a more natural way for controlling prosthetic, but it needs lot of training to be able to activate some areas of the chest rather than others. This approach is interesting and could be discussed in case of total arm amputation. At the

© Springer International Publishing Switzerland 2016
Y. Tan et al. (Eds.): ICSI 2016, Part I, LNCS 9712, pp. 637–644, 2016.
DOI: 10.1007/978-3-319-41000-5_64

EPFL, a prosthesis with a haptic feedback has been developed. Thanks to this prosthesis, the patient can feel the hardness of an object. This is achieved by electrical stimulation of the median and radial nerves. The method consists in surgically connect excitation electrodes to the mentioned nerves [7]. In this way, the patient would be able to regulate his grasp corresponding to what he feels. That sounds more human like and a more intuitive way to use a prosthetic hand. Even if this method allows the patient to somehow feel the grasp, its remains complicated to achieve because of the delicacy of surgery, also because of the haptic method used. In fact the electrical stimulation is a controversial approach [8–11]. Traditional techniques for classify and recognize EMG pattern, like relative wavelet packet energy [12] have various limitations and a lot of computational complexity, so almost bionics system especially those who are available to users through market are built around a non intuitive myoelectrique interface. In fact, they use encoding procedures, frequency or amplitude modulation to classify postures. It induces non intuitive way to use systems that are originally designed to be as intuitive and humanoid as possible. Those myoelectric interfaces are used because they are more efficient and sturdy regarding to EMG interference than the experimental interfaces, which can analyze and classify myoelectric pattern.

This paper is about the conception and the achievement of a prototype of multi-degrees of freedom myoelectric prosthesis. The EMG interface is designed to be a low cost interface and sufficiently efficient to classify a range of myoeletric excitation.

2   Proposed Prosthesis

The proposed prototype is designed in order to be anthropomorphic and easy to produce with an approachable cost, unlike of most prosthesis at disposal on market which are unreachable for majority of hand amputees, especially in developing countries. However, current developed systems are efficient and look humanoid but they remains complicated and made with too expensive composite material. We designed a humanoid prosthesis prototype built around a modular approach. The proposed system comprises eleven (11) parts including five (05) fingers and a carp. Every finger is divided in two segments and have two (02) joints. The system is made of PLA plastic and was 3D printed (Photo 1).

Photo 1. The 3D printed prototype provided by servos and all wires

Each finger is animated separately using a different motor. The motor ensure finger bending using a thin and stiff wire, extending relays on elastic wire. In fact the elastic cable is inserted on the back side of the finger and pull it back when the motor permit it (during extent phase). So the achieved system is able to accomplish a wide range of gesture. Our structure stands for an experimental platform that allows the test of myoeletric interfaces. In fact, we obtained a structure which can accomplish a wide range of posture whatever the technique used in EMG classification. The following step consists to bend over the multi-DOF myoelectric interface.

3 Multi Degree of Freedom Myoeletric Interface

The goal of any bionic system is to allow amputee to perform some daily task those have become laborious given their disability, especially when it comes to the design of bionic hand prosthesis, they have to be dexterous accurate and instinctively commanded at the same time. That's why the myoelectric interface has to be able to decode and classify all the myoelectric excitation sensed over the muscle. The system designed is represented in Fig. 1 below:

Fig. 1. Overall scheme of the MDOF Interface, shows how the system is set. Once signal is detected, EMG is processed then classified in order to command the prosthesis.

The myoelectric acquisition chain is almost based on our previous work [13], with some significant differences as the use of a multiplexer and an ADC. Our myoelectric interface is divided in two main steps, the first one consists into detection and signal processing and the second one is the classification of myoelectric excitations. Those steps are summarized in follow.

Detection. The electrode EMG are simple electrode Ag\AgCl, identical to the electrode used in ECG. Our goal in this work is not to check what type of electrode is best to acquire myoelctric signal, however most studies bearing on myoelectric signals acquisition appeal to the electrode Ag\AgCl [14–18].

Multiplexing. In order to use the same circuit for all fingers we have incorporated a multiplexer.

Pre-amplification. This stage is necessary to amplify the signal without saturate amplifier (AD620).

Filtering. This step consists to extract the myoelectric signal from interference, by using band-pass filter serialized with a rejection filter to remove 50 Hz frequency.

Digitalization. In order to decode and analyze signal by processer we used an ADC with 11 bit resolution incorporated in a PIC 16f876A.

The following figure recaps the steps of the EMG processing (Fig. 2):

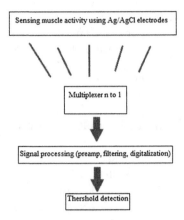

Fig. 2. Diagram of the multiplexed myoelectric acquisition system, which use digital threshold detection.

4 Classification Approach

The chosen method is in fact, a multichannel EMG using, pervious works showed the feasibility of this approach for more intuitive and natural control for artificial prosthesis [19].

To achieve our classifier, we considered the point of view that an EMG signal sensed over a muscle area is different from another one picked up from another muscle area, this way by putting electrode over some areas of the bending and extending muscles of the forearm we can classify different move of fingers. In fact, we have used the same process for several region, the following pictures illustrate this point of view (Photo 2):

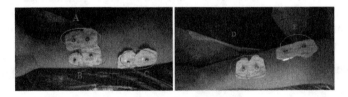

Photo 2. Couples of electrode on different area of the ventral and dorsal face of the forearm muscle (A index finger bending detection, B middle finger bending detection, C ring finger bending detection, D index finger extending detection, E middle finger extending detection).

Every finger move is sensed by analyzing EMG signal picked up from specific area on the forearm through a couple of electrode, the results are compiled at the output of the corresponding lane. So far, to summarize our interface has five (05) different lanes:

- lane0, includes index finger in bending;
- lane1, includes middle finger in bending;
- lane2, includes ring finger and pinkie finger in bending;

- lane3, includes index finger in extending;
- lane4, includes middle finger and ring finger and pinkie finger in extending;

Each lane has digital threshold detection. In fact, we built one circuit used by five (05) different lanes through multiplexer.

5 Evaluation of the Multi DOF Interface

The figure below shows two screenshots from scope displaying EMG signal picked up when only index finger (lane0) was bending during a period of 04 s in each case (Fig. 3):

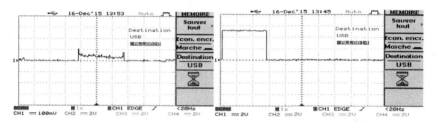

Fig. 3. (Left) EMG signal once processed and before balance, figure shows that amplitude of signal is decreasing a little bit from 100 mV to 80 mV, during a period of 03 s that corresponds to time duration attributed to each lane by the system through multiplexing in this test. (Right) the output of the PIC after digitalization and balance during another bending of index finger only, the amplitude of the signal reaches 5 V. The data out coming from PIC is used to set servomotor of lanes.

All lanes have specific threshold detection experimentally adjusted. This threshold can be different from a person to another as he can also be different for the same person. It is depending from several parameters like electrode position, skin and muscle development. The following figure shows the output of the PIC for lane0 during bending of middle, ring and pinkie fingers only, for duration of (04) s (Fig. 4):

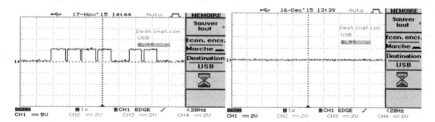

Fig. 4. (Left) output of PIC when multiplexer is set only for lane0, 5v is given for bending index finger and 0v is given if index finger is extending or a different finger is bending. (Right) output for lane0 for bending all fingers except index finger, figure shows that signal delivered from PIC to servo of lane0 is equal to zero, the lane0 is not affected by interference from the other lanes.

The figure below shows results of experimentation made in order to assess the reliability of the EMG interface. We focused on bending of index, middle and annular fingers, indeed each finger was bended 10 times with 3 s rest between each flexion. We noticed chronologically if the movement was detected or not by the EMG interface (Fig. 5).

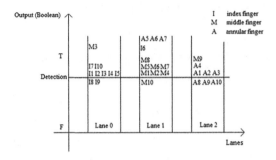

Fig. 5. EMG detection for bending of index, middle and annular fingers (lane0, lane1, lane2 respectively) the figure shows that 70 % of index flexions were detected in the right lane, 20 % were not detected and 10 % were detected in the wrong lane (middle finger lane). Middle finger bending was rightly detected with ratio of 70 % and 20 % of failure were detected in the wrong lanes, 10 % were not detected at all. The annular finger flexion was successfully detected with a ratio of 40 % only, 30 % of movements were detected in the middle finger lane and 30 % were misdetected. We can easily observe that fatigue induce misdetection in all cases, we also notice that owing to proximity between flexors of middle and annular fingers, there are interferences during their detection.

6 Discussion

The work briefly described through this paper consists to design a myoelectric interface that is able to run multi degree of freedom hand prosthesis, without additional constraints for user, with an approachable cost. Almost prosthesis available nowadays for users, even those very expensive are provided by a myoelectric interface that detects EMG activity over muscle. In fact, they capture myoelecric excitation corresponding to muscle contraction. If those contractions are sensed in a specific sequence, interface classifies them and decodes a corresponding posture for prosthesis. That involves some training and induces muscle fatigue when prosthesis is used over long duration. To summarize, we achieved a low cost 3D printed prototype of anthropomorphic hand prosthesis with five (05) fingers. Each finger is separately motorized and has two (02) joints. To run this structure we performed a multi degree of freedom myoelectric interface, which decodes and classify EMG excitation for five (05) different gestures without any constraint. So far the systems can decode bending of index, middle and pinkie fingers and extending of index and middle fingers. The approach achieved through this work brings some interesting results. Although, there are some points to enhance. First, the system has to be able to decode more hands posture and has to adjust the threshold detection by itself, that's why in further work, we will interest in incorporating a neural or fuzzy-neural approach. It is also very interesting to incorporate a haptic feedback that allows user to interact with prosthesis with more human like. Multiplexing the myoelectric circuit is profitable in terms of cost and engineering but it slows somehow the dynamic of the prosthesis, it will be interesting to compromise between swiftness of the system and his cost in a sequent work. The achieved work shows likewise that an efficient bionic system and with an accessible cost can be developed.

References

1. TouchBionics Active Prosthesis. http://www.touchbionics.com
2. Kuiken, T.: Consideration of nerve-muscle grafts to improve the control of artificial arms. Technol. Disabil. **15**, 105–111 (2003)
3. Miller, L., Lipschutz, R., Stubblefield, K., Lock, B.A., Huang, H., Williams, T., Weir, R., Kuiken, T.: Control of a six degree of freedom prosthetic arm after targeted muscle reinnervation surgery. Arch. Phys. Med. Rehabil. **89**, 2057–2065 (2008)
4. Kuiken, T., Dumanian, G., Lipschutz, R., Miller, L., Stubblefield, K.: The use of targeted muscle reinnervation for improved myoelectric prosthesis control in a bilateral shoulder disarticulation amputee. Prosthet. Orthot. Int. **28**, 245–253 (2004)
5. Kuiken, T., Li, G., Lock, B.A., Lipschutz, R., Miller, L., Stubblefield, K., Englehart, K.: Targeted muscle reinnervation for real-time myoelectric control of multifunction artificial arms. J. Am. Med. Assoc. **301**, 619–628 (2009)
6. Zhou, P., Kuiken, T.: Eliminating cardiac contamination from myoelectric control signals developed by targeted muscle reinnervation. Physiol. Meas. **27**, 1311 (2006)
7. Raspopovic, S., Capogrosso, M., Petrini, F.M., Bonizzato, M., Rigosa, J., Pino, J.D., Carpaneto, J., Controzzi, M., Boretius, T., Fernandez, E., Granata, G., Oddo, C.M., Citi, L., Ciancio, L.A., Cipriani, C., Carrozza, M.C., Jensen, E., Guglielmelli, T., Stieglitz, P.M., Rossini, S.: Restoring natural sensory feedback in real-time bidirectional hand prosthsis. SciTransl. Med. **6**, 222 (2014)
8. Shannon, G.F.: A comparison of alternative means of providing sensory feedback on upper limb prosthesis. Med. Biol. Eng. **14**, 284–294 (1976)
9. Kaczmarek, K.A., Webster, J.G., Rita, P.B.Y., Tompkins, W.G.: Electrotactile and vibrotactile displays for sensory substitution systems. IEEE Trans. Biomed. Eng. **38**, 1–16 (1991)
10. Pylatiuk, C., Kargov, A., Schulz, S.: Design and evaluation of a low-cost force feedback system for myoelectric prosthetic hands. Am. Acad. Orthotists Prosthetists **18**, 5–61 (2006)
11. Hu, X., Whang, Z., Ren, X.: Classification of surface EMG signal using relative wavelet packet energy. Comput. Methods Programs Biomed. **79**, 189–195 (2005)
12. Johansson, R.S., Westling, G.: Responses in glabrous skin mechanoreceptors during precision grip in humans. Exp. Brain Res. **66**, 128–140 (1987)
13. Benchabane, S.I., Saadia, N.: Achievement of a myoelectric clamp provided by an optical shifting control for upper limb amputations. In: Tan, Y., Shi, Y., Buarque, F., Gelbukh, A., Das, S., Engelbrecht, A. (eds.) ICSI-CCI 2015. LNCS, vol. 9141, pp. 180–188. Springer, Heidelberg (2015)
14. Zipp, P.: Effect of electrode geometry on the selectivity of myoelectric recordings with surface electrodes. Eur. J. Appl. Physiol. Occup. Physiol. **50**, 35–40 (1986)

15. Hermens, H.J., Freriks, B., Disselhorst-Klug, C., Rau, G.: Development of recommendations for SEMG sensors and sensor placement procedures. J. Electromyogr. Kinesiol. **10**, 361–374 (2000)
16. Roy, S., De Luca, G., Cheng, M., Johansson, A., Gilmore, L., De Luca, J.: Electro-mechanical stability of surface EMG sensors. Med. Bio. Eng. Comput. **45**, 447–457 (2007)
17. Ajiboye, A.B., Weir, R.: A heuristic fuzzy logic approach to EMG pattern recognition for multifunctional prosthesis control. IEEE Neural Syst. Rehabil. **13**, 280–291 (2005)
18. Hargrove, L., Zhou, P., Englehart, K., Kuiken, T.: The effect of ECG interference on pattern recognition based myoelectric control for targeted muscle reinnervated patients. IEEE Trans. Biomed. Eng. **56**, 2197–2201 (2009)
19. Jiang, N., Englehart, K.B., Parker, A.: Extracting simultaneously and proportionally neural control information for multiple DOF prosthesis from the surface electromyographie signal. IEEE Trans. Biomed. **56**, 1070–1080 (2009)

Suspicious Face Detection Based on Key Frame Recognition Under Surveillance Video

Xiaohui Zheng[1,2], Yi Ning[2,3], Xianjun Chen[3,4(✉)],
and Yongsong Zhan[1,4]

[1] Guangxi Colleges and Universities Key Laboratory of Intelligent Processing
of Computer Image and Graphics, Guilin University of Electronic Technology,
Guilin 541004, Guangxi, China
[2] Guangxi Experiment Center of Information Science,
Guilin University of Electronic Technology, Guilin 541004, Guangxi, China
[3] Guangxi Colleges and Universities Key Laboratory of Cloud Computing
and Complex Systems, Guilin University of Electronic Technology,
Guilin 541004, Guangxi, China
hingini@126.com
[4] Guangxi Key Laboratory of Trusted Software,
Guilin University of Electronic Technology, Guilin 541004, Guangxi, China

Abstract. Surveillance video is characterized by large amount of data and redundancy, which makes the suspicious face detection to be a problem. To solve the problem above we proposed suspicious face detection based on key frame. Surveillance video has the type of fixed background, so this paper used the frame difference method with low computational complexity and small computation to extract the key frame. We proposed a new method combined DPM with skin color detection to detect the suspicious face in the key frames. To solve the speed bottleneck of the traditional DPM, we proposed to use the fast HOG LUT feature extraction and the near optimal cost sensitive decision making improving the traditional method. Meanwhile we used YCrCb + otsu skin color segmentation. Since the otsu is easily affected by the illumination, we proposed an improved skin color detection. Experiment results show that the proposed algorithms are robust and accurate for real-time.

Keywords: Suspicious face detection · DPM algorithm · Skin color model · Key frame extraction · Surveillance video

1 Introduction

With the rapid development of Chinese economy, the security awareness of people is gradually increasing, and surveillance video is widely used in many fields. In particular, surveillance video about robbery and theft acts in banks and shopping malls has caused widespread concern. Usually criminals do some makeup, or deliberately obscured themselves when they do criminal. In this paper, we define faces which wear sunglasses or masks as suspicious face. The existence of the above problems makes it difficult to carry out the face detection based on video, especially the tedious and boring retrieval work, and it is time-consuming. Therefore, it is of great importance to detect

© Springer International Publishing Switzerland 2016
Y. Tan et al. (Eds.): ICSI 2016, Part I, LNCS 9712, pp. 645–652, 2016.
DOI: 10.1007/978-3-319-41000-5_65

the suspicious face based on the key frame recognition, and the accuracy of the detection directly affects the subsequent visual processing.

Deformable component model (DPM) is one of the most popular methods of object detection. It was first used Pascal VOC [1] challenge, and as the basis for the VOC 2007–2011 Pa champion system. DPM has been widely applied to the related tasks and achieved the leading performance, such as the joint body pose estimation [2] and face detection [3, 4]. In object detection, the speed and accuracy are equally important, although DPM has a great advantage in dealing with a large number of challenging data sets, but the speed bottleneck has limited its practical application. Researchers have done a lot of research to improve the speed of DPM, such as cascade [5], coarse-to-fine [6], and FFT [7]. Although the above method can improve the detection rate to a certain extent, but the effect is not good.

The structure of this paper is as follows: the second chapter introduces the key frame extraction technology. The third chapter introduces the DPM algorithm and its improvement. Skin color detection and its improved method will be given in the fourth chapter. The fifth chapter is the results of the experiment. The sixth chapter gives the conclusion and future work development direction.

2 Key Frame Extraction

2.1 Key Frame Extraction Based on Frame Difference Method

The frame difference algorithm with low complexity and small computation can detect the moving objects in the camera effectively, in the case of the fixed background. Assuming that the surveillance video is V, f_i is the i-th frame, $i \in (1, \cdots, N)$, and the total frames of the video are N, $V = \{f_1, f_2, \cdots, f_N\}$. Inter frame difference method is that all pixels of one frame minus correspond pixels of next one frame,, then add all the differences, and set a threshold K, if the difference between the f_m frame and the previous frame f_{m-1} exceeds the threshold, considering that there is a moving object and the frame f_m is viewed as the key frame. As shown in formula (1):

$$\begin{cases} f_m - f_{m-1} \geq k; m \in (2, \ldots, N). \ldots . f_m - key\,frame \\ f_m - f_{m-1} < k; m \in (2, \ldots, N). \ldots . f_m - redundant\,frame \end{cases} \quad (1)$$

Adopt frame difference method to extract key frames from surveillance video.

3 Deformable Part Model Based on HOG LUT Feature Extraction and Near Optimal Cost Sensitive Strategy

At present, although DPM algorithm has a high robustness against complex background environment, illumination changes and rotation angle of the object. But it still cannot meet the real-time requirements, based on this we proposed to improve the DPM.

3.1   Lut Hog

The traditional HOG feature extraction reduces the speed of the DPM algorithm. However, many researchers who study DPM have ignored this content, and many study work accelerated computation capacity by GPU, but the algorithm itself has not been improved. In this paper we proposed an improved algorithm by LUT, and it can achieve the same speed with under a single CPU threshold.

In this paper, we mainly introduce how to generate the same HOG features at the same time dynamically reduce the computational cost. We use look up table (LUT) to accelerate the first two steps of HOG extraction. The pixels in the image are represented by "unit8" integers, and LUT uses simpler and more effective array index to compute. They are capable of generating a finite gradient direction and dimension, so they can be computed and stored as a model initialization component in advance. In space polymerization, LUT makes the calculation of linear interpolation weights simpler and more effective, due to the linear weight number is the size of HOG interval.

Take the pixel level feature map as an example. Because the pixel value range is between [0,255], and the range of the gradient in the direction of x and y is [−255,255]. Firstly, we need compute three look up tables T_1, T_2 and T_3, and their size are 511*511. The three look up tables store the index of relative sensitive and insensitive directions, as well as the possible gradient combinations of the x and y directions. At run time, it is not necessary to directly calculate the sensitive and insensitive direction of each pixel and the pixel gradient. We only need look up the content of T_1, T_2 and T_3, which greatly reduced the computation.

The LUT based on HOG computation is very simple and easy to implement. Under the same hardware conditions, the implementation of LUT is 6 times faster than the implementation in, which can eliminate the time bottleneck of the computation of the HOG feature.

3.2   Near Optimal Cost Sensitive Strategy

DPM is based on the sliding window to scan the Pyramid image in the detection process, and it needs to determine the range and position of the object, so the classifier should deal with feature of different scales and locations. The input image is scaled by the sampling method, and the whole image is scanned by a fixed size window. In the process of scanning, because it is based on multi-scale, the object area will produce multiple overlapping detection judgment, so most time of detection is spent on the calculating the high-dimensional appearance, large operation, and the speed of detection is lower. Based on the fact that the early acceptance strategy can reduce the amount of computation, we use the early rejection and early acceptance strategy, to the maximum extent to reduce the computational complexity of scanning window, thus speed up the detection time.

DPM is based on the total score of the root filter and the part filter to judge the matching result. DPM consists of a root filter ω_0 and n component filter, where the t component is determined by the filter ω_t and the deformation term d_t. Assume that the object γ is composed of $\{p_0, p_1, \cdots, p_n\}$, where p_0 is the position of the root filter, and the p_t $(t \geq 1)$ is the position of the t part filter. The root filter and the part filter are connected by a graph structure. Cumulative score function $s(\gamma)$ is defined as below:

$$s(\gamma) = \omega_0^T \varphi_a(p_0, I) + \sum_{t=1}^{n} \omega_t^T(p_t, I) - d_t^T \varphi_d(p_t, p_0). \tag{2}$$

φ_a is the appearance of HOG feature and φ_d is the quadratic function of deformation. The hybrid component can represent the object of different postures.

For the hypothesis γ in detection, the root position p_0 is known, and the part position p_t is obtained by subtracting the displacement deformation from the appearance of the part.

$$p_t = \arg \max_p \omega_t^T \varphi_a(p, I) - d_t^T \varphi_d(p, p_0). \tag{3}$$

P indicates the possible position of the part. Because the parts are directly connected to the root, and their position is assumed to be obtained by the fixed root position. As the sliding window scan every pixel, by computing $s(\gamma)$ to judge matching or not, the calculation is large. For this, we need a method with little computation cost to judge the position of object, so we proposed to use the near optimal cost sensitive strategy.

Near optimal cost sensitive strategy includes a bilateral threshold sequence which execute early acceptance or early rejection based on Cumulative score function $s(\gamma)$ at each stage, where the cost sensitive includes two factors:

(1) The misclassification cost: C_{FN} represents the cost of positive samples judged as negative one and C_{FP} represents the cost of negative samples judged as positive one.
(2) The computational cost: Compute the expected running time in sliding window, which is controlled by parameter λ. The parameter $\Theta = (C_{FN}, C_{FP}, \lambda)$ is according to different visual tasks given or search some certain false positive rate and false negative rate (FNR and FPR).

The "near optimal", which is referred to in the near optimal cost sensitive strategy, is the least summing of the misclassification cost and the computational cost. The specific process is as follows:

The accumulation term p_t is divided into N stages, which are expressed in terms of $G_i(x)(1 \leq i \leq N)$. For the input sample x, the cumulative score of the i phase is defined as $g(x)$. The decision of N stages of p_t is defined as $\prod_N = (\tau_1, \ldots, \tau_I, \ldots, \tau_N)$. τ_i Indicates the bilateral threshold, $\tau_i = (\tau_i^-, \tau_i^+)$, and $\tau_i = (\tau_i^-, \tau_i^+)(i = 1, \ldots, n-1)$, $\tau_i^- = \tau_i^+ (i = n, \ldots, N)$, where n indicates the number of actual use of the decision stage.

In the test, three possible operation are performed in the i_th stage of the cumulative score:

1. Reject x, if $g(x) < \tau_i^-$,
2. Accept x, if $g(x) \geq \tau_i^+$,
3. Otherwise continue computing $G_{i+1}(x)$.

$R(\prod_N; \Theta)$ is the global risk function of $\prod_N$, and $L(.)$ is expected loss, and $C(.)$ is expected computational cost. The optimal cost is the minimum value of the formula (4).

$$R(\prod_N; \Theta) = L(\prod_N; C_{FP}, C_{FN}) + \lambda.C(\prod_N). \tag{4}$$

In this paper, the global optimal solution is solved by fast dynamic programming. The specific process is as follows:

(1) Calculate the optimal allocation risk at each stage, and store the optimal allocation risk in the previous stage as a function of the risk of the latter stage.
(2) For the further seek for the optimal allocation decision, firstly seek for the optimal allocation of the local minimum value at all stages, in accordance with the order of the stage descend, and then go back one by one.

4   Skin Color Detection

Skin color information is one of the very few notable features that distinguish human beings from other organisms. After a large number of scientific studies have found that, for people of different age, gender and race, if not considering the impact of brightness, skin color has an excellent clustering. Based on clustering feature, skin color feature is successfully applied to the detection of skin color. Illumination change can affect the color information of the image, which brings difficulty to the skin color segmentation. Based on this, we presented to improve it.

In order to improve the accuracy of skin color segmentation, the maximum difference method is introduced into the segmentation process. The image background and object are separated by the gray level of image. When the inter variance of background and the object is bigger, greater the difference is. If the part of the object is divided into the background or part of the background is divided into object, the difference will become small. In this paper, we applied maximum difference method to Cb and Cr of YCbCr, which can make the object and background owing maximum distinguish. The specific steps are as follows:

(1) Input the test image, and convert the RGB color space to YCbCr color space, and construct single channel Cr and Cb image.
(2) Separate the Cb and Cr, and apply the maximum difference operation to the two component channels respectively.
(3) Compute the histogram, and make nonlinear stretch to the histogram image. Redistribute the pixel of image pixels, which makes the distribution of pixels in a certain range of gray level tending to be balanced.
(4) Compute the average pixel value of the normalized histogram and the variance.
(5) Based on the maximum variance of Cb and Cr, do optimal image segmentation.

Otsu is the traditional adaptive threshold binarization, and it has a good performance in unimodal internal class variance. However, if there is shadow, it will appear mutimodal class variance, which leads to the result of image segmentation unideal.

In order to solve this problem, we proposed a local binarization algorithm. Using the size of the human eye as the template, the face image is divided into 15 rectangular windows, the size of which is 1/3 width and 1/5 length of the original image. We use otsu algorithm to compute the best threshold of each rectangular window. The specific process is as follows:

Step 1:   The original image S_{image} is divided into 15 partitions S_i, and the width is 1/3 of the original image, the length is 1/5 of the original image

Step 2:   Calculate the gray histogram of the image S_i, and normalize it

Step 3:   Set the threshold, the gray level is divided into 2 parts, C_1 and C_2. The probability of two parts is w_0 and w_1

Step 4:   The cumulative gray value μ_0 and μ_1 are calculated according to the probability of each part

Step 5:   The accumulated gray values μ_T of all the gray levels are in the range of 0-T (255)

Step 6:   Calculate the inter class variance σ.

Between the two classes, the maximum value of the difference represents the biggest distinguish. The threshold t denotes the best partition threshold, then to improve the inaccurate of detection caused by the uneven distribution of light.

5   Experimental Results

In this paper, we propose an improved DPM, which is mainly aimed at the feature extraction and detection process, and the experimental results show that the speed is obviously improved. Compared with other accelerated DPM versions, including cascade, branch-bound, coarse-to-fine, and FFT. In addition to coarse-to-fine, all the other methods of DPM use the default settings and models of realeased4. The Hog interval is 8, and the number of the parts of each component is 8, and the number of components in each class is 6. For coarse-to-fine DPM, its setting references to the proposed settings of [6]. Table 1 is the comparison of the above other methods and proposed method in this paper on the VOC Pascal 2007 data set, and the average feature extraction time and the detection time are listed.

Table 1. The comparison of algorithm processing speed(ms/frame).

	The time of feature extraction	Detection time
DPM	0.46	11.77
Branch-Bound (DPM)	0.46	2.75
Cascade (DPM)	0.46	0.99
FFT (DPM)	0.48	0.98
Coarse to Fine (DPM)	0.67	0.99
Proposed method	0.07	0.49

In order to verify the validity of the proposed detection algorithm, the proposed method is first to detect the suspicious face of the key frames from the video. Meanwhile, in order to further verify the effectiveness of the proposed algorithm in complex environment, we collected the test images from the Internet. As shown in Fig. 1, the first line and the second line are the detection results of key frames from our video. The experimental results show that the improved algorithm can effectively detect the suspicious face in the case of deflection. The improved DPM algorithm is proved to be more robust. The third line and the fourth line are the detection results of the test images are collected from the Internet, this paper proposed the algorithm based on DPM and skin color detection is still able to detect faces with sunglasses and masks in multi-face condition, and the picture shows that the target is not in a prominent position. We can see the validity of the proposed suspicious face detection.

Fig. 1. The detection results of suspicious face.

6 Conclusion

In order to detect the suspicious face under the surveillance video effectively, this paper proposed suspicious face detection based on key frame recognition under surveillance video. We proposed an improved DPM algorithm which used LUT and near optimal cost sensitive strategy. It solved the time bottleneck caused by the traditional feature extraction and detection. After verification, the improved method has a good performance and time consuming. We applied YCbCr + otsu skin color segmentation into skin color detection. But it is vulnerable to the illumination changes, we further proposed the local binarization Otsu, it successfully solved the illumination problem and enhanced the performance of the algorithm.

In the next step, the proposed algorithm will be further optimized and applied to the detection of real-time surveillance video. Because of the diversity of the object, the errors of detection exist. The next step is to combine the detection of suspicious face with suspicious behaviour and voice detection.

Acknowledgments. This research work is supported by the grant of Guangxi science and technology development project (No: 1598018-6), the grant of Guangxi Experiment Center of Information Science of Guilin University of Electronic Technology, the grant of Guangxi Colleges and Universities Key Laboratory of cloud computing and complex systems of Guilin University of Electronic Technology (No: 15210), the grant of Guangxi Colleges and Universities Key Laboratory of Intelligent Processing of Computer Image and Graphics of Guilin University of Electronic Technology (No: GIIP201403), the grant of Guangxi Key Laboratory of Trusted Software of Guilin University of Electronic Technology(No: KX201513).

References

1. Everingham, M., Van Gool, L., Williams, C.K., Winn, J., Zisserman, A.: The pascal visual object classes (voc) challenge (2010)
2. Yang, Y., Ramanan, D.: Articulated pose estimation with flexible mixtures-of-parts. In: Computer Vision and Pattern Recognition (CVPR), June 2011
3. Yan, J., Zhang, X., Lei, Z., Yi, D., Li, S. Z.: Structural models for face detection. In: Automatic Face and Gesture Recognition (FG), April 2013
4. Zhu, X., Ramanan, D.: Face detection, pose estimation, and landmark localization in the wild. In: Computer Vision and Pattern Recognition (CVPR), June 2012
5. Felzenszwalb, P.F., Girshick, R.B., McAllester, D.: Cascade object detection with deformable part models. In: Computer vision and pattern recognition (CVPR), June 2010
6. Pedersoli, M., Vedaldi, A., Gonzalez, J.: A coarse-to-fine approach for fast deformable object detection. In: Computer Vision and Pattern Recognition (CVPR), June 2011
7. Dubout, C., Fleuret, F.: Exact acceleration of linear object detectors. In: Fitzgibbon, A., Lazebnik, S., Perona, P., Sato, Y., Schmid, C. (eds.) ECCV 2012, Part III. LNCS, vol. 7574, pp. 301–311. Springer, Heidelberg (2012)
8. Prisacariu, V., Reid, I.: fastHOG-a real-time GPU implementation of HOG. Department of Engineering Science, 2310 (2009)

Author Index

Printed in the United States
By Bookmasters